Y0-BTD-382

Quantum Mechanics in Mathematics, Chemistry, and Physics

Quantum Mechanics in Mathematics, Chemistry, and Physics

Edited by

Karl E. Gustafson
University of Colorado
Boulder, Colorado

and

William P. Reinhardt
University of Colorado
and
Joint Institute for Laboratory Astrophysics
National Bureau of Standards and University of Colorado
Boulder, Colorado

PLENUM PRESS • NEW YORK AND LONDON

Library of Congress Cataloging in Publication Data

American Mathematical Society.
Quantum mechanics in mathematics, chemistry, and physics.

"Proceedings of a special session in mathematical physics organized as a part of the 774th meeting of the American Mathematical Society, held March 27-29, 1980, in Boulder, Colorado."
Bibliography: p.
Includes index.
1. Quantum theory—Congresses. 2. Quantum chemistry 1. Gustafson, Karl E. II. Reinhardt, William P. III. Title.
QC173.96.A48 1981 530.1'2 81-5846
ISBN 0-306-40737-X AACR2

Proceedings of a Special Session in Mathematical Physics
organized as a part of the 774th Meeting of the American
Mathematical Society, held March 27 – 29, 1980, in Boulder, Colorado.

A Division of Plenum Publishing Corporation
233 Spring Street, New York, N.Y. 10013

Printed in the United States of America

PREFACE

This volume grew from a Special Session in Mathematical Physics organized as a part of the 774th Meeting of the American Mathematical Society in Boulder, Colorado, 27-29 March, 1980. The organizers attempted to include a mix of mathematicians, physicists and chemists. As interest in the session increased and as it became clear that a significant number of leading contributors would be here, we were offered the opportunity to have these proceedings published by Plenum Press.

We would like first to express our thanks to Plenum Press, to the American Mathematical Society, and to the University of Colorado Graduate School, and in particular, respectively, to James Busis, Dr. William LeVeque, and Vice Chancellor Milton Lipetz, for their help in this undertaking. We would also like to thank Burt Rashbaum and Martha Troetschel of the Department of Mathematics and Karen Dirks, Donna Falkenhein, Lorraine Volsky, Gwendy Romey, and Leslie Haas of the Joint Institute for Laboratory Astrophysics for their excellent help in the preparation of these proceedings.

The session took on an international character, representing the countries Federal Republic of Germany, India, Belgium, Peoples Republic of China, Switzerland, Iran, Mexico, German Democratic Republic, England, and the United States. In all there were finally 37 speakers and all have contributed to this volume. The success of the meeting is above all due to them.

We chose to mix, rather than separate, the talks and disciplines, in order to promote interaction and appreciation. The contributions are presented here in the same order as they were given at the meeting.

Thus this volume is in some respects an accident, born of a mixing process which began in the pure state of a special session at a regional meeting of a mathematical society and which in its eventual chaos pulled in thirty-seven mathematicians, chemists, and physicists to a final three-day reaction amid a swirling snow-storm that would not stop until the encounter was over.

Boulder, November, 1980
Departments of Mathematics and Chemistry

Karl E. Gustafson
William P. Reinhardt

CONTENTS

TOTAL CROSS SECTIONS IN NON-RELATIVISTIC SCATTERING THEORY

Volker Enss

Institut für Mathematik
Ruhr-Universität
D-4630 Bochum 1, F.R. Germany

and

Barry Simon

Departments of Mathematics and Physics
Princeton University
Princeton, N.J. 08544, U.S.A.

ABSTRACT

Using time-dependent geometric methods we obtain simple explicit upper bounds for total cross sections σ_{tot} in potential- and multiparticle-scattering. σ_{tot} is finite if the potential decays a bit faster than r^{-2} (in three dimensions) or if weaker direction dependent decay requirements hold. For potentials with support in a ball of radius R bounds are given which depend on R but not on the potential.

We obtain upper bounds on σ_{tot} for large coupling constant λ, the power of λ depending on the falloff of the potential. For spherically symmetric potentials the variable phase method gives also a lower bound growing with the same power of λ.

In the multiparticle case for charged particles interacting with Coulomb forces the effective potential between two neutral clusters decays sufficiently fast to imply finite total cross sections for atom-atom scattering.

We reexamine the definitions of classical and quantum cross sections to discuss some puzzling discrepancies.

1. OUTLINE

The total scattering cross section in quantum mechanics is a simple measure for the strength of a potential when it influences a homogeneous beam of particles with given energy and direction of flight. It can be easily measured in experiments, therefore various approximation schemes have been developed for its calculation. On the other hand relatively little attention has been paid to a mathematically rigorous treatment, probably because it is a rather special quantity derived from basic objects like the scattering amplitude or the scattering operator S. Moreover various assumptions and estimates were motivated by technical rather than physical reasons. In contrast to the conventional time independent approach Amrein and Pearson [1] used time dependent methods to obtain new results. In Amrein, Pearson,and Sinha [2] this was extended to prove finiteness of the total cross section in the multiparticle case if all pairs of particles which lie in different clusters interact with short range forces.

In our approach we add geometric considerations to the previous ones. The main bounds are derived by following the localization of wave packets as they evolve in time. This method is both mathematically simple and physically transparent. Nevertheless it allows to recover or improve most results with simpler proofs. We need not average over directions but we keep the direction of the incident beam fixed. The main defect of the geometric method so far is that we have to average over a small energy range; our bounds blow up in the sharp-energy limit. Consequently we get poor bounds for the low energy behavior or (connected by scaling) for obstacle scattering with the radius going to zero.

In Section 3 we determine the decay requirements for infinitely extended potentials which guarantee finite total cross sections both for the isotropic and anisotropic cases. They are close to optimal. We obtain explicit bounds which have the correct small coupling and high energy behavior. The Kupsch-Sandhas trick is used in the next section to give a bound independent of the potential if the latter has its support inside a ball of radius R. The bound has the correct large R behavior.

One of our main new results combines the two bounds to establish a connection between the decay of the potential at infinity and the rate of increase of the total cross section in the strong coupling limit (Section 5). The variable phase method gives lower bounds with the same rate of increase for spherically symmetric potentials.

The main advantage of time dependent (and geometric) methods is that two cluster scattering is almost as easy to handle as two particle (= potential-) scattering. One has to use a proper effective potential between the clusters which may decay faster than the pair potentials due to cancellations. For a system of charged particles interacting via Coulomb pair potentials the effective potential between neutral clusters (atoms) decays fast enough to give a finite total cross section for atom-atom scattering (including rearrangement collisions and breakup into charged clusters). This new result is derived in Section 7.

In quantum mechanics textbooks usually the classical total cross section is defined first and then the quantum total cross section is derived by analogy. Therefore it is puzzling that both quantities differ considerably even if the quantum corrections should be small. E. g. the quantum cross section is twice as big as the classical one for scattering from big hard spheres ("shadow scattering"), even when $\hbar \to 0$.

In Section 2 we examine the limits involved in the derivation of the quantum total cross section and show that it is basically a pure wave- (and not particle-) concept. This suggests our definition of the quantum total cross section (2.5), which agrees with the traditional one for suitable potentials. (Or one might use (2.5) as an equivalent expression for σ_{tot} which is convenient for estimates.) This point of view explains naturally the discrepancies; we discuss some aspects of the classical limit in Section 6.

For detailed references to earlier and related work see [1,2, 8, 11]. We restrict ourselves here to three dimensions, the results for general dimension as well as various refinements and extensions can be found in [8].

One of us (V.E.) would like to thank the Institute for Advanced Study, Princeton, for its hospitality and support under the Albert Einstein visiting professorship endowed by the Federal Republic of Germany and for a travel grant provided by Deutsche Forschungsgemeinschaft. Another of us (B.S.) acknowledges partial support by the National Science Foundation under Grant No. MCS 78-01885.

2. THE DEFINITION OF CLASSICAL AND QUANTUM TOTAL CROSS SECTIONS

When scattering experiments are performed with microscopic particles like atoms, electrons, nuclei, then (in contrast to billard balls) it is practically impossible to observe the time evolution of individual projectiles. We have to restrict ourselves to very few observables which can be measured well enough,

e. g. the direction of flight of the particle when it has passed the target. This direction is asymptotically constant, thus there is enough space and time available to measure it with arbitrary precision. In classical physics where the possibility to prepare particles with a given trajectory is not restricted by basic principles, the scattering angle depends strongly on the impact parameter. If the latter cannot be controlled the next best thing is to use a homogeneous beam of incoming particles and to observe the distribution of the outgoing particles over the scattering angles. This is the *classical differential cross section*. Let the incoming beam consist of particles flying in the direction $\hat{e}$ with momentum p and a given density (= number of particles per unit area orthogonal to $\hat{e}$); then one defines:

$$\sigma_{\text{class}}(p,\hat{e};d\Omega) = \frac{\text{number of particles deflected into } d\Omega}{\text{density of particles}}$$

where $d\Omega$ does not contain $\hat{e}$. Integrating over the outgoing directions yields the *classical total cross section*:

$$\sigma_{\text{tot,class}}(p,\hat{e}) = \int_{S^2} \sigma(p,\hat{e};d\Omega)$$

$$= \frac{\text{number of deflected particles}}{\text{density of particles}}$$

(If one thinks of an experiment running forever one should understand the numerators and denominators per given time interval.) Note that the idealization of a beam of finite density which is homogeneous in the plane perpendicular to the beam direction $\hat{e}$, necessarily involves infinitely many particles for two reasons. First one would need infinitely many particles per unit area, but this is compensated by the denominator in the definition of the cross section. The second infinity is more delicate which comes from the infinite extension of the beam. If the target has finite size (potential of compact support) then only the particles which hit the target can be deflected, the infinitely many particles which miss the target go on into the forward direction $\hat{e}$ and won't be counted. (The infinite extension of the beam allows to specify the beam independent of the size and localization of the target.) Excluding *one single* direction from the observation we have singled out the finitely many particles of interest (for finite density) out of the infinitely many incoming. This prevents us from measuring the total cross section exactly if the incoming beam cannot be prepared with all particles having the same direction. The (idealized) concept of the total cross section requires for its definition that there are beams of incoming particles with a sharp direction. On the other hand it is irrelevant whether beams with sharp energy (or modulus of the momentum p) are available or not. We will use this freedom below.

Quantum mechanical scattering states for potentials vanishing at infinity are known to behave asymptotically like classical wave packets. Therefore it is reasonable to extend the notion of cross sections to quantum mechanics. However, a further limit is involved because there are no states with a sharp direction in the quantum mechanical state space. Let the z-axis be in the beam direction $\hat{e}$, then a sharp direction would mean that $p_x = p_y = 0$. By the uncertainty principle this implies infinite extension of the states in the x-y-directions. Thus the infinite extension of the state perpendicular to $\hat{e}$, which might look unnecessary in the classical case, is forced upon us in quantum scattering. We will have to handle wave functions which are constant in the plane perpendicular to $\hat{e}$, therefore the quantum cross section behaves like a quantity characteristic for classical waves rather than classical particles for any $\hbar > 0$. A classical particle approximation would require a wave packet well concentrated compared to a length typical for the potential. Thus it is no longer mysterious that in the classical limit ($\hbar \to 0$) the quantum cross section need not converge to the classical one (e. g. shadow scattering off hard spheres).

Another peculiarity of the classical cross section is its discontinuity under small changes of the potential. Consider e. g.

$$V_b(x,y,z) = (a+bx)\ \chi_{[-r,r]}(z)\ \chi_{[-R,R]}(x)\ \chi_{[-R,R]}(y)$$

for some parameters a,b,r,R where $r \ll R$. If the beam direction is along the z-axis (near the z-axis) for $b = 0$ the total cross section is zero (tiny) but for any $b \neq 0$ is jumps to $4R^2$ ($\approx 4R^2$). If one could easily count the particles which have been influenced by V(e. g. time delay for $a > 0$) the discontinuity of $\sigma_{tot,class}$ at $b = 0$ would disappear and it would always have the size of the geometric cross section $4R^2$. For such a potential with $b = 0$ the quasiclassical limit $\hbar \to 0$ of the quantum cross section does not converge at all!

Following the above considerations about the quantum cross section as a wave limit we use for its definition "plane wave packets" which are chosen to describe waves with a sharp direction of propagation $\hat{e}$ parallel to the z-axis, but they are normalized wave packets in the longitudinal direction, thus being as close as possible to a Hilbert space vector. For a given direction $\hat{e}$ the plane wave space $h_{\hat{e}}$ is isomorphic to (and henceforth identified with) $L^2(\mathbb{R}, dz)$. The configuration space wave function is

$$g(x,y,z) = g(z) \text{ with } \int |g(z)|^2\, dz = 1. \qquad (2.1)$$

In momentum space we denote by $\tilde{g}(k)$ the one-dimensional Fourier-transform

$$\tilde{g}(k) = (2\pi)^{-1/2} \int dz\, e^{-ikz}\, g(z), \tag{2.2}$$

corresponding to the three-dimensional Fourier transform

$$\hat{g}(\vec{k}) = \tilde{g}(k_z)\,(2\pi)\,\delta(k_x)\,\delta(k_y). \tag{2.3}$$

Since a beam should hit the target from one side only we assume:

$$\operatorname{supp} \tilde{g}(k) \subset (0,\infty), \tag{2.4}$$

which implies in (2.3) $k_z = |\vec{k}| =: k$.

The scattering operator S is the unitary operator which maps incoming states to the scattered outgoing waves, it is close to one on states which are weakly scattered. $(S-1)g$ corresponds to the scattered part of the wave g. The probability to detect a scattered particle is then $\|(S-1)g\|^2$ where the norm is that of the Hilbert space $\mathcal{H} = L^2(\mathbb{R}^3)$. Thus we *define* as the quantum mechanical *total cross section*

$$\int_0^\infty \sigma_{tot}(k,\hat{e})\, |\tilde{g}(k)|^2\, dk = \|(S-1)g\|^2, \tag{2.5}$$

where $g \in h_{\hat{e}}$ with (2.4). We will show below that for a class of potentials with suitable decay properties S-1 extends naturally from an operator on $\mathcal{H}$ to a bounded map from $h_{\hat{e}}$ into $\mathcal{H}$, then the definition makes sense. We average over the energy of the incident beam but keep the direction fixed. (See also the similar construction in [14].) Certainly we have to verify that our definition agrees with the conventional one given below.

Within the time independent theory of scattering for potentials with sufficiently fast decay the solutions of the Lippman Schwinger equation have the asymptotic form

$$\phi(\vec{k},\vec{x}) \sim \exp(i\vec{k}\cdot\vec{x}) + f(k;\hat{x}\leftarrow\hat{k})\,\frac{\exp(i\,k|x|)}{|x|}\quad,$$

$f(k;\hat{x}\leftarrow\hat{k})$ is the continuous on shell scattering amplitude. Equivalently the kernel of S-1 in momentum space is

$$(S-1)(\vec{k}',\vec{k}) = \frac{i}{2\pi m}\,\delta(k'^2/2m - k^2/2m)\, f(k;\hat{k}'\leftarrow\hat{k})$$

where $\hat{k} = \vec{k}/k$, $k = |\vec{k}|$, etc. Then

$$\sigma_{tot}(k,\hat{e}) = \int d\Omega'\, |f(k;\hat{\omega}'\leftarrow\hat{e})|^2. \tag{2.6}$$

The physical motivation for this choice as given in most textbooks on quantum mechanics uses the "obvious" fact that $\exp(i\,\vec{k}\cdot\vec{x})$ describes an incoming homogeneous beam of particles with momentum k, direction $\hat{k}$ and density one (or $(2\pi)^{3/2}$) particle per unit area, similarly for the outgoing spherical wave.

More careful authors give the following time dependent justification. Let $\phi^{in}(\vec{k})$ be the (square integrable) wave function of a single incoming particle with momentum support well concentrated around a mean value $\vec{q}$. The corresponding outgoing state has a momentum space wave function

$$\phi^{out}(\vec{k}') = (S\,\phi^{in})(\vec{k}') = \int d^3k\ \delta(\vec{k}-\vec{k}')\ \phi^{in}(\vec{k}) + $$

$$+ \frac{i}{2\pi m}\int d^3k\ \delta(k'^2/2m-k^2/2m)\ f(k;\hat{k}'\leftarrow\hat{k})\ \phi^{in}(\vec{k}) . \qquad (2.7)$$

The "scattering into cones" papers [6, 9] show that the asymptotic direction of flight is $\hat{k}$ for the incoming and $\hat{k}'$ for the outgoing state. The first summand in (2.7) is then identified as "not deflected" and for continuous (or not too singular) f's the second term gives the deflected part. Although this splitting is natural it cannot be justified by observations for directions lying in the support of $\phi^{in}(\vec{k})$. Under this assumption the probability $w(\phi^{in})$ that a particle with incoming wave function ϕ^{in} will be deflected, is

$$w(\phi^{in}) = \int d^3k' \left| \frac{i}{2\pi m}\int d^3k\ \delta(k'^2/2m-k^2/2m) f(k;\hat{k}'\leftarrow\hat{k})\,\phi^{in}(\vec{k})\right|^2 =$$

$$= \|(S-1)\,\phi^{in}\|^2 .$$

To represent a homogeneous beam one translates the incoming state by a vector $\vec{a}$ in the plane orthogonal to the mean direction $\vec{q}$, $\phi^{in}_{\vec{a}}(\vec{k}) = e^{-i\,\vec{a}\cdot\vec{k}}\phi^{in}(\vec{k})$, and one sums up the contributions for different $\vec{a}$'s. $\int d^2a$ represents a homogeneous beam with particle density one per unit area. The resulting number of deflected particles is then

$$\int d^2a\ w(\phi^{in}_{\vec{a}}) = \sigma_{tot}(\phi^{in}) \qquad (2.8)$$

$$= \int d^3k(\hat{k}\cdot\hat{q})^{-1}\int d\Omega' |f(k;\hat{\omega}'\leftarrow\hat{k})|^2\ |\phi^{in}(\vec{k})|^2 .$$

In the limit $|\phi^{in}(\vec{k})|^2 \to \delta(\vec{k}-\vec{q})$ expression (2.6) for $\sigma_{tot}(q,\hat{q})$ is recovered and $|\phi^{in}(\vec{k})|^2 \to \delta(k_\perp)|\tilde{g}(k)|^2$ yields $\int dk\ \sigma_{tot}(k;\hat{e})|\tilde{g}(k)|^2$, the left hand side of (2.5). Note that the summation over $\vec{a}$'s is incoherent, we have added probabilities and not states, because we are interested only in interactions between the target and single particles, interference between particles in the beam has to be eliminated.

Let us now calculate the cross section according to our definition.

$$\|(S-1)g\|^2 = \int d^3k' \left| \frac{i}{2\pi m} \int d^3k \; \delta(k'^2/2m - k^2/2m) \right.$$

$$\left. f(k;\hat{k}'\leftarrow\hat{k})\,(2\pi)\,\delta(k_\perp)\,\tilde{g}\,(k) \right|^2$$

$$= \int dk \int d\Omega' \,|f(k;\hat{\omega}'\leftarrow\hat{e})|^2 |\tilde{g}(k)|^2 .$$

Thus our definition coincides with the conventional one if the scattering amplitude is continuous (or not too singular).

At first glance it seems strange that the incoherent superposition in (2.8) yields the same result as the coherent superposition of wave packets with strong correlations which forms the plane wave-packets. The following heuristic argument easily explains the phenomenon. Since $(S-1)g \in L^2(\mathbb{R}^3)$ the action of S-1 "localizes", it essentially annihilates the parts of the state which lie beyond some radius r. Let $R \gg r$ and use in the incoherent case (2.8) the normalized wave function

$$g(z)\;(2R)^{-1}\;\chi_{[-R,R]}\,(x)\;\chi_{[-R,R]}\,(y)$$

whose (3-dimensional) Fourier transform $\phi^{in}(\vec{k})$ obeys $|\phi^{in}(\vec{k})|^2 \to |\tilde{g}(k)|^2\,\delta(k_\perp)$ as $R \to \infty$ ($\tilde{g}$ is the 1-dim. Fourier transform). Then

$$w(\phi^{in}_{\vec{a}}) = \|(S-1)\phi^{in}_{\vec{a}}\|^2 \approx \begin{cases} (2R)^{-2}\,\|(S-1)g\|^2 & \text{for } |a_{1,2}| \le R \\ 0 & \text{otherwise} \end{cases}$$

and

$$\int d^2a\; w(\phi^{in}_{\vec{a}}) \approx \int_{-R}^{R} da_1 \int_{-R}^{R} da_2 \;\|(S-1)\phi^{in}_{\vec{a}}\|^2 \approx \|(S-1)g\|^2 .$$

The sharp direction - limit forces us to use states which are eventually constant in an area much larger than the localization region of S-1. Up to negligible boundary terms all contributions become parallel and the properly normalized coherent and incoherent superpositions do not differ.

In Section 6 we will return to the comparison of the wave picture and particle picture when we discuss the classical limit. There we will explain why it is natural, although it looks unnatural, that the quantum cross section of a hard sphere is twice the corresponding one for classical particles ("shadow scattering"). In the same section we will explain why classical cross sections are generally infinite for potentials with unbounded support although the quantum cross sections may be finite.

3. THE BASIC ESTIMATE FOR σ_{tot}

We assume in this section that the potential $V(\vec{x})$ is a perturbation of the kinetic energy $H_o = -\frac{1}{2}\Delta$ with H_o-bound smaller than 1 (we have set $\hbar=1$ and the particle mass m=1, therefore momenta and velocities coincide). If the potential is of short range (we will impose stronger decay requirements shortly) then the isometric wave operators

$$\Omega^{\mp} = s - \lim_{t\to\pm\infty} e^{i\,Ht}\, e^{-i\,H_o t}$$

exist and are complete, the S-operator

$$S = (\Omega^-)^* \,\Omega^+$$

is unitary and on states in the domain of H_o the following "interaction picture" representation holds:

$$S-1 = (\Omega^-)^* \,[\Omega^+ - \Omega^-]$$

$$= (\Omega^-)^* \int_{-\infty}^{\infty} dt\, e^{i\,Ht}\,(iV)\,e^{-i\,H_o t}. \qquad (3.1)$$

Cook's estimate gives

$$\|(S-1)\Phi\| \leq \int_{-\infty}^{\infty} dt\, \|V\, e^{-i\,H_o t}\Phi\| . \qquad (3.2)$$

Let $\Phi_R \in \mathcal{H} = L^2(\mathbb{R}^3)$ be an approximating sequence of states which tends to the plane wave packet g as $R \to \infty$. For a suitable class of potentials we will show that

$$\lim_{R\to\infty}\ \sup_{R'>R} \int_{-\infty}^{\infty} dt\, \|V\, e^{-i\,H_o t}(\Phi_{R'} - \Phi_R)\| = 0 \qquad (3.3)$$

which implies by (3.2) convergence in $\mathcal{H}$ of $\lim_{R\to\infty} (S-1)\Phi_R =: (S-1)g$ and the finite bound

$$\|(S-1)g\| \leq \int_{-\infty}^{\infty} dt\, \|V\, e^{-i\,H_o t} g\| . \qquad (3.4)$$

It is convenient to use product functions

$$\phi_R(\vec{r}) = g(z)\ f_R(x,y) \tag{3.5}$$

because the free time evolution factorizes:

$$(e^{-i\,H_o t}\ \phi_R)(\vec{r}) = (e^{-i\,h_o t}\ g)(z)\ f_R(t;x,y)\ ,$$

where

$$h_o = -\frac{1}{2}\,d^2/dz^2,\ \ f_R(t;x,y) = [\exp\{-\frac{i}{2}(\frac{d^2}{dx^2} + \frac{d^2}{dy^2})t\}f_R](x,y)\ .$$

In particular the plane wave packet space $h_{\hat{e}}$ is left invariant:

$$e^{-i\,H_o t}\ h_{\hat{e}} = e^{-i\,h_o t}\ h_{\hat{e}} = h_{\hat{e}}\ .$$

For Gaussian $f_R(x,y) = \exp[-(x^2+y^2)/R^2]$ it is well known that

$$|f_R(t;x,y)| \leq 1 \qquad \forall\ t,x,y\ ,$$

and for any L,T,ε there is an R_o such that for $R > R_o$

$$||f_R(t;x,y)|^2 - 1| < \varepsilon \ \text{if}\ x^2+y^2 < L^2, |t| < T\ . \tag{3.6}$$

Then for the convergence (3.3) it is necessary and sufficient to show the finiteness of the bound

$$\int_{-\infty}^{\infty} dt \| V\ e^{-i\,H_o t}\ g\|$$

$$= \int_{-\infty}^{\infty} dt \{\int dz \int dx\ dy |V(x,y,z)|^2\ |[e^{-i\,h_o t}\ g](z)|^2\}^{1/2} \tag{3.7}$$

$$\geq \int_{-\infty}^{\infty} dt \{\int d^3r\ |V(\vec{r})|^2\ |[e^{-i\,H_o t}\ g f_R](\vec{r})|^2\}^{1/2}\ . \tag{3.8}$$

The contribution to this integral is arbitrary small if for some L,T we have $|t| > T$ or $x^2+y^2 > L^2$ whereas inside this region the wave functions of $e^{-i\,H_o t}(\Phi_{R'} - \Phi_R)$ are small for $R' > R$ big enough. Any cutoff f_R which fulfills (3.6) gives the same result.

In the bound (3.7) the potential enters only through the function

$$W(z) = \{\int dx\ dy\ |V(x,y,z)|^2\}^{1/2} \tag{3.9}$$

where we have fixed the z-axis in the beam direction $\hat{e}$, and the *one*-dimensional estimate

$$\|(S-1)g\| \leq \int_{-\infty}^{\infty} \{\int dz\, W^2(z)|[e^{-i h_o t} g](z)|^2\}^{1/2} dt$$
$$= \int_{-\infty}^{\infty} dt\, \|W e^{-i h_o t} g\| < \infty \qquad (3.10)$$

will imply finite total cross sections.

Here we have taken multiplication by $V(\vec{x})$ as a map from $h_{\hat{e}}$ into $\mathcal{H}$ and we have used that

$$\|V g\|_{L^2(\mathbb{R}^3)} = \|W g\|_{L^2(\mathbb{R})} . \qquad (3.11)$$

We collect a few well known facts about wave packets in the following

Lemma.
Let $G(z)$ have the (one-dimensional) Fourier transform $\tilde{G}(k) \in C_o^\infty(\mathbb{R})$, supp $G \subset (-\delta,\delta)$. Define

$$\tilde{G}_v(k) := \tilde{G}(k-v) . \qquad (3.12)$$

Then for $v > 2\delta$ there are C_m independent of v such that

$$\int_{|z| \leq \frac{v}{2} t} dz\, |[e^{-i h_o t} G_v](z)|^2 \leq C_m (1+|vt|)^{-m} . \qquad (3.13)$$

Proof. Using the stationary phase method (see e.g. Theorem XI.14 in [12]) one easily shows that for $|z|/|t| > \delta$

$$|[e^{-i h_o t} G](z)| \leq C_m' (1+|z|)^{-m} .$$

Next observe that

$$[e^{-i h_o t} G_v]^{\sim}(k) = e^{i t v^2/2} e^{i(k-v)vt}[e^{-i h_o t} G](k-v) ,$$

and

$$|[e^{-i h_o t} G_v](z)| = |[e^{-i h_o t} G](z-vt)| .$$

Thus for $|z-vt| / |t| > \delta$

$$|[e^{-i h_o t} G_v](z)| \leq C_m'(1 + |z-vt|)^{-m} .$$

For $|z| \leq v|t|/2$ this implies (3.13) . □

Remark. With some obvious modifications the Lemma and the results below hold for an extremely wide class of "free" Hamiltonians $H_o(\vec{p})$ with velocity operator $\vec{\nabla}_p H_o(\vec{p})$. Only the constant $C(\delta)$ in (3.18) will change.

Let $F(|z| \leq R)$ be the operator of multiplication with the characteristic function of the indicated region. The real function $\psi \in C_o^\infty(\mathbb{R})$ should obey $0 \leq \psi(q) \leq 1$ and $\psi(q) = 1$(resp. 0) for $|q| \leq \delta$(resp. $\geq 2\delta$). Denote by $\psi(K)$ multiplication of wave functions $\tilde{g}(k)$ with $\psi(k)$, then $\psi(K)$ is in z-space convolution with a smooth kernel of rapid decay.

A one-dimensional potential W is called a *short range potential* if

$$\|W\ \psi(K)\ F(|z| \geq R)\| =: h(R) \in L^1(\mathbb{R}_+, dR), \tag{3.14}$$

or equivalently

$$\|W\ F(|z| \geq R)\ \psi(K)\| \in L^1(\mathbb{R}_+, dR) .$$

Going back to the three dimensional potential V from which W was derived in (3.9) and using (3.11) for V as a map from $h_{\hat{e}}$ into $\mathcal{H}$ we require (depending on the direction $\hat{e}$):

$$\|V\ \psi(K)F(|z| \geq R)\|_{h_{\hat{e}},\mathcal{H}} =: h(R) \in L^1(\mathbb{R}_+, dR) \tag{3.15}$$

with the corresponding norm

$$\|V\|_{\hat{e}} = h(0) + \int_o^\infty h(R)\ dR . \tag{3.16}$$

We will discuss simple sufficient conditions for (3.15) below, first we will complete our estimate (3.10).

Observe that $\psi(K-v)$ depends on v in z-space only through phase factors which commute with F and W, thus $\|W\ \psi(K-v)F(|z| \geq R)\| = h(R)$ for all v. Let $\tilde{g}$ have support in $(v-\delta, v+\delta)$, $v > 2\delta$, then $g = \psi(K-v)g$ and with the Lemma we obtain

$$\|(S-1)g\| \leq \int_{-\infty}^{\infty} dt\ \|W\ e^{-i\ h_o t}\ g\|$$

$$\leq \int_{-\infty}^{\infty} dt \|W\ \psi(K-v)\ F(|z| \geq v|t|/2)\ e^{-i\ h_o t}\ g\|$$

$$+ \int_{-\infty}^{\infty} dt \|W\ \psi(K-v)\ F(|z| \leq v|t|/2)\ e^{-i\ h_o t}\ g\| \leq$$

$$\leq \frac{1}{v} \int_{-\infty} d(vt) \; \{2 \; h(v|t|/2) + 2h(0) \; C_2 (1+|vt|)^{-2}\}$$

$$\leq C \; v^{-1}\{h(0) + \int_0^\infty dR \; h(R)\} = C \; v^{-1} \|V\|_{\hat{e}} \tag{3.17}$$

where the constant C depends on the shape of the wave function $\tilde{g}$, but it is independent of v and W.

Let us take for g a function with supp $\tilde{g} \subset (v-\delta, v+\delta)$ and $\tilde{g}(k)=1$ for $|k-v| < \delta/2$; furthermore we introduce a coupling constant λ, then we can sum up our results in the following

Theorem. For the pair H_o, $H = H_o + \lambda V$ and incident beam direction $\hat{e}$ the total cross section is bounded by

$$\int_{v-\delta/2}^{v+\delta/2} \sigma_{tot} \; (k,\hat{e})dk \leq C(\delta)(\lambda/v)^2 \; \|V\|^2_{\hat{e}} \; . \tag{3.18}$$

Note that the bound can be calculated explicitly and that it depends on the beam direction for non-isotropic forces. It is correct or close to optimal in its dependence on several properties as we will discuss now.

There are some simple sufficient conditions for $\|V\|_{\hat{e}} < \infty$. Assume that V is locally square integrable and continuous outside a ball of radius ρ. Then $\|V\|_{\hat{e}}$ is finite if

$$\int_\rho^\infty d\zeta \; \sup_{|z|>\zeta} \; \{\int dx \int dy \; V^2(x,y,z)\}^{1/2} < \infty \; . \tag{3.19}$$

If singularities may occur at arbitrary distances we use the fact that $\psi(K)$ maps $L^2(\mathbb{R})$ into $L^\infty(\mathbb{R})$ in z-space and the kernel decays rapidly. Therefore the decay of local L^2-norms is sufficient and we obtain

$$\|V\|_{\hat{e}} \leq \text{const} \int_o^\infty d\zeta \quad \sup_{|z'|\geq\zeta} \quad \int_{|z-z'|\leq 1} dz \quad \{\int dx \int dy \; V^2(x,y,z)\}^{1/2} . \tag{3.20}$$

We will get a finite total cross section if the potential is bounded at infinity by

$$|x|^{-(1/2)-\varepsilon} |y|^{-(1/2)-\varepsilon} |z|^{-1-\varepsilon}, \; \varepsilon > o \; . \tag{3.21}$$

The total cross section is finite for all directions if the

decay is like

$$|\vec{r}|^{-2} \; (\ln|\vec{r}|)^{-1-\varepsilon}, \; \varepsilon > 0 . \tag{3.22}$$

Up to a square root of the logarithm (see [11]) this is optimal. Using the variable phase method of Calogero [4] and Babikov [3] one proves that the total cross section is infinite for some spherically symmetric potential with $|\vec{r}|^{-2}\,(\ln|\vec{r}|)^{-1/2}$ - decay (see the remark after Prop. 2.3 and Appendix 2 in [8]).

If the coupling constant λ is small or the energy high (i.e. (λ/v) small) then the Born approximation converges and it gives the same $(\lambda/v)^2$ - behavior as our bound (3.18).

Also in the strong coupling limit $\lambda \to \infty$ (v fixed) there is for any $\mu < 2$ a spherically symmetric potential with $\|V\|_{\hat{e}} < \infty$ such that the total cross section increases at least like $(\lambda)^{\mu}$. For potentials with faster decay, however, we will prove a slower increase in λ in Section 5.

The main drawback of our geometric method is that we do not get estimates for sharp energy: our bound $C(\delta)$ in (3.18) remains bounded but does not decrease like $O(\delta)$ as $\delta \to 0$. Related to this we get poor estimates on the low energy behavior. The reason for this limitation is our estimate

$$\| \int_{-\infty}^{\infty} dt \; e^{i\,Ht} \; V \; e^{-i\,H_0 t} \; g \| \tag{3.23}$$

$$\leq \int_{-\infty}^{\infty} dt \; \| V \; e^{-i\,H_0 t} \; g \| \, . \tag{3.24}$$

For a small momentum spread δ(and similarly for potentials with small support) the size of the wave packet g in z-space becomes large compared to the size of the region where V is strong; the main contribution to the integral (3.23) comes from a time interval $\sim \delta^{-1}$. In the continuous spectral subspace for H, away from zero-energy resonances, one expects a growth

$$\| \int_{-\delta^{-1}}^{\delta^{-1}} dt \; e^{+i\,Ht} \; V \| \sim \delta^{-1/2}$$

rather than the δ^{-1} of our estimate. The cancellations in (3.23) which are lost in (3.24) would be necessary to get good bounds for sharp energies (or small obstacles).

To sum up our strategy in this section was as follows:

according to our definition (2.5) of σ_{tot} we have to estimate $\|(S-1)g\|$ for plane wave packets g. It is bounded by

$$\int_{-\infty}^{\infty} dt \; \|V \, e^{-i H_o t} g\| \; . \qquad (3.25)$$

This expression is particularly convenient because it uses *freely* evolving wave packets and the potential V, but it does not use the full Hamiltonian H. The same properties are shared by the first order Born approximation which will give better approximations in the parameter range of its applicability. Our bound, however, is a universal upper bound.

The travelling plane wave packet $e^{-i H_o t}$ g is mainly localized in a region where $z \approx v\,t$, $v \in \text{supp}\, \tilde{g}$, the velocity(=momentum) support of g; the tails into the classically forbidden region decay rapidly. Thus one has to control that V as a map from suitable plane wave packets localized in $|z| \geq R$ into the Hilbert space is of short range (has a norm integrable in R). This is exactly what the $\|\cdot\|_{\hat{e}}$-norm controls (the factor $\psi(K)$ is simply a regularization which smoothes out local singularities; it does not affect the decay properties of the potential). At each time the effect of the potential on the plane wave packet is independent of the mean velocity v, but the time necessary for the wave packet to pass the potential behaves like v^{-1} . Thus the $(\lambda/v)^2\|V\|_{\hat{e}}^2$ -bound is quite natural.

For potentials V with stronger singularities like the Rollnik class which are form bounded perturbations of H_o one can use the intertwining property of the wave operators to get the estimate

$$\|(S-1)g\|$$
$$\leq \int_{-\infty}^{\infty} dt \|\psi[(2H^{a.c.})^{1/2} - v] \; V \; \psi[(2H_o)^{1/2} - v] \, e^{-i H_o t} g\|$$

for states with momentum support around v . Here we have used that $g = \psi(K-v)g = \psi[(2H_o)^{1/2} - v]g$ for these states.

If the interaction term

$$\psi[(2H^{a.c.})^{1/2} - v)] \; V \; \psi[(2H_o)^{1/2} - v]$$

is of short range as a map from plane wave packets into the Hilbert space *uniform* in v, then all the above results remain true. There is another way to handle even stronger local singularities with the Kupsch-Sandhas trick, explained in the next section. But then the high energy decay in (3.18) is lost.

4. POTENTIALS OF COMPACT SUPPORT

If the support of the potential V is contained in a ball of radius R then the classical total cross section can at most be πR^2 no matter what the potential is. We have just counted all particles entering the interaction region as potentially being deflected. Similarly in the quantum case we get a uniform bound (independent of the potential) by estimating the part of the plane wave packet which can possibly be influenced by the potential. The technical trick used for this estimate is due to Kupsch and Sandhas [10].

Let $j(|\vec{r}|) = 1$ (resp. 0) if $|\vec{r}| \leq 1$ (resp. ≥ 2) be a smooth cutoff function and define

$$j_R(|\vec{r}|) = j(|\vec{r}| / R). \tag{4.1}$$

For any $R < \infty$ and any $\Phi \in \mathcal{H}$ we have

$$\| j_R \, e^{-i H_o t} \Phi \| \to 0 \text{ as } |t| \to \infty , \tag{4.2}$$

and the same is true if Φ is replaced by a plane wave packet g (see the Lemma in Section 3). Therefore

$$\begin{aligned} \Omega &= \text{s-lim}\, e^{i Ht} e^{-i H_o t} \\ &= \text{s-lim}\, e^{i Ht} (1-j_R) e^{-i H_o t} \end{aligned} \tag{4.3}$$

and with the Cook argument

$$\begin{aligned} \Omega^+ - \Omega^- &= i \int_{-\infty}^{\infty} dt\, e^{i Ht} \{H(1-j_R) - (1-j_R) H_o\} e^{-i H_o t} \\ &= -i \int_{-\infty}^{\infty} dt\, e^{i Ht} [H_o, j_R] e^{-i H_o t} \\ &= i \int_{-\infty}^{\infty} dt\, e^{i Ht} \{\tfrac{1}{2} (\Delta j_R) + (\vec{\nabla} j_R) \cdot \vec{\nabla}\} e^{-i H_o t} . \end{aligned} \tag{4.4}$$

Here we have used that $V(1-j_R) = 0$ if V has support inside a ball of radius R, no matter how bad the singularities of V may be. If for the description of hard cores or other severe local singularities an identification operator is used to define the wave operators then the second line of (4.3) and (4.4) are still true for big enough R.

Inserting the "potential" $\{\frac{1}{2}\Delta j_R + (\vec{\nabla} j_R)\cdot\vec{\nabla}\}$ in the estimates of the preceding section, e.g. (3.19) with $\rho = 0$, one easily obtains

Theorem. Let $H = H_o + V$, V any potential with support contained in a ball of radius $R \geq R_o$, then for $v > 2\delta$

$$\int_{v-\delta}^{v+\delta} \sigma_{tot}(k,\hat{e})\, dk \leq \text{const. } R^2, \tag{4.5}$$

where the constant depends on δ and R_o but is independent of v,R and V.

Except for the value of the constant in (4.5) (see Section 6 for estimates) the bound is saturated for large R by hard cores giving $2\pi R^2$. The energy decay has disappeared because the gradient applied to g in (3.17) yields an increase proportional to v. The remarks following (3.24) showed why the small R behavior of our simple bound is not optimal. The correct behavior as $R\to 0$ should be a constant because there are point interactions with non trivial scattering, see [8] for a discussion.

5. STRONG COUPLING BEHAVIOR

For strong coupling, when the Born approximation does not converge, the traditional time independent method yields finiteness of the total cross section but no control on its size because a Fredholm alternative is used to solve the Lippman Schwinger equation. Recently Amrein and Pearson [1] gave a bound independent of λ for potentials of compact support and increasing as λ^2 otherwise. Martin [11] proved a λ^4-bound for spherically symmetric Rollnik potentials. Actually the increase in the coupling constant λ will depend on the decay at infinity of the potential.

Theorem. Let $V(\vec{r})$ obey for some $\alpha > 2$, $r := |\vec{r}| \geq R_o$:

$$|V(\vec{r})| \leq \text{const. } (1 + r)^{-\alpha} \tag{5.1}$$

or

$$|V(\vec{r})| \leq \text{const. } e^{-\mu r} \tag{5.2}$$

then for given direction $\hat{e}$ and Hamiltonian $H_o + \lambda V$

$$\int_{v-\delta}^{v+\delta} \sigma_{tot}(k,\hat{e};\lambda)\, dk \leq D(\delta) \begin{cases} (\lambda/v)^{\gamma}, \gamma = 2/(\alpha-1) & (5.3) \\ \text{or} & \\ \ln^2(\lambda/v) & (5.4) \end{cases}$$

where $v > 2\delta$, $(\lambda/v) > 2$.

Remark. The power γ in (5.3) is correct because there are spherically symmetric potentials decaying like (5.1) with a lower bound increasing like (5.3). For $\alpha > 3$ trace class methods give similar results (see Appendix 2 and 3 in [8]).

Proof. Using j_R (4.1) we get in (4.4) in addition to the commutator a tail term $\lambda V(1-j_R)$. Combining the bounds (3.18) for the tail part and (4.5) for the inner part which is independent of V and λ, we obtain

$$\int_{v-\delta}^{v+\delta} \sigma_{tot}(k,\hat{e};\lambda)\,dk \le D(\delta)[R^2 + (\lambda/v)^2 \| V(1-j_R)\|_{\hat{e}}^2] \ .$$

We minimize the bound by choosing $R = (\lambda/v)^{1/(\alpha-1)}$ in case (5.1) and $R = (\mu')^{-1} \ln(\lambda/v)$, $\mu' < \mu$ in case (5.2).

□

6. THE CLASSICAL LIMIT

So far we have chosen our units such that Planck's constant $\hbar=1$. We reinsert it to study the classical limit $\hbar\to 0$ for the pair $H_o = -(\hbar^2/2)\Delta$, $H = H_o + V$. The wave number is $\vec{k}=\hbar^{-1}\vec{p}$ for the physical momentum $\vec{p}$(= velocity). Scaling the time this corresponds to scattering for the pair $H_o = -(1/2)\Delta$, $H = H_o + \hbar^{-2}V$; thereby the S-operator and its kernel in $\vec{k}$-space are not changed. With the physical momentum $\vec{p}$ fixed the wave vector k diverges as $\hbar^{-1}$ in the classical limit $\hbar\to 0$ and the coupling constant λ grows as $\hbar^{-2}$. In terms of the quantites of the previous section we have $v \sim \hbar^{-1}$ ($\hbar$ v = const is the physical velocity), thus $(\lambda/v) \sim \hbar^{-1}$ diverges and the classical limit is a strong coupling limit. We have seen that the total cross section then generally diverges unless the potential has compact support. Therefore infinitely extended potentials will in general have infinite classical total cross sections.

Fix now an obstacle or potential of compact support and let F be its area as seen from the fixed incident beam direction $\hat{e}$. The classical cross section can be determined with any beam which covers F, the particles passing outside F will miss the target and they do not contribute. Similarly in the quantum case the part of the plane wave packet which is at time $t = 0$ far away from F will hardly be scattered by the potential. The main contribution to the total cross section comes from the part that covers F, the outside part does not contribute in the classical limit $\hbar\to 0$. This gives simple estimates of σ_{tot} in the quasiclassical regime and moreover allows to give bounds on the constant in (4.5) for the large R behavior.

For simplicity we treat obstacles inside a ball of radius R, the changes necessary to treat other shapes are obvious.

Similar to (3.5) we use a family of smooth cutoff functions in the plane perpendicular to $\hat{e}$. Let $\phi \in C^\infty(\mathbb{R})$ be monotone with $\phi[u] = 1$ (resp. 0) if $u \leq 0$ (resp. ≥ 1). Define for $R,s > 0$

$$f_{R,s}(x,y) := \phi[\{(x^2+y^2)^{1/2} - R - 2s\}/s], \tag{6.1}$$

this implies

$$f_{R,s}(x,y) = \begin{cases} 1 & \text{for } x^2+y^2 \leq (R+2s)^2 \\ 0 & \text{for } x^2+y^2 \geq (R+3s)^2 . \end{cases} \tag{6.2}$$

Now split the plane wave packet g as

$$g = g\, f_{R,s} + g(1-f_{R,s}) \tag{6.3}$$

then the Hilbert space norm of the first summand is bounded by

$$\| g\, f_{R,s} \|^2 \leq \| g \|^2\, \pi(R+3s)^2 \tag{6.4}$$

where the norm of g is in $L^2(\mathbb{R})$.

Now consider the normalized sequence of wave packets

$$\tilde{g}_\hbar(k) = (\hbar)^{1/4}\, \tilde{G}[\sqrt{\hbar}(k-v/\hbar)]$$

for $\tilde{G} \in C_0^\infty(\mathbb{R})$, v a given physical velocity. Then in the limit $\hbar \to 0$ $|g_\hbar(z)|^2\, dz$ converges to $\delta(z)\, dz$ and the distribution of the physical momentum $p = \hbar\, k$ converges to $\delta(p-v)dp$.

With the free time evolution generated by $h_0 = -(\hbar^2/2)d^2/dz^2$ the estimate (3.13) of the Lemma in Section 3 can be changed to

$$\int_{|z| \leq vt/2} dz\, |[e^{-i h_0 t} g_\hbar](z)|^2 \leq C_m \hbar^m (1+|vt|)^{-m}. \tag{6.5}$$

As in Section 4 we use again the Kupsch-Sandhas trick for the estimate of

$$\begin{aligned} &\| (S-1) g_\hbar (1-f_{R,s}) \| \\ &\leq \int_{-\infty}^{\infty} dt\, \| \{(1/2)(\Delta j) + (\vec{\nabla} j)\cdot\vec{\nabla}\}\, e^{-i H_0 t}\, g_\hbar (1-f_{R,s}) \| \end{aligned} \tag{6.6}$$

where we choose $j \in C_0^\infty$ with $j(\vec{r}) = 1(0)$ if $|\vec{r}| \leq R(\geq R+s)$ and $|\vec{\nabla}j|$, Δj are proportional to s^{-1}, s^{-2}. The support of the "bounded potential" in curly brackets in (6.6) is contained in a ball of radius R+s. With the estimate (6.5) the time integration in (6.6) has a contribution for $|t| \geq 2(R+s)/v$ which is bounded by const $\hbar^m(s^{-1}+s^{-2})$.

It remains to estimate the tail of

$$e^{-i H_0 t}(1-f_{R,s}(x,y)) \tag{6.7}$$

which propagates into the region $x^2+y^2 \leq (R+s)^2$ for the time interval $|t| \leq 2(R+s)/v \leq 4R/v$ (if $s \leq R$) independent of $\hbar$. Now we let s tend to zero slowly as $\hbar \to 0$, e.g. like a small power $\hbar^\varepsilon$, then the momentum distribution shrinks as $\hbar^{(1-\varepsilon)}$, we have to control propagation beyond a distance $\hbar^\varepsilon$ into the region where the "potential" of strength $\hbar^{-2\varepsilon}$ acts. The same kind of estimate as above yields

$$\lim_{\hbar\to 0} \sup_{|t| \leq 4R/v} \| \{(1/2)(\Delta j) + (\vec{\nabla}j)\cdot\vec{\nabla}\} \times e^{-i H_0 t} g_\hbar (1-f_{R,s}) \| = 0. \tag{6.8}$$

Thus the contribution to σ_{tot} from the outer part disappears

$$\lim_{\hbar\to 0} \| (S-1) g_\hbar (1-f_{R,s}) \| = 0.$$

With (6.4) and $\|S-1\| \leq 2$ we obtain

$$\sigma_{tot}(v;\hbar=0) := \lim_{\hbar\to 0} \| (S-1) g_\hbar f_{R,s} \|^2$$

$$\leq 4\pi R^2 .$$

In the classical limit the sharp energy quantum cross section is bounded by four times the geometric classical πR^2 (or 4F for general shapes).

By scaling one can see that the relevant quantity is the dimensionless $k R = p R/\hbar$ which has to be big. Thus $\hbar \to 0$ is equivalent to the high energy or large R limit for given physical $\hbar$. Using this (or an analogous estimate as above with s growing slightly slower than R) we can improve (4.5) in the Theorem of Section 4 ($\hbar$ fixed):

$$\int_{v-\delta}^{v+\delta} \sigma_{tot}(k,\hat{e})dk \leq 2\delta\, 4\pi R^2 + o(R^2). \tag{6.9}$$

The remainder term can be estimated explicitly.

This bound (6.9) is saturated if $S \approx -1$ on $g\, f$. This happens for potentials of the type

$$V(x,y,z) = a\, \chi_{[-r,r]}(z)\, \chi_{R^2}(x^2+y^2)$$

discussed in Section 2. For δ small and suitably adjusted parameters a and r depending on $v/\hbar$ the S operator can approximate any phase factor. In particular for $\hbar \to 0$ it oscillates between $S \approx -1$ and $S \approx 1$, therefore σ_{tot} does not converge as $\hbar \to 0$.

For typical potentials, however, the particles are deflected if they hit the target and only very few of them continue to fly approximately in the forward direction. Then $g\, f$ and $S\, g\, f$ are approximately orthogonal and

$$\int \sigma_{tot}(k) |\tilde{g}(k)|^2 dk \approx 2\pi R^2 \|g\|^2. \qquad (6.10)$$

For general shapes $2\pi R^2$ is replaced by $2F$. This is the well known "shadow scattering"-result which holds e.g. for hard spheres. A short time after the scattering $g\, f$ and $S\, g\, f$ are essentially localized in disjoint regions, thus for these particular beams simple amplitude measurements close to the target can be made. In Section 5 of [8] we propose a characterization of potentials which should be "typical" in the above sense.

To sum up this discussion we have seen that with our definition of the total cross section it is perfectly justified to use beams of finite width. For small but macroscopic targets (k R big enough depending on the admissible error) a beam is even wide enough if it just covers the target. Moreover for typical potentials simple measurements can be carried out near the target which should yield good approximate results.

On the other hand the conventional definition based on counters detecting deflected particles will always require a much wider beam. An extremely well collimated beam of finite width 2ρ (like a laser beam) will typically have the following shape. Up to a finite distance it looks like a plane wave restricted to a tube of radius ρ and asymptotically it looks like a spherical wave restricted to a cone. By the uncertainty principle the momentum- (=velocity-) spread perpendicular to the propagation direction is of the order $\hbar\, \rho^{-1}$ which should be small compared to the average velocity v. The opening angle of the asymptotic cone is then $\hbar(\rho v)^{-1}$. The transition between the two regimes happens near a distance D where the cone is as wide as the tube, i.e. $D \approx \rho^2\, v/\hbar$.

A counter which should detect only deflected particles must be located outside the union of the tube and the cone where the incoming beam would propagate. Consider for example a target of radius $R \leq \rho$. Typically most particles which hit the target will be significantly deflected and are detected easily giving the classical geometric cross section πR^2. The subtle effects come from the "shadow", the particles missing behind the target; their wave function is the negative of the part of the plane wave packet restricted in the perpendicular direction to the radius R at time zero. Its tube region is contained in the bigger tube of the incoming beam and thus never matters. Later it spreads into a cone with angles $\tan \theta \leq \hbar (Rv)^{-1}$. For the main part of this wave to be detectable outside the cone of the incoming wave one has to choose $\rho \gg R$. One can see these "shadow"-particles only if their cone region is wider than the tube of radius ρ, i.e. beyond a minimal distance $d \approx R \rho v/\hbar$ from the target. To get an idea of the order of magnitude take a neutron of energy 100 eV, a target of radius 10^{-5}m and a beam ten times wider, then $d \approx 2 \cdot 10^3$m! Increasing the mass or energy of the projectile or the size of the target will only increase this distance. In the laboratory one will see nothing but the classical cross section for tiny but macroscopic targets if deflected particles are counted. (See also [13] where an approximate calculation for hard spheres is given.) Although both definitions of the total cross section agree asymptotically our definition has the advantage of giving a good approximate value from observations within a reasonable distance of the target.

7. TWO CLUSTER SCATTERING

So far we have studied potential scattering. This is equivalent to two particle scattering if one can separate off the center of mass motion, i.e. if the potential depends only on the relative position of the particles. Similarly one can consider in the multiparticle case scattering of two bounded subsystems like atoms; the relative position and momentum of the centers of mass for the two subsystems corresponds to position and momentum in potential scattering. A "channel" is specified if both the decomposition of the particles into clusters and the bound states for each cluster are given. For each channel (labelled by the index α) there is a subspace $\mathcal{H}_\alpha$ of the state space $\mathcal{H}$ consisting of product wave functions

$$\phi \prod_i \eta_i \tag{7.1}$$

where ϕ is the square integrable function which describes the relative motion of the centers of mass of the clusters, and η_i are the cluster bound state wave functions. The cluster Hamiltonian $H(\alpha)$ which leaves $\mathcal{H}_\alpha$ invariant is obtained from the full Hamiltonian

H as $H(\alpha) = H - I_\alpha$, where I_α is the sum of all potentials which couple particles in different clusters. The channel wave operators are mappings from $\mathcal{H}_\alpha$ into $\mathcal{H}$ defined as

$$\Omega_\alpha^\mp = \lim_{t\to\pm\infty} e^{i\,Ht}\, e^{-i\,H(\alpha)t} \tag{7.2}$$

and the full wave operators

$$\Omega^\mp = \bigoplus_\alpha \Omega_\alpha^\mp \tag{7.3}$$

are isometric mappings into $\mathcal{H}$ from $\oplus\,\mathcal{H}_\alpha$ which is interpreted as space of outgoing or incoming configurations, respectively. The scattering operator

$$S = (\Omega^-)^* \,\Omega^+$$

maps incoming configurations into outgoing ones. (See Section XI.5 of [12] for details.) For the channel α and a given incoming state $\Psi_\alpha \in \mathcal{H}_\alpha$ one has

$$\|(S-1)\Psi_\alpha\| \le \|(\Omega_\alpha^+ - \Omega_\alpha^-)\Psi_\alpha\| \tag{7.4}$$

$$\le \int_{-\infty}^{\infty} dt\, \|I_\alpha\, e^{-iH(\alpha)t}\, \Psi_\alpha\| \tag{7.5}$$

similar to the two particle case. (7.4) is an equality if asymptotic completeness holds.

Fix now a *two* chuster channel α, then the incoming wave function in $\mathcal{H}_\alpha$ is of the form

$$\Psi_\alpha = \phi_\alpha\, \eta_1\, \eta_2 \tag{7.6}$$

and ϕ_α is a function of one variable, the relative coordinate of the two centers of mass. (As usual we have separated off the total center of mass motion.) The total cross section is now defined analogously. In addition to the clusters which are scattered elastically and deflected one also counts all excitations, breakup and rearrangement collisions. In our wave-limit approach we use incoming plane wave packets of the channel α described by

$$g_\alpha := g\, \eta_1\, \eta_2 \tag{7.7}$$

where η_i are the corresponding bound state wave functions (or one if the cluster consists of a single particle) and g is the familiar plane wave packet for the relative center of mass motion, obtained as a limit of square integrable functions as discussed in Section 3. We now define

$$\int_0^\infty \sigma_{tot}\ (k,\hat{e};\alpha)\ |\tilde{g}(k)|^2\ dk\ :=\ \|(S-1)g_\alpha\|^2, \tag{7.8}$$

and we use (7.5) as a simple estimate.

$H(\alpha)$ acts trivially on η_1 and η_2 and reduces on g to $h_\alpha = -(1/2\ m_\alpha)\ d^2/dz^2$ where m_α is the reduced mass of the two clusters. Therefore

$$\|I_\alpha\ e^{-i\ H(\alpha)t}\ g_\alpha\| = \|V_\alpha\ e^{-i\ h_\alpha t}\ g\| \tag{7.9}$$

with the effective potential between the clusters

$$V_\alpha(\vec{r}) = [\int |I_\alpha(\vec{r},\zeta_1,\zeta_2)|^2\ |\eta_1(\zeta_1)|^2\ |\eta_2(\zeta_2)|^2\ d\zeta_1\ d\zeta_2]^{1/2} \tag{7.10}$$

Here $\vec{r}$ is the separation of centers of mass and ζ_i are the inner-cluster coordinates (of dimension 3(k-1) if k particles belong to the cluster). The analysis of Section 3 immediately applies and all we have to do in the multiparticle case is to control the decay of $V_\alpha(\vec{r})$ for a proof of finite total cross sections. On the other hand the analysis of Sections 4 - 6 cannot be used directly because effective potentials won't have compact support, and a growing coupling constant will change I_α and the bound state wave functions η_i simultaneously. This makes it difficult to control the strong coupling or classical limits.

Typical bound state wave functions (except at thresholds) have exponential decay [5]. If there are bound states with slow decay we will omit in the following the corresponding channels. If all pair potentials contributing to I_α

$$V_{ij}(\vec{r}_i - \vec{r}_j) = V_{ij}(\vec{r} + \vec{\zeta}_{1,i} - \vec{\zeta}_{2,j}) \tag{7.11}$$

decay faster than $|\vec{r}_i - \vec{r}_j|^{-2}$ as specified in (3.22), then the convolution in (7.10) preserves this property and the total cross section is finite. Of particular physical interest, however, is the case of long range pair potentials like the Coulomb force between charged particles which may nevertheless give rise to an effective potential of short range. This is the case for atom-atom scattering. If both clusters are neutral then the contribution with slowest decay is the dipole-dipole potential which behaves as $|\vec{r}|^{-3}$.

Let i label the particles in one cluster and j label those in the other. Consider as a typical example pair potentials of the type

$$e_i e_j |\vec{r}_i - \vec{r}_j|^{-1} + V_{ij}(\vec{r}_i - \vec{r}_j) \in L^2_{\ell oc}$$
$$V_{ij}(\vec{u}) = O(|\vec{u}|^{-2-\varepsilon}) \text{ for } |\vec{u}| \geq R_o \ . \tag{7.12}$$

If both clusters are neutral: $\Sigma e_i = \Sigma e_j = 0$, then $V_\alpha(\vec{r})$ satisfies (3.20). This proves the following

Theorem. Let N charged particles interact with pair potentials which fulfill (7.12). Let α be a channel with two neutral clusters whose bound state wave functions have rapid decay. Then

$$\int_{v-\delta}^{v+\delta} \sigma_{tot}(k,\hat{e};\alpha)dk \leq C(\delta)\ v^{-2}\|V_\alpha\|^2_{\hat{e}} \tag{7.13}$$

is finite and $C(\delta)$ is independent of the channel.

One expects that a similar result holds if one cluster is neutral and does not have a permanent dipole moment, but we cannot prove that ([7] and Section 6 of [8]).

REFERENCES

1. W.O. Amrein and D.B. Pearson, J. Phys. A 12, 1469 (1979).
2. W.O. Amrein, D.B. Pearson, and K.B. Sinha, Nuovo Cimento 52A, 115 (1979).
3. V.V. Babikov, Sov. Phys. Uspekhi 92, 271 (1967).
4. F. Calogero, The Variable Phase Approach to Scattering, Academic Press, New York 1967.
5. P. Deift, W. Hunziker, B. Simon, and E. Vock, Commun. Math. Phys. 64, 1 (1978).
6. J.D. Dollard, Commun. Math. Phys. 12, 193 (1969).
7. V. Enss and B. Simon, Phys. Rev. Lett. 44, 319 and 764 (1980).
8. V. Enss and B. Simon, Commun. Math. Phys., in press.
9. T.A. Green and O.E. Lanford, J. Math. Phys. 1, 139 (1960).
10. J. Kupsch and W. Sandhas, Commun. Math. Phys. 2, 147 (1966).
11. A. Martin, Commun. Math. Phys. 69, 89 (1979) and 73, 79 (1980).
12. M. Reed and B. Simon, Methods of Modern Mathematical Physics, III Scattering Theory, Academic Press, New York 1979.

13. R. Peierls, Surprises in Theoretical Physics, Princeton University Press, 1979.
14. T. Kato, Scattering Theory, in: Studies in Mathematics Vol 7, Studies in Applied Mathematics, A. H. Taub ed., The Mathematical Association of America, 1971; page 90-115.

CLASSICAL-QUANTUM CORRESPONDENCE IN NON-LINEAR SYSTEMS

Eric J. Heller, Ellen B. Stechel, & Michael J. Davis

University of California
Department of Chemistry
Los Angeles, CA 90024

ABSTRACT

Quantum mechanical autocorrelation functions are surprisingly accurate using classical dynamics with quantum initial conditions, for parameters appropriate to molecular vibration. The accuracy generally decreases with increasing time; thus the classically determined Fourier transform power spectra (molecular absorption spectrum) are most accurate at low and intermediate spectral resolution, by the time-frequency uncertainty relation. Spectral band widths and other absorption features are given a simple classical interpretation and easily calculated by running a modest number of classical trajectories.

One of the most useful tools available for the analysis of dynamical systems is the Fourier Transform Power Spectrum. Such diverse areas as fluid dynamics, electronic networks, vibration analysis, stochastic processes, molecular physics and many more find a common link in the thoery and practice of Fourier analysis. In the optical spectroscopy of polyatomic molecules, which is the motivation for the present work, attention has mainly been focussed on the frequency domain (i.e., absorption or emission spectra). The challenge to the spectroscopist was (and to a large extent still is) to obtain fundamental structure, harmonic force constance and anharmonicities (within perturbation theory) by measuring line spectra. Most of this work is carried out at low total energies, and line positions (corresponding to transitions between two quantum eigenstates) and sometimes intensities are used to derive the parameters which characterize the potential surface (or surfaces) on which the dynamics takes place. The texts by Herzberg[1] have for years been the primary general reference for these types of spectroscopy.

The Fourier transform of these molecular spectra into the time domain has received rather less attention than the spectra themselves. In part, we feel this has been due to a theoretical vacuum as to the physical meaning of the time domain results, particularly in the case of electronic spectra, where the potential surface on which the nuclei move changes suddenly upon absorption of a photon. However, it has long been understood[2] that the Fourier transform of any frequency domain spectrum gives ensemble averaged autocorrelation functions of the appropriate transition moment, e.g.

$$\varepsilon(\omega) \propto \int_{-\infty}^{\infty} e^{i\omega t} \langle \underline{\mu}\,\underline{\mu}(t) \rangle \, dt \tag{1}$$

where $\varepsilon(\omega)$ is the absorption intensity at frequency ω, $\underline{\mu}$ is the appropriate transition moment operator, $\underline{\mu}(t)$ is that operator propagated in time according to the Heisenberg picture, and the brackets imply an ensemble average. The case of linewidth-lifetime relations for lines broadened by interaction of a molecule with a medium has received wide attention[3] and here it is fair to say that there is a good understanding of the physical processes leading to (typically) exponential decay and Lorenzian lineshapes. In infrared and microwave transitions, molecules do not typically change electronic state and the selection rules give the bulk of the intensity to relatively few spectra lines; this makes the time domain results rather uninteresting except for the exponential decay of the individual lines just mentioned. Electronic spectra are very much richer in this respect, and tens or hundreds or billions of lines can have significant intensity in a typical electronic spectrum.

The focus of our work has been to interpret and predict the <u>patterns</u> of intensity appearing in the spectra, and we wish to literally "defocus" attention on individual lines by seeking the physical meaning of the patterns. The spectroscopists' natural tendency is to concentrate attention on the highest resolution features available in his spectrometer; these features are the hard-won fruits of clever and meticulous effort. By looking for physical meaning in the <u>patterns</u> of intensity we can defocus the spectrometer and examine the spectrum at lower resolution. In the time domain, events which occur on a time scale T are evident on a resolution $\sim 2\pi/T$ in the frequency domain, and to look at higher resolution may obscure the forest for the trees! The information which we are able to glean from this somewhat different approach to spectroscopy is more dynamic than static. Thus, rather than extract explicit potential surface parameters from measurements of line positions, we obtain information on several different types of rate processes which occur even in isolated molecules and which place indirect constraints on the possible form of the potential surfaces.

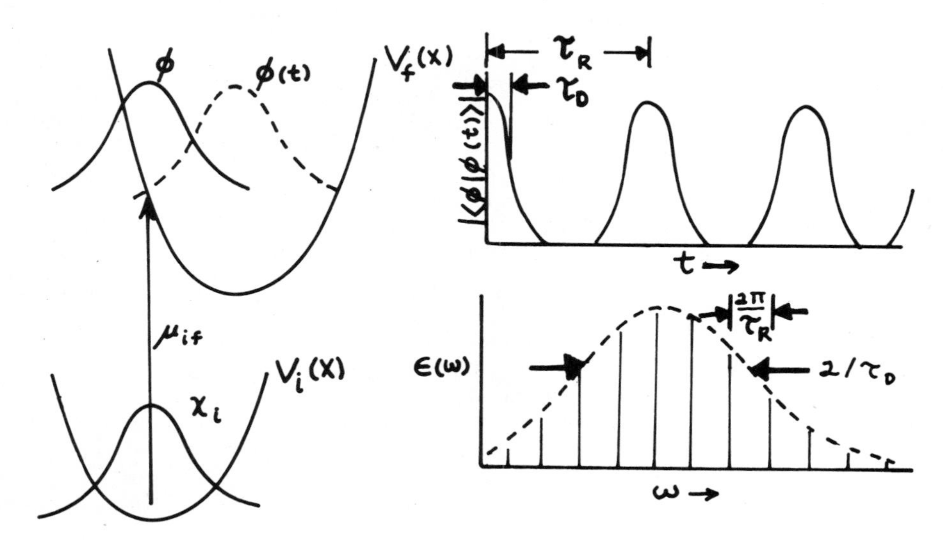

Fig. 1. A diatomic molecule absorbs a photon and the potential of interaction between the two atoms changes from $V_i(x)$ to $V_f(x)$. The moving wavepacket $\phi(t)$ oscillates and gives rise to periodic behavior in $|\langle\phi|\phi(t)\rangle|$ of period τ_R, and structure in ω-space of spacing $2\pi/\tau_R$. Likewise the initial, nearly Gaussian decay time τ_D produces the Gaussian bandwidth $2\tau_D^{-1}$ in ω-space.

The physical meaning of an electronic absorption spectrum is much clearer when Eq. 1 is recast into the form[4]

$$\varepsilon(\omega) = C\cdot\omega \int_{-\infty}^{\infty} e^{i\omega t} \langle\phi|\phi(t)\rangle \tag{2}$$

where $\phi = \mu_{if}(\underline{R})\ \chi_i(\underline{R})$ is the electronic transition moment between surface i and f multiplied by the initial vibrational wavefunction $\chi_i(\underline{R})$ on surface i, and C is a constant. The function $\phi(t)$ is simply ϕ propagated for a time t on the upper "f" surface, according to the time dependent Schrödinger equation

$$i\hbar\dot{\phi}(t) = H_f\phi \tag{3}$$

The function ϕ is physically a compact wavepacket because typically the absorption is out of the ground or low lying vibrational states of the "i" potential surface, and $\phi(t)$ is a moving version of this wavepacket, whose average properties we know from Eherenfest's theorem follow classical dynamics. Thus, we may take an experimental absorption spectrum, divided by the frequency ω, and Fourier transform it to give the overlap between ϕ and $\phi(t)$. For a one dimensional system (corresponding to vibrational motion of a diatomic) the situation is shown in Fig. 1. There is an initial decay of $\langle\phi|\phi(t)\rangle$ followed by periodic buildup in decays caused by approximately periodic (for a harmonic system, exactly periodic) motion of $\phi(t)$. The Fourier transform gives the spectrum, where the overall envelope is governed by the initial fast decay in time, and the line spacing is the reciprocal of the frequency of the periodic motion in time. We shall have more to say about this below, in connection with the two dimensional system which we consider next. The "one dimensional" effects of a rapid falling away of $\phi(t)$ from the region of ϕ, and periodic motion of $\phi(t)$ will be evident also in our two dimensional example, which possesses other more interesting dynamics as well.

Consider the potential

$$V(s,u) = \frac{1}{2}\,\omega_s^2\,s^2 + \frac{1}{2}\,\omega_u^2\,u^2 + \lambda s u^2 \qquad (4)$$

for $\omega_s = 1.0$, $\omega_u = 1.1$, $\lambda = -0.11$. Contours of this potential, together with an outline of the wavepacket

$$\phi(s,u) = (\omega_s\omega_u/\pi^2)^{1/4}\,\exp[-\omega_s/2\,(s-4)^2 - (\omega_u/2)\,u^2] \qquad (5)$$

are shown in Fig. 2. The wavepacket is a displaced ground state eigenfunction of the same potential as Eq. 4 except $\lambda = 0$. The very early time evolution of the overlap $\langle\phi|\phi(t)\rangle$ can easily be guessed on the basis of classical intuition and Eherenfest's theorem. The wavepacket ϕ finds itself perched on the side of a hill, the direction of steepest descent pointing along the s axis toward the equilibrium position $s = 0$, $u = 0$ at the bottom of the well. Clearly, $\phi(t)$ will at first move away in this direction, and the overlap $\langle\phi|\phi(t)\rangle$ will decay to a very small value on a time scale significantly less than a vibrational period. The steepness of the potential in the s-direction will determine the lifetime of this initial decay, which in turn determines the breadth of the overall envelope of the Fourier transform spectrum.

To make this idea a little more quantitative, note that the wavepacket ϕ can be written as a product of two wavepackets, one in each of the two dimensions, i.e.,

$$\phi = \phi_s\phi_u\ . \qquad (6)$$

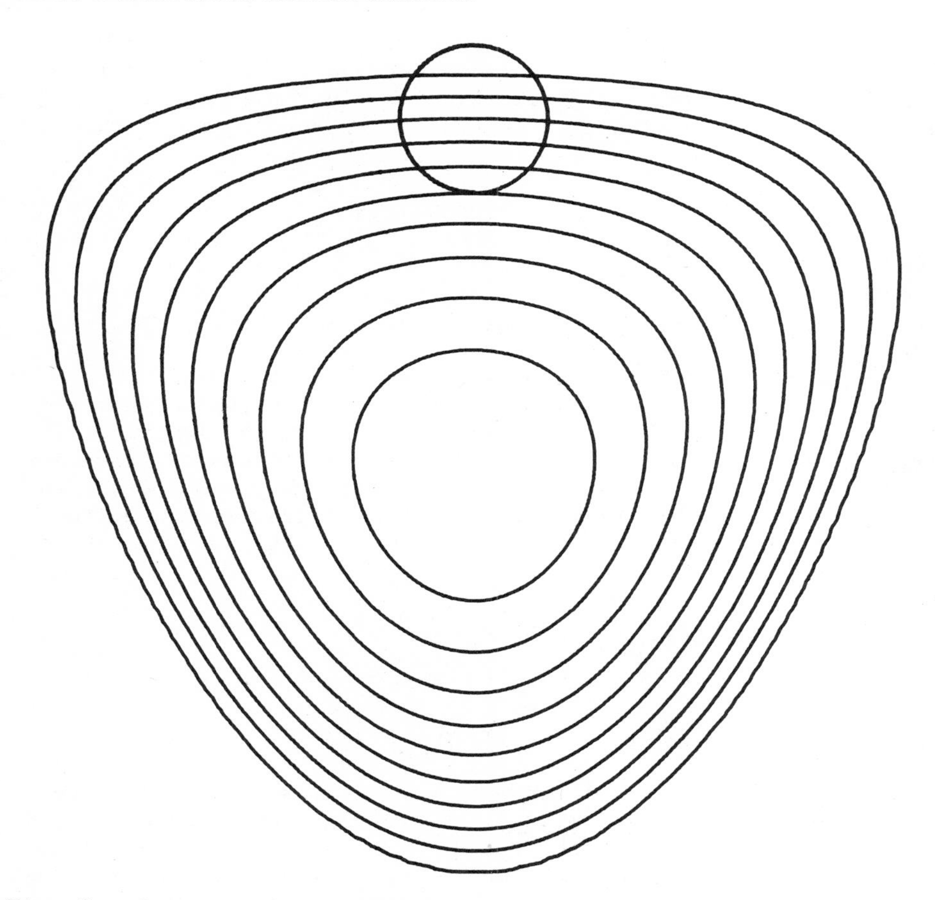

Fig. 2. Contour of the potential function, Eq. 4, and a displaced wavepacket.

Since the potential is separable all along the s axis near u = 0, i.e.,

$$\left.\frac{\partial^2 V}{\partial s \partial u}\right|_{u=0} = 0 \tag{7}$$

the wavepacket $\phi(t)$ remains separable to a good approximation, especially at early times, thus

$$\phi(t) = \phi_s(t)\ \phi_u(t). \tag{8}$$

Further, the initial decay of $\langle\phi|\phi(t)\rangle$ is due almost entirely to the motion in the s-coordinate, so we focus attention on this degree of freedom, which for u = 0 is just a harmonic oscillator. It is very instructive to examine the motion of $\phi_s(t)$ in phase space. Using a Wigner transform[5] the phase space density corresponding to the quantum density matrix

$$\rho_s = |\phi_s\rangle\langle\phi_s| \tag{9}$$

is just

$$\rho_s(s,p_s) = \frac{1}{\pi} \exp[-\omega_s(s-4)^2 - 1/\omega_s\, p_s^2]. \tag{10}$$

In the s,p_s phase space, this density is centered around s = 4, $p_s = 0$. The situation is shown in Fig. 3 along with the phase orbit that passes through this point. In phase space, the moving density $\rho_s(s,p_s;t)$ will evolve exactly as in classical mechanics (the correspondence is perfect for harmonic oscillators) which implies in this case that the circular "blob" of density just travels around the orbit classically with no change in shape. As it begins to move, the classical overlap

$$\int ds\,dp_s\,\rho_s(s,p_s)\ \rho_s(s,p_s;t) \equiv \mathrm{Tr}(\rho_s\rho_s(t)) \tag{11}$$

will decay. Also, it is trivial to show that

$$|\langle\phi_s|\phi_s(t)\rangle| = [\mathrm{Tr}(\rho_s\rho_s(t))]^{1/2} \tag{12}$$

so that, except for a phase, we are able to glean the desired decay in time from the <u>classical motion</u>. To simplify matters further, let us approximate the path of the center of $\rho_s(t)$ by a straight line tangent to the orbit at $p_s = 0$, which is a valid approximation just where the overlap is large. Note that for even larger displacements, which would imply a steeper slope in the vicinity of the displaced wavepacket, the circle representing the phase orbit has even larger radius and the straight line tangent approximation would be even better. The motion of the center of the density in phase space of $\rho_s(t)$ is just the classical motion, which is approximately

$$p_s(t) = \dot{p}_s \cdot t = -\left.\frac{\partial V}{\partial s}\right|_{s=4,\ u=0} \cdot t \equiv -V_s' \cdot t \tag{13a}$$

$$s(t) = s_0 = 4.0. \tag{13b}$$

Thus we have

$$\rho_s(t) = \frac{1}{\pi} \exp[-\omega_s(s-4)^2 - (p_s + V_s' \cdot t)^2/\omega_s] \tag{14}$$

and

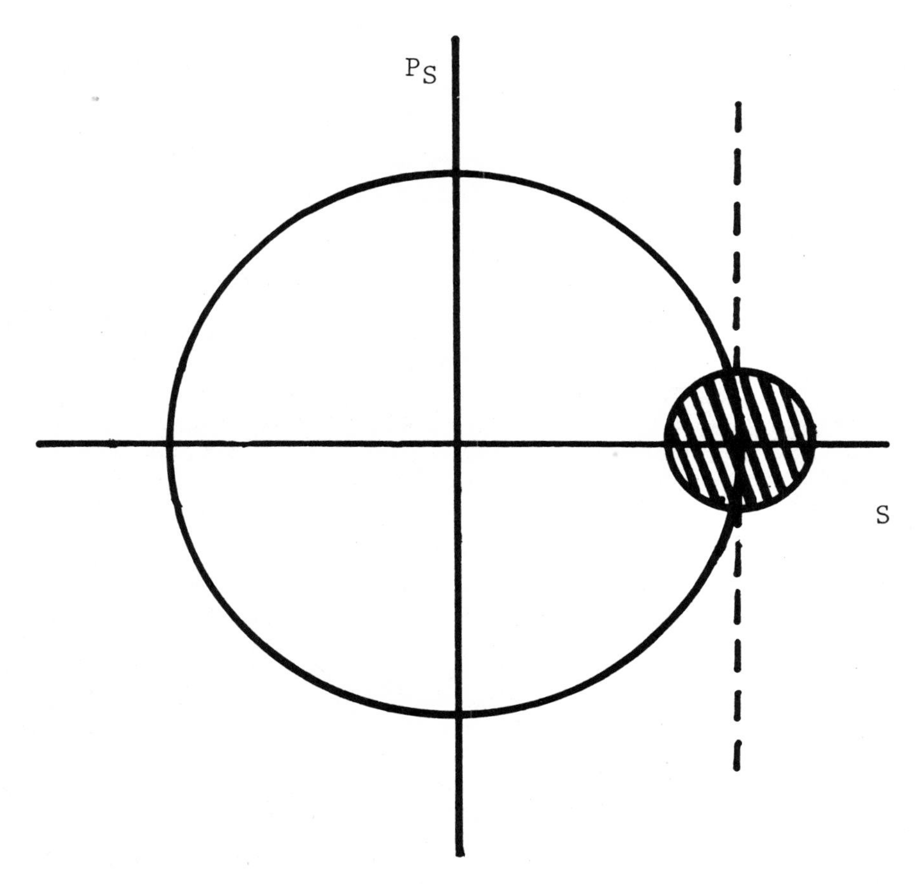

Fig. 3. Phase space map of wavepacket displaced in s-coordinate, with classical phase orbit (circle) shown. A tangent line (dashed) is a reasonable approximate path for the center of the moving wavepacket near where the overlap is large with the fixed wavepacket.

$$[\mathrm{Tr}(\rho_s \rho_s(t))]^{1/2} = |\langle\phi|\phi(t)\rangle| \\ = e^{-V_s'^2 t^2/4\omega_s} . \tag{15}$$

If we ignore the phase of $\langle\phi|\phi(t)\rangle$ in the Fourier transform and instead transform the magnitude, we have

$$\varepsilon(\omega) = C\cdot\omega \, \frac{\sqrt{4\pi\omega s}}{V'_s} \, e^{-(\omega_s/V'^2_s)\omega^2} . \tag{16}$$

Except for an overall shift in the spectrum, the neglected phase should not be too severe for such short times, for we can write

$$<\phi|\phi(t)> \equiv e^{i\theta(t)} \, |<\phi|\phi(t)>| \tag{17}$$

with $\theta \approx \theta_o + \theta_1 \cdot t = \theta_1 \cdot t$. Equation 16 tells us that the overall spectral envelope is a Gaussian of dispersion $V'_s/\sqrt{2\omega_s}$. Thus, steep slopes (large V'_s) or broad wavepackets (small ω_s) will cause a broader envelope. This is in accord with the idea of the reflection approximation.[6] In Fig. 4a the absolute value of the overlap is shown as a function of time. The initial decay in time we have just been discussing is labeled D_1 on the figure, which was computed numerically by a convergent basis set expansion. In Fig. 4c the spectrum is shown at various resolutions, and the Gaussian envelope predicted by Eq. 16 is seen to be quite reasonable.

The initial decay D_1 is the first event in time of four which we will be discussing. Two of the four are decays, and two are recurrences. The recurrences follow on the heels of their corresponding decays, that is, in the order of increasing time, the events are D_1, R_1, D_2, R_2. In the two dimensional example we are discussing, these four events are about all that can be usefully extracted from the time domain. We have just discussed the first event, D_1, its physical basis and application for the spectrum, and we now proceed to analyze the remaining three events.

Whatever happens to $<\phi|\phi(t)>$ after the initial D_1 decay, it cannot alter the Gaussian envelope we have just discussed. This is a general property of Fourier transform which is central to our point of view. It applies that any level in the "decay-recurrence hierarchy" in that once a decay has occured certain average spectral features are "frozen" in, independent of the subsequent dynamics. Mathematically, this can be seen in the Fourier transform of an arbitrary function f(t) by introducing a cutoff function $c_T(t)$, where $c_T(t) \approx 1$ for $|t| < T$ and $c_T(t) \to 0$ for $|t| > T$. From the Fourier convolution theorem, we have

$$\begin{aligned} \hat{f}_T(\omega) &= \frac{1}{\sqrt{2\pi}} \int_{-\infty}^{\infty} e^{i\omega t} \, f(t) \, c_T(t)dt \\ &= \int_{-\infty}^{\infty} d\omega' \, \hat{f}(\omega') \, \hat{c}_T(\omega - \omega'). \end{aligned} \tag{18}$$

The function $c_T(t)$ renders the behavior of f(t) for $|t| > T$ moot, yet the spectrum $\hat{f}_T(\omega)$, convolved or "smoothed" with $\hat{c}_T$, is fully

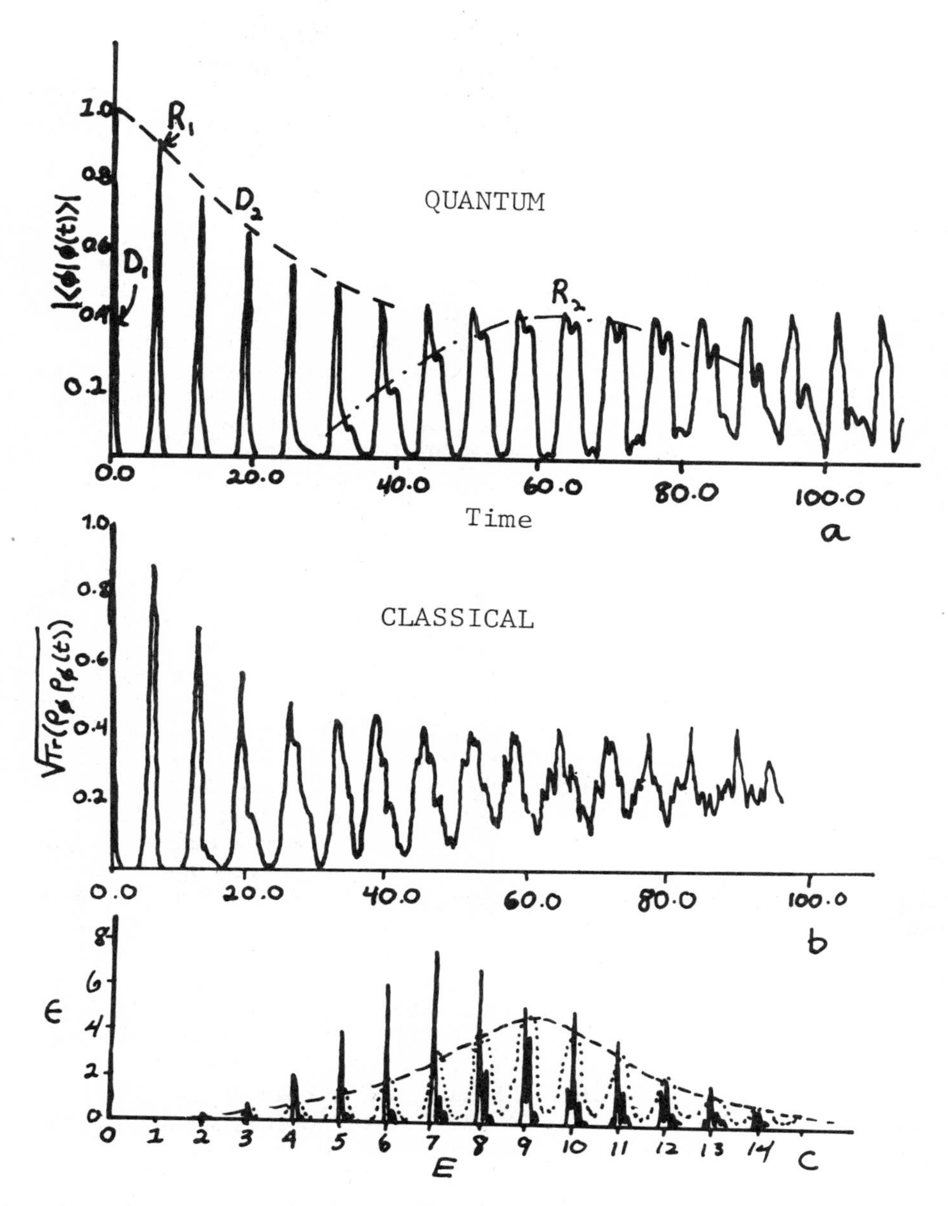

Fig. 4. a) Quantum $|\langle\phi|\phi(t)\rangle|$ for the potential of Eq. 4 and Fig. 1, with labels for the "events" D_1, R_1, D_2, R_2.

b) Same as a) but computed with classical trajectories. See text.
c) Quantum spectrum at various resolutions. Dashed line: Gaussian envelope, showing effects of D_1 only. Dotted line: effect of D_1, R_1 and D_2 only. Solid line: Spectrum seen at high resolution with D_1, R_1 D_2, R_2 effects noticable.

determined by f(t) for $|t| < T$.

Returning to our example, we see that in the vicinity of $t = 2\pi$ there is a partial recurrence (R_1) arising from the return of the wavepacket to the starting position. The recurrence is not perfect because $\phi(2\pi) \neq \phi(0)$. Subsequent peaks at multiples of 2π suffer further reduction in intensity. These peaks are all classed as "R_1" recurrences. The decline of one peak to the next for the first few periods constitutes the second "D_2" decay phenomenon. Finally the subsequent growth in the peaks (their area as well as their height) that starts after about four periods and reaches a maximum at about 10 periods in this is the second (R_2) recurrence. Because of the time sequence $\tau_{D_1} < \tau_{R_1} < \tau_{D_2} < \tau_{R_2}$, we can say that R_1 imposes structure on the spectrum consistent with the envelope imposed by D_1; D_2 imposes broadening on the otherwise sharp single lines which would appear at $E = 1,2...$ in the absence of D_2 and R_2, and that R_2 imposes structure on the broadened sub-envelopes implied by R_1 and D_2. (If we examine the spectrum, Fig. 4c, we see evidence of all these features.) Ignoring the details of each cluster of lines, we note that there is a buildup of intensity at $E = 1,2...$ caused by R_1 recurrence of period 2π. The spacing is $\Delta E = 2\pi/\tau_{R_1} = 1.0$. At higher energy, the enhanced intensity near $E = 1,2...$ takes the form of clusters of lines, and we may speak of an envelope for each of these smaller clusters; the finite width of this envelope is attributable to D_2, which is damping out the R_1 recurrences. (The fact that these envelopes are getting broader with higher energy has an interesting significance, discussed below. It is not evident from our qualitative arguments about D_2 decay why the lines should be getting broader but the idea of D_2 decay does give the correct average width.) If the D_2 decay continued indefinitely, the "lines" at $E = 1,2...$ would be intrinsically broadened, with no structure visible at any resolution. When a true continuum is present as in photodissociation, D_2 decay may indeed be permanent and the lines intrinsically broadened. This was recently shown to be the case in the dissociation of symmetric triatomic molecules.[7] However, in the present example there is a second recurrence R_2 reaching a maximum at about 10 vibrational periods; this recurrence induces structure in the otherwise broadened bands, and the structure consists of individual lines spaced by $2\pi/\tau_{R_2} \sim 0.1$, in accordance with the ~ 10 vibrational periods required to reach the maximum R_2 recurrence.

What is the physical significance of the D_2 decay? So far, we have noticed that motion in the s-coordinate has been responsible for D_1 and R_1. The more subtle and longtime events D_2 and R_2 are due to motion in the u-coordinate, and its coupling to the s-coordinate. We will examine the situation classically. Consider a trajectory started out near s = 4, u = 0. For some time in the s-coordinate, we will have, approximately,

$$s(t) = 4 \cos \omega_s t = S_o \cos \omega_s t. \tag{19}$$

The full Hamiltonian reads

$$H = p_s^2/2 + p_u^2/2 + V(s,u) \tag{20}$$

where V(s,u) is given by Eq. 4. If now we substitute Eq. 19 into Eq. 20 and examine the u-motion only, we have

$$H_u = \frac{1}{2} p_u^2 + (\frac{1}{2} \omega_u^2 + \lambda s_o \cos \omega_s t) u^2. \tag{21}$$

If we set $\tau = \frac{1}{2} \omega_s t$, $\sqrt{a} = 2\omega_u/\omega_s$, and $q = 4\lambda s_o/\omega_s^2$, we have the classical equation of motion

$$\frac{d^2u}{d\tau^2} + (a - 2q \cos 2\tau)u = 0, \tag{22}$$

which is Mathieu's[8] Equation. This is the equation of a quadratically forced Harmonic oscillator, and depending on the parameters λ, s_o, ω_s, ω_u the oscillator may become unstable and absorb energy from the forcing caused by motion in the s-coordinate. Eq. 22 does not include the effect of the loss of energy in the s-coordinate as the u-coordinate absorbs it, but it is valid for reasonably short times before the s-coordinate has lost much energy. In our particular example with a displacement of $s_o = 4.0$, the Mathieu Equation is unstable. Returning to the quantum mechanics, we note that the effective Hamiltonian in the u-coordinate given by Eq. 21 is Harmonic, albeit time dependent. Nonetheless, we can still bank on the close correspondence between classical and quantum mechanics for Harmonic potential, and instability in the classical forced oscillator motion implies a decay in the overlap $\langle\phi_u|\phi_u(t)\rangle$. This is the origin of the D_2 decay. If the coupling λ were set to 0, no such decay would occur.

The linearized, separable dynamics has led to an understanding of the D_1, R_1, and D_2 features but it fails to give the R_2 recurrence when the motion about u = 0 is unstable. This is because the linearized dynamics in the u-coordinate (Mathieu Equation) is only first order in the s-coordinate, and if unstable shows no further recurrences. However, this does not imply that the R_2 recurrence is a quantum effect, because we need only to run full, non-linear classical trajectories to discover the classical origin of R_2. When a classical trajectory with small displacement in the u-coordinate is initiated with s = 4, the value of u grows steadily and unstably as oscillation continues, but eventually a maximum value of u is reached, and the process reverses itself as energy flows back from the u-coordinate into the s-coordinate. This last feature of the classical mechanics is the R_2 recurrence. After the last recurrence is complete, the whole sequence $D_1 \rightarrow R_2$ is repeated *ad infinitum*.

We can make the classical-quantum correspondence much more definite by using a Wigner distribution, as already done in connection with the overall spectral envelope due to s-motion; Eq. 10, as follows: first Wigner transform the initial wavefunction ϕ, giving the density

$$\rho_\phi = \frac{1}{\pi^2} \exp[-\omega_s(s - s_o)^2 - 1/\omega_s(p_s - p_{so})^2 - \omega_u(u - u_o)^2 - 1/\omega_u(p_u - p_{uo})^2]. \tag{23}$$

Second, approximate this density by a biased random sampling of points in phase space, so that as the number of points is increased, the moments of the approximate ρ_ϕ approaches those of ρ_ϕ itself. Third, propagate each trajectory in the approximate ρ_ϕ to obtain $\rho_\phi(t)$. Fourth, calculate

$$\mathrm{Tr}(\rho_\phi \rho_\phi(t)) = N \sum_j \rho_\phi(p_{sj}(t),\ p_{uj}(t),\ s_j(t),\ u_j(t)), \tag{24}$$

where the sum is over all trajectories, and where N is a normalizing factor. Fifth, take the square root and hope that

$$|\langle\phi|\phi(t)\rangle| \simeq \sqrt{\mathrm{Tr}\rho_\phi \rho_\phi(t)} \tag{25}$$

where the left hand side is the quantum amplitude. (For harmonic systems Eq. 25 will be exactly true; for anharmonic systems approximately true with the approximation becoming worse at longer times.) Sixth, Fourier transform $\sqrt{\mathrm{Tr}\rho_\phi \rho_\phi(t)}$ and examine the resulting spectrum (i.e., hope that the neglect of the phase still leaves us with useful information about the spectrum.) This is an *ad hoc* set of approximations but the results are encouraging. Examine Fig. 4b, which is the result of $\sqrt{\mathrm{Tr}\rho_\phi \rho_\phi(t)}$ determined classically in the way just described, and compare it with Fig. 4a which was computed quantum mechanically. It is seen that the time dependence of the amplitude determined classically has the correct qualitative features including both decays and both recurrences.

The neglect of phase and the use of classical mechanics has the result that the structure in the Fourier transform near $\omega = 0$ is representative of the spectral bands of the largest intensity in the overall spectrum. Thus, to examine other regions in the spectral band, we move ρ_ϕ along the s-coordinate to larger (smaller) displacements to examine bands at higher (lower) energies. Since the bandwidth of the clusters and the structure under the bands are independent of band intensity, the width and structure (i.e., information on D_2 and R_2) are still available for our inspection.

Results for two types of displaced wavepackets are shown in Fig. 5. In Fig. 5a, the full spectrum (insert) was generated by displacing the wavepacket to s = u = 3.2, on the potential surface of Eq. 4. This is really a type of local mode displacement, since it involves a linear combination of normal modes. The resulting spectrum is best labelled by a local mode overtone quantum number V. Various "local mode" overtone bands (V = 4, 7, 12) are selected for higher resolution scrutiny, and these are shown in the first row of spectra below the insert. The finite width of each individual line is an artifact of truncating the integration time in the Fourier transform. Using the same truncation time, the classical spectra using the procedures 1-6 discussed above are shown below the corresponding quantum spectra. In this case, the bandwidth narrowing as energy increases is correctly reproduced by the classical spectra, and even the spacing of the combination lines under the band is well represented. The bands narrow in this case because of the presence of nearly periodic trajectories near $E \approx 12$ with $2 = u$, $p_s = p_u = 0$ initial conditions. The nearly periodic behavior of the $\sqrt{\mathrm{Tr}\rho_\phi \rho_\phi(t)}$ at $E \approx 12$ is reflected in the Fourier transform spectra as a narrowed band.

Fig. 5b shows an analogous normal mode spectrum for $\omega_s = 1.818181$, $\omega_u = 1.0$, $\lambda = 0.1$, and s = 2.7, u = 0 initially. Here the band envelopes broaden at higher energies. Again, the agreement is surprisingly good between the classical and quantum results.

In this article we have tried to discuss the physical origins of the features seen in a molecular absorption spectrum in terms of the time domain dynamics of wavepackets on the potential surface. If much of our interpretation has been in classical terms, it is because trajectories are easier to visualize than solutions of the time dependent Schrödinger equation, and because trajectories contain so much information that is apparently correct as regards dominant features in the absorption spectra and the time dependence. Although the example we have discussed here is fairly specific, the idea of a decay-recurrence heirarchy, and the mechanisms for the various decays and recurrences are much more general. More details on this new, dynamical approach to molecular spectroscopy may be found in papers which are now in press.[9]

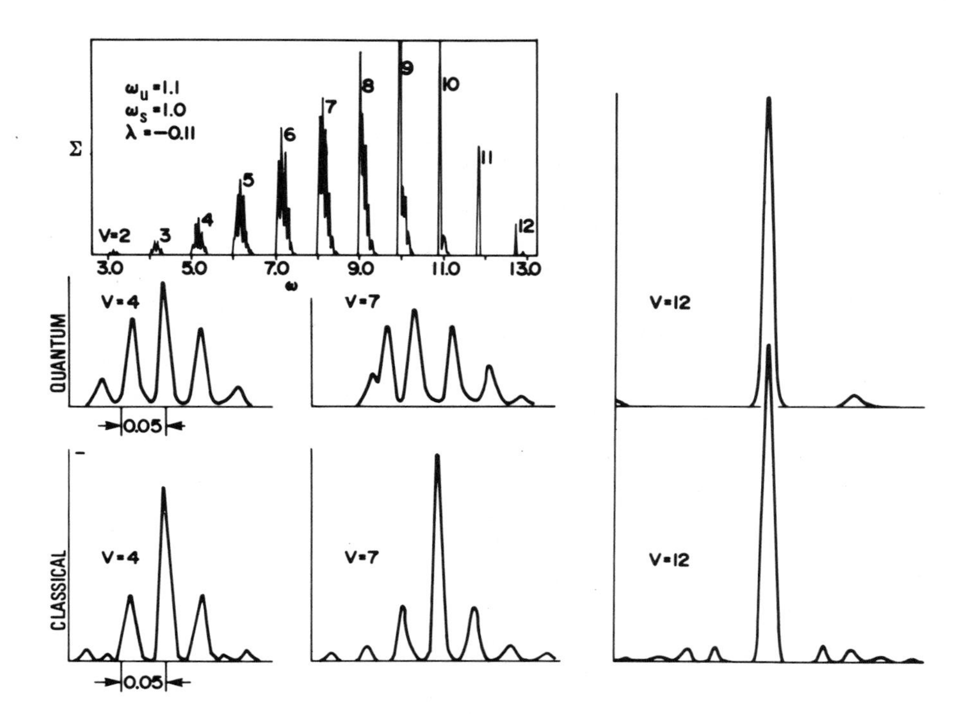

Fig. 5a. Quantum and classical Franck-Condon spectra for the Hamiltonian and wave packet described in text. The insert shows the full spectrum. Details for the bands v = 4, 7, and 12 are shown for the quantum (top row) and classical (bottom row) cases.

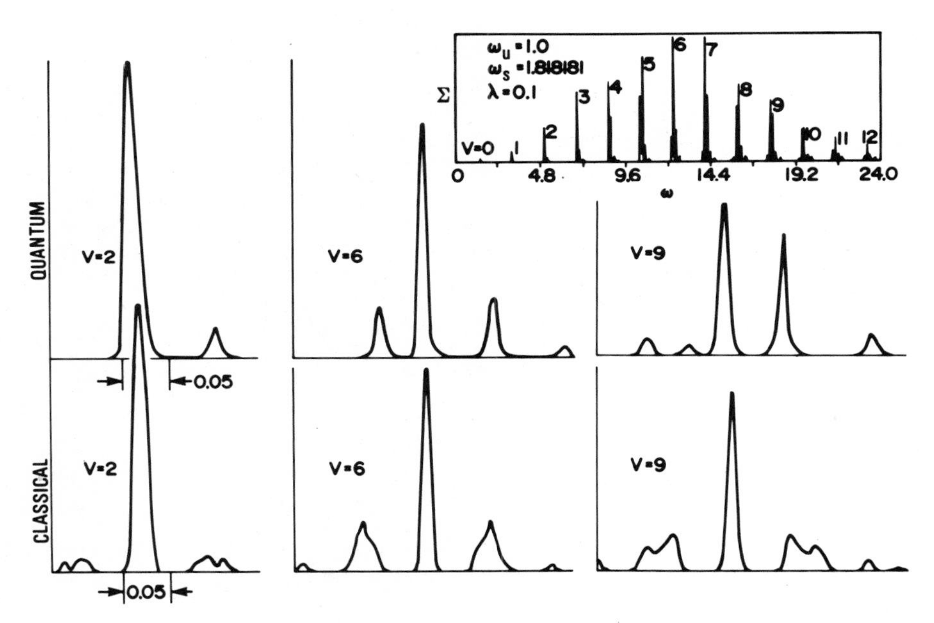

Fig. 5b. Same as Fig. 5a, except as noted in text.

1. G. Herzberg, Molecular Structure and Molecular Spectra II. Infrared and Raman Spectra of Polyatomic Molecules. (Van Nostrand, New York, 1945); III. Electronic Spectra and Electronic Structure of Polyatomic Molecules. (Van Nostrand, New York, 1966).
2. R. G. Gordon, "Correlation Function for Molecular Motion", Adv. Mag. Res. 3:1 (1968).
3. D. W. Oxtoby, Adv. Chem. Phys. 40:1 (1979).
4. See, e.g. A. Messiah, Quantum Mechanics Vol. I (Wiley, New York, 1958) p. 216.
5. See, e.g. J. E. Moyal, Proc. Cambridge Philos. Soc. 45:99 (1949); K. Imre, E. Ozizmir, M. Rosenbaum, and P. F. Zweifel, J. Math. Phys. 8:1097 (1967).
6. E. J. Heller, J. Chem. Phys. 68:2066 (1978).
7. E. J. Heller, J. Chem. Phys. 68:3891 (1978); K. C. Kulander and E. J. Heller, J. Chem. Phys. 69:2439 (1978).
8. W. J. Cunningham, Introduction to Non-Linear Analysis (McGraw-Hill, New York, 1958). Chaps. 9, 10.
9. a) E. J. Heller, E. B. Stechel, and M. J. Davis, "Molecular Spectra, Fermi Resonances, and Classical Motion", J. Chem. Phys., to be published; b) M. J. Davis, E. B. Stechel, and E. J. Heller, "Quantum Dynamics in Integrable and Non-integrable Regions" Chem. Phys. Lett., to be published; c) M. J. Davis and E. J. Heller, "Molecular Overtone Bandwidths from Classical Trajectories", J. Phys. Chem., to be published.

REAL AXIS ASYMPTOTICS AND ESTIMATES OF HAMILTONIAN RESOLVENT KERNELS

F. H. Brownell

University of Washington
Seattle, Washington 98195

ABSTRACT

The three space Newtonian potential of a spherically symmetric charge distribution of finite total charge then has summability Fourier transform proportional to $|\vec{z}|^{-2} g(|\vec{z}|)$ over $\vec{z} \in R_3$; secondly, this g is assumed to have an even analytic extension from $[0,+\infty)$ to the strip $|\mathscr{I}[s]| \leq b$ for some $b \in (0,+\infty)$ such that also there $\left(\sup (1+|\mathscr{R}[s]|)^{\eta} |g(s)|\right) < +\infty$ for some $\eta \in (0,1]$. For such potentials, both the usual Schrödinger Hamiltonian and the free electron state second quantized Dirac Hamiltonian are Fourier transformed to momentum space, and then by spherical harmonics reduced from three dimensional to one dimensional radial integral operators H_q on $(0,+\infty)$. Then we obtain a Faddeev integral operator representation of the resolvent $(\lambda I - H_q)^{-1}$, the kernel being explicitly constructed by contour deforming $(0,+\infty)$ off the real axis to the side opposite λ. From this construction and the resulting estimates and asymptotics, we find $\lambda_0 \in (0,+\infty)$, H_q has no point spectrum in $(\lambda_0, +\infty)$ and is absolutely continuous there, and $((\lambda I - H_q)^{-1} u, u)$ is sub $\frac{1}{4}$ order Hölder continuous in λ up to both sides of the real axis for a dense linear manifold of u; some speculative ramifications of the latter for psuedo-eigenvalues are discussed.

Here we consider both the usual Schrödinger Hamiltonian and the Dirac Hamiltonian modified by second quantization with respect to free electron states as in [3] and [4], the latter being considered as a reasonable physically realistic model. Both Hamiltonians are considered for atomic potentials close to the Coulomb one, and both have rather simple and quite analogous representations as three dimensional integral operators after Fourier transformation to momentum space. These three dimensional integral operators are then reduced by means of spherical harmonics to radial one dimensional integral operators H_q, quite trivially in the Schrödinger case and with a bit of work in the Dirac case (see [11]).

The new material here reported represents the resolvents $(\lambda I - H_q)^{-1}$ by other one dimensional integral operators, and from the explicit construction of the latter we determine the asymptotic behavior as λ approaches suitably restricted intervals of the real axis, from above and below in the complex plane. Here H_q is self-adjoint and the real axis contains its spectrum. This resolvent integral operator construction essentially follows Faddeev

[8]. But his work was for the three dimensional analogue, and more importantly the potential Fourier transform entering his work was assumed Hölder continuous. Such continuity breaks down drastically with an isolated second degree denominator zero (see eq. 6) below) for our potentials close to the Coulomb one. But our potential transforms are actually analytic away from this single singular point, and the key to our construction is the exploitation of this analyticity by contour deformation of the original $(0,+\infty)$ radial integration to the broken line contours $E_{\mp}=[0,(1\pm i)b]\cup[(1\pm i)b,+\infty\pm ib)$, taking the contour on the opposite side of the real axis from λ in $(\lambda I-H_q)^{-1}$ here.

Before sketching the above resolvent kernel construction and results, we now indicate their consequences for four different problems. First, as reported two years ago at the New York meeting, for the analogues of Faddeev's wave operator formulas the limits so specified do not exist, due to infinitely often bounded oscillation of the resolvent kernel at the points in question. This break down agrees with our potentials being long range (close to the Coulomb one), for which in the Schrödinger case the nonexistence of the ordinary wave operator is well known. But for modified wave operators, originally due to Dollard, then extended by Kato & Alsholm and others, and currently the focus of much recent work by Ikebe and associates ([9], [12]), despite considerable inconclusive fumbling I have found no way to alter Faddeev's formulas in light of the above asymptotics to obtain anything significant.

Secondly we eliminate the point spectrum of H_q from the upper part of the real axis. In fact, as constitutes the third problem, for the intervals in question all the H_q are shown to have absolutely continuous spectrum. Here is found a $\lambda_o \in (0,+\infty)$ such that each k_-, k_+ with $\lambda_o < k_- < k_+ < +\infty$ and each integer $q \geq 0$ have a dense linear manifold $\mathcal{M}$ in the Hilbert space such that $|((\lambda I - H_q)^{-1}u,u)|$ is bounded over $\lambda \in \Omega_+[k_-,k_+] \cup \Omega_-[k_-,k_+]$ for each $u \in \mathcal{M}$, defining $\Omega_\pm[k_-,k_+] = \{\lambda \in C_1 \mid \mathcal{R}[\lambda] \in [k_-,k_+]$ and $\pm \mathcal{J}[\lambda] \in (0,1]\}$. Hence all H_q and such $[k_-,k_+]$ have the H_q spectrum be absolutely continuous on $[k_-,k_+]$ by the frequently used Gustafson & Johnson criterion [5]. In the Schrödinger case for our potentials, these second and third problem conclusions, even extended by replacing λ_o by 0, are already known by the results of Rejto [6] and Weidmann [7] and others by quite different O.D.E. methods; but for our Dirac case they are apparently new.

For the fourth and final problem, with λ_o, k_-,k_+,q as above, we show the existence of a dense linear manifold $\mathcal{M}'$ such that not only does the above third problem boundedness hold but also each $u \in \mathcal{M}'$ has $((\lambda I - H_q)^{-1}u,u)$ be sub $\frac{1}{4}$ order Hölder continuous over $\lambda \in \Omega_+[k_-,k_+]$ and over $\lambda \in \Omega_-[k_-,k_+]$ separately; that is, for each $\gamma \in (0,\frac{1}{4})$ and each $u \in \mathcal{M}'$ there exists a constant $M_{\gamma,u} \in (0,+\infty)$ having

$$1) \qquad |((\lambda' I - H_q)^{-1}u,u) - ((\lambda I - H_q)^{-1}u,u)| \leq M_{\gamma,u} |\lambda' - \lambda|^{\gamma}$$

over all $\lambda', \lambda \in \Omega_+[k_-,k_+]$ and also over all $\lambda', \lambda \in \Omega_-[k_-,k_+]$. I have some ideas for, but have not yet carried out, extending such results below λ_o to the transition point to the at most countable point spectrum at the H_q spectrum bottom. If such can be carried out in the Dirac case, it appears possible then to give a mathematically complete analysis of photon emission with associated psuedo-eigenvalues for the three component space model of section 11 of [4] (p. 377-378), to which we return below (see [1], [2] for such for a drastically simplified, quite artificial model). These psuedo-eigenvalues (the term here is ascribed to the talk of B. Friedman [13]) are essentially the same phenomena as resonances, whose theory Simon developed in [10] in the different context of several body scattering.

To indicate some of the details for the above construction and conclusions, the initial Schrödinger operator (after scaling physical units) is

$$-2^{-1}\nabla^2 w(\vec{y}) - a\,\Phi(\vec{y})w(\vec{y}) \qquad alm.\ \mu_3\ ev.\ \vec{y} \in R_3, \tag{2}$$

with domain say the Hermite functions w. Here $a = Na_o$, the fine structure constant $a_o \doteq (137)^{-1}$, the real constant $N \neq 0$ (the positive integer atomic number for a physical atom).

Concerning the relative electrostatic potential Φ we assume: first (ν signifying the relative charge distribution)

$$\Phi(\vec{x}) = \int_{R_3} |\vec{\xi} - \vec{x}|^{-1}\, d\nu(\vec{\xi}) \tag{3}$$

where ν is a fixed Borel measure on R_3 having $\nu(R_3) = 1$ and being invariant under rotations about the origin (as suffices for [11]); for such ν defining over $\vec{x} \in R_3$

$$g(|\vec{x}|) = \int_{R_3} e^{-i(\vec{x}\cdot\vec{\xi})} d\nu(\vec{\xi}), \tag{4}$$

clearly making g be continuous real valued on $[0,+\infty)$ with $g(0) = 1 \geq |g(r)|$ there, secondly we assume that ν in 3) has the even extension of g from $[0,+\infty)$ to $(-\infty,+\infty)$ possess an analytic continuation over the complex plane strip $|\mathcal{I}[s]| \leq 2b_1$ for some $b_1 \in (0,+\infty)$ such that for some $\eta \in (0,1]$

$$\left[\sup_{|\mathcal{I}[s]|\leq 2b_1} (1+|\mathcal{R}[s]|)^{\eta} |g(s)|\right] < +\infty. \tag{5}$$

Consider the examples: the Coulomb case, $\nu = \nu_0$ unit weight at the origin, $\Phi_0(\vec{x}) = |\vec{x}|^{-1}$, $g_0(r) \equiv 1$, the first assumption holds but not the second (5) fails); the truncated Coulomb case, $\nu = \nu_1$ uniform weight distribution on the sphere $|\vec{\xi}| = k > 0$,

$$\Phi_1(\vec{x}) = (\max |\vec{x}|, k)^{-1}, \quad g_1(s) = \begin{bmatrix} (ks)^{-1}\sin(ks) & \text{if } s \neq 0, \\ 1 & \text{if } s = 0 \end{bmatrix}$$

extends even entire, 5) holds with $\eta = 1$ for each $b_1 \in (0,+\infty)$, both potential assumptions hold.

With μ_3 denoting three dimensional Lebesgue measure, next under $L_2(R_3)$ fourier transform 2) goes over to

$$
6) \begin{cases} H_S = A_S + B_S, \\ [A_S u](\vec{x}) = 2^{-1}|\vec{x}|^2 u(\vec{x}) = f_S(|\vec{x}|)\, u(\vec{x}), \\ [B_S u](\vec{x}) = -\left(\frac{a}{2\pi^2}\right) \int_{R_3} \frac{g(|\vec{x}-\vec{y}|)}{|\vec{x}-\vec{y}|^2}\, u(\vec{y})\, d\mu_3(\vec{y}), \end{cases}
$$

alm. μ_3 ev. over $\vec{x} \in R_3$, first for Hermite u and then for $u \in \mathfrak{D}_S$, the including domain below, with $f_S(r) = 2^{-1} r^2$. The corresponding free electron state second quantized Dirac Hamiltonian H_D is ([11], 3.14))

$$
7) \begin{cases} H_D = A_D + B_D, \\ [A_D u]_\gamma(\vec{x}) = \sqrt{1+|\vec{x}|^2}\, u_\gamma(\vec{x}) = f_D(|\vec{x}|)\, u_\gamma(\vec{x}), \\ [B_D u]_\gamma(\vec{x}) = -\left(\frac{a}{2\pi^2}\right) \int_{R_3} \left(\frac{g(|\vec{x}-\vec{y}|)}{|\vec{x}-\vec{y}|^2}\right) \frac{\left\{ \begin{array}{l} \left[f_1(|\vec{x}|) f_1(|\vec{y}|) + (\vec{x}\cdot\vec{y}) + i(-1)^\gamma [\vec{x}\otimes\vec{y}]_3\right] u_\gamma(\vec{y}) + \\ + \left[(-1)^\gamma [\vec{x}\otimes\vec{y}]_2 + i[\vec{x}\otimes\vec{y}]_1\right] u_{\gamma+}(\vec{y}) \end{array} \right\}}{2\sqrt{f_D(|\vec{x}|) f_1(|\vec{x}|) f_D(|\vec{y}|) f_1(|\vec{y}|)}}\, d\mu_3(\vec{y}) \end{cases}
$$

over $\gamma \in \{1,2\}$ and alm. μ_3 ev. $\vec{x} \in R_3$. here $[\vec{x} \otimes \vec{y}]$ denotes the usual R_3 cross product, $f_D(r) = \sqrt{1+r^2}$, $f_1(r) = 1 + f_D(r)$, and for $\gamma \in \{1,2\}$ also $\gamma+$ denotes the $\gamma' \in \{1,2\}$ having $\gamma' \neq \gamma$. For 6) the Hilbert space is $X_S = L_2(R_3)$, $\mathfrak{D}_S = \{u \in X_S \mid w(x) = f_S(|\vec{x}|)\, u(\vec{x})$ has $w \in X\}$ is the domain for all three of A_S, B_S, H_S, this B_S is symmetric, A_S is self-adjoint with $(kI+A_S)^{-1}$ bounded Hermitian for $k \in (0,+\infty)$, and from $0 = \lim_{k\to+\infty} \|B_S(kI+A_S)^{-1}\|$ also H_S is self-adjoint, having for large $k > 0$

6)' $$(kI+H_S)^{-1}=(kI+A_S)^{-1}+\sum_{n=1}^{\infty}(kI+A_S)^{-1}\left[-B_S(kI+A_S)^{-1}\right]^n .$$

Correspondingly for 7) the Hilbert Space $X_D = L_2(R_3) \oplus L_2(R_3)$, a $u \in X_D$ being a pair of $u_\gamma \in L_2(R_3)$, $\gamma = 1,2$; also

$$\mathfrak{D}_D=\{u\in \mathfrak{X}_D \mid w_\gamma(\vec{x})=f_D(|\vec{x}|)\,u_\gamma(\vec{x}) \quad \text{has} \quad w\in \mathfrak{X}_D\}$$

is the domain for A_D, B_D, H_D, this B_D is symmetric, A_D is self-adjoint with A_D^{-1} bounded Hermitian, and (see [3], [11])

8) $$\|B_D A_D^{-1}\| \le 2|a| .$$

Hence in the Dirac case adding the assumption

9) $$|a| = |N|\,a_0 < \frac{1}{2},$$

then in 7) also H_D is self-adjoint, having the bounded Hermitian inverse

7)' $$H_D^{-1}=A_D^{-1}+\sum_{n=1}^{\infty}A_D^{-1}\left[-B_D A_D^{-1}\right]^n .$$

Here we make throughout the two assumptions 3) and 5) about Φ, $\mathcal{V}$, g, and 9) is additionally assumed in the Dirac case. These are the hypotheses under which our resolvent kernel construction is carried out and the above four problem conclusions reached.

The spherical harmonic reduction in [11] springs from the identity that, for integer $q \ge 0$ and nonnull $\vec{x} \in R_3$,

$$10)\begin{cases} \int_{R_3} F(|\vec{x}|,|\vec{y}|,(\vec{x}\cdot\vec{y}))\, Y_q(|\vec{y}|^{-1}\vec{y})\,d\mu_3(\vec{y}) = {}_qw_F(|\vec{x}|)\, Y_q(|\vec{x}|^{-1}\vec{x}), \\ {}_qw_F(s) = 2\pi \int_0^{+\infty} r^2 \left\{ \int_{-1}^{+1} F(s,r,srt)\, P_q(t)\,dt \right\} dr, \end{cases}$$

assuming the complex measurable F has $\int_{R_3} |F(|\vec{x}|,|\vec{y}|,(\vec{x}\cdot\vec{y}))|\, d\mu_3(\vec{y}) < +\infty$. Here Y_q is any spherical harmonic of degree q, and P_q is the q degree Legendre polynomial. Under this [11] reduction, in both the Schrödinger and Dirac cases the Hilbert space is $X_0 = L_2((0,+\infty),\tilde{\mu})$ with the $\tilde{\mu}$ measure having $\tilde{\mu}(E) = \int_E r^2\, d\mu_1(r)$ over Lebesgue measurable subsets E of $(0,+\infty)$, μ_1 being one dimensional Lebesgue measure. Arising from 10), for integer $q \geq 0$

$$11)\qquad K_q(x,y) = \int_{-1}^{+1} \frac{g\left(\sqrt{x^2+y^2-2xyt}\right)}{x^2+y^2-2xyt}\, P_q(t)\,dt$$

is defined over $x, y \in (0,+\infty)$ having $x \neq y$. Then with domain $\mathcal{D}_s = \{u \in X_0 \mid w(x) = f_s(x)u(x) \text{ has } w \in X_0\}$, 6) goes over to

$$12)\begin{cases} {}_sH_q = {}_sA + {}_sB_q, \\ [{}_sAu](x) = f_s(x)\,u(x), \\ [{}_sB_qu](x) = -\left(\frac{a}{2\pi}\right) \int_0^{+\infty} \{2K_q(x,y)\}\, u(y)\, y^2\, dy \end{cases}$$

alm. $\tilde{\mu}$ ev. over $x \in (0,+\infty)$; with domain

${}_D\mathfrak{D} = \{u \in \mathfrak{X}_0 \mid w(x) = f_D(x)u(x)$ has $w \in X_0\}$, 7) goes over to

$$
13) \begin{cases} {}_D H_{q,\pm} = {}_D A + {}_D B_{q,\pm} , \\ [{}_D A u](x) = f_D(x)\, u(x) , \\ [{}_D B_{q,\pm} u](x) = -\left(\frac{a}{2\pi}\right) \int_0^{+\infty} \frac{[f_1(x) f_1(y) K_q(x,y) + xy\, K_{q\pm 1}(x,y)]}{\sqrt{f_D(x) f_1(x) f_D(y) f_1(y)}}\, u(y) y^2 dy \end{cases}
$$

alm. $\tilde{\mu}$ ev. over $x \in (0,+\infty)$, with $q \geq 1$ also required in the case of ${}_D H_{q,-}$ and ${}_D B_{q,-}$ due to the K_{q-1} entrance. With 8), here $\| {}_D B_{q,\pm}\, {}_D A^{-1} \| \leq \| B_D A_D^{-1} \| \leq 2|a|$, and the same self-adjointness follows in 12) and 13) as in 6) and 7).

Notice, as a notational convenience, that 12) and 13) have the common form

$$
14) \begin{cases} H_q = A + B_q , \\ [Au](x) = f(x)\, u(x) , \\ [B_q u](x) = -\int_0^{+\infty} G_q(x,y)\, u(y)\, y^2 dy \end{cases}
$$

with domain $\mathfrak{D} = \{u \in \mathfrak{X}_0 \mid w(x) = f(x)u(x)$ has $w \in \mathfrak{X}_0\}$ and the obvious f and G_q definitions.

Next for 14) following the Faddeev form, for all nonreal complex λ the known bounded resolvent $(\lambda I - H_q)^{-1}$ on X_0 will be found to have the representation (at least for a suitable dense linear manifold of u)

15) $$\left[(\lambda I - H_q)^{-1} u\right](x) = \frac{u(x)}{\lambda - f(x)} - \int_0^{+\infty} \frac{F_q(x,y;\lambda)\, u(y)}{(\lambda - f(x))(\lambda - f(y))} y^2 dy$$

alm. $\tilde{\mu}$ ev. over $x \in (0,+\infty)$, where for $x,y \in (0,+\infty)$ having $x \neq y$

16) $$F_q(x,y;\lambda) = G_q(x,y) + J_q(x,y;\lambda) ,$$

17) $$J_q(x,y;\lambda) + \int_0^{+\infty} G_q(x,s) \frac{J_q(s,y;\lambda)}{\lambda - f(s)} s^2 ds = -\int_0^{+\infty} G_q(x,s) \frac{G_q(s,y)}{\lambda - f(s)} s^2 ds .$$

For $R[\lambda]$ sufficiently negative the usual (note 6)', 7)')

18) $$(\lambda I - H_q)^{-1} = (\lambda I - A)^{-1} + \sum_{n=1}^{\infty} (\lambda I - A)^{-1} \left[B_q (\lambda I - A)^{-1}\right]^n$$

makes 15)-17) clear there, whence our construction following of the 17) solution J_q yields 15) as stated by analytic continuation in λ.

To obtain this 17) solution, as in the foregoing introductory remarks the 17) left side $(0,+\infty)$ integration will be contour deformed to E_+ and E_- respectively in the two cases $I[\lambda] > 0$ and $I[\lambda] < 0$, $E_+ = [0,(1-i)b] \cup [(1-i)b, +\infty - ib)$, $E_- = [0,(1+i)b] \cup [(1+i)b, +\infty + ib)$, $b = (\min 1, b_1)$ with b_1 in 5).

19)

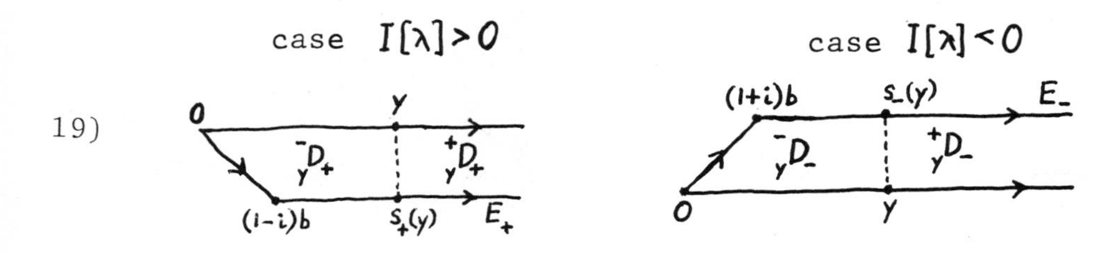

Conveniently for $\mathcal{R}[z] \geq 0$ we denote $S_{\pm}(z)$ as the unique $z' \in E_{\pm}$ having $\mathcal{R}[z'] = \mathcal{R}[z]$; also $D_{\pm} = \{z \in C_1 | 0 < \mathcal{R}[z] \text{ and } 0 < \frac{I[z]}{I[S_{\pm}(z)]} < 1\}$ and for $y \in (0,+\infty)$ take ${}_{y}^{-}D_{\pm} = \{z \in D_{\pm} | \mathcal{R}[z] < y\}$ and ${}_{y}^{+}D_{\pm} = \{z \in D_{\pm} | \mathcal{R}[z] > y\}$, as 19) indicates.

This 17) left side integral contour deformation requires appropriate anayltic continuation of the integrand factor off the real axis. For the G_q factor from 14), first note $s = \sqrt{x^2 + y^2 - 2xyt}$ in 11) yields

$$11)' \quad K_q(x,y) = K_q(y,x) = (xy)^{-1} \int_{|x-y|}^{x+y} s^{-1} g(s) P_q\left(2^{-1}\left[\frac{x}{y} + \frac{y}{x}\right] - \frac{s^2}{2xy}\right) ds$$

over $x \neq y$ both in $(0,+\infty)$, exhibiting the $\log\left(|x-y|^{-1}\right)$ singularity as $|y-x| \to 0$ with $g(0) = 1 = P_q(1)$ and g analyticity in 5); thus defining

$$20) \quad K_q(z',z) = K_q(z,z') = (z'z)^{-1} \int_{[z'-z,\, z'+z]} s^{-1} g(s) P_q\left(2^{-1}\left[\frac{z'}{z} + \frac{z}{z'}\right] - \frac{s^2}{2zz'}\right) ds$$

over complex z', z having $0 < \mathcal{R}[z] < \mathcal{R}[z']$ and $|I[z]| \leq b_1$ and $|I[z']| \leq b_1$ yields separate analyticity in z' and z there and so extends 11). With the additional restriciton $|I[z]| \leq \mathcal{R}[z]$ sufficing for the analytic extension $f_D(z)$ for f_D in 7), and likewise for z', thus the $G_q(z,z')$ kernels in 14), 13), 12) are correspondingly extended and likewise separately analytic. From the above log singularity in 20), and the consequent jump discontinuities on the verticals $\mathcal{R}[z'-z] = 0$ when $z' \neq z$ due to jumps in the angle, thus $G_q(z,z')$ exhibits the same behavior; explicitly for $\tau \neq t$ both in $[-b_1, b_1]$ and $x \in (0,+\infty)$ and $z = x + it$ and $z' = x + i\tau$

$$\left[\lim_{\tilde{x}\to x^+} G_q(\tilde{x}+it, z')\right] - \left[\lim_{\tilde{x}\to x^-} G_q(\tilde{x}+it, z')\right] =$$

21)
$$= -\left(\frac{t-\tau}{|t-\tau|}\right) Q_q(z,z')$$

$$= -\left\{\left[\lim_{\tilde{x}\to x^+} G_q(z, \tilde{x}+i\tau)\right] - \left[\lim_{\tilde{x}\to x^-} G_q(z, \tilde{x}+i\tau)\right]\right\}.$$

Here, in the various cases, over $z', z \in \Sigma = \{z \in C_1 \mid \mathcal{R}[z] > 0$ and $|\mathcal{G}[z]| \leq \mathcal{R}[z]\}$,

22)
$$\begin{cases} {}_S Q_q(z,z') = ia\,(zz')^{-1} P_q\left(2^{-1}\left[\frac{z}{z'} + \frac{z'}{z}\right]\right), \\ {}_D Q_{q,\pm}(z,z') = \left(\frac{ia}{2}\right) \dfrac{\left[(zz')^{-1} f_1(z) f_1(z') P_q\left(2^{-1}\left[\frac{z}{z'} + \frac{z'}{z}\right]\right) + P_{q\pm 1}\left(2^{-1}\left[\frac{z}{z'} + \frac{z'}{z}\right]\right)\right]}{\left[f_D(z) f_1(z)\right]^{1/2} \left[f_D(z') f_1(z')\right]^{1/2}}. \end{cases}$$

and thence in all cases

23)
$$z^2 Q_q(z,z) = ia = iNa_0 \neq 0.$$

From this complex extension behavior of G_q, and from the explicit series representation of J_q indicated by 18) for very negative $\mathcal{R}[\lambda]$, it appears natural to require similar extension behavior for the 17) solution J_q. In fact we now impose the following extension specification.

24) $J_q^{\pm}$ solution of 17) specification

For $y \in (0,+\infty)$, when $I[\lambda] > 0$ the $J_q^{+}(\cdot,y;\lambda)$ is analytic in ${}^{-}_{y}D_{+} \cup {}^{+}_{y}D_{+}$ and has its restriction to ${}^{-}_{y}D_{+}$ possess a continuous extension ${}^{-}J_q^{+}(\cdot,y;\lambda)$ to $\overline{{}^{-}_{y}D_{+}}$ and its restriciton to ${}^{+}_{y}D_{+}$ possess a continuous extension ${}^{+}J_q^{+}(\cdot,y;\lambda)$ to $\overline{{}^{+}_{y}D_{+}}$, when $I[\lambda] < 0$ the $J_q^{-}(\cdot,y;\lambda)$ is analytic in ${}^{-}_{y}D_{-} \cup {}^{+}_{y}D_{-}$ and has its restriction to ${}^{-}_{y}D_{-}$ possess a continuous extension ${}^{-}J_q^{-}(\cdot,y;\lambda)$ to $\overline{{}^{-}_{y}D_{-}}$ and its restriction to ${}^{+}_{y}D_{-}$ possess a continuous extension ${}^{+}J_q^{-}(\cdot,y;\lambda)$ to $\overline{{}^{+}_{y}D_{-}}$, and in addition:

i) these $J_q^{\pm}(\cdot,y;\lambda)$ boundary values on $(0,y)\cup(y,+\infty)$ satisfy 17);

ii) $$0 = \left[\lim_{x\to+\infty} x^{\alpha}\left(\sup_{\substack{z\in \overline{D_{\pm}},\\ \mathcal{R}[z]=x}} \left|{}^{+}J_q^{\pm}(z,y;\lambda)\right|\right)\right]$$ for each $\alpha\in(0,2)$.

Our construction of $J_q^{\pm}$ forming the unique solution of 24) comes from the following lemma, with hypotheses stated after 9), with $E_{\pm}$ and $D_{\pm}$ in 19) and thereafter, and with Q_q in 21), 22).

25) lemma

For some $y \in (0,+\infty)$ and $\lambda \in \{\lambda \in C_1 \mid \mathcal{I}[\lambda] > 0\}$ let $J_q^{+}(\cdot,y;\lambda)$ satisfy all the 24) properties except possibly i) (satisfaction of 17) on $(0,y)\cup(y,+\infty)$); then 17) holds over $(0,y)\cup(y,+\infty)$ if and only if over all $z \in \overline{D_+}$ having $0 < R[z] \neq y$ holds

$$J_q^+(z,y;\lambda) = \tag{26}$$
$$= -\int_{E_+} \frac{G_q(z,s)}{\lambda - f(s)} \left[G_q(s,y) + J_q^+(s,y;\lambda)\right] s^2 ds + \int_{[s_+(z),z]} \frac{Q_q(z,s)}{\lambda - f(s)} \left[G_q(s,y) + J_q^+(s,y;\lambda)\right] s^2 ds +$$
$$+ \int_{[s_+(y),y]} \frac{G_q(z,s)}{\lambda - f(s)} \left[Q_q(s,y) - \left\{{}^{-}J_q^+(s,y;\lambda) - {}^{+}J_q^+(s,y;\lambda)\right\}\right] s^2 ds;$$

moreover, then 26) holds over the z there stated if and only if over all $z \in E_+$ having $0 < R[z] \neq y$ holds

$$J_q^+(z,y;\lambda) + \int_{E_+} \frac{G_q(z,s)}{\lambda - f(s)} J_q^+(s,y;\lambda) s^2 ds = \tag{27}$$
$$= -\int_{E_+} G_q(z,s) \frac{G_q(s,y)}{\lambda - f(s)} s^2 ds + \int_{[s_+(y),y]} \frac{G_q(z,s)}{\lambda - f(s)} \left[Q_q(s,y) - \left\{{}^{-}J_q^+(s,y;\lambda) - {}^{+}J_q^+(s,y;\lambda)\right\}\right] s^2 ds.$$

The $I[\lambda] < 0$ case with J_q^-, E_-, D_- goes similarly.

To see 25), first notice the $I[z] = 0$ special case of 26) (so $z = x \in (0,+\infty)$) is clearly equivalent to 17) by contour deformation in 19), taking into account the 21) and 24) jumps on the verticals $R[s] = R[z]$ and $R[s] = y$. Thus the general 26) implies 17), and the converse implication follows by unique analytic continuation of boundary values, after first rewriting (by means of 21)) the sum of the 26) right side first two terms in an equivalent form assuring analyticity in z. Also 27) is the $z \in E_+$ special case of 26), and thus similarly 27) is equivalent to 26) and 25) is shown.

Next for y and λ as in 25) defining

$$v_q^+(z;y,\lambda) = {}^-J_q^+(z,y;\lambda) - {}^+J_q^+(z,y;\lambda) \tag{28}$$

over $z \in [s_+(y), y]$, then 26) implies (by taking right and left hand limits along this vertical) that

$$v_q^+(z;y,\lambda) - \int_{[z,y]} Q_q(z,s)\frac{v_q^+(s;y,\lambda)}{\lambda - f(s)} s^2 ds = -\int_{[z,y]} Q_q(z,s)\frac{Q_q(s,y)}{\lambda - f(s)} s^2 ds \tag{29}$$

over such z, and hence $v_q^+(y;y,\lambda) = 0$ initially.

Thus consider the following three stage construction:

30) first stage

$v_q^+(\cdot;y,\lambda)$ continuous on $[s_+(y),y]$ satisfies 29) there;

31) second stage

$u_q^+(\cdot;y,\lambda)$ continuous on $E_+ \dot{-} (\{0\} \cup \{s_+(y)\})$ satisfies there

$$u_q^+(z;y,\lambda) + \int_{E_+} G_q(z,s)\frac{u_q^+(s;y,\lambda)}{\lambda - f(s)} s^2 ds =$$

$$= -\int_{E_+} G_q(z,s)\frac{G_q(s,y)}{\lambda - f(s)} s^2 ds + \int_{[s_+(y),y]} G_q(z,s)\frac{[Q_q(s,y) - v_q^+(s;y,\lambda)]}{\lambda - f(s)} s^2 ds, \tag{32}$$

$u_q^+(\cdot\,;y,\lambda)$ having finite right and left hand limits at $s_+(y)$ and right hand at 0, and $u_q^+(s_+(x);y,\lambda) = o(x^{-\alpha})$ as $x \to +\infty$ for each $\alpha \in (0,2)$;

33) <u>third stage</u>

for each $x \in (0,y) \cup (y,+\infty)$, $w_q^+(\cdot\,;y,\lambda)$ continuous on $[s_+(x), x]$ satisfies there

34)
$$w_q^+(z;y,\lambda) - \int_{[s_+(z),z]} Q_q(z,s) \frac{w_q^+(s;y,\lambda)}{\lambda - f(s)} s^2 ds =$$
$$= \int_{[s_+(z),z]} Q_q(z,s) \frac{G_q(s,y)}{\lambda - f(s)} s^2 ds - \int_{E_+} G_q(z,s) \frac{[G_q(s,y) + u_q^+(s;y,\lambda)]}{\lambda - f(s)} s^2 ds +$$
$$+ \int_{[s_+(y),y]} G_q(z,s) \frac{[Q_q(s,y) - v_q^+(s;y,\lambda)]}{\lambda - f(s)} s^2 ds ,$$

so by 32) having initially for such x

35)
$$w_q^+(s_+(x);y,\lambda) = u_q^+(s_+(x);y,\lambda) .$$

Granting this three stage construction, then defining

36)
$$J_q^+(z,y;\lambda) = w_q^+(z;y,\lambda)$$

over $z \in \overline{D}_+$ having $0 < R[z] \neq y$ yields the desired J_q^+.

Note here, starting with a J_q^+ satisfying all the 24) properties, then v_q^+ defined by 28) satisfies 29) and 30), and w_q^+ and u_q^+ defined by 36), 35) then satisfy 32) by 27) and 34) by 26), and hence such J_q^+ must coincide with the supposed unique construction result 36). Conversely for this construction result $\tilde{J}_q^+$ in 36), accordingly defining $\tilde{v}_q^+$ in 28), and then in 34) taking right and left limits along the $[s_+(y), y]$ vertical by 21) and subtracting 29), we find over $z \in [s_+(y), y]$

$$37) \quad \tilde{v}_q^+(z;y,\lambda) - v_q^+(z;y,\lambda) = \int_{[s_+(y),z]} \frac{Q_q(z,s)}{\lambda - f(s)} \left[\tilde{v}_q^+(s;y,\lambda) - v_q^+(s;y,\lambda)\right] s^2 ds ,$$

whence by the obvious estimates $\tilde{v}_q^+(z;y,\lambda) = v_q^+(z;y,\lambda)$ there; hence the known 34) yields 26) for this 36) constructed $\tilde{J}_q^+$, which hence satisfies all the 24) properties and is the desired J_q^+.

Thus we have reduced solution of 24) to carrying out the to be verified unique three stage construction 30), 31), 33). For the first and third stages 30), 33), the quite similar Volterra integral equations 29) and 34) on verticals always have unique continuous solutions, in fact such solutions being represented explicitly by their uniformly convergent iteration series. But in this quite satisfactory analysis of the first and third stages, interesting asymptotics arise as $I[\lambda] \to 0$ under the coincidences $\mathcal{R}[\lambda] \to f(x)$ or $\mathcal{R}[\lambda] \to f(y)$ (see the second paragraph below).

Thus there remains for our concern only the second stage 31) with its integral equation 32) (not of Volterra type), which with

E_+ in 19) exhibits the great advantage over the original 17) that for $r_- \in (0,+\infty)$ the 32) left side integrand factor $|\lambda - f(s)|^{-1}$ is bounded over $s \in E_+$ and $\lambda = f(r) + ih$ and $r \in [r_-,+\infty)$ and $h \in [0,+\infty)$, whereas in the 17) left side integral the same factor is not bounded over $s \in (0,+\infty)$ and such λ. Two fairly simple solution methods appear available for 32), both after introduction of a suitable Banach space: first showing the 32) left side integral operator to be compact, and then using the Fredholm alternative; secondly showing this operator T to have norm < 1, whence $(I+T)^{-1} = I + \sum_{n=1}^{\infty} (-T)^n$ yields the solution u_q^+ explicitly. For the first method we use the Banach space of continuous functions on E_+ vanishing at ∞ with sup norm, operator compactness (after multiplying 32) by $(1+|z|)^\alpha$ for $\alpha \in (0,2)$) follows by Arzela-Ascoli, and a unique 32) solution results, since the homogeneous equation has no nontrivial solution by going back from this analogue of 32) or 27) to the equivalent analogue of 17) and using symmetry with $I[\lambda] > 0$. Unfortunately I have not succeeded in rescuing this last step from collapse when $I[\lambda] = 0$, and hence do not use this first method for $I[\lambda] \to 0$ asymptotics. For the second method, as one of several alternatives, take the similar Banach space of measurable functions u on E_+ with norm $\left(\operatorname{ess\,sup}_{x \in (0,+\infty)} (1+|s_+(x)|)^\alpha |u(s_+(x))|\right) < +\infty$, and for $\alpha \in (0,2)$ our operator norm < 1 follows from a lengthy direct estimate, involving K_q estimation on $E_+ \otimes E_+$ by contour deforming 20) to use $|P_q(t)| \le 1$ over $t \in [-1,+1]$; thus for $a \neq 0$ in 2), 6), 7), 9),

14) we estimate $r_a \in (0,+\infty)$, independent of q and having $r_a \to 0$ as $a \to 0$, such that over $r \in (r_a,+\infty)$ and $h \in [0,+\infty)$ and $\lambda = f(r) + ih$ the above norm < 1 and the unique 32) solution u_q^+ is determined by action of the resulting geometric series inverse operator above.

To return to the $I[\lambda] \to 0^+$ asymptotics in the first and third stages (second preceding paragraph), we see in the first stage 29) with integration over $s \in [z,y] \subseteq [s_+(y),y]$ that the integrand denominator $[\lambda - f(s)] \to 0$ as $s \to y$ and $\lambda = f(r) + ih$ has real $r \to y$ and $h \to 0^+$, the only singularity arising. But differentiating 29) (or similarly 34)) once with respect to $t = I[z]$, multiplying by the elementary integrating factor for a first order linear ordinary differential equation, and integrating back again, we find a new Volterra integral equation with the singularity removed. After further straightforward manipulation, thus for 29) the solution v_q^+ has

$$38)\qquad v_q^+(z;y,\lambda) - Q_q(z,y) = {}_{+}\varphi_q(z;y,\lambda)Q_q(z,y)\left\{\frac{e^{L_+(z,\lambda)}}{e^{L_+(y,\lambda)}}\right\},$$

where for any constants y_1, y_2, r_-, r_+, $0 < y_1 < r_- < r_+ < y_2 < +\infty$, there exists $\delta_1 \in (0,y_1)$ having $Q_q(y+it,y) \neq 0$ over $(y,t) \in [y_1,y_2] \otimes [-\delta_1,+\delta_1]$, for such δ_1 this ${}_{+}\varphi_q(y+it;y,\lambda)$ is sub 1 order Hölder continuous in (t,y,λ) over $\Sigma = \{(t,y,\lambda) \mid t \in [-\delta_1,0],\ y \in [y_1,y_2],\ \lambda = f(r)+h$ with $r \in [r_-,r_+]$ and $h \in [0,1]\}$, and 38) holds over

$z = y + it$ and $(t,y,\lambda) \in \Sigma_0 = \{(t,y,\lambda) \in \Sigma \mid \lambda \neq f(y)\}$.

In this 38) the before mentioned integrating factor introduces the e^{L_+} factors, where

$$39)\quad L_+(z,\lambda) = -ia \int_{[s_+(z),z]} (\lambda - f(s))^{-1} ds = \frac{ia}{f'(x)} \log_+(\lambda - f(x) - if'(x)t) + W_+(z,\lambda)$$

over $(z,\lambda) \in {}_0\mathcal{Z}_+ = \{(z,\lambda) \in \mathcal{Z}_+ \mid \lambda \neq f(z)\}$ with

$\mathcal{Z}_+ = \{(z,\lambda) \in C_1 \otimes C_1 \mid \mathcal{I}[\lambda] \geq 0,\ \mathcal{R}[z] > 0,\ 0 \leq -\mathcal{I}[z] \leq \mathcal{R}[z]\}$,

here denoting $x = \mathcal{R}[z]$ and $t = \mathcal{I}[z]$ and for complex $\xi \neq -i\mathcal{I}[\xi]$ defining $\log_+ \xi = \ln|\xi| + i\Theta_+(\xi)$ with $\xi = |\xi| e^{i\Theta_+(\xi)}$ and $-\frac{\pi}{2} < \Theta_+(\xi) < \frac{3\pi}{2}$. Also, for any y_1, y_2, r_-, r_+ as in following 38), in 39) the $W_+(z,\lambda)$ is sub 1 order Hölder continuous in $(z,\lambda) \in \tilde{\Sigma} = \{(z,\lambda) \in \mathcal{Z}_+ \mid \mathcal{R}[z] \in [y_1,y_2]$ and $\lambda = f(r) + ih$ with $r \in [r_-,r_+]$ and $h \in [0,1]\}$. Note 39) yields in 38) with $z = x + it$

$$40)\quad e^{L_+(z,\lambda)} = \exp\left(W_+(z,\lambda) + \frac{ia}{f'(x)} \ln|\lambda - f(x) - if'(x)t| - \frac{a}{f'(x)} \Theta_+(\lambda - f(x) - if'(x)t)\right),$$

exhibiting the bounded infinite oscillation mentioned in the introduction. The third stage 34) solution goes very similarly, with the same e^{L_+} factors in 40) entering.

From this three stage 30), 31), 33) explicit solution of 17), 24) and for the resulting J_q^+ in 36)(and similarly for J_q^-), the following additional properties are determined, with $r_a \in (0,+\infty)$ found above (paragraph before 38)) independent of q and with

η in 5):

I) with $\Omega_+ = \{\lambda \in C_1 | \mathscr{I}[\lambda] > 0\}$, $J_q^+(x,y;\lambda)$ is continuous complex valued over $(x,y,\lambda) \in (0,+\infty)\otimes(0,+\infty)\otimes\Omega_+$ and separately analytic in λ there;

II) with 16) and this J_q^+, this 15) holds over $\lambda \in \Omega_+$ and at least over $u \in {}_1\mathcal{M}_\rho$ for each $\rho \in (0,1)$, where ${}_1\mathcal{M}_\rho$ is the dense linear manifold of X_0 of 14) consisting of all complex valued measurable functions u on $(0,+\infty)$ having $\int_0^{+\infty} |u(x)|^2 (1+x)^{2(1+\rho)} dx < +\infty$;

III) over $(x,y,\lambda) \in (0,+\infty)\otimes(0,+\infty)\otimes\Omega_+$ holds the symmetry

41) $$J_q^+(y,x;\lambda) = \overline{J_q^-(x,y;\overline{\lambda})} = J_q^+(x,y;\lambda),$$

and likewise over $(y,\lambda) \in (0,+\infty)\otimes\Omega_+$ and $z \in {}_y^-D_+ \cup {}_y^+D_+$

41)' $$J_q^+(z,y;\lambda) = \overline{J_q^-(\overline{z},y;\overline{\lambda})};$$

IV) we have the single coincidence asymptotics that for each pair $(x_1,x_2) \in (0,+\infty)\otimes(0,+\infty)$ having $x_1 \neq x_2$ and $x_2 > r_d$ there holds

42) $$\begin{cases} 0 = \lim\limits_{\substack{(x,y,\lambda)\to(x_1,x_2,f(x_2)),\\ \mathscr{I}[\lambda]\geq 0}} \left| F_q^+(x,y;\lambda) - {}_1\psi_q^+(x,y;\lambda)e^{-L_+(y,\lambda)} \right| \\ 0 = \lim\limits_{\substack{(x,y,\lambda)\to(x_2,x_1,f(x_2)),\\ \mathscr{I}[\lambda]\geq 0}} \left| F_q^+(x,y;\lambda) - {}_2\psi_q^+(x,y;\lambda)e^{-L_+(x,\lambda)} \right| \end{cases}$$

with 16) and 40) and with $_{p}\psi_{q}^{+}(x,y;\lambda)$ sub $\frac{1}{2}$ order Hölder continuous in $\vartheta[\lambda]\geq 0$ neighborhoods of such points;

V) with η in 5), we have the domination that, for each integer $q \geq 0$, each r_1, r_2 having $r_2 < r_1 < r_2 < +\infty$, each $\gamma \in (0,1)$, some $\eta' \in (0,\eta)$, some $\delta \in (0, 2^{-1}[r_1 - r_2])$, there exists a constant $M \in (0,+\infty)$ such that

$$\left(\sup_{z\in[s_+(x),x]} |J_q^+(z,y;\lambda)|\right) \leq$$

43)
$$\leq M\left(1+\xi_{x,y}^{-1}\right)^{1+\gamma}\left(1+\xi_{x,y}\right)^{-2-\eta'}\left\{1+\chi_{(r-\delta,r+\delta)}(x)\left[1+\ln\left(1+|y-r|^{-1}\right)\right]\left[1+\ln\left(1+|x-y|^{-1}\right)\right]\right\}$$

over $x, y \in (0,+\infty)$ and $\lambda = f(r) + ih$ with $r \in [r_1, r_2]$ and $h \in (0,+\infty)$ having $x \neq y \neq r \neq x$, here denoting $\xi_{x,y} = (\max x,y)$.

Now to return to the four problems in the introduction, first for Faddeev's wave operator formulas, as per [8] (8.4)-.7) p. 60, also 8.19), .26) p. 63,64) with F_q^+ in 16) here for the spherical harmonic reduced radial analogue we would want the h', $h \to 0^+$ simultaneous limit of

44.1)
$$\int_0^{+\infty} \frac{F_q^+(x,y;f(y)+ih')}{f(x)-[f(y)+ih]}\, u(y) y^2 dy\,,$$

and similarly for the wave operator adjoint would want the $h \to 0^+$ limit of

44.2)
$$\int_0^{+\infty} \frac{F_q^+(x,y;f(x)+ih)}{[f(x)+ih]-f(y)}\, u(y) y^2 dy\ .$$

But 42) 40) show for $x > r_a$ and $\rho \in (0,1)$ that the $[x+\rho,\ x+\rho^{-1}]$ portion of 44.2) becomes as $h \to 0^+$

45) $$o(1)+\left[e^{-ia[f'(x)]^{-1}(\ln h)}\right]\left[e^{a\pi[2f'(x)]^{-1}-W_+(x,f(x)+ih)}\right]\int_{x+\rho}^{x+\rho^{-1}}\frac{\psi^+_{2q}(x,y;f(x)+ih)}{f(x)-f(y)+ih}u(y)y^2$$

for continuous u, for which $\ln h \to -\infty$ and the bounded infinite oscillation of the first factor yields no finite limit as asserted. In 44.1) the same 42), 40) by Riemann-Lebesgue yield annihilation of any compact interval portion in $(x,+\infty) \cap (r_a,+\infty)$. Thus in both the Faddeev formulas break down as asserted.

Next for the second problem, we will see that an elementary estimate (independent of the more complicated Gustafson-Johnson criterion) from II) and 43) shows that no $r_1 \in (r_a,+\infty)$ can have $\lambda_1 = f(r_1)$ satisfy $\lambda_1 \in \sigma_p(H_q)$. For if such r_1 with $\lambda_1 = f(r_1) \in \sigma_p(H_q)$ existed, then the linear manifold $\mathcal{N}$ of continuous functions on $(0,+\infty)$ having compact support $\subseteq (0,r_1) \cup (r_1,+\infty)$ has $\mathcal{N}$ be dense in X_0 and $\mathcal{N} \subseteq {}_1\mathcal{M}_\ell$ of II); hence $H_q w_1 = \lambda_1 w_1$, $\|w_1\| = 1$ would have $v_1 \in \mathcal{N}$ with $\|v_1 - w_1\| < \frac{1}{8}$, whence $|(w_1,v_1)| \geq \frac{7}{8}$ and $(\bar{\lambda}_h I - H_q)^{-1} w_1 = h^{-1} i w_1$ for $\lambda_h = \lambda_1 + ih$ and $h > 0$,

46) $$\left(\tfrac{7}{8}\right)h^{-1} \leq |h^{-1} i (w_1,v_1)| = |([\lambda_h I - H_q]^{-1} v_1, w_1)|,$$

in which $\int_{r_1-\delta}^{r_1+\delta} |f(r_1)+ih-f(y)|^{-1}|w_1(y)|y^2 dy = O(h^{-\frac{1}{2}})$ as $h \to 0^+$ (by Schwarz-Hölder) readily shows by II), 15), 16), 43) that the 46) right side would also be $O(h^{-1/2})$, yielding a contradiction and showing that $\lambda_1 \in \sigma_p(H_q)$ is impossible.

For the third and fourth problems, since 1) implies the boundedness desired for the third problem, thus only 1) need be indicated. Here for given k_-, k_+ having $f(r_a) < k_- < k_+ < +\infty$, simplifying somewhat (it appears both useful and possible to enlarge $\mathcal{M}'$ substantially, as we expect to do in our final detailed publication) the manifold $\mathcal{M}'$ for 1) is taken to be the set of all continuous functions u on $[0,+\infty)$ which each have a $\rho \in (0,1)$ such that $f(\rho) < k_- < k_+ < f(\rho^{-1})$, u vanishes on $[2\rho^{-1}, +\infty)$, and u coincides with a polynomial on $[\rho, \rho^{-1}]$ (whence both u and $\bar{u}$ have unique analytic continuations v and $\tilde{v}$ on the strip $\rho \leq \mathcal{R}[z] \leq \rho^{-1}$). Thus by II) and 15), 16) with $\mathcal{M}' \subseteq {}_1\mathcal{M}_\beta$, over $\lambda \in \Omega_+(k_-, k_+)$ in 1)

$$47)\quad ([\lambda I - H_q]^{-1}u, u) = \int_0^{+\infty} \frac{\overline{u(x)}u(x)}{\lambda - f(x)} x^2 dx - \int_0^{+\infty}\int_0^{+\infty} \frac{\overline{u(x)}u(y)F_q^+(x,y;\lambda)}{(\lambda - f(x))(\lambda - f(y))} x^2 y^2 dy\,dx$$

in which with $D_\rho = (0,\rho) \cup (\rho^{-1}, 2\rho^{-1})$

$$48)\quad \int_0^{+\infty} u(s)w(s)ds = \int_{D_\rho} u(s)w(s)ds + \int_\rho^{\rho^{-1}} u(s)w(s)ds$$

and likewise for $\overline{u(s)}$. With $0 < \delta < |\mathcal{J}[s_+(\rho)]|$ and the contour $C_\delta = [\rho, \rho - i\delta] \cup [\rho - i\delta, \rho^{-1} - i\delta] \cup [\rho^{-1} - i\delta, \rho^{-1}]$, with 48) deform the resulting $\int_\rho^{1/\rho}$ part of the 47) first integral to $\int_{C_\delta}$, and likewise in the 47) second integral deform the $\int_{D_\rho}\int_\rho^{1/\rho}$ part to $\int_{D_\rho}\int_{C_\delta}$ and $\int_\rho^{1/\rho}\int_{D_\rho}$ to $\int_{C_\delta}\int_{D_\rho}$, using 24) and 41). Thus all parts of 47) are seen sub 1 Hölder continuous in $\lambda \in \Omega_+(k_-, k_+)$ except for $\int_\rho^{1/\rho}\int_\rho^{1/\rho}$, for which by 24), 28), 21) and Fubini used twice

$$\int_\rho^{\rho^{-1}}\int_\rho^{\rho^{-1}} \frac{\overline{u(x)}u(y)F_q^+(x,y;\lambda)}{(\lambda-f(x))(\lambda-f(y))}x^2y^2\,dy\,dx = \iint(\text{same})\,dx\,dy =$$

$$49)\quad = \int_\rho^{\rho^{-1}} \frac{y^2u(y)}{\lambda-f(y)}\left\{\int_{[y-i\delta,y]}\left[\frac{-Q_q(z,y)+v_q^+(z;y,\lambda)}{\lambda-f(z)}\right]\tilde{v}(z)z^2dz\right\}dy +$$

$$+\int_{C_\delta}\frac{z^2\tilde{v}(z)}{\lambda-f(z)}\left\{\int_\rho^{\rho^{-1}}\frac{F_q^+(z,y;\lambda)u(y)}{\lambda-f(y)}y^2dy\right\}dz,$$

in which latter $\int_{C_\delta} = \int_{[\rho,\rho-i\delta]} + \int_{[\rho-i\delta,\rho^{-1}-i\delta]} + \int_{[\rho^{-1}-i\delta,\rho^{-1}]}$

In the 49) first term using 38) and a number of lemmas for the elementary oscillation 40), and in the $\int_{[\rho-i\delta,\rho^{-1}-i\delta]}$ part of the second term using like 42)

$$50)\quad J_q^+(z,y;\lambda) = {}_3\psi_q^+(z,y;\lambda) + {}_4\psi_q(z,y;\lambda)e^{-L_+(y,\lambda)}$$

with ${}_p\psi_q^+(z,y;\lambda)$ sub $\frac{1}{2}$ order Hölder continuous there, we find all parts of 49) to be sub $\frac{1}{4}$ order Hölder continuous over $\lambda\in\Omega_+(k_-,k_+)$, and the desired 1) follows.

Returning to the speculative remarks following 1) on psuedo-eigenvalues, these apparently arise when two separate Hamiltonian systems are allowed to interact mildly. The first Hamiltonian H_1 should have at least two true eigenvalues ${}_1\lambda_1$ and ${}_1\lambda_2$ with ${}_1\lambda_1 < {}_1\lambda_2$; the second H_2 should have $(b_2,+\infty)$ in its continuous spectrum (corresponding to an escaping free particle); then a shift downward of the first system from ${}_1\lambda_2$ to ${}_1\lambda_1$ can have the resulting energy carried away by the escaping particle of the second system. $H = H_1 + H_2 + H'$ being the total Hamiltonian of the combined system, H' the interaction, u_0 being the initial state vector, the time flow $u_t = U_t u_0$ with $U_t = e^{-itH}$ being represented by the usual loop integral of the resolvent of H, P being the orthogonal projection for the state where the first system is in level ${}_1\lambda_1$ and the second system escaping particle has momentum $\vec{z} \in V$ with bounded open $V \subseteq R_3$, then (simplifying somewhat) the expectation value

$$(P U_t u_0 : U_t u_0) = \int_V |w(\vec{z},t)|^2 d\mu_3(\vec{z}), \tag{51}$$

$$w(\vec{z},t) = 2^{-1}\left[h_-(|\vec{z}|+{}_1\lambda_1;\vec{z}) + h_+(|\vec{z}|+{}_1\lambda_1;\vec{z})\right] + \tag{52.1}$$
$$+ \lim_{\rho\to 0^+} (2\pi i)^{-1} \int_{\lambda_0}^{+\infty} \chi_{[\rho,+\infty)}\big(|r-(|\vec{z}|+{}_1\lambda_1)|\big)\left\{\frac{e^{it[(|\vec{z}|+{}_1\lambda_1)-r]}}{r-(|\vec{z}|+{}_1\lambda_1)}\right\}\left[h_-(r;\vec{z})-h_+(r;\vec{z})\right]dr,$$

$$h_\pm(r;\vec{z}) = \left[\lim_{\gamma\to 0^\pm} h(r+i\gamma;\vec{z})\right], \tag{52.2}$$

52) following by shrinking the foregoing loop integrals to both sides of the real axis. In 52) would follow

$$w(\cdot,t) \xrightarrow[t\to\pm\infty]{} h_\pm(|\cdot|+{}_1\lambda_1;\cdot) \tag{53)'}$$

with $L_2(V)$ norm convergence, and hence in 51)

$$\lim_{t\to\pm\infty} (P U_t u_0 : U_t u_0) = \int_V |h_\pm(|\vec{z}|+{}_1\lambda_1;\vec{z})|^2 d\mu_3(\vec{z}), \tag{53}$$

if in the 52.1) integral $[h_-(r;\vec{z}) - h_+(r;\vec{z})]$ could be replaced by $\varphi(r)$ <u>independent</u> of $\vec{z} \in V$ with merely $\varphi \in L_2([\lambda_0,+\infty))$ known, as is easily seen by standard L_2 Fourier and Hilbert transform results. But here with $\varphi(r;\vec{z}) = [h_-(r;\vec{z}) - h_+(r;\vec{z})]$ apparently genuinely dependent on $\vec{z} \in V$, this strong version is apparently lost; the most convenient sufficient condition appears to be

54) $$\left|\varphi(r;\vec{z}) - \varphi(|\vec{z}|+{}_1\lambda_1;\vec{z})\right| \leq \left|(|\vec{z}|+{}_1\lambda_1)-r\right|^{\gamma}|\psi(\vec{z})|$$

with $\psi \in L_2(V)$ and constant $\gamma \in (0,1]$, then 53)′ being easily seen ([8], lemma 10.2), p. 86). Here 54) follows from γ order Hölder continuity of $h(\lambda;\vec{z})$ over $(\lambda,\vec{z}) \in \Omega_+(k'_-,k_+)\otimes V$ and separately over $\Omega_-(k'_-,k_+)\otimes V$ with suitable $k'_- < k_-$, which after chasing through several other transformations is a naturally expected consequence of 1) with $[k_-,k_+]$ suitably enlarged below as there stated. These considerations motivate our pursuit of 1) here.

Finally, supposing 53) can be so established, the explicitly known $|h_\pm(|\vec{z}|+{}_1\lambda_1;\vec{z})|$ there has very strongly peaked maxima at $|\vec{z}|+{}_1\lambda_1={}_1\lambda'_2$, with ${}_1\lambda'_2={}_1\lambda_2+{}_1\omega_2$ and the shift ${}_1\omega_2$ known with $|{}_1\omega_2|$ small, yielding a strong peak in escaping particle energy at ${}_1\lambda'_2-{}_1\lambda_1$. This ${}_1\lambda'_2$ is the above designated psuedo-eigenvalue associated with the original H_1 true eigenvalue ${}_1\lambda_2$.

References

1) F. Brownell, "Perturbation Theory and an Atomic Transition Model," Arch. Rat. Mech. & Anal., 10, (1962), p. 149-170.
2) F. Brownell, "Psuedo-Eigenvalues, Perturbation Theory, and the Lamb Shift Computation," p. 393-423, "Perturbation Theory and its Applications in Quantum Mechanics," ed. C.H. Wilcox, Wiley, 1966.
3) F. Brownell, "Second Quantization and Recalibration of the Dirac Hamiltonian of a Single Electron Atom without Radiation," Journ. d'Anal. Math., 16, (1966), p. 1-422.

4) F. Brownell, "A Limbotic Reformulation of Quantum Electrodynamics and the Lamb Shift Basis," LD00012 monograph, University Microfilm, Ann Arbor, 1973.
5) K. Gustafson & G. Johnson, "On the Absolutely Continuous Subspace of a Self-Adjoint Operator," Helv. Phys. Acta, 47, (1974), p. 163-166.
6) P. Rejto, "On a Theorem of Titchmarsh-Kodaira-Weidmann Concerning Absolutely Continuous Operators II," Indiana Univ. Math. J., 25, (1976), p. 629-658.
7) J. Weidmann, "Zur Spektraltheorie von Sturm-Liouville Operatoren," Math. Zeits, 98, (1967), p. 268-302; particularly th 5.1), corr. 5.2), A) & $B)_3$, p. 293.
8) L. Faddeev, "Mathematical Aspects of the Three-Body Problem in Quantum Mechanical Scattering Theory," Israel Program Scientific Translations, 1965; original Russian: Trudy Mat. I. Steklov, 69, (1963), p. 1-122.
9) T. Ikebe & Y. Saito, "Limiting Absorption Method and Absolute Continuity for the Schrödinger Operator," J. Math Kyoto U., 12, (1972), p. 513-542.
10) B. Simon, "Resonances in n-body quantum systems with dilation analytic potentials and the foundations of time-dependent perturbation theory," Annals Math., 97, (1973), p. 247-274.
11) F. Brownell, "Spherical Harmonic Integrals and Dirac Hamiltonian Radial Reduction," Journal of Integral Equations, to appear.
12) H. Izozaki, On Long Range Stationary Wave Operators," Publ. R.I.M.S. Kyoto, 13, (1977), p. 589-626.
13) B. Friedmann, "Two Theorems on Wave Propagation," Bull. A.M.S., 62, (1956), p. 589, #741.

STATISTICAL INFERENCE IN QUANTUM MECHANICS

Jean-Paul Marchand

Department of Mathematics
University of Denver
Denver, Colo. 80208

ABSTRACT

This lecture is a brief account of a new theory of statistical inference which is applicable to classical and quantum physics and generalizes the concept of coarse-graining. The mathematical setting is non-commutative probability theory on von Neumann algebras.

HEURISTICS

Inference, as we perceive it here, aims at determining the most likely state w_0 of a system S about which only partial information I is available. The choice of w_0 is guided by the principles of

(a) Compatibility: w_0 reflects all of I, and

(b) Least Reason: w_0 reflects only I.

Before establishing the formalism in its mathematical generality, we consider two examples.

Example 1: Classical Probability

Let $\frac{d\mu}{dx}$ be the probability density for the position of a one-dimensional particle enclosed in a box $[0, L]$. The "coarse-grained" measurement of the discrete partition $\{\Delta_i\}$ of $[0, L]$ results in the expectations $\mu(\Delta_i)$. The partial measurement therefore defines the restriction $\mu_{\mathcal{B}}$ of μ to the Boolean σ-subalgebra $\mathcal{B} = \{\Delta_i\}''$ generated by $\{\Delta_i\}$, and the principle of compatibility implies that the restriction of μ_o to $\mathcal{B}$ equals $\mu_{\mathcal{B}}$. On the other hand, the principle of least reason implies that $\frac{d\mu_o}{dx}$ is constant on the Δ_i's. Hence we have

$$\text{(a)} \quad \mu_o|_{\mathcal{B}} = \mu_{\mathcal{B}} ,$$

$$\text{(b)} \quad \frac{d\mu_o}{dx} \text{ is } \mathcal{B}\text{-measurable.}$$

(a) and (b) define μ_o essentially uniquely (Figure 1).

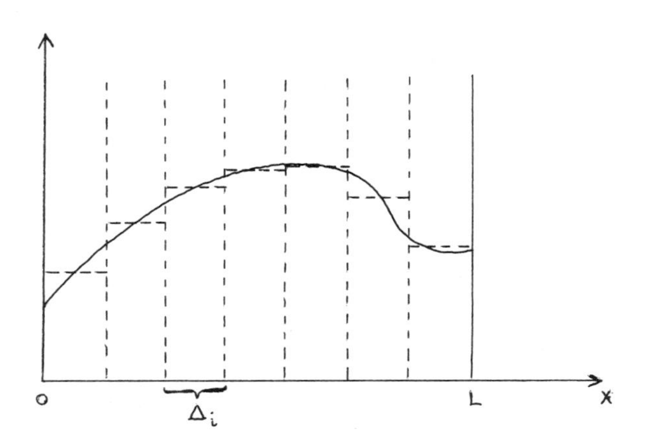

Figure 1: One-dimensional particle in [0, L]

——— : True density $\frac{d\mu}{dx}$

------ : Inferred density $\frac{d\mu_o}{dx}$.

Example 2: Quantum Mechanics

Let S be a quantum system, $\mathcal{A}$ the von Neumann algebra of operators in a Hilbertspace H generated by the observables of S, and $\mathcal{B} = \{P_i\}''$ the abelian subalgebra generated by a partition $\{P_i\}$ of the identity. If w is the state of S, then the partial measurement of $\{P_i\}$ results in the expectations $w(P_i)$.

If we assume, for simplicity, that S has no superselection rules, then $\mathcal{A} = B(H)$, the algebra of all bounded operators in H, and the inferred state w_o ca be written $w_o(A) = \tau(W_o A)$, where, in analogy to the classical case, the unique density operator $W_o = \frac{dw_o}{d\tau}$ may be considered as a generalized (bounded) Radon-Nikodym derivative of w_o with respect to the trace τ. The two principles now imply

(a) $w_o(P_i) = w(P_i)$,

(b) $W_o = \frac{dw_o}{d\tau} \in \mathcal{B}$.

These conditions have the unique solution (if it exists!)

$$W_o = \sum_i \frac{w(P_i)}{\tau(P_i)} P_i \, . \qquad (1)$$

The state w_o defined by (1) is called the coarse-grained state with respect to the measurement of $\{P_i\}$. We note, however, that its existence is limited by the requirement that the projections P_i have finite trace. This excludes, in particular, the position measurement

$$\mathcal{B} = \{Q\}'' = \{E(\Delta)\}'' ,$$

since the spectral measure $\{E(\Delta)\}$ of the position operator Q does not contain (non-trivial) finite-dimensional projections.

It turns out that these limitations can be avoided by generalizing the notion of a priori state in a way which is, mathematically and physically, reasonable. Let us briefly reflect on the feasibility and significance of such a generalization.

In example 1 we have tacitly assumed that the a priori measure dx on [0, L] is Lebesgue (equidistribution). This assumption is physically too restrictive. Suppose, for instance, that the box is in a gravitational field. It is then natural to include this information in the a priori measure $d\nu$ of the system, and the only mathematical condition for the existence of a $\mathcal{B}$-measurable Radon-Nikodym derivative $\frac{d\mu_o}{d\nu}$ is that $\mu_o|_{\mathcal{B}}$ is absolutely continuous w.r.t. $\nu|_{\mathcal{B}}$. (An instant's reflection shows that this technical condition is quite natural: If a dice tossing experiment results in a non-zero probability for the occurrence of face i, it would be awkward to assume an a priori distribution ν for which $\nu(i) = 0$!) .

In example 2 we encounter a similar situation: The choice of the unitarily invariant trace functional τ reflects an a priori equidistribution. But for a particle located on the infinite real line R this choice is inappropriate if it can be expected that the measurement yields non-zero expectations for finite intervals Δ . It could in fact be argued that a particle always interacts with some field and that the a priori situation must therefore be described by a normalizable state. (Note that τ is normalizable only if $\dim H < \infty$.)

Our general idea is to formulate a theory of inference for arbitrary normalizable a priori states v. The only mathematical problem is the precise meaning of the generalized Radon-Nikodym derivative $\frac{dw}{dv}$. Once this hurdle has been overcome, the inferred state can be defined by the postulates

(a) $w_o|_{\mathcal{B}} = w_{\mathcal{B}}$ (compatibility) ,

(b) $\frac{dw_o}{dv} \in \mathcal{B}$ (least reason) .

INFERENCE PRINCIPLE

Mathematically, the key to our reformulation is provided by the following result[1]:

Sakai Theorem: Let v, w be states on a von Neumann algebra $\mathcal{A}$ such that $w \leq \lambda v$, $\lambda \geq 1$. Then there is a unique $T \in \mathcal{A}^+$, $1 \leq \|T\|^2 \leq \lambda$, such that $w(A) = v(TAT)$, $\forall A \in \mathcal{A}$.

In analogy to the classical theory, the positive square root of the Sakai operator T may be viewed as a non-commutative generalization of the (bounded) Radon-Nikodym derivative $\frac{dw}{dv}$. If $\mathcal{A}$ is abelian, we have in fact

$$w(A) = v\left(\frac{dw}{dv} A\right), \quad \frac{dw}{dv} = T^2 .$$

Our generalized inference principle can now be described as follows: Let $\mathcal{A}$ and $\mathcal{B} \subset \mathcal{A}$ be the von Neumann algebras generated by the observables and the measured observables of S, respectively. Let v be an a priori state on $\mathcal{A}$ reflecting our information about S prior to measurement, and suppose that the state $w_{\mathcal{B}}$ measured on $\mathcal{B}$ satisfies $w_{\mathcal{B}} \leq \lambda\, v|_{\mathcal{B}}$. Finally, let $T_o \in \mathcal{B}$ be the Sakai operator defined by $w_{\mathcal{B}}(B) = v(T_o B T_o)$, $\forall B \in \mathcal{B}$. Then the $(\mathcal{B},v)$-inference w_o is defined on $\mathcal{A}$ by

$$w_o(A) = v(T_o A T_o) ,$$

or, in other words, w_o is the unique extension of $w_{\mathcal{B}}$ whose Sakai operator T relative to v is in $\mathcal{B}$.

The following properties of w_o can be easily verified[2] :

(1)	$w_o\|_{\mathcal{B}} = w_{\mathcal{B}}$	(compatibility)
(2)	$T_o \in \mathcal{B}$	(least reason)
(3)	$\mathcal{B} = \mathcal{A} \Rightarrow w_o = w_{\mathcal{B}}$	(complete measurement)
(4)	$\mathcal{B} = \{I\}'' \Rightarrow w_o = v$	(no measurement)
(5)	$w_{\mathcal{B}} = v\|_{\mathcal{B}} \Rightarrow w_o = v$	(measurement confirms a priori guess)

Moreover, $(\mathcal{B},v)$-inference generalizes the classical concept (example 1) and coarse-graining (example 2).

OTHER CHARACTERIZATIONS OF $(\mathcal{B},v)$-INFERENCE

We have defined the inferred state w_o by the property that its generalized Radon-Nikodym derivative relative to the a priori state v is in $\mathcal{B}$. We shall now show that w_0 can also be characterized by minimal Bures distance from v and by the property that it does not distinguish between an observable and its (suitably defined) conditional expectation with respect to $\mathcal{B}$.

Bures Distance and Uhlmann Transition Probability

Let v, w be relatively bounded and T be the corresponding Sakai operator. Then

$$P(v,w) = |v(T)|^2$$

is the Uhlmann transition probability[3] and

$$d(v,w) = \left(2\left[1 - v(T)\right]\right)^{1/2}$$

the Bures distance[4] between v and w. (For a very general proof that d defines a metric on the set N of states on $\mathfrak{A}$ cf. Gudder[5] .)

Theorem[6] : w_0 is characterized as the extension of $w_{\mathcal{B}}$ with maximum Uhlmann transition probability and minimum Bures distance from the a priori state v.

Sketch of Proof: Let $w(A) = v(TAT)$ be an extension of $w_{\mathcal{B}}$. Then $v(TBT) = w_{\mathcal{B}}(B) = v(T_0 B T_0)$, $B \in \mathcal{B}$. For $B = T_0^{-1}$ this implies $v(T\,T_0^{-1}\,T) = v(T_0)$ and the Schwarz inequality yields

$$\begin{aligned} |v(T)|^2 &= \left|v\left[T_0^{1/2}(T_0^{-1/2}T)\right]\right|^2 \\ &\leq v(T_0).v(T\,T_0\,T) = |v(T_0)|^2 . \end{aligned}$$

Hence,

$$P(v,w) \leq P(v,w_0) \quad ; \qquad d(v,w) \geq d(v,w_0) .$$

The uniqueness of w_0 follows from Schwarz' equality.

We note that the characterization of inference by minimal Bures distance allows a generalization of the original concept to situations where the partial measurement defines expectations on a subset $\mathcal{S}$ of $\mathfrak{A}$ which is not necessarily a von Neumann subalgebra of $\mathfrak{A}$ itself. An example of this (measurements at various times) will be briefly discussed at the end.

Conditional Expectation

In the classical case it can be shown that the inferred measure μ_0 does not distinguish between a random variable f and its conditional expectation $E(f|\mathcal{B})$ w.r.t. the measured σ-subring $\mathcal{B}$, i.e.

$$\int E(f|\mathcal{B})\, d\mu_0 = \int f\, d\mu_0 .$$

Hence the question arises whether this property can be generalized to the non-commutative case. In Gudder et al.[7] we have shown that the proper generalization of $E(f \mid \mathcal{B})$ is as follows:

Let $L^2(\mathcal{A},v)$ and $L^2(\mathcal{B},v)$ be the Hilbertspace completions of $\mathcal{A}$ and $\mathcal{B}$ relative to the inner product $\langle A_1, A_2 \rangle = v(A_1^+ A_2)$, $\mathcal{P}$ the projection $L^2(\mathcal{A},v) \rightarrow L^2(\mathcal{B},v)$, and $\pi_v(A)$ the linear extension of the map $\pi_v(A)\, A' = A\, A'$ from $\mathcal{A}$ to $L^2(\mathcal{A},v)$. If the <u>$(\mathcal{B},v)$-conditional expectation of A</u> is defined as

$$E_v(A \mid \mathcal{B}) = \mathcal{P}\ \pi_v(A)\Big|_{L^2(\mathcal{B},v)} ,$$

then the following holds:

<u>Theorem</u>[7] : w_0 is the unique extension of $w_{\mathcal{B}}$ for which

$$w_0\left[E(A \mid \mathcal{B})\right] = w_0(A) , \quad \forall\ A \in \mathcal{A} .$$

Moreover, $E(A \mid \mathcal{B})$ has all the expected properties of a conditional expectation.

Relative Entropy

It is natural to question the connection between maximum likelihood and maximum entropy. If the a priori weight is τ (example 2), it can be shown that the inferred state w_0 (defined by (1)) is the unique extension of $w_{\mathcal{B}}$ for which the von Neumann entropy

$$S_{\tau}(w) = - \tau\left[\frac{dw}{d\tau} \log \frac{dw}{d\tau}\right]$$

is maximal. It is therefore reasonable to search for a relative entropy $S_v(w)$ which can be defined for arbitrary a priori states v and and which characterizes w_0 as a maximum entropy state. The following result provides a partial answer.

<u>Theorem</u>[6] : Let $N_v(\mathcal{A})$ be the class of those extensions w of $w_{\mathcal{B}}$ which admit Sakai operators T relative to v, and let $F(T) = \sum_0^\infty C_n (T-\lambda)^n$ $\lambda \neq 0$. Then a relative entropy of the form $v(F(T))$ characterizes w_0 as a maximum entropy state in $N_v(\mathcal{A})$ if and only if $C_n = 0$ for $n \geq 3$, or equivalently if, up to a trivial normalization,

$$S_v(w) = v(T) . \qquad (2)$$

Various relative entropies have been proposed in the literature (for a list cf. Benoist et al.[8]), but since none of them satisfies (2), they do not characterize inference by maximal entropy. The spin system may serve as an example.

Example: The Spin

Let $\mathfrak{A} = B(H)$, dim H = 2 , $\mathfrak{B} = \{\sigma_1\}''$, $w(\sigma_1) = b$, $v(A) = \tau(VA)$ $V = Z^{-1}\exp(-\beta\sigma_3) = \frac{1}{2}(\sigma_0 + a\sigma_3)$. This represents a spin system in interaction with a heat reservoir at temperature β^{-1} of which only the σ_1-component has been measured. The inferred state w_0 has the density matrix[2]

$$W_0 = \tfrac{1}{2}(\sigma_0 + b\sigma_1 + a(1-b^2)^{1/2}\sigma_3) .$$

For arbitrary $b \in [0, 1]$ (and $a > 0$), $W_0(b)$ can be represented as an ellipse in the first quadrant of the (σ_1, σ_3)-plane of the Liouville sphere. In Figure 2 this curve is compared with the curves of the maximum entropy states relative to the entropies S^N and S^{UL} defined by Naudts and Umegaki-Lindblad (cf.[8]).

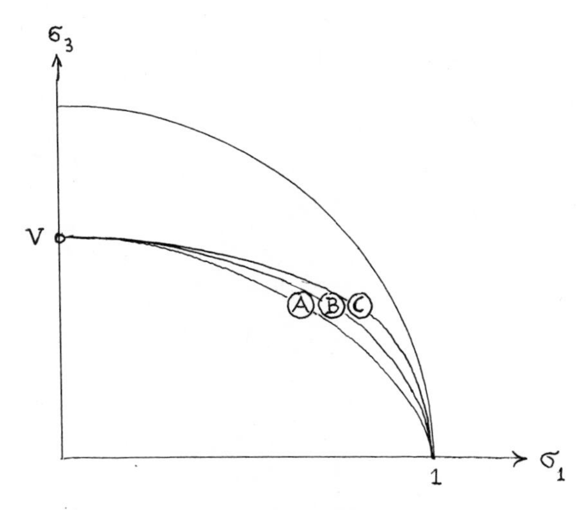

Figure 2: Inference versus Maximum Entropy

A: Inferred states (ellipse)

B: Maximum S^N-states

C: Maximum S^{UL}-states.

APPLICATIONS

I conclude this review with an outline of the main applications which have so far been considered. If the reader finds my indications enigmatic, I refer him to the literature.

Classical Probability[2]

$$\{\Omega, \Sigma, \nu\} \;;\quad \Sigma' \subset \Sigma \;;\quad \mu_{\Sigma'} \prec \nu|_{\Sigma'} ;\quad \mathfrak{A} \approx L^{\infty}(\Omega, \Sigma, \nu) \;;$$
$$\mathcal{B} \approx L^{2}(\Omega, \Sigma', \nu|_{\Sigma'}) \;;\quad w_{o}(A) = \int_{\Omega} A(\omega)\, d\mu_{o} \;;$$
$$\frac{d\mu_o}{d\nu} \in L^{1}(\Omega, \Sigma', \nu|_{\Sigma'}) \,.$$

Quantum-mechanical Position[2]

$$\mathfrak{A} = B(H) \;;\quad \mathcal{B} = \{E(\Delta)\}'' \;;\quad v(\lambda) = v(E(\lambda)) \;;$$
$$w(\lambda) = w\,(E(\lambda)) \;;$$
$$w_{o}(A) = \iint d\lambda\, d\rho \left(\frac{w'(\lambda)\, w'(\rho)}{v'(\lambda)\, v'(\rho)}\right)^{1/2} \frac{\partial^2}{\partial\lambda\,\partial\rho}\, v\big[E(\lambda)\, A\, E(\rho)\big] \,.$$

Coupled Systems[9]

$$H = H_1 \otimes H_2 \;;\quad \mathfrak{A} = B(H) \;;\quad \mathcal{B} = B(H_1) \otimes I_2 \;;$$
$$V_1 = \tau_2(V) \;;\quad w_{\mathcal{B}} = (A_1 \otimes I_2) = \tau_1(W_1 A_1) \;;$$
$$w_o(A) = v(T_o\, A\, T_o) \;:$$
$$T_o = T_1 \otimes I_2 \;;\quad T_1 = V_1^{-1/2}(V_1^{1/2}\, W_1\, V_1^{1/2})^{1/2}\, V_1^{-1/2}$$

Measuring Process[10]

H_1 : Apparatus A ; H_2 : Probe P ;

v : Correlation between the states of A and P ;

Rest as for coupled systems.

Measurement at Two Times[11]

$$\mathfrak{A} = B(H) \;;\quad \mathcal{B} = \mathcal{B}_o \cup \mathcal{B}_t \;;\quad \mathcal{B}_t = U_{-t}\, \mathcal{B}\, U_t \quad (U_t \text{ evolution}) \,.$$

Here $\mathcal{B}$ is not a subalgebra of B(H). But w_o can still be defined via minimal Bures distance!

REFERENCES

1. S.Sakai, Bull.AMS 71, 149 (1965)
2. J.P.Marchand, Found.Phys. 7, 35 (1977)
3. A.Uhlmann, Rep.Math.Phys. 9, 273 (1976)
4. D.Bures, Trans.AMS 135, 199 (1969)
5. S.Gudder:"Expectation and Transition Probability", Preprint Univ.of Denver 1980
6. J.P.Marchand & W.Wyss, J.Stat.Phys. 16, 349 (1977)
7. S.Gudder & J.P.Marchand, Rep.Math.Phys. 12, 317 (1977)
8. R.Benoist, J.P.Marchand & W.Wyss, Lett.Math.Phys. 3, 169 (1979)
9. R.Benoist & J.P.Marchand, Lett.Math.Phys. 3, 93 (1979)
10. R.Benoist, J.P.Marchand & W.Yourgrau, Found.Phys. 7, 827 (1977)
11. J.P.Marchand, to appear in the memorial volume for Wolfgang Yourgrau, ed.van der Merwe (1980)

SCHRÖDINGER OPERATORS WITH OSCILLATING POTENTIALS

Allen Devinatz*

Department of Mathematics
Northwestern University
Evanston, IL 60201

Peter Rejto**

School of Mathematics
University of Minnesota
Minneapolis, MN 55455

§1. INTRODUCTION

Due to the pioneering contributions of Kato-Kuroda [10], [11] and to the more recent works of Agmon [1] and Enss [6], a spectral and scattering theory for Schrödinger operators with short range potentials is now well established. An interesting example of a potential which does not belong to this class is the Wigner-von-Neumann [17] potential. This potential is the sum of a short range potential and of an oscillating one which is of the form,

$$p_o(x) = c\,\frac{\sin b\,|x|^{\alpha}}{|x|^{\beta}}\ , \quad \alpha, \beta > 0 \tag{1.1}$$

where $c = -8$, $b = 2$ and $\alpha = \beta = 1$.

For radially symmetric potentials of the form $V(|x|) = p_o(x) + V_S(|x|)$, where V_S is short range, the existence and asymptotic completeness of the usual Moller wave operators have been established by Dollard and Friedman [5] and by Ben-Artzi and Devinatz [2], for suitable values of α and β . These results

*Supported by NSF Grant MCS 79-02538-A01

**Supported by NSF Grant MCS 78-02199-A01

have been improved by Bourgeous [3], Devinatz [4] and White [25]. When V_S is not necessarity radially symmetric the existence of the Moller wave operators has been established in [4].

A natural problem which arises in connection with the work discussed above is to give a spectral and scattering theory for Schrödinger operators of the form

$$H \equiv -\Delta + p_o(|x|) + V_S(x) , \tag{1.2}$$

where $p_o(|x|)$ is a radially symmetric potential whose prototype is (1.1) and V_S is a short range potential, generally not radially symmetric. This in turn natually suggests the consideration of other types of Schrödinger operators, e.g.,

$$H \equiv -\Delta + p_o(|x|) + V_L(x) + V_S(x) , \tag{1.3}$$

where $V_L(x)$ is a long range potential of non-oscillatory type. Depending on the nature of V_L one can consider questions of existence and uniqueness of modified wave operators or questions concerning the spectral structure and an eigenfunction expansion theorem for the Schrödinger operator. A number of interesting papers have appeared in the literature along these lines and as a sample of recent work we refer to Mochizuki and Uchiyami [14], and Saito [21].

In this paper we shall make a contribution to one such problem. We shall give results and outline our methods. Full details will appear elsewhere.

§2. Preliminaries

Once the existence of the Moller wave operators has been established an interesting technique for establishing asymptotic completeness is to first establish a limiting absorption principle. (See [1] for a discussion and bibliography.)

Let us suppose that $H \mid \mathfrak{C}_0^\infty(R^3)$ has a self-adjoint closure. Here H is given by (1.2). Let $\mu \in \mathbb{C}$ with $\mathcal{I}m\mu \neq 0$. As is well known, the resolvent $R(H,\mu) \equiv (H-\mu I)^{-1}$ exists as a bounded operator. Suppose λ is real and not an eigenvalue of H. Roughly speaking, a limiting absorption principle says that

$$\lim_{\epsilon \downarrow 0} R(H, \lambda \pm i\epsilon) \equiv R^{\pm}(H,\lambda) \tag{2.2}$$

exists in a suitable operator topology. Of course, if λ is in the spectrum of H, this limit cannot exist in the uniform operator topology, or indeed in the strong or weak operator topology of $\mathfrak{B}(\mathfrak{L}^2(\mathbb{R}^3))$, the latter symbol designating the algebra of bounded operators on $\mathfrak{L}^2(\mathbb{R}^3)$.

To be specific about the meaning of the limit in (2.2) we define a weight function

$$m_s(x) = (1+|x|)^{-s}, \quad s > 1. \tag{2.3}$$

The limit in (2.2) means there exist operators

$$R^{\pm}(H,\lambda) : m_s^{1/2} L^2(\mathbb{R}^3) \to m_s^{-1/2} L^2(\mathbb{R}^3)$$

so that

$$\lim_{\epsilon \downarrow 0} m_s^{1/2} R(H, \lambda \pm i\epsilon) m_s^{1/2} = m_s^{1/2} R^{\pm}(H,\lambda) m_s^{1/2} \quad \text{in } \mathfrak{B}(\mathfrak{L}^2(\mathbb{R}^3)), \tag{2.2'}$$

where the uniform operator topology is taken for $\mathfrak{B}(\mathfrak{L}^2(\mathbb{R}^3))$. In other words we have

$$\lim_{\epsilon \downarrow 0} \| m_s^{1/2} [R(H, \lambda \pm i\epsilon) - R^{\pm}(h,\lambda)] m_s^{1/2} \| = 0, \tag{2.2''}$$

where $\|\cdot\|$ is the operator norm.

The first step in getting a limiting absorption principle for the operator H is to get it for the operator

$$H(p) \equiv -\Delta + p_0(x) \tag{2.4}$$

We shall suppose that $H(p) \,|\, \mathfrak{C}_0^{\infty}(\mathbb{R}^3)$ has a self-adjoint closure in $\mathfrak{L}^2(\mathbb{R}^3)$. Then $R(H(p),\mu)$ exists for $\mathcal{I}m\,\mu \neq 0$.

Let $\mathcal{I}$ be a compact interval on the real line which does not contain an eigenvalue of $H(p)$. For a given angle $0 < \gamma \leq \pi$, let

$$(2.5) \qquad \mathcal{R}_{\pm}(\mathcal{I}) = \{\mu \in \mathbb{C} : \operatorname{Re}\mu \in \mathcal{I},\ 0 < \operatorname{Im}\mu < \gamma\} .$$

The principle step in obtaining a limiting absorption principle for $H(p)$ is to prove the inequality

$$(2.6) \qquad \sup_{\mu \in \mathcal{R}_{\pm}(\mathcal{I})} \| m_s^{1/2} R(H(p),\mu)\, m_s^{1/2} \| < \infty .$$

The object of this report is to give results and to outline the principal steps in obtaining such an inequality.

§3. FORMULATION OF THE RESULTS

For simplicity, in this report we shall limit ourselves to long range potentials of the form

$$(3.1) \qquad p_0(|x|) = \frac{c \sin b\,|x|^{\alpha}}{|x|} , \quad \alpha > 0 .$$

We shall use the notations of the previous section.

THEOREM 3.1. <u>Let</u> $\mathcal{I}$ <u>be a compact interval in the half-line</u> $\mathbb{R}^+ \equiv (0,\infty)$. <u>If</u> $\alpha \neq 1$, <u>or if</u> $\alpha = 1$ <u>and</u> $\mathcal{I} \subset (0, b^2/4)$, <u>then the inequality</u> (2.6) <u>is valid</u>. <u>If</u> $\alpha = 1$ <u>there exists a constant</u> γ_0 <u>so that if</u> $\mathcal{I} \subset (b^2/4, \infty)$ <u>and</u>

$$(3.2) \qquad \frac{|c|}{\left|\frac{b^2}{4} + \omega\right|^{1/2}} \int_1^{\left|\frac{b^2}{4\omega} + 1\right|^{1/2}} \frac{d\sigma}{(\sigma^2 - 1)^{1/2}} < \gamma_0 ,$$

<u>where</u> $\omega = \inf\{\mu - b^2/4 : \mu \in I\}$, <u>then again the inequality</u> (2.6) <u>is valid</u>.

For the purposes of the next theorem we introduce the family of operators

$$(3.3) \qquad F^{\pm}(\mu) \equiv m_s^{1/2} R(H(p),\mu)\, m_s^{1/2} , \quad \mu \in \mathcal{R}_{\pm}(\mathcal{I}) .$$

THEOREM 3.2. _Under the assumptions of Theorem 3.1, each of the two operator valued functions of definition (3.3) admits a continuous extension to the closure of $\mathcal{R}_+(\mathcal{J})$ and $\mathcal{R}_-(\mathcal{J})$, respectively, with respect to the uniform operator topology._

§4. OUTLINE OF THE PROOFS OF THEOREM 3.1 AND THEOREM 3.2

Our proof of Theorem 2.1 will make essential use of the well known fact [2], [24] that the operator $H(p)$ admits a complete family of reducing subspaces on each of which it acts like an ordinary differential operator.

To describe these operators, to each potential p in $\mathfrak{L}^2_{\ell oc}(\mathbb{R}^+)$ assign the operator

$$(4.1)\qquad L(p)f(\rho) = -f''(\rho) + p(\rho)f(\rho) \ , \quad f \in \mathfrak{C}_0^\infty(\mathbb{R}^+) \ .$$

Next define

$$(4.2)\qquad p(j)(\rho) = p_o(\rho) + \frac{j(j+1)}{\rho^2} \ , \quad j = 0,1,2,\ldots$$

Then we know [2], [24] that $H(p)(j)$, the part of the operator $H(p)$ over the j - th reducing subspace is unitarily equivalent to a self-adjoint extension of the operator $L(p(j))$;

$$(4.3)\qquad H(p)(j) \sim L(p(j)) \ .$$

For the boundary condition defining this self-adjoint extension we refer to the Appendix of [2]. It follows from the explicit form of this unitary transformation that

$$(4.4)\qquad m_s^{1/2}\, H(p)(j)\, m_s^{1/2} \sim m_s^{1/2}\, L(p(j))\, m_s^{1/s} \ .$$

Since the norm of an orthogonal sum equals the supremum of the norms and since unitary equivalence preserves norms, these relations yield

$$(4.5)\qquad \| m_s^{1/2} R(H(p),\mu) m_s^{1/2} \| = \sup_{j \geq 0} \| m_s^{1/2} R(L(p(j)),\mu) m_s^{1/2} \| \ .$$

So the conclusion (2.6) is equivalent ot

$$(4.6) \qquad \sup_{\mu \in \mathfrak{R}_{\pm}(\mathcal{J})} \ \sup_{j \geq 0} \| m_s^{1/2} R(L(p(j)),\mu)\, m_s^{1/2} \| < \infty \, .$$

The considerations of [2] show that for each fixed j the supremum over $\mathfrak{R}_{\pm}(\mathcal{J})$ is finite. Hence (4.6) is equivalent ot

$$(4.7) \qquad \limsup_{j \to \infty} \ \sup_{\mu \in \mathfrak{R}_{\pm}(\mathcal{J})} \| m_s^{1/2} R(L(p(j)),\mu)\, m_s^{1/2} \| < \infty \, .$$

In spite of the simplicity of the potentials $p(j)$ given by (4.2) we have not found it possible to directly obtain the inequality (4.7). Instead, our proof of this inequality will make essential use of the fact that we can construct a family $\{q(j,\mu)\}$ of approximate potentials. The properties of these approximate potentials are described by the next theorem. We first introduce the functions

$$(4.8) \qquad n_s(\rho) \equiv n_s(j,\mu)(\rho) \equiv \max \{ |p(j) - q(j,\mu)|(\rho) \, , \ m_s(\rho) \} \, .$$

As usual for a given Hilbert space $\mathfrak{H}$ we denote by $\rho(A)$ the resolvent set for an operator A defined in $\mathfrak{H}$, and by $\mathfrak{D}(\mathfrak{A})$ the domain of A.

We now define the domain of $L(q(j,\mu))$ by

$$(4.9) \qquad \mathfrak{D}(L(q(j,\mu))) = \{ f \in \mathfrak{L}^2 : f, f' \text{ are locally AC} \, , \ \lim_{\rho \to 0} f(\rho) = 0 \, , \ L(q(j,\mu))\, f \in \mathfrak{L}^2 \} \, .$$

Using this definition we have the following theorem.

THEOREM 4.1. *Let the assumptions of Theorem* 3.1 *hold. Then there are regions* $\mathfrak{R}_{\pm}(\mathcal{J})$ *of the form* (2.5), *an integer* j_o *and potentials* $q(j,\mu)$ *with the following properties*:

For each $\mu \in \mathfrak{R}_{\pm}(\mathcal{J})$ *and for each* $j \geq j_o$

$$(4.10) \qquad \mu \in \rho(L(q(j,\mu))) \, ;$$

$$(4.11) \qquad (p(j) - q(j,\rho))\, R(L(q(j,\mu),\mu) \in \mathfrak{B}(\mathfrak{L}^2(\mathbb{R}^+) \, .$$

Furthermore,

$$\limsup_{j\to\infty} \sup_{\mu \in \mathcal{R}_\pm(\mathcal{J})} \| n_s^{1/2} R(L(q(j,\mu)),\mu)\, n_s^{1/2} \| < \infty . \tag{4.12}$$

$$\limsup_{j\to\infty} \sup_{\mu \in \mathcal{R}_\pm(\mathcal{J})} \| [I + n_s^{-1/2}(p(j) - q(j,\mu))\, R(L(q(j,\mu)))\, n_s^{1/2}]^{-1} \| < \infty . \tag{4.13}$$

Assuming the validity of this theorem let us show how this leads to the estimate (4.7). Using (4.9) and (4.10) we have the resolvent equation

$$R(L(p(j),\mu) - R(L(q(j,\mu)),\mu) = R(L(p(j)),\mu)[q(j,\mu) - p(j)]\, R(L(q(j,\mu)),\mu) ,$$

which is valid on a dense set in $\mathfrak{L}^2$. Hence, multiplying on the left and on the right by $n_s^{1/2}$ we have the equality

$$n_s^{1/2} R(L(p(j)),\mu) n_s^{1/2} \{ I + n_s^{-1/2} [p(j) - q(j,\mu)]\, R(L(q(j,\mu)),\mu)\, n_s^{1/2} \} = n_s^{1/2} R(L(q(j,\mu)),\mu)\, n_s^{1/2} ; \tag{4.14}$$

which is also valid on a dense set in $\mathfrak{L}^2$. This equality in conjunction with conclusions (4.12) and (4.13) gives the inequality

$$\limsup_{j\to\infty} \sup_{\mu \in \mathcal{R}_\pm(\mathcal{J})} \| n_s^{1/2} R(L(p(j)),\mu)\, n_s^{1/2} \| < \infty . \tag{4.15}$$

Since $m_s \leq n_s$ we have

$$\| m_s^{1/2} R(L(p(j)),\mu)\, n_s^{1/2} \| \leq \| n_s^{1/2} R(L(p(j)),\mu)\, n_s^{1/2} \| .$$

Using the fact that a bounded operator and its adjoint have the same norm we have

$$\begin{aligned} \| m_s^{1/2} R(L(p(j)),\mu)\, m_s^{1/2} \| &\leq \| n_s^{1/2} R(L(p(j)),\mu)\, m_s^{1/2} \| \\ &= \| m_s^{1/2} R(L(p(j)),\bar{\mu})\, n_s^{1/2} \| \\ &\leq \| n_s^{1/2} R(L(p(j)),\bar{\mu})\, n_s^{1/2} \| \end{aligned}$$

$$= \| n_s^{1/2} R(L(p(j)),\mu) n_s^{1/2} \|$$

This inequality combined with (4.15) gives estimate (4.7).

The proof of Theorem 3.2 follows from Theorem 3.1 with the help of a result obtained in [2]. Let $\mathcal{J}$ be a compact interval as described in Theorem 3.1. As was shown in [2] there exist kernels $k^{\pm}(j,\mu)(\xi,\ )$ which for each j are defined and continuous on $\mathcal{R}_{\pm}(\mathcal{J}) \times \mathbb{R}^{+} \times \mathbb{R}^{+}$, and for $\mathcal{J}m\,\mu \neq 0$ is Green's kernel for $R(L(p(j)),\mu)$. Thus if $\lambda \in \mathcal{J}$ and $f \in \mathfrak{L}^2(\mathbb{R}^+)$ with compact support

$$R^{\pm}(H(p(j)),\lambda)\, f(\xi) \equiv \int_{\mathbb{R}^+} k^{\pm}(j,\lambda)(\xi,n)\, f(n)\, dn \tag{4.16}$$

are well defined. Thus for $\mu \in \mathcal{R}_{\pm}(\mathcal{J})$ we may write

$$[R^{\pm}(L(p(j)),\mu) - R^{\pm}(L(p(j)),\lambda)]\, f(\xi) = \tag{4.17}$$

$$\int_{\mathbb{R}^+} [k^{\pm}(j,\mu) - k^{\pm}(j,\lambda)](\xi,\eta)\, f(\eta)\, d\eta \,.$$

It is clear that the left side remains bounded and goes to zero as $\mu \to \lambda$.

Since $m_s \in \mathfrak{L}^1(\mathbb{R}^+)$, it follows by the previous remarks that

$$\| m_s^{1/2} [R^{\pm}(L(p)(j),\mu) - R^{\pm}(L(p)(j),\lambda)]\, m_s^{1/2} f \| \to 0 \quad \text{as} \quad \mu \to \lambda \,.$$

Thus on a dense set of elements f in $\mathfrak{L}^2(\mathbb{R}^3)$ there are operators $R^{\pm}(H(p),\mu)$ defined for $\mu \in \mathcal{R}_{\pm}(\mathcal{J})$ so that

$$\lim_{\mu \to \lambda} \| m_s^{1/2} [R^{\pm}(H(p),\mu) - R^{\pm}(H(p),\lambda)\, m_s^{1/2} f \| = 0 \,. \tag{4.18}$$

From Theorem 3.1 we know that the operator valued functions $F^{\pm}(\mu)$, given by (3.3) are bounded on $\mathcal{R}_{\pm}(\mathcal{J})$. This combined with (4.18) shows immediately that if $\lambda \in \mathcal{J}$, the stong limits

$$s\text{-}\lim_{\mu \to \lambda} F^{\pm}(\mu) \equiv F^{\pm}(\lambda) \tag{4.19}$$

and the extensions are continuous on $\mathcal{R}_{\pm}(\mathcal{J})$, respectively.

The fact that the existence of the strong limit implies the existence of the limit in the operator norm may be proved, _mutatis mutandis_, in the same way as in [1, pp. 164-165].

§5. CONSTRUCTION OF APPROXIMATE POTENTIALS

In this section we shall indicate how to construct approximate potentials. We wish these approximations to be valid uniformly in j and μ in the sense of Theorem 4.1. We shall construct these approximate potentials with the help of approximate solutions to the differential equation

$$f''(\rho) + (\mu - p(j)(\rho))\, f(\rho) = 0\,, \quad \mathcal{I}m\,\mu \neq 0\,. \tag{5.1}$$

Note that if μ is real, the coefficient of this equation vanishes at some real point, the turning point. For each complex μ we define a turning point τ' as the zero of this coefficient;

$$\mu - p(j)(\tau') = 0\,.$$

Near the turning point we seek approximations to the solutions of (5.1). Following Langer [7], [8], [12], [13], [15], we seek these approximate solutions with the help of the Airy equation

$$f''(z) - z(f(z) = 0\,. \tag{5.2}$$

Since the Airy equation has a turning point at zero, and the equation (5.1) has a turning point at τ', we make a change of independent variable $\varphi = \varphi(\rho)$, as yet unspecified, so that $\varphi(\tau') = 0$. The function $Ai(-\varphi(p))$ is a solution to (5.2) in the variable $-\varphi$. However, instead of working with this latter function it is much more convenient to work with an adjusted function

$$y(\rho) = \varphi'(\rho)^{-1/2}\, Ai(-\varphi(\rho))\,, \quad \rho \in \mathbb{R}^+\,. \tag{5.3}$$

In order to determine the function φ and the approximate potential we shall proceed on a formal basis. Using the fact that $Ai(-\varphi(\rho))$ satisfies (5.2) in the variable $-\varphi$, it is an elementary exercise to verify that

$$y''(\rho) + [\tfrac{1}{2}\{\varphi,\rho\} + \varphi'(\rho)^2\varphi(\rho)]\,y(\rho) = 0\,, \quad \rho \in \mathbb{R}^+\,. \tag{5.4}$$

where $\{\varphi,\rho\}$ is the Schwarzian derivative [8],

$$\{\varphi,\rho\} = \frac{\varphi'''(\rho)}{\varphi'(\rho)} - \frac{3}{2}\left(\frac{\varphi''(\rho)}{\varphi'(\rho)}\right)^2\,. \tag{5.5}$$

If we would choose φ so that

$$\mu - p(j)(\rho) = \varphi'(\rho)^2\varphi(\rho) + \tfrac{1}{2}\{\varphi,\rho\}\,, \quad \varphi(\tau') = 0\,, \tag{5.6}$$

then the function $y(\rho)$ of (5.3) would be an exact solution to (5.1). However, only under very special circumstances can a solution to (5.6) be given in terms of elementary functions, and hence it would be extremely difficult to obtain sufficient knowledge about φ to be useful. Instead we solve an approximate equation by dropping the term $p_o(\rho)$ on the left and the Schwarzian derivative on the right in (5.6). More precisely, we follow Langer [8], [12], [13], [15] and solve the non-linear initial value problem

$$\varphi_\ell'(\rho)^2\varphi_\ell(\rho) = \mu - (j(j+1) + \tfrac{1}{4})\rho^{-2}\,, \quad \varphi(\tau) = 0\,. \tag{5.7}$$

Here, τ is the solution to the equation $\mu - (j(j+1) + \frac{1}{4})\rho^{-2} = 0$; i.e.

$$\tau = \frac{\nu}{\sqrt{\mu}}\,, \text{ where } \nu^2 = j(j+1) + \tfrac{1}{4}\,, \tag{5.8}$$

and the principal square root of μ has been chosen. The term $-(1/4)\rho^{-2}$ has been in order to get a good approximate solution near the point zero. A formal solution to (5.7) is

$$\varphi_\ell(\rho) = \left\{\frac{3}{2}\int_\tau^\rho \left(\mu - \frac{\nu^2}{\rho^2}\right)^{1/2}\right\}^{2/3}\,. \tag{5.9}$$

The subscript ℓ indicates that we expect φ_ℓ to be an approximate solution to φ not only near the turning point but actually over an entire left interval of the form

$$\mathcal{I}_\ell = (0, \delta j)\,, \tag{5.10}$$

where the constant δ is to be chosen in an appropriate way.

With a suitable solution φ_ℓ of (5.7) we obtain a function $y_\ell(\rho)$ given by (5.3) which satisfies the equation (5.4), with φ_ℓ replacing φ. Hence we are led to take the approximate potential on $\mathcal{I}_\ell$ as

$$(5.11) \qquad q_\ell(j,\mu)(\rho) = \mu - \frac{1}{2}\{\varphi_\ell, \rho\} - \varphi_\ell'(\rho)^2\, \varphi_\ell(\rho) \ .$$

Near the point at infinity we seek an approximation to the solutions of equation (5.1) with the help of solutions of the equation

$$(5.12) \qquad f''(z) + f(z) = 0 \ .$$

Similarly to definition (5.3) we define

$$(5.13) \qquad y(\rho) = \varphi'(\rho)^{-1/2} \exp(i\,\varphi(\rho)) \ , \quad \rho \in \mathbb{R}^+ \ .$$

An elementary calculation shows that $y(\rho)$ satisfies the differential equation

$$(5.14) \qquad y''(\rho) + [\tfrac{1}{2}\{\varphi,\rho\} + \varphi'(\rho)^2]\, y(\rho) = 0 \ , \quad \rho \in \mathbb{R}^+ \ .$$

If we would choose φ so that

$$(5.15) \qquad \mu - p(j)(\rho) = \frac{1}{2}\{\varphi,\rho\} + \varphi'(\rho)^2 \ ,$$

then the function $y(\rho)$ given by (5.13) would be an exact solution to the equation (5.1). However, as before it is too difficult to get a solution for this equation for which one has a good control on the function φ. Hence we seek a solution to an approximate equation. Taking a que from the JWKB method [15], [16], [22], we seek a solution in the form

$$(5.16) \qquad \varphi_r(\rho) = \int_{\delta j}^{\rho} (\mu - \psi(\sigma))^{1/2}\, d\sigma \ .$$

Inserting the definitions (5.5) and (5.16) into equation (5.15) we find the non-linear differential equation,

$$(5.17) \qquad p(j) = \psi + \frac{1}{4}\,\frac{\psi''}{\mu - \psi} - \frac{5}{16}\left(\frac{\psi'}{\mu - \psi}\right)^2 \ .$$

By reasoning somewhat analogous to the reasoning employed in the construction of φ_ℓ , it is possible to choose an appropriate φ . Normalizing the original potential so that $b = 1$, in case $\alpha = 1$, ψ turns out to be

$$\psi = p(j) - r(j,\mu) + \{r(j,\mu)^2 - \tfrac{1}{4}p_o\}^{1/2} , \tag{5.18}$$

where

$$r(j,\mu) = \tfrac{1}{2}[p(j) - \mu + \tfrac{1}{4}] . \tag{5.19}$$

In case $\alpha \neq 1$, ψ is somewhat more complicated. As on the left interval $\mathcal{I}_\ell$ we are lead to define the approximate potential on the right interval

$$\mathcal{I}_r = (\delta j , \infty) \tag{5.20}$$

by means of the potential term in the differential equation (5.14). That is to say we take the approximate potential on $\mathcal{I}_r$ as

$$q_r(j,\mu)(\rho) = \mu - \tfrac{1}{2}\{\varphi_r , \rho\} - \varphi_r'(\rho)^2 . \tag{5.21}$$

Finally we define the approximate potential on $\mathbb{R}^+$ by

$$q(j,\mu)(\rho) = \begin{cases} q_\ell(j,\mu)(\rho) , & \rho \in \mathcal{I}_\ell \\ q_r(j,\mu)(\rho) , & \rho \in \mathcal{I}_r \end{cases} . \tag{5.22}$$

§6. ESTIMATES FOR THE APPROXIMATE RESOLVENTS

The conclusion (4.10) of Theorem 4.1 is obtained by constructing Weyl kernels from certain "eigensolutions" to the differential operators $L(q(j,\mu))$, $\mathcal{I}m\,\mu \neq 0$, and showing that these Weyl kernels are kernels for the resolvents of $L(q(j,\mu))$, provided μ is close enough to the real axis. These eigensolutions are specifically described by the following lemma.

LEMMA 6.1. *There exists a* $\gamma_o \leq \pi$ *so that for each non-negative integer* j *and for each non-real* μ *in the right half of the complex plane with* $|\mathrm{Arg}\,\mu| < \gamma_o$, *there are functions* k_ℓ *and* k_r *defined on* $\mathbb{R}^+$ *such that*

$$(6.1) \qquad k''_{\ell,r}(\rho) + (\mu - q(j,\mu)(\rho))\, k_{\ell,r}(\rho) = 0 ,$$

$$(6.2) \qquad \lim_{\rho \to 0} k_\ell(\rho) = 0 , \quad k_r(\rho) \in \mathfrak{L}^2(\mathcal{I}_r) ,$$

<u>and there is a</u> j_o <u>so that for</u> $j \geq j_o$ <u>the Wronskian</u> $W(k_\ell , k_r)$ <u>of</u> k_ℓ <u>and</u> k_r <u>is different from zero</u>.

Using this lemma it makes sense to form the Weyl kernel

$$(6.3) \qquad K(j,\mu)(\xi,\eta) = \frac{1}{W(k_\ell , k_r)} \begin{cases} k_\ell(\eta)\, k_r(\xi) , & \eta \leq \xi \\ k_\ell(\xi)\, k_r(\eta) , & \eta \geq \xi \end{cases} .$$

It turns out that this function is the kernel of the resolvent of $L(q(j,\mu))$. The first thing to show is that it is the kernel of a bounded operator on $\mathfrak{L}^2(\mathbb{R}^+)$. For this purpose we use a result of Schur-Holmgren-Carleman [7, pp. 103 ff] which implies the following: Let $K(\cdot,\cdot)$ be a kernel, K the corresponding integral operator, and $t(\cdot)$ a positive measurable function, and

$$\|K\|_t = \max \{ \sup_\xi t^{-1}(\xi) \int |K(\xi,\eta)|\, t(\eta)\, d\eta ,$$

$$\sup_n t^{-1}(\eta) \int |K(\xi,\eta)|\, t(\xi)\, d\xi \} .$$

Then

$$(6.4) \qquad \|K\| \leq \|K\|_t ,$$

where $\|\cdot\|$ is the usual norm in $\mathfrak{B}(\mathfrak{L}^2)$.

We use this Schur-Holmgren-Carleman result for

$$(6.5) \qquad t(\rho) = \begin{cases} |k_\ell(\rho)\, z_\ell(\rho)| , & \rho \in \mathcal{I}_\ell \\ |k_r(\rho)\, z_n(\rho)| , & \rho \in \mathcal{I}_r \end{cases} ,$$

and

$$z_\ell(\rho) = \varphi'_\ell(\rho)^{-1/2}\, \mathrm{Ai}(-\varphi_\ell(\rho) \exp \frac{2\pi i}{3}) ,$$

$$z_r(\rho) = \varphi'_r(\rho)^{-1/2} \exp(-i\, \varphi_r(\rho)) .$$

Of course, it must be shown that $t(\rho)$ is a positive function. That the operator $K(j,\mu)$ corresponding to the Weyl kernel (6.3) is a bounded operator is a consequence of the Schur-Holmgren-Carleman result and the following lemma.

LEMMA 6.2. *Under the hypotheses of Lemma* 6.1, *for* $j \geq j_o$,

$$\|K(j,\mu)\|_t < \infty . \tag{6.6}$$

The proof of (4.10) is then completed by showing that the operator $K(j,\mu)$ is actually the resolvent of $L(q(j,\mu))$. This latter fact is similar to a proof used elsewhere [18].

The proofs of the inequalities (4.11) and (4.12) of Theorem 4.1 involve, among other things, uniformity considerations in μ and j. A crucial role in these uniformity considerations is played by certain uniform estimates in μ and j for the Weyl kernel (6.3). In order to describe these estimates we first define some terms. We set

$$w(\rho) \equiv w(j,\mu)(\rho) = \begin{cases} |\mu - \nu^2\rho^2|^{-1/4}, & \rho \in \mathcal{J}_\ell \\ |\varphi_r'(\rho)|^{-1/2}, & \rho \in \mathcal{J}_r \end{cases} \tag{6.7}$$

and

$$v(\rho) \equiv v(j,\mu)(\rho) = \begin{cases} -\mathrm{Re}\, i\nu \int_1^{\rho\tau^{-1}} (1 - \frac{1}{\sigma^2})^{1/2}\, d\sigma, & \rho \in \mathcal{J}_\ell , \\ -\mathrm{Re}\, i\nu \int_1^{\delta j \tau^{-1}} (1 - \frac{1}{\sigma^2})^{1/2}\, d\sigma, & \rho \in \mathcal{J}_r . \end{cases} \tag{6.8}$$

THEOREM 6.3. *There exists a* $\gamma > 0$ *and a* $j_o \geq 0$ *so that for every compact interval* $\mathcal{J}$ *as described in Theorem* 3.1, *there is an* $\mathcal{R}_\pm(\mathcal{J})$ *so that for all* $j \geq j_o$, $\mu \in \mathcal{R}_\pm(\mathcal{J})$ *and* $\xi, \eta \in R^+$.

$$|K(j,\mu)(\xi,\eta)| \leq \gamma\, w(\xi)\, w(\eta) \exp[-|v(\eta) - v(\xi)|] , \tag{6.9}$$

where $K(j,\mu)$ is given by (6.3).

Let us now set

$$G(j,\mu) = n_s^{1/2}\, K(j,\mu)\, n_s^{1/2} \tag{6.10}$$

and

$$t(\rho) = w(\rho)\, n_s^{1/2}(\rho) \quad , \tag{6.11}$$

where recall n_s is given by (4.8). Note also that $t(\rho)$ depends on j and μ. Recall also that we have already noted that $K(j,\mu)(\xi,\eta)$ is the kernel of $R(L(q(j,\mu)),\mu)$ when $\mu \in \mathcal{R}_{\pm}(\mathcal{S})$. The inequalities (4.11) and (4.12) of Theorem 4.1 are now easy consequences of the following theorem and the Schur-Holmgren-Carleman result.

THEOREM 6.4. *Let the hypotheses of Theorem* 3.1 *hold with* $\gamma_o = \gamma^{-1}$, *where* γ *is the constant of Theorem* 6.3. *If* t *is given by* (6.11) *then*

$$\limsup_{j\to\infty} \sup_{\mu \in \mathcal{R}_{\pm}(\mathcal{S})} \|G(j,\mu)\|_t < 1 . \tag{6.12}$$

REFERENCES

1. S. Agmon, Spectral properties of Schrödinger operators and scattering theory, Ann. Scuol. Norm. Sup. Pisa, 2 (1975) 151-218.
2. M. Ben-Artzi and A. Devinatz, Spectral and scattering theory for the adiabatic oscillator and related potentials, J. Math. Phys. 111 (1979) 594-607.
3. B. Bourgeois, Quantum-mechanical scattering theory for long-range oscillatory potentials, Thesis, University of Texas at Austin, 1979.
4. A. Devinatz, The existence of wave operators for oscillating potentials, J. Math. Phys. (to appear).
5. J. Dollard and C. Friedman, Existence of the Moller wave operators for $V(r) = \lambda r^{-\beta}\sin(\mu r^{\alpha})$, Ann. Phys. 111 (1978) 251-266.
6. V. Enss, Asymptotic completeness for quantum mechanical potential scattering, Comm. Math. Phys. 61 (1978) 285-29 .

7. A. Erdelyi, Asymptotic solutions of differential equations with transition points on singularities, J. Math. Phys. 1 (1960), 16-26.
8. ________, Asymptotic Expansions, Dover Publications, New York, 1956.
9. K.O. Friedrichs, Spectral theory of operators in Hilbert Space, Springer-Verlag, 1973.
10. T. Kato, Perturbation Theory for Linear Operators, Springer-Verlag, 1966 and 1976.
11. T. Kato, Scattering theory and perturbation of continuous spectra, Actes, Congrés Intern. Math. 1970, Tome 1, 135-140.
12. R.E. Langer, On the asymptotic solutions of ordinary differential equations with an application to Bessel functions of large order, Trans. Amer. Math. Soc., 33 (1931) 23-64.
13. ________, The asymptotic solutions of ordinary linear differential equations of the second order with special reference to a turning point, Trans. Amer. Math., 67 (1949) 461-490.
14. K. Mochizuki and J. Uchiyami, Radiation conditions and spectral theory for 2-body Schrödinger operators with "oscillating" long range potentials I, J. Math. Kyoto Univ. 18-2 (1978) 377-408.
15. F.W.T. Olver, Asymptotics and Special Functions, Academic Press, 1974.
16. Pauli Lectures on Physics, C.P. Enz, editor, Vol. 5. Wave Mechanics, The MIT Press (1977). See Section 27, The WKB-method.
17. M. Reed and B. Simon, Methods of Modern Mathematical Physics, Vol. IV, Analysis of Operators, Academic Press 1978. See Example 1 in Section XIII.13.
18. P. Rejto, On the resolvents of a family of non-self adjoint operators, Lett. Math. Phys. 1 (1976) 111-118.
19. ________, An application of the third order JWKB-approximation method to prove absolute continuity I. The construction. Helv. Phys. Acta. 50 (1977), 479-494. Part II. The estimates,..., 495-508.
20. ________, Uniform estimates for the resolvent kernels for the parts of the Laplacian, Univ. Minn. Tech. Summ. Report, 1970.
21. Y. Saito, Spectral Representation for Schrödinger Operators with Long-Range Potentials, Springer-Verlag Lecture Notes in Mathematics, 727, (1979).
22. Y. Sibuya, Global Theory of a Second Order Linear Differential Equation with a Polynomial Coefficient, North-Holland/American Elsevier, 1975.
23. W. Wasov, Asymptotic Expansions for Ordinary Differential Equations, Wiley-Interscience, 1965.
24. J. Weidmann, Zur Spectraltheorie von Sturm-Lionville-Operatoren, Math. Zeit., 98 (1967), 268-302.
25. D. White, Scattering theory for oscillatory potentials, Thesis, Northwestern University, 1980.

TIME DELAY AND RESONANCE IN SIMPLE SCATTERING

Kalyan B. Sinha

Indian Statistical Institute, 7 Sansanwal
Marg, New Delhi 110029, India and
Department of Mathematics, University of Colorado
Boulder, Colorado, USA 80309

Abstract: The concept of time delay is used to understand resonances in simple scattering. In a simple model of one dimensional Stark Hamiltonian, we test this idea and show that for small field strength one has spectral concentration near the eigenvalue of the unperturbed Hamiltonian. Time-delay for the model also exhibits a similar feature.

1. Introduction

The notion of a 'resonance' in scattering has been of great interest to theoretical physicists. The oldest and standard method of understanding has been the method of analytic continuation of suitable matrix elements of the resolvent of the total Hamiltonian or the S-matrix elements and looking for certain complex poles of these analytically continued elements as 'resonances'. For a mathematically rigorous treatment of these ideas using dilatation analyticity, the reader is referred to Simon [1].

Here we shall use the concept of time-delay in a scattering process to understand resonances. The energy λ_r in whose neighborhood, the time-delay attains a "large" value shall be termed the "resonance energy". In other words, this reflects the intuition that the particle with energy close to λ_r spends a relatively long time in the vicinity of the target before finally moving away.

In the second section, we describe the phenomenon of spectral concentration and related decay of stationary states. Though no general theory for relating time-delay and resonances is attempted, in the last section we deal with the simple example of 1-dimensional Stark-Hamiltonian in presence of a rank-one perturbation to illustrate the ideas.

2. Spectral Concentration and Decay

Let $H, H(\mu) \equiv H + \mu W$ be self adjoint and let $H\psi_0 = \lambda_0\psi_0$. Then if there exists $\lambda_0(\mu)$ and $\Gamma(\mu)$ with the property $\lambda_0(\mu) \to \lambda_0$ and $\Gamma(\mu) \to 0$ as $\mu \to 0$ so that

$$(\psi_0, E^{\mu}(\lambda_0(\mu) + \Gamma(\mu)a, \lambda_0(\mu) + \Gamma(\mu)b)\psi_0) \to \pi^{-1}\int_a^b \frac{dh}{1 + h^2}, \qquad \dots (1)$$

we say that the spectrum of $H(\mu)$ is concentrated near $\lambda_0(\mu)$ to the order $\Gamma(\mu)$. In the above, we have written E^{μ} for the spectral family of the selfadjoint operator $H(\mu)$ and $(a,b]$ is an interval in $\mathbb{R}$. $\lambda_0(\mu)$ is said to be an approximate eigenvalue of $H(\mu)$ (or a resonance energy of the system) and $\Gamma(\mu)$ the width of the resonance.

Whenever (1) is satisfied, one arrives at a decay result of the following type:

$$(\psi_0, \exp[-i(H(\mu) - \lambda_0(\mu))\Gamma(\mu)^{-1}t]\psi_0) \to \exp(-|t|) \qquad \cdots (2)$$

as $\mu \to 0$, uniformly for t in compact intervals of $\mathbb{R}$. For a proof of (2) , see Davies [2] and for models leading to (1) or similar structures, see Sinha [3] and Howland [4] .

One may derive a decay result for dilatation-analytic potential as in the following

Theorem 1: Let $H_0 = -\Delta$ and V and W be sums of dilatation analytic potentials in $\mathbb{R}^3$. Let λ_0 be a non-threshold, non-degenerate eigenvalue, with normalized eigenvector ψ_0, of $H = H_0 + V$ embedded in the continuum. For μ small and real, let $\lambda_0(\mu) = \lambda_0(\mu) - i\Gamma(\mu)$ be the resonance energy of $H_0 + V + \mu W \equiv H + \mu W \equiv H(\mu)$ near λ_0 . Let E^{μ} be the spectral family of $H(\mu)$ and let $E_0 \equiv E^0(\{\lambda_0\})$. Then

$$(\psi_0, \exp[-i(H(\mu) - \lambda_0(\mu))\Gamma(\mu)^{-1}t]\psi_0) \to \exp(-|t|)$$

as $\mu \to 0$, uniformly for t in compacts.

Proof (Sketch): Proceeding as in the proof of Theorem 6.1 of Simon [1] , one gets for φ in the analytic domain,

$$(\varphi, E^{\mu}(\lambda_0(\mu) + \Gamma(\mu)a, \lambda_0(\mu) + \Gamma(\mu)b)\varphi)$$

$$= 1/\pi \ \mathrm{Im}\left[(\varphi(\theta), E(\theta,\mu)\varphi(\theta)) \int_{\lambda_0(\mu)+\Gamma(\mu)a}^{\lambda_0(\mu)+\Gamma(\mu)b} \frac{dx}{\lambda(\mu) - x}\right] + o(1).$$

By a change of variable: $x = \lambda_0(\mu) + \Gamma(\mu)h$ and observing that $(\varphi(\theta), E(\theta,0)\varphi(\theta)) = (\varphi, E_0\varphi)$, we get

$$(\varphi, E^{\mu}(\lambda_0(\mu) + \Gamma(\mu)a, \lambda_0(\mu) + \Gamma(\mu)b)\varphi) \to \left(\pi^{-1}\int_a^b \frac{dh}{1 + h^2}\right)(\varphi, E_0\varphi)$$

as $\mu \to 0$. Thus we arrive at the weak convergence of $E^{\mu}(\lambda_0(\mu) + \Gamma(\mu)a, \lambda_0(\mu) + \Gamma(\mu)b)$ to $\left(\pi^{-1}\int_a^b \frac{dh}{1 + h^2}\right)E_0$ as $\mu \to 0$ and then by a theorem of Davies [2] , we have $(\psi_0, \exp[-i(H(\mu) - \lambda_0(\mu))\Gamma(\mu)^{-1}t]\psi_0) \to \exp(-|t|)$ as $\mu \to 0$.

3. Time-delay and Resonance

In this section, we shall attempt to relate the notion of spectral concentration as described in section 1 and time delay in an associated scattering system. But first we recall some of the facts about time delay in a simple scattering system. Let (H,H_0) be a simple scattering system so that $\Omega_\pm = \underset{t\to\pm\infty}{\text{s lim}}\exp(iHt)\exp(-iH_0t)$ exist and are complete. Time delay for such a system is the difference between the time of flight across a 'large' sphere by a particle in state f (at time $t = 0$) under the influence of interaction $H - H_0$ and that in the absence of any interaction. A mathematical expression for it is:

$$\mathcal{T}(f) = \lim_{r\to\infty}\int_{-\infty}^{\infty}(e^{-iH_0t}f,[\Omega_-^*F_r\Omega_- - F_r]e^{iH_0t}f)dt\ , \qquad (4)$$

whenever the limit exists. Here we have denoted by F_r, the multiplication by the characteristic function of the r-ball in $\mathbb{R}^3$. Writing $\mathcal{T}_r(f)$ for the expression in (4) before taking the limit, one knows from Jauch, Sinha and Misra [5] that there exists a family $\{\mathcal{T}_r(\lambda)\}$ of trace-class operators in the multiplicity-space $\mathcal{K}_0 \equiv L^2(S^{(2)})$ of the spectral representation of $H_0 \equiv -\Delta$, such that $\mathcal{T}_r(f) = \int_0^\infty (f_\lambda, \mathcal{T}_r(\lambda)f_\lambda)d\lambda$, for f such that $\|f_\lambda\|_0$ has compact support in $(0,\infty)$.

We write $R_z^0 \equiv (H_0 - z)^{-1}$ and $R_z = (H - z)^{-1}$ and define Krein's displacement function ξ as

$$\xi(\lambda) = \pi^{-1}\lim_{\varepsilon\downarrow 0}\text{Im}\,\ell n\Delta(\lambda + i\varepsilon)\ , \text{ where} \qquad (5)$$

$$\Delta(z) = \det[I + (R_{z_0} - R_{z_0}^0)(R_{z_0} - z)^{-1}] \quad \text{for some}$$

fixed $z_0 \in \rho(H) \cap \rho(H_0)$. $\Delta(z)$ is called the perturbation determinant. For some of its properties and those of ξ, see [6]. Here we need to observe that if $R_z - R_z^0$ is trace-class for some $z \in \rho(H) \cap \rho(H_0)$, $\Delta(z)$ and $\xi(\lambda)$ are well defined, and moreover $\xi(\lambda)(1+\lambda^2)^{-1} \in L^1(d\lambda)$ (see [6]). We have the following theorem, proof of which can be found in [5].

<u>Theorem 2</u>: Let $V \equiv H - H_0$ be such that $R_z - R_z^0$ is trace class. Then $\text{Tr}_0\mathcal{T}_r(\lambda)$ converges in the sense of distribution to $\tau(\lambda)$ and one has $\tau(\lambda) = 2\pi\xi'(\lambda)$.

We call $\tau(\lambda)$ the average time delay at the energy λ and the above relation between τ and ξ can be looked upon as a variant of Eisenbud-Wigner relation. We also note from [6] and [7] that under the assumptions of Theorem 2, one has that $S(\lambda) - I_0$ is trace class and $\xi(\lambda) = (-i/2)\mathrm{Tr}_0 \ln S(\lambda)$, where $S = \{S(\lambda)\}$ is the scattering operator.

Next consider the following Stark-like model in 1-dim; for $\mu > 0$:

$$\begin{aligned} H_0 &= -d^2/dx^2 & H_0(\mu) &= -d^2/dx^2 + \mu x \\ H &= -d^2/dx^2 + V(x) , & H(\mu) &= -d^2/dx^2 + V(x) + \mu x . \end{aligned} \tag{6}$$

Both (H,H_0) and $(H(\mu),H_0(\mu))$ are simple scattering systems if V is short range, say in $L^1(\mathbb{R})$. Also H_0 and H are bounded below while $H_0(\mu)$ and $H(\mu)$ are not for all $\mu > 0$. For the two scattering systems, we have two average time delay functions $\tau(\lambda)$ and $\tau(\lambda,\mu)$ respectively. While $\tau(\lambda)$ is defined for $0 < \lambda < \infty$, $\tau(\lambda,\mu)$ is defined for $-\infty < \lambda < \infty$. We have the following results from Sinha [8] which show that the scattering system $(H(\mu),H_0(\mu))$ "converges" to (H,H_0) as $\mu \to 0$.

<u>Theorem 3</u>: Let V be such that $(1 + x^2)^{1/4} V \in L^1(\mathbb{R})$. Then $R_z(\mu) - R_z^0(\mu)$ is trace class for all $\mu > 0$ and $\mathrm{Im}\, z \neq 0$, and $R_z(\mu) - R_z^0(\mu) \to R_z - R_z^0$ in trace-norm as $\mu \to 0+$.

Combining this theorem with theorem 2, one has as in [8]

<u>Corollary 4:</u> $\tau(\lambda,\mu) \to \tau(\lambda)$ for $\lambda \geq 0$ and $\tau(\lambda,\mu) \to 2\pi\delta_{\lambda_0}$ as $\mu \to 0+$ in the sense of distribution, if we furthermore assume that H has just one (simple) eigenvalue $\lambda_0 < 0$.

Thus time delay $\tau(\lambda,\mu)$ is "large" in the neighborhood of λ_0. We next study the detailed structure of $\tau(\lambda,\mu)$ in the neighborhood of λ_0 for small $\mu > 0$. But for this part, we simplify the model further, and assume V is a (non-local) rank 1 perturbation, i.e. $Vf = \alpha(h,f)h$ with $h \in \mathscr{S}(\mathbb{R})$. Nevertheless the conclusions of Theorem 3 and Corollary 4 remain valid.

In the following, we denote by $\tilde{h}$ the Fourier transform of h in $L^2(\mathbb{R})$ and the next three lemmas collect some of the relevant properties of the model.

Lemma 5: $\lambda_0 < 0$ is an eigenvalue of H if and only if $\alpha\int|\tilde{h}(p)|^2(\lambda_0 - p^2)^{-1}dp = 1$, and in such a case $\varphi(p) = c(\lambda_0 - p^2)^{-1}\tilde{h}(p)$ is the normalized eigenfunction with $c^2\int|\tilde{h}(p)|^2(\lambda_0 - p^2)^{-2}dp = 1$.

Lemma 6: Let $f \in L^2(\mathbb{R})$. The spectral representative of f in the spectral representation of $H_0(\mu)$ is:

$$f^0(\lambda,\mu) = (2\pi\mu)^{-1/2}\int dp\, \exp[-i\mu^{-1}(p^3/3 - \lambda p)]\tilde{f}(p). \cdots \quad (7)$$

Also, if $\lambda < 0$, then $h^0(\lambda,\mu) \to 0$ as $\mu \to 0$ and if $\lambda > 0$ and if $\tilde{h}(p)|_{p^2=\lambda} = 0$, then $h^0(\lambda,\mu) \to 0$ as $\mu \to 0$.

The spectral representation for $H^0(\mu)$ is obtained by taking the Airy transform of $\tilde{f}$ as in (7) and if $\lambda < 0$, then the exponent inside the integral has no critical point and since $\tilde{h} \in \mathscr{S}(\mathbb{R})$, one has the above result. Now we note that the perturbation determinant in this example is given as $\Delta(z,\mu) = 1 + \alpha(h, R_z^0(\mu)h)$ and let $\lambda_0 < 0$ be a simple eigenvalue for H as in section 1.

Lemma 7: Assume that $h^0(\lambda_0,\mu) \neq 0$ for some μ positive and small. Then $H(\mu)E^{\mu}(\lambda_0 - \delta, \lambda_0 + \delta)$ is absolutely continuous and is unitarily equivalent to $H^0(\mu)E^{0\mu}(\lambda_0 - \delta, \lambda_0 + \delta)$ for sufficiently small δ. Furthermore, as $\varepsilon \to 0+$

$$\Delta(\lambda' \pm i\varepsilon,\mu) \to 1 + \alpha\, P\!\!\int \frac{|h^0(\lambda,\mu)|^2}{\lambda - \lambda'}\, d\lambda \pm i\pi\alpha|h^0(\lambda',\mu)|^2$$

$$\equiv \Delta_{\pm}(\lambda',\mu), \quad (8)$$

for all $\lambda' \in \lambda_0 - \delta, \lambda_0 + \delta)$ and μ small and positive.

The assumption $h^0(\lambda_0,\mu) \neq 0$ implies by continuity that $h^0(\lambda',\mu) \neq 0$ in a small region around λ_0 and that rules out presence of singular continuous spectrum in the same interval. For details of the method, see section 8-3 in [7] and Friedrichs and Rejto [9] .

Next we rescale the energy variable in the average time-delay $\tau(\lambda,\mu)$ for the scattering system $(H(\mu),H^0(\mu))$ as in (1) and obtain an exactly similar behavior. Set

$$\gamma(\mu) = c^2\left[⨍ \frac{|h^0(\lambda,\mu)|^2}{\lambda - \lambda_0} d\lambda - \int \frac{|\tilde{h}(p)|^2}{p^2 - \lambda_0} dp \right]$$

and $\Gamma(\mu) = \pi c^2 \; |h^0(\lambda_0,\mu)|^2$.

Theorem 8: Let $H(\mu)$, $H^0(\mu)$, h satisfy the assumptions of the preceding lemmas. Then

$$\Gamma(\mu)\tau(\lambda_0 - \gamma(\mu) + \Gamma(\mu)k,\mu) \to 2(1 + k^2)^{-1} \quad \text{as} \quad \mu \to 0+ \; .$$

Sketch of proof: We use the relations $\tau(\lambda,\mu) = 2\,\xi'(\lambda,\mu)$, (5) and (8) to obtain

$$\Delta_+(\lambda',\mu) = \alpha \left[⨍ \frac{|h^0(\lambda,\mu)|^2}{\lambda - \lambda'} d\lambda - \int \frac{|\tilde{h}(p)|^2}{p^2 - \lambda_0} dp \right] + i\pi\alpha|h^0(\lambda',\mu)|^2$$

$$= \alpha \left[⨍ \frac{|h^0(\lambda,\mu)|^2}{\lambda - \lambda'} d\lambda - \int \frac{|\tilde{h}(p)|^2}{p^2 - \lambda'} dp + (\lambda' - \lambda_0) \int \frac{|\tilde{h}(p)|^2 dp}{(p^2 - \lambda')(p^2 - \lambda_0)} \right]$$

$$+ i\pi\alpha|h^0(\lambda',\mu)|^2 \; .$$

Next we note that if we write $\lambda' = \lambda_0 - \gamma(\mu) + \Gamma(\mu)k$ and let

$$\mu \to 0+ \; , \; \int \frac{|\tilde{h}(p)|^2 dp}{(p^2 - \lambda')(p^2 - \lambda_0)} \to \int \frac{|\tilde{h}(p)|^2 dp}{(p^2 - \lambda_0)^2} = c^{-2} \quad \text{by Lemma 5 and}$$

$$⨍ \frac{|\tilde{h}^0(\lambda,\mu)|^2}{\lambda - \lambda'} d\lambda - \int \frac{|\tilde{h}(p)|^2 dp}{p^2 - \lambda'} \to \gamma(\mu)c^{-2} \; .$$

Thus the average time delay in the rescaled variable k shows a behavior similar to that of spectral concentration as in section 1. Another way of looking at this result is to say that Theorem 8 shows the detailed structure of $\tau(\lambda,\mu)$ in the neighborhood of the embedded eigenvalue λ_0 and exhibits the approach to $2\pi\delta_{\lambda_0}$ as

$\mu \to 0+$ (Corollary 4).

The above result is formally consistent with a Levinson's theorem for the (H, H_0) scattering system with one simple eigenvalue $\lambda_0 < 0$ (see [10] for a proof of Levinson's theorem). In this case, we have $\xi(0) = (-i/2\pi)\mathrm{Tr}_0 \ln S(0) = 1$. Now, since $S(\lambda,\mu) \to I$ as $\lambda \to -\infty$, we have $\xi(-\infty,\mu) = 0$. In the relation $\int_{-\infty}^{0} d\lambda\ \tau(\lambda,\mu) = 2\pi[\xi(0,\mu) - \xi(-\infty,\mu)] \to 2\pi$ as $\mu \to 0+$, we make the change of variable $\lambda = \lambda_0 - \gamma(\mu) + \Gamma(\mu)k$ as before to have the L.H.S $= \int dk\ \Gamma(\mu)\tau(\lambda_0 - \gamma(\mu) + \Gamma(\mu)k,\mu)$. By Theorem 8, this expression converges as $\mu \to 0+$ to $2\int_{-\infty}^{\infty} dk(1 + k^2)^{-1} = 2\pi$, demonstrating the consistency.

The author is indebted to Dr. Ph. Martin for a stimulating discussion in this subject.

References:

[1] Simon, B., Ann. Math. 97, 247 (1973).
[2] Davies, E.B., Lett. Math. Phys. 1, 31 (1975).
[3] Sinha, K.B., Lett. Math. Phys. 1, 251 (1976).
[4] Howland, J.S., Amer. J. Math. 91, 1106 (1969).
[5] Jauch, J.M., Sinha, K.B., Misra, B., Helv. Phys. Acta 45, 398 (1972).
[6] Birman, M.S., Krein, M.G., Soviet Math Doklady 3, 740 (1962).
[7] Amrein, W.O., Jauch, J.M., Sinha, K.B., Scattering Theory in Quantum Mechanics, W.A. Benjamin, Reading, Mass., 1977.
[8] Sinha, K.B., Rep. Math. Phys. 14. 65 (1978).
[9] Friedrichs, K.O., Rejto, P.A., Comm. Pure Appl. Math. 15, 219 (1962).
[10] Dreyfus, T., J. Math. Anal. Appl. 64, 114(1978).

A STUDY OF THE HELMHOLTZ OPERATOR

Karl Gustafson
Department of Mathematics
University of Colorado
Boulder, CO 80302

Guy Johnson
Department of Mathematics
Syracuse University
Syracuse, NY 13210

ABSTRACT

The operator $R^o_\lambda = (-\Delta - (\lambda + i0))^{-1}$, $0 \leq \lambda$, is a right inverse of the Helmholtz operator $-\Delta-\lambda$ on the weighted L^1 space $L^{1,-1}(\mathbb{R}^3)$. It is shown to be a bounded operator into $L^{1,s}(\mathbb{R}^3)$ for each $s < -3$. Comparisons are made with its operation on L^2 spaces. The operator $\text{Im}R^o_\lambda$ maps $L^{1,-1}(\mathbb{R}^3)$ into the null space of $-\Delta-\lambda$. Bounds for the growth at infinity of the derivatives of $\text{Im}R^o_\lambda f$ are obtained.

1. INTRODUCTION

This paper presents some properties of the operator $R^o_\lambda = (-\Delta - (\lambda + i0))^{-1}$, $0 \leq \lambda$, in the context of weighted L^1 spaces. The motivation for a study of this operator is its importance as a basis, either directly or indirectly, for potential scattering theory. One may note, for example, the work of Ikebe[1], Shenk[2], Rejto[3], Alsholm and Schmidt[4], and Agmon[5] in which explicit properties are used. For compelling reasons it has been customary to study scattering theory in an L^2 setting. One can mention its physical origins as a problem in Hilbert space and the corresponding role that spectral theory has played. However, as will be observed in the results to follow, it is quite natural to investigate the properties of R_λ^o using $L^{1,s}$ spaces and they relate to the L^2 properties in an interesting way.

We will restrict attention to function spaces on $\mathbb{R}^3$. The starting point for our study is the integral representation

$$(1.1) \qquad R_z^o f(x) = \int \frac{e^{i\sqrt{z}|x-y|}}{4\pi|x-y|} f(y)dy$$

for the resolvent $R_z^o = (-\Delta-z)^{-1}$ where Δ is the Laplacian in three dimensions, and the branch of $\sqrt{z}$ is chosen so that $\mathrm{Im}\sqrt{z} > 0$ for $0 < \arg z < 2\pi$. For $0 \leqq \lambda$

$$(1.2) \qquad R_{\lambda \pm i0}^o f(x) = \int \frac{e^{\pm i\sqrt{\lambda}|x-y|}}{4\pi|x-y|} f(y)dy$$

and we will write $R_\lambda^o = R_{\lambda+i0}^o$.

Weighted L^p spaces are defined by

$$L^{p,s}(\mathbb{R}^3) = \{f : (1 + |x|^2)^{s/2} f \in L^p(\mathbb{R}^3)\}$$

for $1 \leqq p \leqq \infty$ and $s \in \mathbb{R}$, with norms

$$\|f\|_{p,s} = \|(1 + |x|^2)^{s/2} f\|_{L^p(\mathbb{R}^3)} .$$

We will use these for $p = 1$ and $p = 2$. In certain calculations equivalent weight functions of the form $(1+|x|^\alpha)^{s/\alpha}$ will be convenient. The inequality

$$(1.3) \qquad 2^{(\alpha-2)/2\alpha}(1 + |x|^\alpha)^{1/\alpha} \leqq (1 + |x|^2)^{1/2} \leqq (1 + |x|^\alpha)^{1/\alpha},$$

$0 < \alpha < 2$, makes this possible.

In section 2, $L^{1,-1}(\mathbb{R}^3)$ is established as a domain for R_z^o and and $L^{1,s}(\mathbb{R}^3)$, $s < -3$, as range spaces. The corresponding operator norms are bounded uniformly with respect to $z \in \mathbb{C}$ and $R_z^o f \to R_\lambda^o f$ in $L^{1,s}(\mathbb{R}^3)$ as $z \to \lambda \geqq 0$, $\mathrm{Im} z \geqq 0$. In section 3 it is shown that $L^{1,-1}(\mathbb{R}^3)$ contains the appropriate L^2 domains for R_λ^o. Section 4 is concerned with the calculation of estimates for the growth at infinity of derivatives of $\mathrm{Im}R_\lambda^o f$.

In some of the derivations to follow use will be made of ideas from classical potential theory.

2. AN L^1 DOMAIN AND L^1 RANGE SPACES FOR R_z^o

The integrand in (1.1) and (1.2) satisfies

$$\left|\frac{e^{i\sqrt{z}|x-y|}}{4\pi|x-y|} f(y)\right| \leqq \frac{|f(y)|}{4\pi|x-y|} \tag{2.1}$$

and one may take limits with respect to z using the Lebesgue dominated convergence theorem, provided the function on the right is integrable. Since it depends on x we make use of the following result from potential theory.

Lemma 2.1 If $\int|x-y|^{-1}|f(y)|dy < \infty$ for at least one $x_o \in \mathbb{R}^3$, then it is in $L^1_{loc}(\mathbb{R}^3)$. Moreover, this is the case if and only if $f \in L^{1,-1}(\mathbb{R}^3)$.

Proof.
$$\int_{|x|<R} dx \int \frac{|f(y)|}{|x-y|} dy = \int \frac{|f(y)|}{|x_o-y|} \left[|x_o-y| \int_{|x|<R} \frac{dx}{|x-y|} \right] dy$$
$$\leqq M_R \int \frac{|f(y)|}{|x_o-y|} dy < \infty$$

since the integral in the brackets is the potential of a homogeneous ball. Thus the integral is a locally integrable function of x. For the second statement we make use of the first. If $f \in L^{1,-1}(\mathbb{R}^3)$, then
$$\int_{|y|<1} \frac{|f(y)|}{|x-y|} dy \leqq \frac{2}{|x|-1} \int_{1<|y|} \frac{|f(y)|}{1+|y|} dy \leqq \frac{2}{|x|-1} \|f\|_{1,-1} < \infty$$

for $1 < |x|$, and

$$\int_{1\leq|y|} \frac{|f(y)|}{|y|}\,dy \leq 2 \int_{1\leq|y|} \frac{|f(y)|}{1+|y|}\,dy \leq 2\,\|f\|_{1,-1} < \infty .$$

Thus $$\int \frac{|f(y)|}{|x-y|}\,dy = \int_{|y|<1} \frac{|f(y)|}{|x-y|}\,dy + \int_{1\leq|y|} \frac{|f(y)|}{|x-y|}\,dy$$

is the sum of two locally integrable functions. Conversely, if $\int |x_o-y|^{-1}|f(y)|dy < \infty$, then

$$\int \frac{|f(y)|}{1+|y|}\,dy \leq \max(|x_o|,1) \int \frac{|f(y)|}{|x_o-y|}\,dy < \infty$$

which implies $\|f\|_{1,-1} < \infty$. This completes the proof of the lemma.

Therefore $R_z^o f$ is a locally integrable function for any $f \in L^{1,-1}(\mathbb{R}^3)$. For such f, $R_z^o f \to R_\lambda^o f$ almost everywhere as $z \to \lambda \geq 0$, $\operatorname{Im} z \geq 0$. Actually $R_z^o f$ is in an L^1 space as stated in

<u>Theorem 2.1</u> If $f \in L^{1,-1}(\mathbb{R}^3)$ and $s < -3$, then $R_z^o f \in L^{1,s}(\mathbb{R}^3)$ and R_z^o is a bounded operator with

(2.2) $$\|R_z^o f\|_{1,s} \leq c_s \|f\|_{1,-1}$$

where c_s is independent of f and z. Also $R_z^o f \to R_\lambda^o f$ in $L^{1,s}(\mathbb{R}^3)$ as $z \to \lambda \geq 0$, $\operatorname{Im} z \geq 0$.

Proof. The function $u(x) = (1+|x|^\alpha)^{-1/\alpha}$, $0 < \alpha < 2$, is a positive superharmonic function on $\mathbb{R}^3$ with greatest harmonic minorant zero. According to the Riesz decomposition theorem (see Helms[6]) it is the potential of the density $-\Delta u$ where the Laplacian is taken in the distribution sense. We find

$$-\Delta u(x) = (\alpha+1)|x|^{\alpha-2}(1+|x|^\alpha)^{-(2\alpha+1)/\alpha} + c\delta$$

since u is differentiable except at x = 0. The fact that u(0) is finite implies c = 0. Use of (1.3) and $s = -\alpha-3$ yields

$$(2.3) \qquad (1+|x|^2)^{s/2} \leqq c_s[-\Delta u(x)]$$

where $c_s = -2^{s(s+5)/(s+3)}/(s+2)$, $-5 < s < -3$. From (2.1) and (2.3) we obtain the dominant

$$(2.4) \qquad (1+|x|^2)^{s/2}|R_z^o f(x)| \leqq c_s \int \frac{-\Delta u(x)}{4\pi|x-y|}|f(y)|dy$$

which is in $L^1(\mathbb{R}^3)$ because, again using (1.3),

$$(2.5) \qquad \int\left[\int \frac{-\Delta u(x)}{4\pi|x-y|}dx\right]|f(y)|dy = \int(1+|y|^\alpha)^{-1/\alpha}|f(y)|dy \leqq \|f\|_{1,-1} \quad .$$

In view of (2.5) the integration of (2.4) gives $\|R_z^o f\|_{1,s} \leqq c_s\|f\|_{1,-1}$ for $-5 < s < -3$. For $s \leqq -5$, the inequality is valid with $c_s = 1/3$. The last statement of the theorem is a consequence of (2.4) and the almost everywhere convergence. This completes the proof.

The example $f(x) = -\Delta(1+|x|)^\beta$, $-1 < \beta < 0$, may be used to show that (2.2) is not valid for $s = -3$. A precursor of Theorem 2.1 may be found in Lemma 3.1 in Ikebe[1].

3. COMPARISON WITH L^2 SPACES

In this section we will show that $L^{1,-1}(\mathbb{R}^3)$ is an extension of the appropriate L^2 domains for R_λ^o. The first result compares the $L^{1,s}$ and $L^{2,s}$ spaces.

<u>Theorem 3.1</u> Let $s,t \in \mathbb{R}$. Then $L^{2,s}(\mathbb{R}^3) \subset L^{1,t}(\mathbb{R}^3)$ if and only if $s > t + (3/2)$.

Proof. If the inequality is satisfied then Schwarz's inequality yields

$$\int(1+|x|^2)^{t/2}|f(x)|dx = \int(1+|x|^2)^{(t-s)/2}(1+|x|^2)^{s/2}|f(x)|dx$$

$$\leqq \left(\int (1+|x|^2)^{t-s} dx\right)^{1/2} \left(\int (1+|x|^2)^s |f(x)|^2 dx\right)^{1/2}$$

or $\|f\|_{1,t} \leqq c_{t-s} \|f\|_{2,s}$ where $c_{t-s} < \infty$ since $t-s < -3/2$. The example $f(x) = (1+|x|^2)^{-(t+3)/2} (\log |x|)^{-1}$, $2 \leqq |x|$, with $f \in C(\mathbb{R}^3)$ has norms $\|f\|_{1,t} = \infty$ and $\|f\|_{2,s} < \infty$ for $s \leqq t + (3/2)$. Thus $L^{2,s}(\mathbb{R}^3) \not\subset L^{1,t}(\mathbb{R}^3)$ for $s \leqq t + (3/2)$. This completes the proof.

In Agmon[5], see his Theorem 4.1, the domain of R_z^o is $L^{2,s}(\mathbb{R}^3)$ for any $s > 1/2$. Setting $t = -1$ in Theorem 3.1 we have $L^{2,s}(\mathbb{R}^3) \subset L^{1,-1}(\mathbb{R}^3)$ if and only if $s > 1/2$. Observe that $L^{1,-1}(\mathbb{R}^3)$ contains the union of these L^2 domains.

The same can be said of the space B introduced as a domain for R_z^o in Agmon and Hörmander[7] and used by Agmon in his Salt Lake City lectures[8]. That is, $L^{2,s}(\mathbb{R}^3) \subset B$ for all $s > 1/2$. We only consider B here as a function space on $\mathbb{R}^3$. As we will show now, it is also true that $B \subset L^{1,-1}(\mathbb{R}^3)$.

The space B consists of those $f \in L^2_{loc}(\mathbb{R}^3)$ for which

$$\|f\|_B = \sum_{j=1}^{\infty} R_j^{1/2} \left(\int_{A_j} |f(x)|^2 dx\right)^{1/2} < \infty$$

where $A_j = \{x \in \mathbb{R}^3 ;\ R_{j-1} < |x| < R_j\}$, $R_o = 0$, and $R_j = 2^{j-1}$, $j = 1, 2, \text{---}$. Using Schwarz's inequality we obtain

$$\begin{aligned}\|f\|_{1,-1} &= \sum_{j=1}^{\infty} \int_{A_j} (1+|x|^2)^{-1/2} |f(x)| dx \\ &\leqq \sum_{j=1}^{\infty} \left(\int_{A_j} (1+|x|^2)^{-1} dx\right)^{1/2} \left(\int_{A_j} |f(x)|^2 dx\right)^{1/2} \\ &\leqq c \sum_{j=1}^{\infty} R_j^{1/2} \left(\int_{A_j} |f(x)|^2 dx\right)^{1/2} = c\|f\|_B .\end{aligned}$$

Therefore $B \subset L^{1,-1}(\mathbb{R}^3)$ and the identity map is continuous.

The space $L^{1,-1}(\mathbb{R}^3)$ is the maximal function space which can serve as domain for R_z^o as an integral operator for all $z \in \mathbb{C}$. It is possible to consider R_z^o on the space of positive measures μ which satisfy $\int(1+|x|^2)^{-1/2}\mu(dx) < \infty$. Again

$$R_z^o\mu(x) = \int \frac{e^{i\sqrt{z}|x-y|}}{4\pi|x-y|}\,\mu(dy)$$

is a function in $L^{1,s}(\mathbb{R}^3)$ for $s < -3$ with

$$\|R_z^o\mu\|_{1,s} \leqq c_s\int(1+|x|^2)^{-1/2}\mu(dx)$$

The L^2 result which corresponds to (2.2) is $\|R_z^of\|_{2,-s-1} \leqq c_s'\|f\|_{2,s}$ for $s > 1/2$ where c_s' is independent of f and z. This may be compared with a result implied by Agmon[5], $\|R_z^of\|_{2,-s} \leqq c_s''\|f\|_{2,s}$ for $s > 1/2$ where c_s'' is independent of f and of z in $\{1/K \leqq |z| \leqq K\}$, $K > 1$.

4. THE OPERATOR $S_\lambda = \text{Im}R_\lambda^o$

The operator $\text{Im}R_\lambda^o$ is of interest because of its relation to the spectral family E_λ^o of $-\Delta$, as in $(d/d\lambda)(E_\lambda^of,g) = (1/\pi)(\text{Im}R_\lambda^of,g)$ for suitable functions f and g. In this section we derive some of its properties as an integral operator on $L^{1,-1}(\mathbb{R}^3)$. First let us verify a familiar property of R_λ^o for the space under consideration. It was derived heuristically and used by Lord Rayleigh[9].

<u>Lemma 4.1</u> If $f \in L^{1,-1}(\mathbb{R}^3)$, then $(-\Delta-z)R_z^of = f$ in the distribution sense.

Proof. The function $h_z(x) = e^{i\sqrt{z}|x|}/4\pi|x|$ satisfies $(-\Delta-z)h_z = \delta$ and thus for any test function $\varphi \in C_o^\infty(\mathbb{R}^3)$, $(-\Delta-z)(h_z * \varphi) = \varphi$. Then

$$\langle(-\Delta-z)R_z^of,\varphi\rangle = \langle R_z^of,(-\Delta-z)\varphi\rangle = \int R_z^of(x)(-\Delta-z)\varphi(x)dx$$

$$= \int \left[\int \int h_z(x-y)(-\Delta - z)\varphi(x)dx \right] f(y)dy = \int \varphi(y)f(y)dy = \langle f, \varphi \rangle .$$

Therefore $(-\Delta - z)R_z^o f = f$ completing the proof.

For $z = \lambda \geqq 0$, we will write $R_\lambda^o = C_\lambda + iS_\lambda$ where

$$(4.1) \qquad C_\lambda f(x) = \int \frac{\cos\sqrt{\lambda}\,|x-y|}{4\pi|x-y|} f(y)dy \ , \ S_\lambda f(x) = \int \frac{\sin\sqrt{\lambda}\,|x-y|}{4\pi|x-y|} f(y)dy$$

for all $f \in L^{1,-1}(\mathbb{R}^3)$. The Lemma 4.1 has the

Corollary If $f \in L^{1,-1}(\mathbb{R}^3)$ and $0 \leqq \lambda$, then $(-\Delta-\lambda)C_\lambda f = f$ and $(-\Delta-\lambda)S_\lambda f = 0$ in the distribution sense.

Proof. We may suppose without loss of generality that f is real valued and then the result follows from $(-\Delta-\lambda)C_\lambda f + i(-\Delta-\lambda)S_\lambda f = (-\Delta-\lambda)R_\lambda^o f = f$.

By an extension of Weyl's lemma, weak solutions of $(-\Delta-\lambda)u = 0$ are essentially equal to C^2 solutions and smooth solutions are real analytic. See, for example, Hellwig[10]. An easy consequence of (4.7) below is that $S_\lambda f$ is continuous. Thus the corollary implies that $S_\lambda f$ is analytic. The next three lemmas prepare the way for proof of the bounds for its derivatives as given in Theorem 4.1.

We note first that $(\sin\sqrt{\lambda}\,|x-y|)/|x-y|$ is analytic in x and is a solution of $(-\Delta-\lambda)u = 0$. The notation

$$\partial_\alpha = \partial^{|\alpha|}/\partial x_1^{\alpha_1}\,\partial x_2^{\alpha_2}\,\partial x_3^{\alpha_3} \ , \ \alpha = (\alpha_1, \alpha_2, \alpha_3) \ , \ |\alpha| = \alpha_1+\alpha_2+\alpha_3$$

where the α_i are nonnegative integers, will be used.

Lemma 4.2 For each α there is a constant c_α depending only on α such that

$$(4.2) \qquad \left| \partial_\alpha \frac{\sin\sqrt{\lambda}\,|x-y|}{4\pi|x-y|} \right| \leqq c_\alpha \lambda^{|\alpha|/2} \min(\lambda^{1/2}, |x-y|^{-1})$$

for all $x, y \in \mathbb{R}^3$ and $\lambda > 0$.

Proof. The use of Leibnitz's rule and $h(r) = e^{ir}/r$ permits one to find bounds for the derivatives of $g(r) = \mathrm{Im}\,h(r) = (\sin r)/r$.

$$(4.3) \qquad |g^{(n)}(r)| \leq r^{-1} \sum_{j=0}^{n} \binom{n}{j} j! \quad , \quad 1 \leq r .$$

With $r = |x-y|$, the x derivatives of $g(r)$ may be written

$$\partial_\alpha g(r) = g^{(|\alpha|)}(r) \left(\frac{\partial r}{\partial x_1}\right)^{\alpha_1} \left(\frac{\partial r}{\partial x_2}\right)^{\alpha_2} \left(\frac{\partial r}{\partial x_3}\right)^{\alpha_3} + \cdots + g^{(1)}(r)\partial_\alpha r .$$

Since for each β, $|\partial_\beta r| \leq m_\beta / r^{|\beta|-1}$ for all $r > 0$ with m_β independent of r, we obtain an inequality of the form

$$(4.4) \qquad |\partial_\alpha g(r)| \leq |g^{(|\alpha|)}(r)| + |g^{(|\alpha|-1)}(r)| \frac{c_2'}{r} + \cdots + |g^{(1)}(r)| \frac{c'_{|\alpha|}}{r^{|\alpha|-1}} .$$

Combining (4.3) and (4.4)

$$(4.5) \qquad |\partial_\alpha g(r)| \leq \frac{c_1}{r} + \frac{c_2}{r^2} + \cdots + \frac{c_{|\alpha|}}{r^{|\alpha|}} \leq \frac{c_1 + \cdots + c_{|\alpha|}}{r} \quad , \quad 1 \leq r ,$$

where the constants $c_1, \ldots, c_{|\alpha|}$ depend only on α. The continuous function $\partial_\alpha g(r)$ is bounded on $0 \leq r \leq 1$, so we obtain an inequality

$$\left| \partial_\alpha \frac{\sin|x-y|}{|x-y|} \right| \leq c'_\alpha \min(1, |x-y|^{-1})$$

for all $x, y \in \mathbb{R}^3$ where c'_α depends only on α. Replacing $|x-y|$ with $\sqrt{\lambda}\,|x-y|$ this becomes

$$\left| \partial_\alpha \frac{\sin\sqrt{\lambda}\,|x-y|}{\sqrt{\lambda}\,|x-y|} \right| \leq \lambda^{|\alpha|/2} c'_\alpha \min(1, \lambda^{-1/2}|x-y|^{-1})$$

and (4.2) follows with $c_\alpha = c'_\alpha / 4\pi$. This completes the proof of the lemma.

Lemma 4.3 For each α and each $f \in L^{1,-1}(\mathbb{R}^3)$

$$(4.6) \qquad \partial_\alpha S_\lambda f(x) = \int \partial_\alpha \frac{\sin\sqrt{\lambda}\,|x-y|}{4\pi|x-y|} f(y)dy .$$

Proof. Notice first that

$$\sup_{|x|\leq R} \inf(\lambda^{1/2}, |x-y|^{-1}) = \begin{cases} \lambda^{1/2} & , \quad |y| < R+\lambda^{-1/2} \\ (|y|-R)^{-1} & , \quad R+\lambda^{-1/2} \leq |y| \end{cases}$$

so that for each α and each λ, applying Lemma 4.2,

$$\sup_{|x|\leq R} \left| \partial_\alpha \frac{\sin\sqrt{\lambda}\,|x-y|}{4\pi|x-y|} \right| \leq c(1+|y|)^{-1}. \tag{4.7}$$

for all $y \in \mathbb{R}^3$. The argument now proceeds by induction. Assume (4.6) to be valid for all α with $|\alpha| = n$ and compute $(\partial/\partial x_j)\partial_\alpha S_\lambda f$ using the mean value theorem, the dominant (4.7) for the derivative $(\partial/\partial x_j)\partial_\alpha$, the fact that $(1+|y|)^{-1}f(y)$ is integrable, and the dominated convergence theorem. These steps will complete the proof.

The continuity of $\partial_\alpha S_\lambda f$ follows from (4.7) and the dominated convergence theorem. From Lemma 4.3 we find directly that

$$(-\Delta-\lambda)S_\lambda f(x) = \int (-\Delta-\lambda) \frac{\sin\sqrt{\lambda}\,|x-y|}{4\pi|x-y|} f(y)dy = 0.$$

The bound of Lemma 4.2 may be used also to derive a bound for $\partial_\alpha S_\lambda f$.

<u>Lemma 4.4</u> For each $R \geq \lambda^{-1/2}$ and all $x \in \mathbb{R}^3$

$$|\partial_\alpha S_\lambda f(x)| \leq c_\alpha \lambda^{|\alpha|/2} \left\{ \lambda^{1/2}(1+|x|+R) \int_{|x-y|<R} \frac{|f(y)|}{1+|y|}\,dy + \left(1 + \frac{1+|x|}{R}\right) \int_{R\leq|x-y|} \frac{|f(y)|}{1+|y|}\,dy \right\}. \tag{4.8}$$

Proof. The condition $R \geq \lambda^{-1/2}$ implies

$$\inf(\lambda^{1/2}, |x-y|^{-1}) \leq \begin{cases} \lambda^{1/2} & , \quad |x-y| < R \\ |x-y|^{-1} & , \quad R \leq |x-y| \end{cases}.$$

Also, for fixed x, the maximum value of $|y|\,|x-y|^{-1}$ on the sphere

$|y-x| = r$ occurs at $y = (1 + r|x|^{-1})x$ where $|y|$ is largest. Thus $|x-y| = r \geqq R$ implies

$$\frac{1+|y|}{|x-y|} \leqq \frac{1+|x|+r}{r} \leqq 1 + \frac{1+|x|}{R} .$$

If $|x-y| < R$ then $1+|y| \leqq 1+|x|+R$. The use of Lemmas 4.3 and 4.2 and the above estimates yield (4.8) completing the proof.

This dominant may be used in two ways to obtain estimates on the growth at infinity. An example will be given which shows these to be sharp for $S_\lambda f$ itself.

Theorem 4.1 For each α, each $\lambda > 0$, and each $f \in L^{1,-1}(\mathbb{R}^3)$

$$\partial_\alpha S_\lambda f(x) = o(|x|) \text{ as } |x| \to \infty, \text{ and} \tag{4.9}$$

$$\left|\frac{\partial_\alpha S_\lambda f(x)}{1+|x|}\right| \leqq 2c_\alpha \lambda^{|\alpha|/2}(1+\lambda^{1/2})\|f\|_{1,-1} \text{ for all } x. \tag{4.10}$$

Proof. According to Lemma 4.4, for all $x \in \mathbb{R}^3$ and $R \geqq \lambda^{-1/2}$

$$\left|\frac{\partial_\alpha S_\lambda f(x)}{1+|x|}\right| \leqq c_\alpha \lambda^{|\alpha|/2}\left\{\lambda^{1/2}\left(1 + \frac{R}{1+|x|}\right)\int_{|x-y|<R} \frac{|f(y)|}{1+|y|}\,dy + \left(\frac{1}{1+|x|} + \frac{1}{R}\right)\int_{R\leqq|x-y|} \frac{|f(y)|}{1+|y|}\,dy\right\} . \tag{4.11}$$

If we take $|x| \geqq R$ in (4.11) it may be written

$$\left|\frac{\partial_\alpha S_\lambda f(x)}{1+|x|}\right| \leqq c_\alpha \lambda^{|\alpha|/2}\left\{2\lambda^{1/2}\int_{|x-y|<R} \frac{|f(y)|}{1+|y|}\,dy + \frac{2}{R}\int_{R\leqq|x-y|} \frac{|f(y)|}{1+|y|}\,dy\right\} . \tag{4.12}$$

The second term on the right may be made small for large R and the first term on the right tends to zero as $|x|$ tends to infinity for fixed R. It follows that

$$\lim_{|x|\to\infty} \frac{\partial_\alpha S_\lambda f(x)}{1+|x|} = 0$$

proving (4.9). For any x, (4.11) implies

$$\left|\frac{\partial_\alpha S_\lambda f(x)}{1+|x|}\right| \leqq c_\alpha \lambda^{|\alpha|/2} \left\{ \lambda^{1/2}(1+R) \int_{|x-y|<R} \frac{|f(y)|}{1+|y|}\, dy + \left(1+\frac{1}{R}\right) \int_{R<|x-y|} \frac{|f(y)|}{1+|y|}\, dy \right\} ,$$

and (4.10) follows on substituting $R = \lambda^{-1/2}$ and using (1.3). This completes the proof.

Example 1 If $f = 1_{(|y|<R)}$, then

$$S_\lambda f(x) = \int_{|y|<R} \frac{\sin\sqrt{\lambda}\,|x-y|}{4\pi|x-y|}\, dy = a(\lambda,R)\, \frac{\sin\sqrt{\lambda}\,|x|}{|x|} \tag{4.13}$$

where $a(\lambda,R) = \lambda^{-3/2}(\sin\sqrt{\lambda}R - \sqrt{\lambda}R\cos\sqrt{\lambda}R) = (\pi/2)^{1/2}(R/\sqrt{\lambda})^{3/2}J_{3/2}(\sqrt{\lambda}R)$.

To verify this use the expansion

$$\int_{|y|<R} \frac{\sin\sqrt{\lambda}\,|x-y|}{|x-y|}\, dy = \sum_{k=0}^{\infty} \frac{(-1)^k(\sqrt{\lambda})^{2k+1}}{(2k+1)!} \int_{|y|<R} |x-y|^{2k} dy . \tag{4.14}$$

The integrals in the sum may be evaluated readily using spherical coordinates. After utilizing the binomial formula and taking advantage of certain cancellations the result

$$\int_{|y|<R} |x-y|^{2k} dy = \frac{4\pi}{(2k+2)(2k+3)(2k+4)} \sum_{j=1}^{k+1} j\binom{2k+4}{2j+1} |x|^{2(k-j+1)} R^{2j+1} \tag{4.15}$$

is obtained. Substituting (4.15) into (4.14) and converting the triangular sum into a square sum yields

$$\int_{|y|<R} \frac{\sin\sqrt{\lambda}\,|x-y|}{|x-y|}\,dy = \frac{4\pi}{\lambda}\sum_{m=o}^{\infty}\frac{(-1)^m}{(2m+1)!}(\sqrt{\lambda}\,|x|)^{2m}$$

$$\cdot \sum_{n=o}^{\infty}\frac{(-1)^n}{(2n+1)!(2n+3)}(\sqrt{\lambda}R)^{2n+3}$$

$$= \frac{4\pi}{\lambda}\frac{\sin\sqrt{\lambda}\,|x|}{\sqrt{\lambda}\,|x|}(\sin\sqrt{\lambda}R - \sqrt{\lambda}R\cos\sqrt{\lambda}R)\,.$$

Example 2 Use will be made of (4.13) to construct an unbounded $S_\lambda f$. This will give a measure of the precision of the estimate (4.9). Let $f = a_n$ on $B_n = \{y \in \mathbb{R}^3 \ \ |y-x_n| < r_n\}$, $n = 1, 2, ---$, where $x_n = (n,0,0)$ and a_n and r_n are to be chosen later but which satisfy $0 \leqq a_n$ and $0 < r_n \leqq 1/2$. Define $f = 0$ on $\mathbb{R}^3 \setminus (\cup B_n)$. Then

$$(4.16) \qquad \int \frac{f(y)}{1+|y|}\,dy = \sum_{n=1}^{\infty} a_n \int_{B_n} \frac{dy}{1+|y|} \leqq \frac{4\pi}{3}\sum_{n=1}^{\infty}\frac{a_n r_n^3}{n} < \infty$$

for suitable choices of a_n and r_n. On the other hand, using translations of (4.13)

$$S_\lambda f(x) = \sum_{n=1}^{\infty} a_n \int_{B_n} \frac{\sin\sqrt{\lambda}\,|x-y|}{4\pi|x-y|}\,dy = \sum_{n=1}^{\infty} a_n a(\lambda, r_n)\frac{\sin\sqrt{\lambda}\,|x-x_n|}{|x-x_n|}\,.$$

In particular for k a positive integer

$$S_\lambda f(x_k) = \sum_{n=1}^{\infty} a_n a(\lambda, r_n)\frac{\sin\sqrt{\lambda}\,|k-n|}{|k-n|}$$

and for the choice $\lambda = \pi^2$, all but one of the terms is zero and

$$S_{\pi^2} f(x_k) = a_k a(\pi^2, r_k)\pi \quad .$$

An elementary estimate gives $(3/10)R^3 \leqq a(\lambda, R) \leqq (1/3)R^3$ for $\sqrt{\lambda}R \leqq 1$. Restricting $\pi r_k \leqq 1$ we obtain

$$\frac{3\pi}{10} a_k r_k^3 \leqq S_{\pi^2} f(x_k) \leqq \frac{\pi}{3} a_k r_k^3 \quad , \; k = 1, 2, \cdots \quad .$$

The convergence of the series in (4.16) implies $S_{\pi^2} f(x_k) = o(|x_k|)$, but for any α, $0 < \alpha < 1$, it is possible to choose a_k and r_k satisfying (4.16) and for which $a_k r_k^3 / k^\alpha$ is unbounded. Therefore, there is no bound of the form $O(|x|^\alpha)$ for $S_\lambda f(x)$ which is valid for all $f \in L^{1,-1}(\mathbb{R}^3)$.

REFERENCES

1. T. Ikebe, Eigenfunction expansions associated with Schrödinger operators and their applications to scattering theory, Arch. Ratl.Mech.Anal., 5:1-34 (1960).
2. N. Shenk, The invariance of wave operators associated with perturbations of -Δ, J.Math.Mech. 17:1005-1022 (1968).
3. P. A. Rejto, On partly gentle perturbations II, J.Math.Anal. Appl. 20:145-187 (1967).
4. P. Alsholm and G. Schmidt, Spectral and scattering theory for Schrödinger operators, Arch.Ratl.Mech.Anal. 40:281-311 (1971).
5. S. Agmon, Spectral properties of Schrödinger operators and scattering theory, Ann.Scuola Norm.Sup.Pisa, Classe Sci. (4)2:151-218 (1975). (section 4)
6. L. L. Helms, Introduction to Potential Theory, Wiley-Interscience, New York (1969).
7. S. Agmon and L. Hörmander, Asymptotic properties of solutions of differential equations with simple characteristics, J.d'Anal. Math. 30:1-38 (1976).
8. S. Agmon, Spectral and scattering theory for elliptic operators, NSF Regional Conference in Mathematics, University of Utah (1978).
9. Lord Rayleigh, The Theory of Sound, Macmillan (1877), Dover, New York (1945). (vol.II page 106)
10. G. Hellwig, Partial Differential Equations, Blaisdell, New York (1964) (pages 96 and 182)

INVARIANT MANIFOLDS AND BIFURCATIONS IN THE TAYLOR PROBLEM

R.D. Richtmyer

Department of Mathematics
University of Colorado
Boulder, Colorado 80309

The calculations of Davey 1962, of Davey, DiPrima, and Stuart 1968, and of Eagles 1971, on the Taylor problem of the flow between concentric rotating cylinders are interpreted as calculations of the finite dimensional invariant unstable manifold in the infinite-dimensional configuration space of the system and subsequent determination of the fixed points and closed orbits of the dynamical system in that manifold. The procedure is not a spectral method or eigenfunction expansion method, although certain of the eigenfunctions of the linearized problem play an important role. In contrast with such methods, the restriction to a finite number of dimensions is not an approximation, because the manifold under study is rigorously invariant. (There are of course plenty of other approximations to worry about.)

In the interest of possible application to more general calculations, we have modified and extended the method in minor ways, as follows:

1. The number of dimensions of the unstable manifold is determined by the eigenvalues of the linearized problems that lie in the right half of the complex plane and increases with the Reynolds number. Here it is arbitrary, not restricted to 6 as in the previous work. With the present program for the Cray 1 computer, calculations up to 14 dimensions are feasible. By certain modifications of the procedure, the number of dimensions could probably be increased somewhat further.

2. The algebraic calculations made by hand by the previous authors, especially by Eagles, and which would be prohibitive in more than about six dimensions, have been mechanized and incorporated in the computer program.

3. Certain consequences of the cylindrical symmetry, which reduce the amount of calculation and storage by a factor around 3.0 , and which were taken into account in the hand calculations referred to above, have been formulated as lemmas and also incorporated in the computer program.

4. In order to be able to calculate the helical vortices a little more realistically, a modification has been introduced to inhibit net axial flow.

5. Multiple shooting or reorthonormalization has been introduced in the numerical solution of the two-point boundary problems, to deal with the stiffness of the radial ordinary differential equations at higher Reynolds numbers.

6. Numerous checks have been built into the program to monitor things like the accuracy of the numerical solution of the ordinary differential equations, the conservation laws, consequences of the symmetry, the biorthogonality of the eigenfunctions and adjoint eigenfunctions of the linearized problem, and certain other orthogonalities.

7. Orbit calculation procedures (based mainly on the Adams-Bashforth-Moulton method) have been provided, for the study of the resulting dynamical system in the invariant manifold, especially for looking for fixed points, closed orbits, and attractors generally.

With those modifications, the method was found suitable for study of the Taylor problem under a moderate range of parameters, though unfortunately not for high enough Reynolds numbers to permit study of the aperiodic phenomena found by Gollub and Swinney and other investigators. The main limitation comes from the number of dimensions of the invariant manifold. It appears we should have to go to at least 32 dimensions. Unfortunately, the amount of computing and storage varies with a high power of the number of dimensions. It is also possible that at high Reynolds number the power series expansions that give the manifold and the dynamical system in it would have to be carried to higher-degree terms than has been done so far.

The method will be described in detail elsewhere (for example in a forthcoming report by the author at the National Center for Atmospheric Research in Boulder). Here, we merely describe briefly the (mostly negative) results that have been obtained so far.

One series of calculations, intended mainly to verify that the computer program could reproduce the results of the earlier authors, was for a narrow gap between the cylinders. A ten-dimensional manifold was studied, based on the first five of the eigenvalues of the linearized problem that cross into the right half of the complex plane; each has multiplicity two. The corresponding eigenfunctions depend on θ as $e^{im\theta}$, where r, θ, z are cylindrical coordinates, and where $m = 0, \pm1, \pm2$. The outer cylinder was at rest, and the gap between the cylinders was equal to .05 times the mean radius. The periodicity in the z direction was such that the axial extent of a single vortex was 1.0047 times the gap. The calculations were carried to a Taylor number T (see, for example, Eagles 1968 for the definitions) up to 3000. The power series expansions were carried to the 5th order. The results can be summarized as follows:

Taylor ring vortices appear at	$T = 1753.1$
They are stable for	$T < 1971.5$
Wavy vortices $m = 1$ appear at	$T = 1971.5$
They are stable for	$T < 1985$
Wavy vortices $m = 2$ appear at	$T = 1970$
They are always stable	
Helical vortices $m = 1$ appear at	$T = 1768.4$
They are stable for	$T > 1770.3$
Helical vortices $m = 2$ appear at	$T = 1784$
They are stable for	$T > 1815$

The values of T given may not be very precise because of the small number of intervals used (24) in the solution of the radial ordinary differential equations. It is perhaps worth noting that for an interval of T from about 1970 to about 1985 both wavy vortex modes, $m = 1$ and $m = 2$, appear to be stable.

This series was discontinued, because the number of dimensions of the unstable manifold increases rather rapidly so that for T above about 1820, the ten-dimensional manifold, although invariant, is only a part of the unstable manifold, hence perturbations can take the solution out of that part.

That difficulty is somewhat less severe for the wide gap cases, hence we made a second series of calculations with gap = 0.5 times the mean radius. Eigenvalue calculations show that for that gap, the eigenvalue with $m = 4$ remains in the left half plane to the highest values of T considered, hence a 14-dimensional manifold based on $m = 0, \pm1, \pm2, \pm3$ was studied. (However, for large T, further eigenvalues appear in the right half plane for $m = 0, \pm1, \pm2$, so that there is here the same difficulty as in the previous series, but it now appears only at higher values of T.) For large values of T, the number of steps in the radial integration was increased to 240, but in this series the power series expansions were carried only to the 3rd order. It was found that the Taylor vortices give way to wavy vortices with $m = 3$ somewhere between $T = 320{,}000$ and 640,000, and the wavy vortices are stable from there to $T \approx 1{,}148{,}000$. For $T \geq 1{,}150{,}000$, no stable modes were found, either by mode analysis of the kind made by the previous authors or by orbit calculations; all orbits escape from the neighborhood of the origin in the configuration space where the Taylor and wavy vortices had been to larger distances, where the power series expansions used in computing the invariant manifold are inaccurate.

We must apparently assume that the bifurcation at $T \approx 1{,}148{,}000$ is subcritical. It may lead to an unstable mode at $T < 1{,}148{,}000$ consisting of pulsating wavy vortices. For $T > 1{,}148{,}000$ there may be an aperiodic behavior (strange attractor), but it was felt that the present method is inadequate for further study without increasing the number of dimensions and the order of the power series expansions far beyond the capabilities of the present computer program.

This research was performed largely at the National Center for Atmospheric Research, sponsored by the National Science Foundation.

Davey, A., 1962, The growth of Taylor vortices in the flow between rotating cylinders, Jour. Fluid Mech. 14: 336-368.

Davey, A., DiPrima, R.C., and Stuart, J.T., 1968, On the stability of Taylor vortices, Jour. Fluid Mech. 31: 17-51.

Eagles, P.M., 1971, On stability of Taylor vortices by fifth-order amplitude expansions, Jour. Fluid Mech. 49: 529-550.

ON THE DIMENSION OF A FINITE DIFFERENCE APPROXIMATION TO DIVERGENCE-FREE VECTORS

R.L. Hartman and K. Gustafson

Department of Mathematics
University of Colorado
Boulder, Colorado 80309

Abstract. The condition $\text{div}\,\vec{u} = 0$ on a region Ω in R^n is approximated by finite differences. Grid functions $u_{i,j}$ satisfying this condition form a vector space. Calculation of the dimension and calculation of a basis for this vector space are discussed. An illustrative example is included.

1. Introduction.

The condition $\text{div}\,\vec{u} = 0$ arises in many situations in mathematics and the physical sciences. Often the first step in numerical solution of partial differential equations involving this condition is the choice of a family of functions satisfying (in some sense) the divergence-free condition, from which a function approximating the true solution will be chosen. See for example Temam [1] .

A new integral approach to the condition $\text{div}\,\vec{u} = 0$ is put forth in Gustafson and Young [3] , and we refer to that paper; see also the survey Gustafson [2] , for further references and discussion of the various schemes in use.

The simplest among these schemes is finite differences. More than once in the course of his discussion, Temam [1] mentions the lack of a 'simple' basis for the vector space family of (step) functions $\vec{u}$ satisfying $\text{div}\,\vec{u} = 0$. With possible reservations about the word 'simple', we discuss such a basis and its dimension.

2. Problem Statement and Reformulation.

Let R be the real numbers and let Z be the integers. Let

$\vec{h} \in R^n$ with $h_i > 0$ for $1 \le i \le n$. Then $\vec{h}$ defines a grid G in R^n by $\vec{x}(\vec{m}) = \sum_{i=1}^{n} m_i h_i \varepsilon_i$ where $(\varepsilon_i)_{i=1}^{n}$ is the standard basis for R^n and $\vec{m} \in Z^n$. Given a vector function $\vec{u} : R^n \to R^n$, approximate the quantity div $\vec{u}(\vec{p})$ by the discrete form $\sum_{j=1}^{n} \frac{u_j(\vec{p} + h_j \vec{\varepsilon}_j) - u_j(\vec{p})}{h_j}$. Given a finite subset $X \subset G$, find a basis for and the dimension of the vector space $U(X)$ of functions $\vec{u} : G \to R^n$ such that:

$$\vec{u} = 0 \quad \text{on} \quad G \setminus X \tag{1}$$

and

$$\text{div } \vec{u}(\vec{p}) = 0 \text{ in discrete form holds for all } \vec{p} \in G. \tag{2}$$

Replacement of u_j by $\frac{u_j}{h_j}$ and $\vec{h}$ by $(1,1,\ldots,1)$ leads to the equation $0 = \text{div } \vec{u}(\vec{m}) = \sum_{j=1}^{n} [u_j(\vec{m} + \vec{\varepsilon}_j) - u_j(\vec{m})]$ and G becomes Z^n. It is sufficient to consider this latter problem.

Reformulation in Terms of Graph Theory.

It is advantageous at this point to interpret $u_j(m)$ as the amount of fluid flow through an arc directed from point $(\vec{m} - \vec{\varepsilon}_j)$ to point $\vec{m}$. For that purpose define (N,T,σ) to be a directed graph where N is a finite set of vertices, T is a finite set of arcs connecting vertices in N and $\sigma : T \times N \to \{-1,0,1\}$ indicates directions along arcs. That is, $\sigma(t,p) = 1$ if arc t is directed toward vertex p, $\sigma(t,p) = -1$ if arc t is directed away from vertex p and $\sigma(t,p) = 0$ if arc t connects vertex p to itself or arc t does not connect vertex p to another vertex. In the case at hand let $N = \{\vec{m} | \vec{m} \in X \text{ or } (\vec{m} + \vec{\varepsilon}_j) \in X \text{ for some } j\}$ and vertices $\vec{m}_1$ and $\vec{m}_2$ of N are connected by an arc if $\vec{m}_2 = \vec{m}_1 \pm \vec{\varepsilon}_j$ for some $1 \le j \le n$. For definiteness define $\sigma(t,\vec{m}) = 1$ if arc t connects vertices $\vec{m} - \vec{\varepsilon}_j$ and $\vec{m}$ for some $1 \le j \le n$. In correspondence to $U(X)$ above, define the flow space (N,T,σ) to be the vector space of functions $u : T \to R$ for which:

$$\sum_{t \in T} u(t)\sigma(t,p) = 0 \quad \text{for each} \quad p \in N \tag{3}$$

Equation (3) states that fluid is conserved at each point of N. There is an isomorphism between $U(X)$ and $U(N,T,\sigma)$ because equation (3) is the graph-theoretic equivalent of equation (2).

Example.

Let $X = \{(0,0),(1,0),(2,0),(2,1),(2,2),(1,2),(0,2),(0,1)\}$. The vertices and directed arcs of the corresponding directed graph are shown in the following figure where "●" denotes a vertex from X and "○" denotes a vertex from $Z^n \setminus X$.

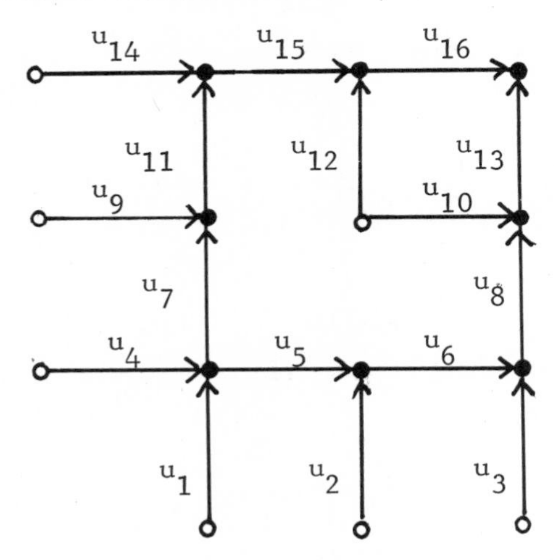

The equations corresponding to (3) are:

$-u_1 = 0$
$-u_2 = 0$
$-u_3 = 0$
$-u_4 = 0$
$u_1 + u_4 - u_5 - u_7 = 0$
$u_2 + u_5 - u_6 = 0$
$u_3 + u_6 - u_8 = 0$
$-u_9 = 0$

$u_7 + u_9 - u_{11} = 0$
$-u_{10} - u_{12} = 0$
$u_8 + u_{10} - u_{13} = 0$
$-u_{14} = 0$
$u_{11} + u_{14} - u_{15} = 0$
$u_{12} + u_{15} - u_{16} = 0$
$u_{13} + u_{16} = 0$

3. Computation of the Dimension.

Relative to the directed graph (N,T,σ) define the set $M \subset N$ to be connected if for any $\{p,q\} \subset M$, there exists a sequence $p = p_0, t_0, p_1, t_1, \cdots, p_{k-1} t_{k-1}, p_k = q$ where arc t_j connects (ignoring direction) vertices p_j and p_{j+1}, $(p_j)_{j=0}^{k} \subset M$ and $k \geq 0$. Define a component of N to be a maximally connected subset. Given any set S, denote its cardinality by $\overline{S}$. The following is a well-known theorem in graph theory. See for example Berge [5].

Theorem 1. Let (N,T,σ) be a directed graph. Then

$$\dim U(N,T,\sigma) = \overline{T} - \overline{N} + \overline{C} \tag{4}$$

where C is the set of components of N.

Sketch of proof. Assume that N is connected. Equation (3) supplies a linear equation $\mathcal{E}_p$ for each vertex $p \in N$. Let $\sum_{p \in N} c_p \mathcal{E}_p = 0$ be a dependence relation among this set of equations. If $\{p,q\} \subset N$ and $t \in T$ connects p and q, then $c_p = c_q$. Connectivity of N thus implies that c_p is a constant function of p hence $\dim U(N,T,\sigma) = \overline{T} - \overline{N} + 1$. In the multi-component case $\dim U(N,T,\sigma) = \sum_{i=1}^{\overline{C}} (\overline{T}_i - \overline{N}_i + 1) = \overline{T} - \overline{N} + \overline{C}$ where the graph (N_i, T_i, σ_i) corresponds to the component N_i.

In the example given above $\overline{T} = 16$, $\overline{N} = 15$ and N is connected hence $\dim U(N,T,\sigma) = 2$.

4. Finding a Basis.

Define the sequence $p_0, t_0, p_1, t_1, \cdots, p_{k-1}, t_{k-1}, p_k$ to be a cycle if:

a) t_i is an arc connecting vertices p_i and p_{i+1},
b) no arc appears in the sequence more than once,
c) $p_k = p_0$ but $k \geq 1$.

The following lemma is a consequence of well-known results from graph theory.

Lemma 2. The graph (N,T,σ) has a cycle involving arc τ if and only if there exists $u \in U(N,T,\sigma)$ with $u(\tau) \neq 0$.

Sketch of proof: If there exists a cycle involving arc τ, set up a unit flow through the arcs of that cycle obtaining $u \in U(N,T,\sigma)$ for which $u(\tau) \neq 0$. Assume that $u \in (N,T,\sigma)$, that $u(\tau) \neq 0$ and that arc τ connects vertices p and q. Set $S(0) = \{p\}$ and inductively define $S(n+1)$ to be all vertices either in $S(n)$ or connected to a vertex in $S(n)$ by an arc in $T \setminus \{\tau\}$. There exists m such that $n \geq m$ implies that $S(n) = S(m)$. Define $S_\infty = S(m)$ and assume that $q \notin S_\infty$. Then $0 = \sum_{v \in S_\infty} \sum_{t \in T} u(t)\sigma(t,v) = u(\tau)\sigma(\tau,p) \neq 0$. This contradiction implies that $q \in S_\infty$ and hence that there exists a cycle involving arc τ.

The following algorithm suggested by Lemma 2 may be applied to yield a basis. This algorithm is amenable to being programmed and the basis vectors' support may usually be arranged to be sparse.

Algorithm:

1) Let (N,T_0,σ_0) be the directed graph (N,T,σ). Set $i = 0$ and go to (2).
2) Choose any arc t of graph (N,T_i,σ_i). Determine whether there exists a cycle in (N,T_i,σ_i) involving arc t and save the cycle if so. Go to (3).
3) Delete arc t from graph (N,T_i,σ_i) to obtain (N,T_{i+1},σ_{i+1}) and increment i. If now $T_i = \emptyset$, then stop; otherwise, go to (2).

Theorem 3. The above algorithm produces a set of cycles $(\gamma_i)_{i=1}^k$. Set up a unit flow through the arcs of each cycle to obtain corresponding elements $(u_i)_{i=1}^k \subset U(N,T,\sigma)$. The elements $(u_i)_{i=1}^k$ form a basis for $U(N,T,\sigma)$.

Sketch of proof: Let C_i be the set of components in graph (N,T_i,σ_i). If there exists a cycle involving arc t, then $\overline{C}_{i+1} = \overline{C}_i$ and by equation (4) $\dim U(N,T_{i+1},\sigma_{i+1}) = \dim(N,T_i,\sigma_i) - 1$. If no cycle exists, then $\overline{C}_{i+1} = \overline{C}_i + 1$ and $\dim U(N,T_{i+1},\sigma_{i+1}) = \dim U(N,T_i,\sigma_i)$. Hence the number of cycles found is $\dim U(N,T,\sigma)$. Since an arc is removed at each step, u_j is not a linear combination of $(u_i)_{i=j+1}^k$ hence $(u_i)_{i=1}^k$ is a linearly independent set.

For the example given above, remove the arc form (0,0) to (1,) noting that this arc is part of the following cycle given in terms of vertices only:

$$(0,0),(1,0),(2,0),(2,1),(2,2),(1,2),(0,2),(0,1),(0,0) .$$

Next, the arc from $(1,1)$ to $(2,1)$ is part of the cycle:

$$(1,1),(2,1),(2,2),(1,2),(1,1) .$$

The resulting two basis vectors are given in the following diagrams (a) and (b) where values of the basis vectors at grid points not pictured are $(0,0)$.

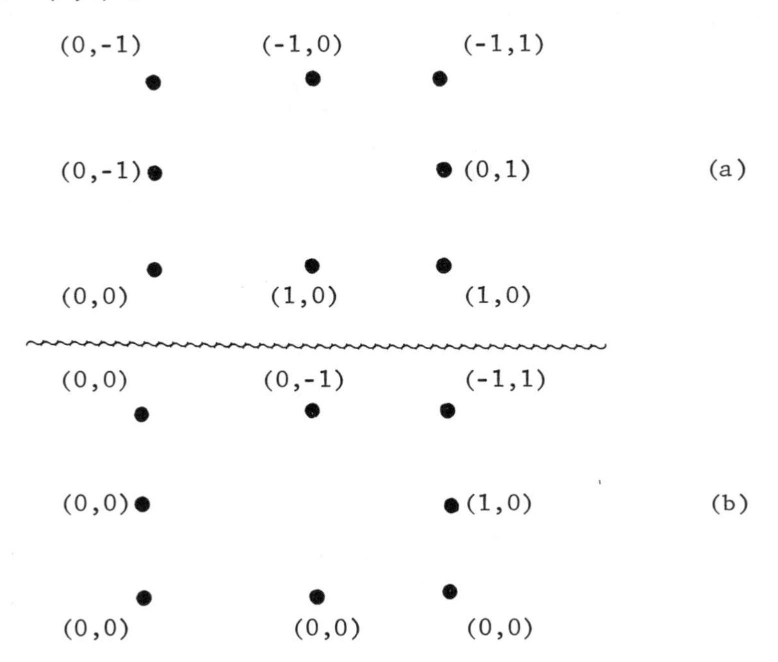

5. Further Applications.

The kind of analysis described above has been applied to discrete Dirichlet and Neumann-like problems. In this application a graphical flow problem is associated with the original problem so that solution of the flow problem is equivalent to solution of the original problem. For further details see [4].

References

1. R. Temam, Navier-Stokes Equations; Theory and Numerical Analysis, Elsevier North-Holland, New York, 1979.
2. K.E. Gustafson, Recent Progress on the Nonlinear Equations of Hartree-Fock, Concentration-Diffusion, Navier-Stokes, Proc. Conf. on Bifurcation Theory, Bielefeld, Oct. 1979, Applications of Nonlinear Analysis in the Physical Sciences, Pitman, London, to appear.

3. K.E. Gustafson and D.P. Young, Computation of Solenoidal (Divergence-free) Vector Fields, to appear.
4. R.L. Hartman, Ph.D. Thesis, University of Colorado, Boulder, Colorado, to appear.
5. C. Berge, Graphs and Hypergraphs, Elsevier North-Holland, New York, 1973.

INTRAMOLECULAR DYNAMICS IN THE QUASIPERIODIC AND STOCHASTIC REGIMES*

D. W. Noid, M. L. Koszykowski, and R. A. Marcus

Oak Ridge National Laboratory, Oak Ridge, TN 37830
Sandia National Laboratories, Livermore, CA 94550
California Institute of Technology, Pasadena, CA 91125

INTRODUCTION

Nonlinear dynamics has been the subject of intense study in recent years. In this review article, we present a summary of our recent work in this field, in which nonlinear dynamics is applied to problems involving molecular behavior. Four aspects are described: (1) semiclassical methods for the calculation of bound-state eigenvalues, (2) classical spectra and correlation functions in the quasiperiodic and "chaotic" regimes, (3) "chaotic" behavior in quantum mechanics, and (4) applications to collisional and laser interactions.

I. SEMICLASSICAL METHODS FOR THE CALCULATION OF BOUND-STATE EIGENVALUES

The calculation of bound-state properties using semiclassical concepts has been of interest for many years.[1-3] This problem has a well-known WKB solution for systems permitting separation of variables. Recently, several methods of calculating eigenvalues for non-separable systems that are quasiperiodic have been presented. The earliest method[4] for systems with smooth potentials involves the use of classical trajectories for the calculation of the phase space path integrals ("actions") used to quantize the system and hence find the semiclassical eigenvalues.

The studies involved systems with incommensurate frequencies[4a-4c]

*Research sponsored by the U. S. Department of Energy under contract W-7405-eng-26 with the Union Carbide Corporation, and DOE at Sandia National Laboratory, and NSF at California Institute of Technology.

and then systems with a 1:1 zeroth order commensurability.[4d] Recently we have been able to extend the applicability of this method to systems that have Fermi resonance[4e] and to systems with more degrees of freedom.[4f] The system that exhibits Fermi resonance has frequencies of the unperturbed problem in a ratio of 1:2 and also a perturbation term that couples the two degrees of freedom resonantly. The Hamiltonian which we used in this calculation is

$$H = \frac{1}{2}(p_x^2 + p_y^2 + \omega_x^2 x^2 + \omega_y^2 y^2) + \lambda(xy^2 - \beta x^3) \ . \qquad \text{(I-1)}$$

A typical trajectory of Hamiltonian (1) is shown in Fig. 1. Introducing a curvilinear Poincaré surface of section (in parabolic coordinates), we were able to calculate the eigenvalues semiclassically.

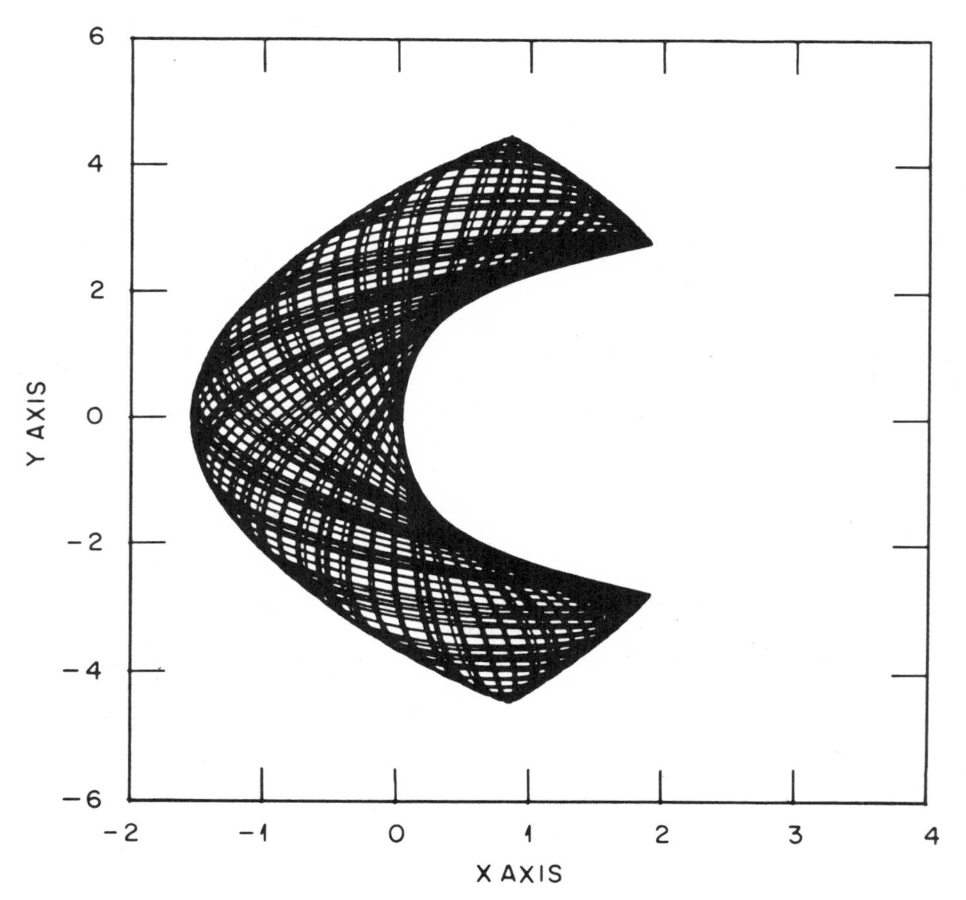

Fig. 1. A quasiperiodic trajectory for Hamiltonian (I-1) which has a Fermi resonance.

The path used in our calculation is shown in Fig. 2 (ξ and η are parabolic coordinates). An example of a p_ξ vs. ξ Poincaré surface of section for η = constant is shown in Fig. 3. The quantum conditions for this type of trajectory were ($\hbar = 1$)

$$J_\xi = \oint p_\xi d\xi = 2\pi(n_1 + \frac{1}{2}) \tag{I-2a}$$

and

$$J_\eta = \oint p_\eta d\eta = 2\pi(n_2 + \frac{1}{2}) \ . \tag{I-2b}$$

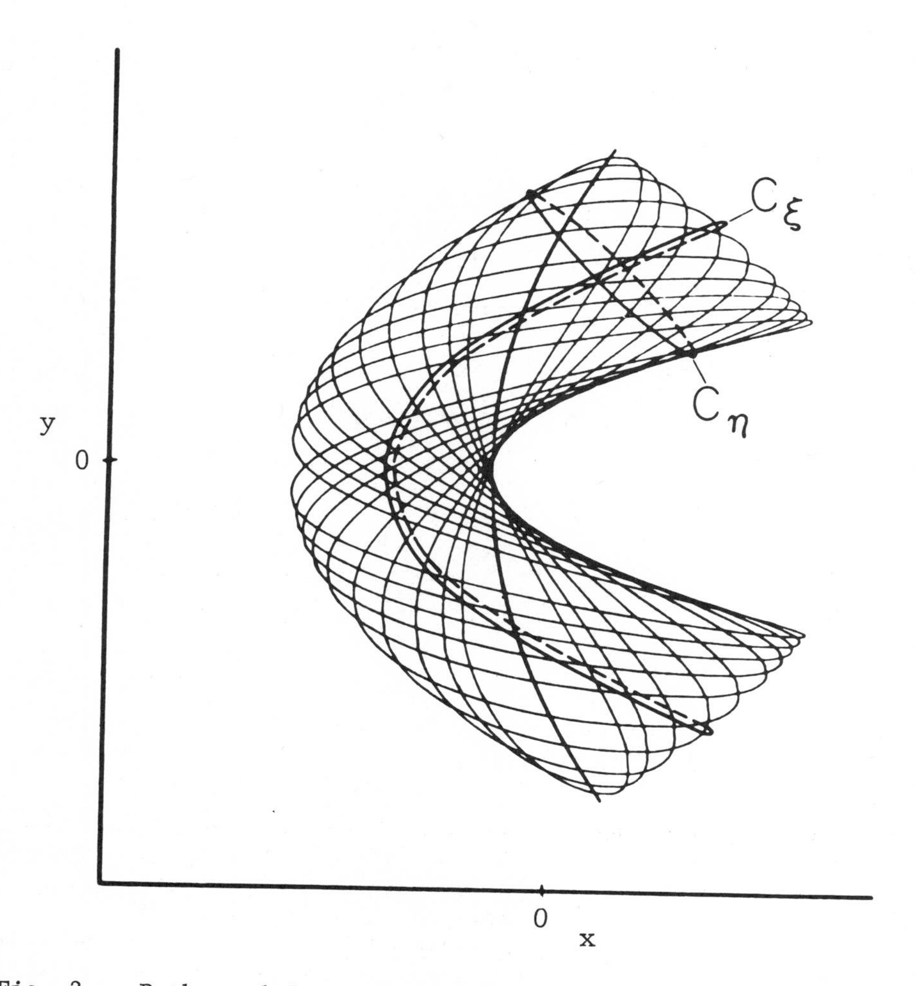

Fig. 2. Path used for quantization of the trajectory in Fig. 1.

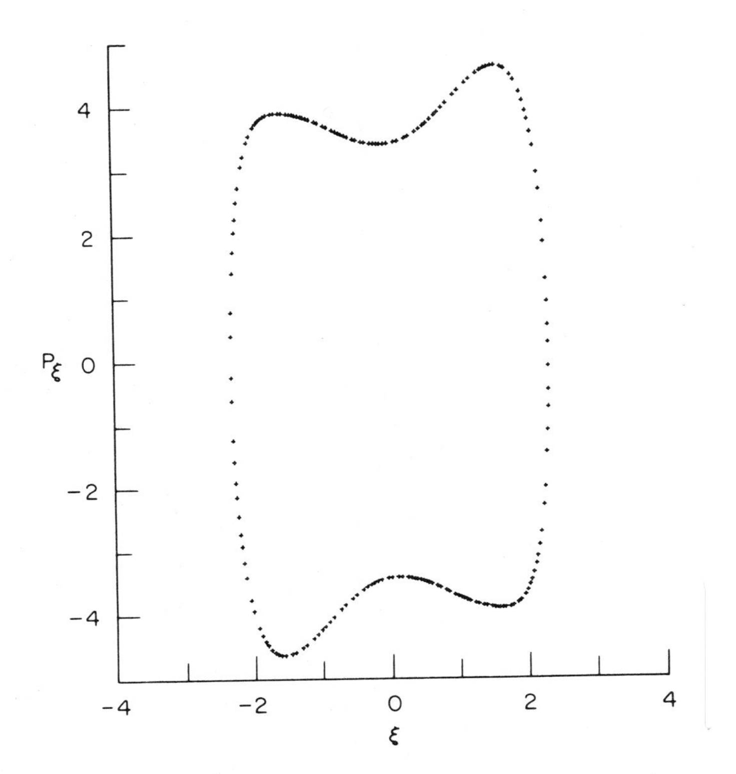

Fig. 3. An example of a P_ξ, ξ surface of section (η = constant, $P_\eta \geqslant 0$).

We found usually very good agreement between quantum and semiclassical eigenvalues. Since our method depends on having quasiperiodic motion, we were not able to calculate eigenvalues for all of the bound states for this Hamiltonian. For example, at $E = 6$ in our units, the trajectories became predominantly "chaotic", as shown in Fig. 4. An interesting feature of Hamiltonian (1) is that for $\beta = -2$, the Hamiltonian-Jacobi equation is separable in parabolic coordinates and hence cannot exhibit chaotic motion. A recently developed

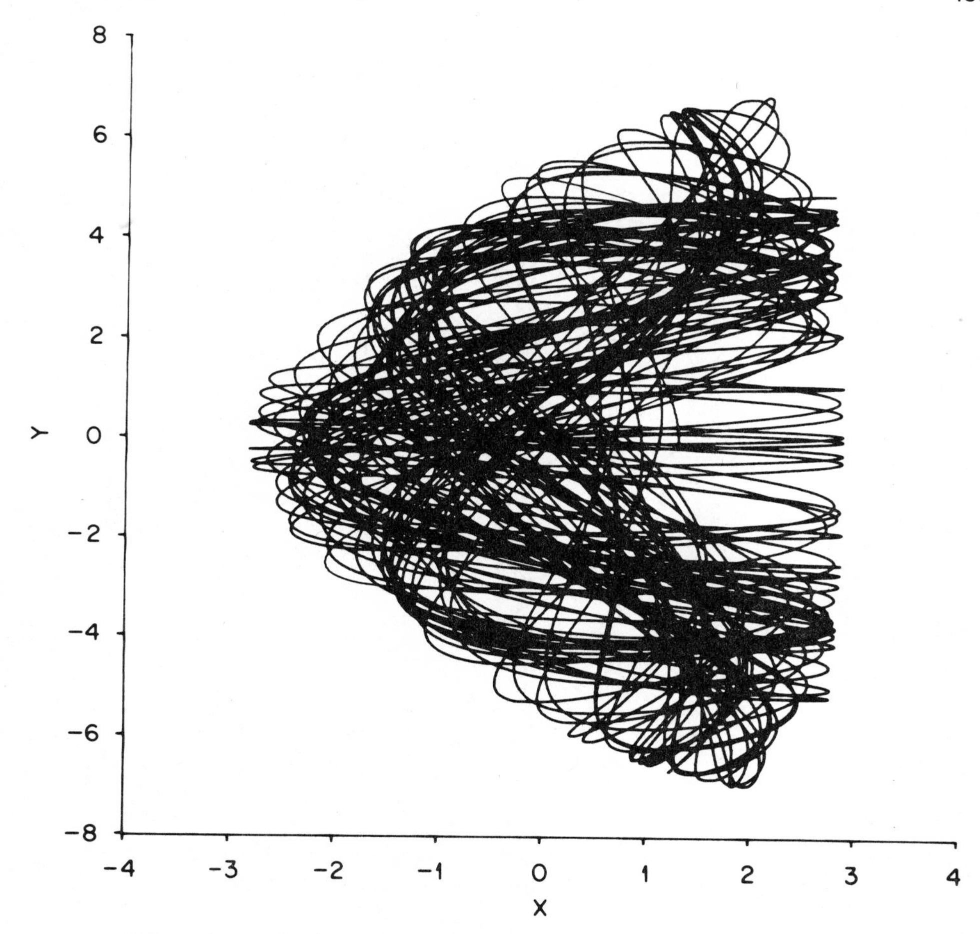

Fig. 4. A chaotic trajectory of Hamiltonian (I-1).

method[5] for the chaotic limit was applied to this model for the separable case, and it was found that for a large range of x, y initial conditions, the prediction is incorrect.[4e]

A method for calculating eigenvalues for systems with more degrees of freedom has also been developed.[4f] This method is an extension of the Poincaré surface of section method. Now, we use a surface of section in 2N-dimensional phase space, with a small but finite width in several coordinate directions. This surface of section with the appropriate path integrals is shown in Fig. 5. A surface of section calculated this way for the Hamiltonian

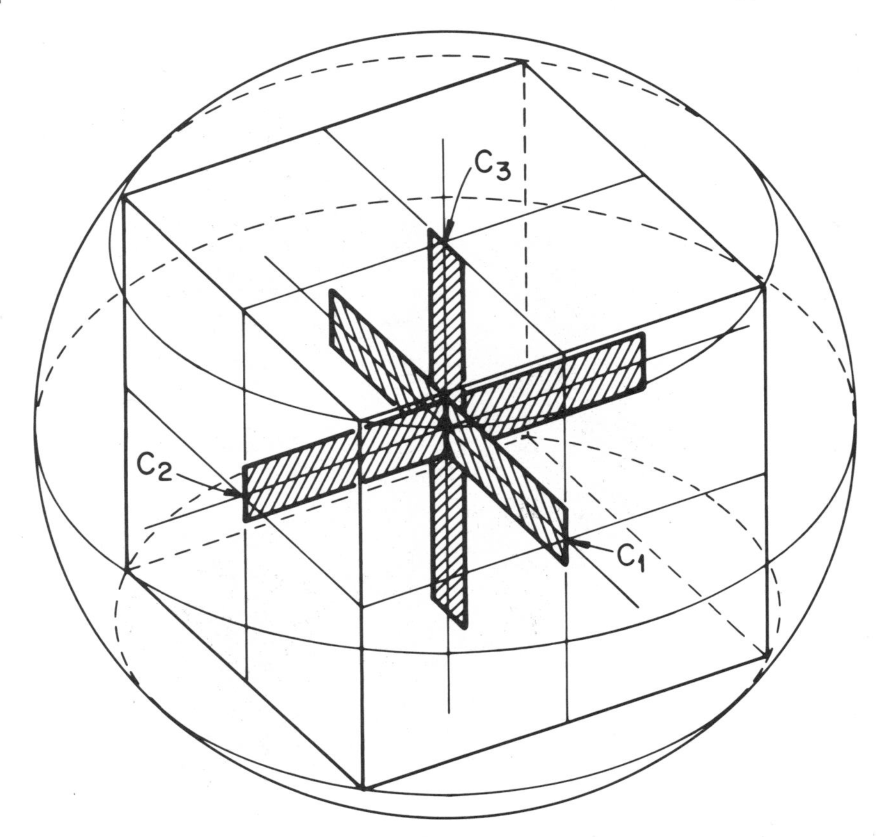

Fig. 5. Diagram showing the three paths for quantization in a system of three coupled oscillators (Hamiltonian (I-3)).

$$H = \frac{1}{2}(p_x^2 + p_y^2 + p_z^2 + \omega_x^2 x^2 + \omega_y^2 y^2 + \omega_z^2 z^2) + \lambda(xy^2 - \eta x^3) + \lambda(yz^2 - \eta y^3) \tag{I-3}$$

is shown in Fig. 6. Upon calculation of the three path integrals $J_x = \oint p_x dx$, $J_y = \oint p_y dy$, and $J_z = \oint p_z dz$, the values of J_i's can be interpolated for $J_i = 2\pi\ (n_i + \frac{1}{2})$, and the energy eigenvalues can be obtained. The agreement with the quantum mechanical eigenvalues was excellent.

In studies of unimolecular reaction rate theory, particularly for comparison with classical trajectory results, it is useful to have a method of calculating the number and density of states for

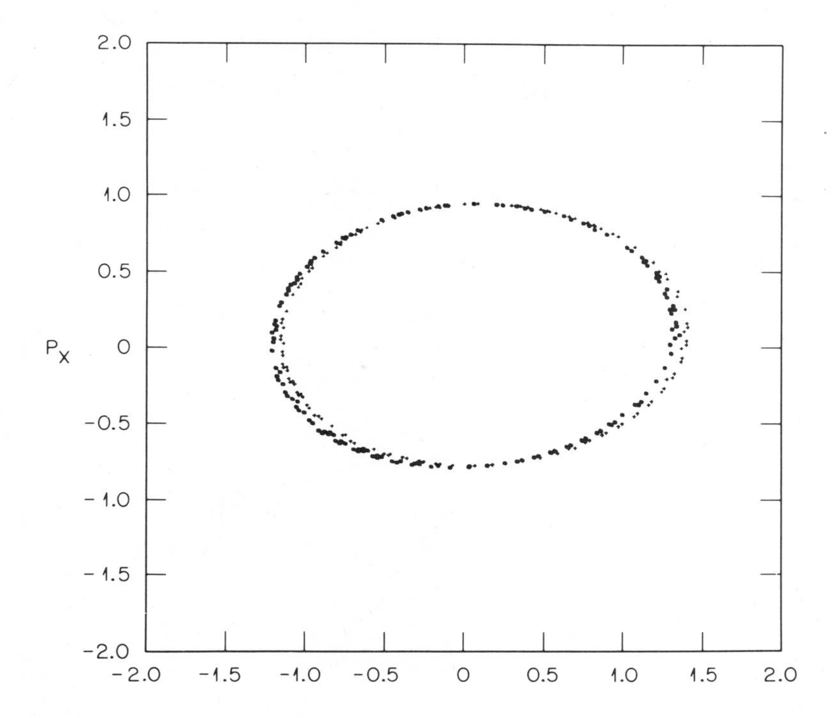

Fig. 6. A Px, x surface of section for Hamiltonian (I-3).

nonseparable systems.[6] The "classical" number of states $N_{c\ell}(E)$ for a system of n-degrees of freedom is simply given by

$$N_{c\ell}(E) = \frac{1}{(2\pi\hbar)^n} \int d\underset{\sim}{p} \int d\underset{\sim}{q}\, \theta(E-H(\underset{\sim}{p},\underset{\sim}{q})) \tag{I-4}$$

where p and q are the n-dimensional vectors of momentum and conjugate coordinate, respectively. θ is the (unit) step-function. (The density of states is just $dN_{c\ell}(E)/dE$.) The integral can be evaluated quite rapidly using efficient Monte Carlo procedures. In Fig. 7, the smooth $N_{c\ell}(E)$ is compared with the corresponding exact quantal result for the case of a model Hamiltonian with two degrees of freedom. The agreement is very good. An analogous problem (Sinai Billiard Problem) has been recently solved by Berry[1b] with similar agreement.

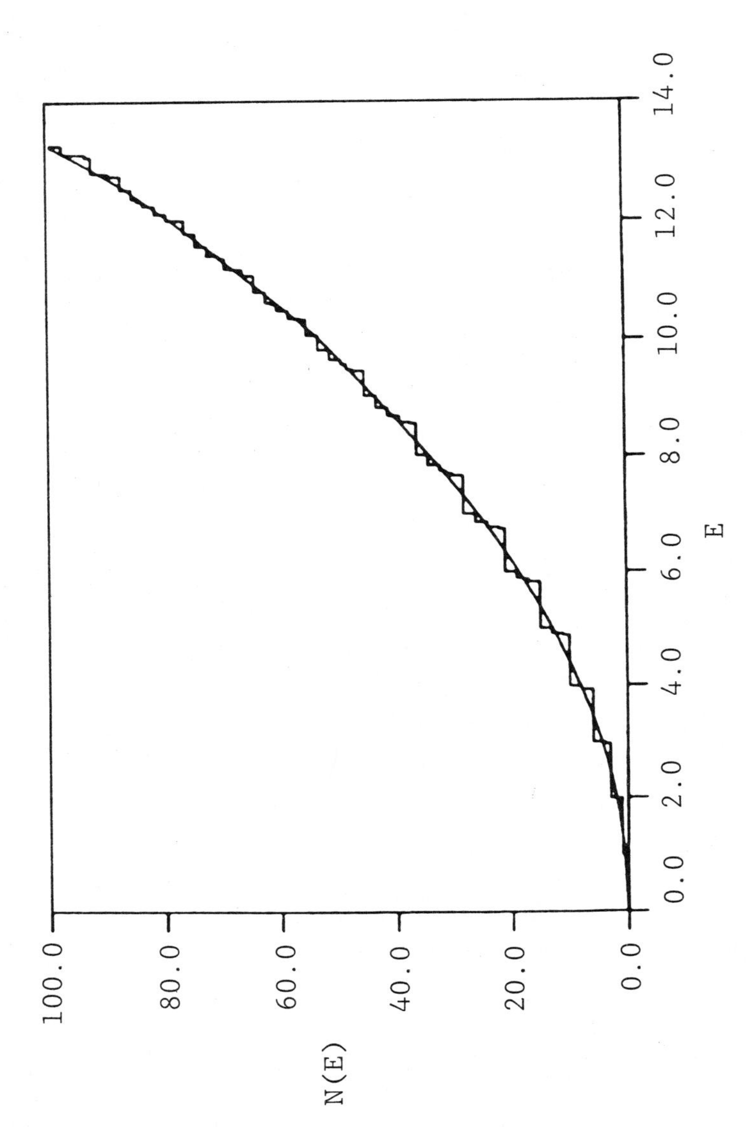

Fig. 7. A plot of the classical and quantum density of states for the Hénon-Heiles Hamiltonian.

II. SPECTRA AND CORRELATION FUNCTIONS

In the quasiperiodic regime, we have demonstrated[6,7,8] that the semiclassical (calculated from trajectories) and quantum (calculated from "exact" wavefunctions) power spectra are directly comparable in both frequency and intensity. More recently, this method of generating a spectrum from classical trajectories has also been used by Hansel[9] and Powell, et al.[10]

Briefly, we obtained the classical autocorrelation function, C(t), by averaging (e.g., the correlation eq. of the dipole) the appropriate ensemble. The classical spectrum, $I(\omega)$ is then given by the Fourier transform of the autocorrelation function

$$I(\omega) = \frac{1}{2\pi}\int_{-\infty}^{\infty} \langle \mu \cdot \mu(t) \rangle e^{i\omega t} dt \ . \tag{II-1}$$

Recently, another spectral method has been introduced by Heller.[11]

The justification of the use of classical trajectories to obtain semiclassical spectra will be restated now. The quantum power spectrum is

$$I(\omega) = \sum_k |\langle k|\mu|n\rangle|^2 \delta(\omega - \omega_{kn}) \tag{II-2}$$

where we have assumed only state $|n\rangle$ is initially populated and $\omega_{kn} = \frac{1}{\hbar}(E_k - E_n)$, or the transition frequency. Introducing the semiclassical wavefunction

$$\langle w|n\rangle = e^{2\pi i n w} \ , \tag{II-3}$$

the matrix element for the dipole operator for quasiperiodic motion becomes

$$\langle k|\mu|n\rangle = \int_0^1 e^{-2\pi i k w} \mu(\underset{\sim}{J},\underset{\sim}{w}) e^{2\pi i n w} dw \ . \tag{II-4}$$

Performing the operation gives

$$\langle k|\mu|n\rangle = \int_0^1 \mu(J_n, w) e^{2\pi i (n-k) w} dw \ , \tag{II-5}$$

which is the $(n-k)^{th}$ Fourier component of the dipole moment μ. Introducing eq. (II-5) into eq. (II-2) results in

$$I(\omega) = \sum_s |\mu_s(n)|^2 \ \delta(\omega - \omega_s(n)) \tag{II-6}$$

where $\omega_s(n)$ is the classical frequency component on the torus $(\tilde{=}\omega_{kn})$. μ_s is the Fourier coefficient for the dipole operator dependent on the action variable, J. Eq. (II-6) is exactly the result previously used.[7]

A typical quasiperiodic spectrum is shown in Fig. 8 and comparison of semiclassical and quantum frequencies for various states is given in Table 2 of Ref. 7 for a three-oscillator system. A comparison of both frequency and intensity is shown in Fig. 9 for a single Morse oscillator.[8] In Fig. 9, it is clear that not only the frequencies and intensities of the "allowed" transitions are predicted, but also those of the overtones are correctly predicted with appropriate interpolation.[8]

At higher energies, the motion becomes chaotic and the spectrum is composed of a broad distribution of sharp peaks.[6,7] A microcanonically averaged spectrum forms a broad envelope of the single trajectory spectrum and is shown in Fig. 10. There have not been definitive comparisons of semiclassical and quantal stochastic spectra,[6] and we are currently studying the comparison.

In contrast to the above correlation functions, the mode energy correlation function (i.e., $\langle E_n E_n(t) \rangle$ where E is the energy in the nth normal mode) is important not to molecular spectra, but to reaction rate theory.[12] Some chemically interesting experiments involve the excitation of a given bond or mode, for example in infrared multiphoton absorption, with subsequent decay of the excitation energy into other modes.

We have numerically evaluated several different correlation functions using microcanonical averages.[12] As expected, the correlation function in the quasiperiodic regime is oscillatory and is shown in Fig. 11.

In the chaotic regime, it has been previously postulated[13] that the correlation function will behave exponentially. The correlation function in the chaotic regime has, in addition to the usual property present in all regimes given by

$$c(o) = \langle a \cdot a(o) \rangle = \langle a^2 \rangle \ , \qquad \text{(II-7)}$$

also the property that

$$c(\infty) = \langle a \cdot a \rangle = \langle a \rangle \langle a(\infty) \rangle = \langle a \rangle^2 \ . \qquad \text{(II-8)}$$

In general, $\langle a^2 \rangle \neq \langle a \rangle^2$ and c(t) will decay to some value. From the present system, we expect[12] $\frac{c(\infty)}{c(o)} \sim 3/4$ in the chaotic regime, which is consistent with the results shown in Fig. 12. We have modeled the

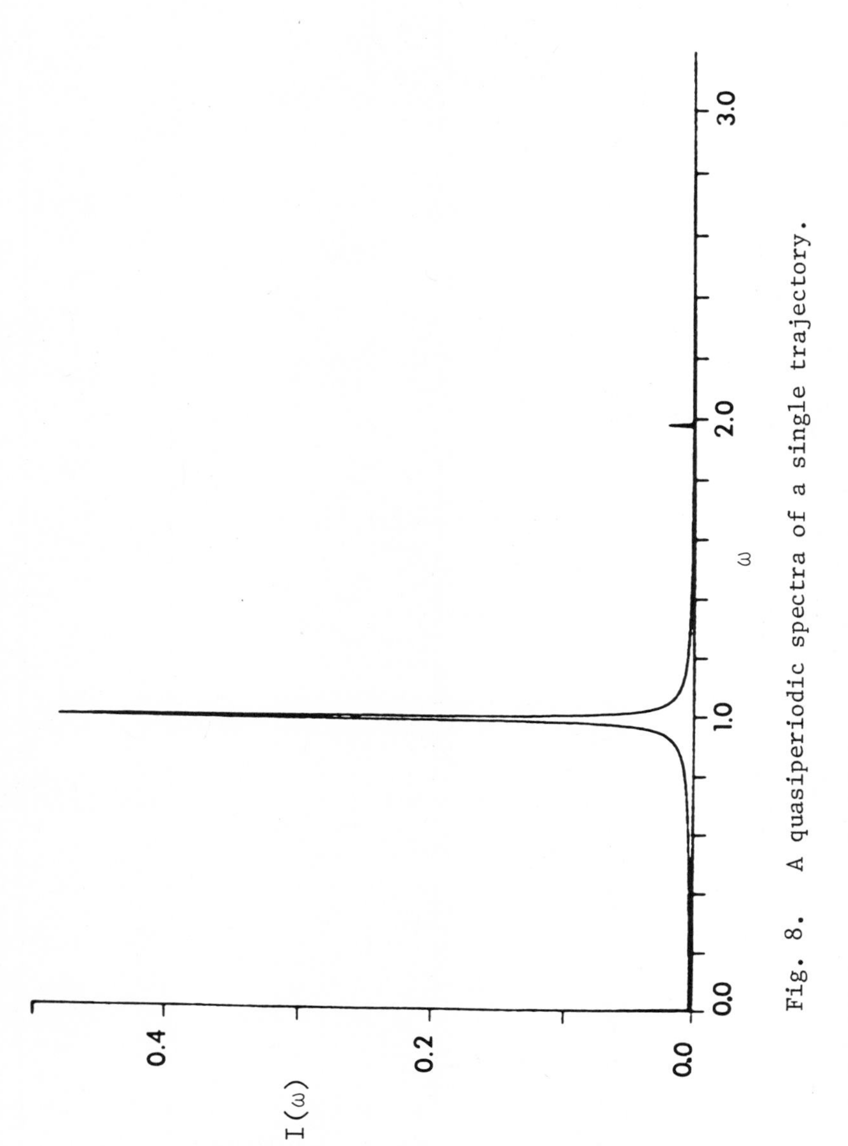

Fig. 8. A quasiperiodic spectra of a single trajectory.

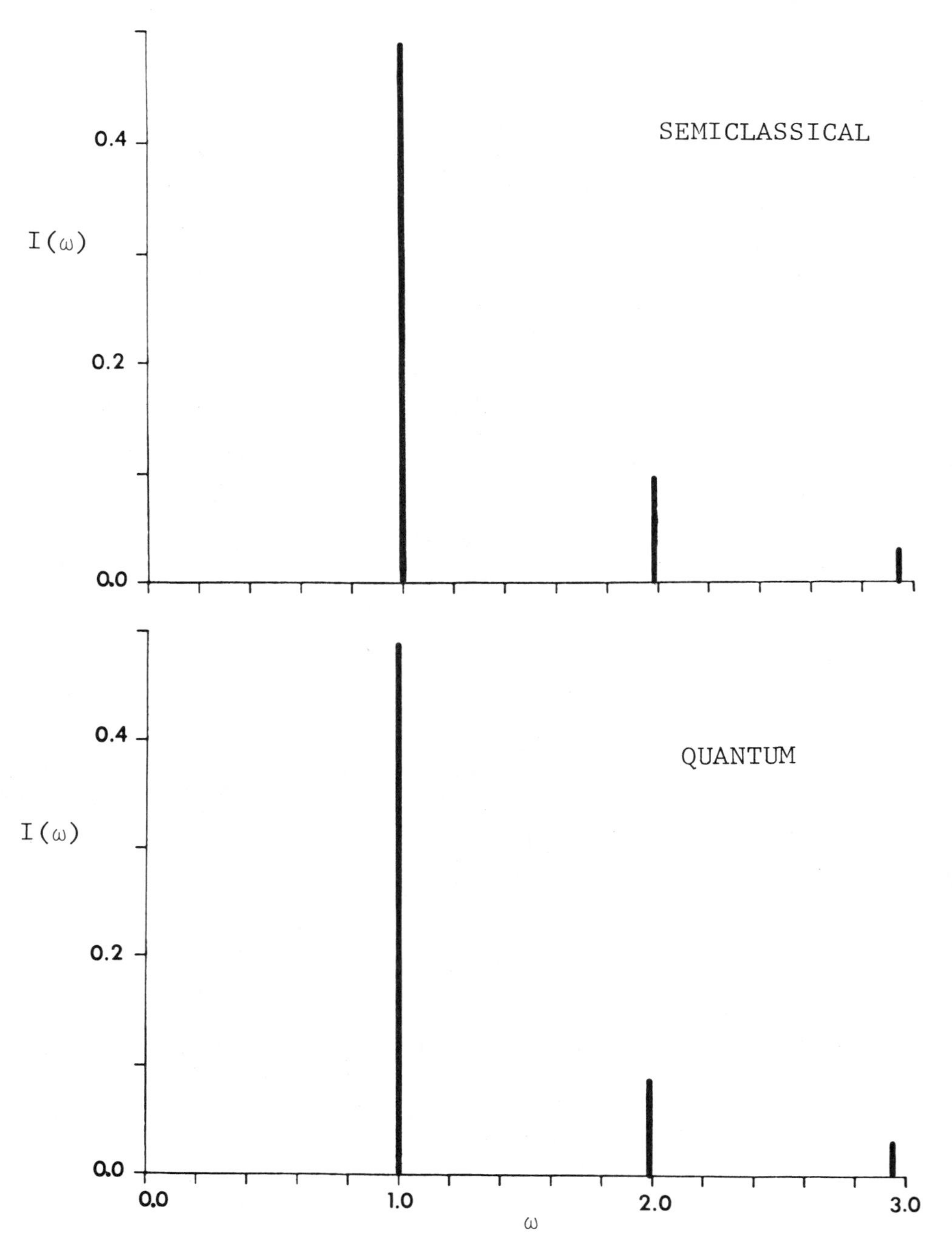

Fig. 9. Quantum and semiclassical spectrum for a Morse oscillator.

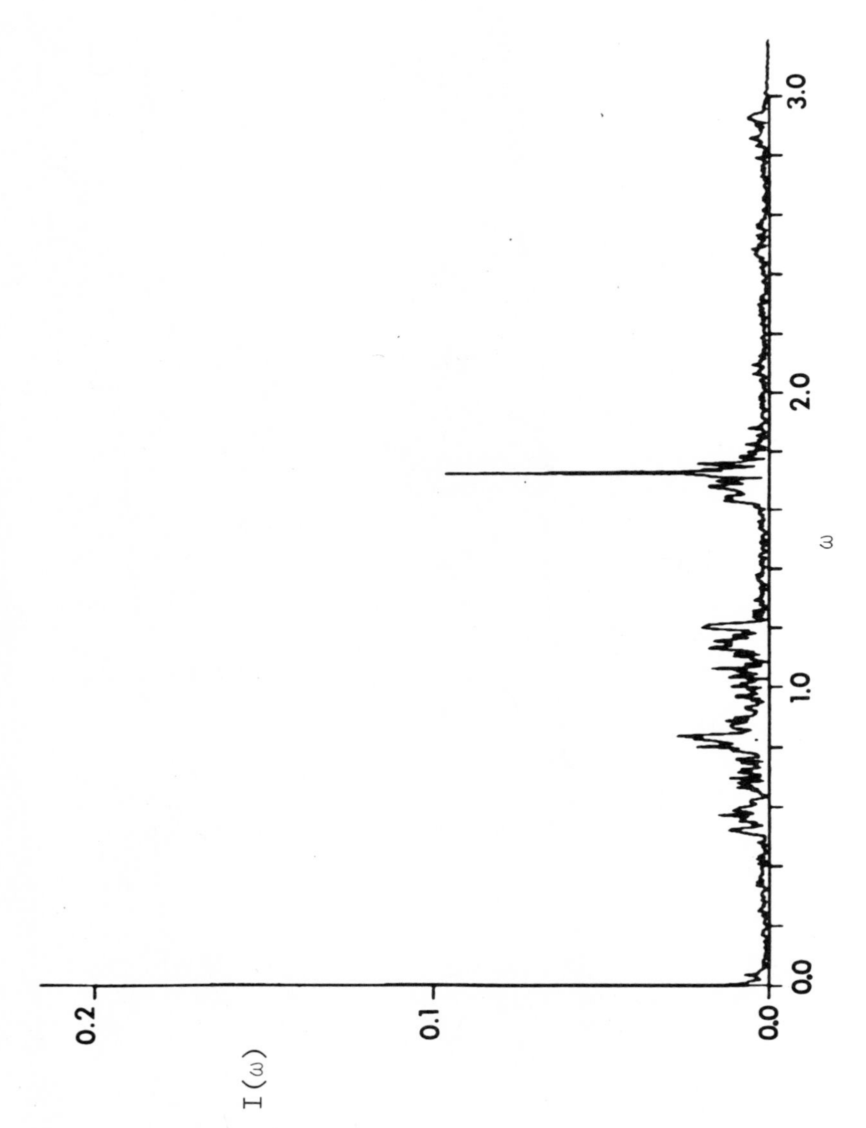

Fig. 10. As in Fig. 8, but for the chaotic regime.

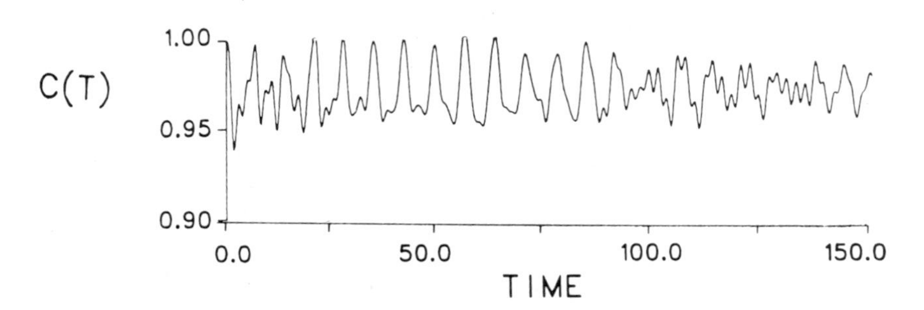

Fig. 11. Correlation function for quasiperiodic motion.

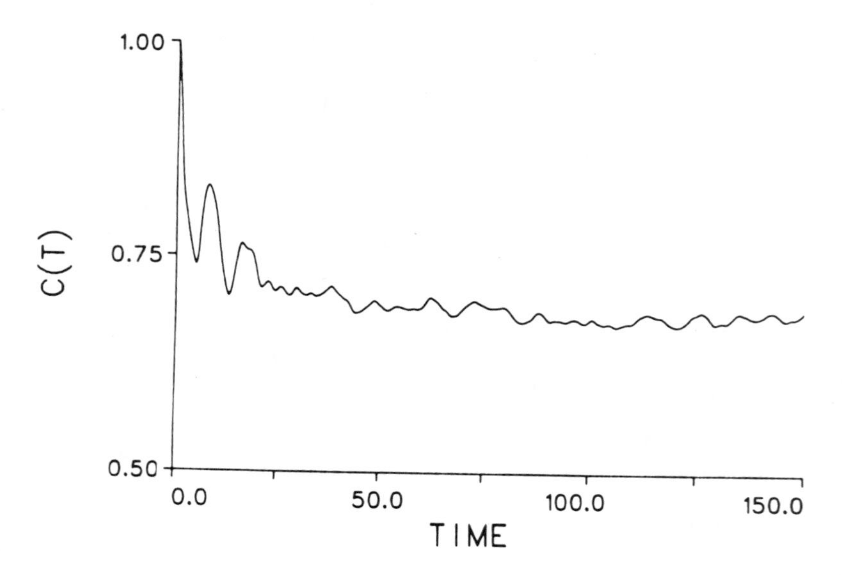

Fig. 12. Correlation function for chaotic motion.

memory functions of Figs. 11 and 12, with the result in the chaotic regime being

$$c(t) = a + (b \cos \omega_1 t + c)e^{-dt} + f \cos \omega_2 t , \quad \text{(II-9)}$$

and in the quasiperiodic regime

$$c(t) = a' + b' \cos \omega_1 t + f' \cos \omega_2 t , \quad \text{(II-10)}$$

with a corresponding result in the memory function.[12] The correlation functions of molecular systems are currently under investigation.[14]

III. CHAOTIC BEHAVIOR IN QUANTUM MECHANICS

The classical motion of nonlinearly coupled oscillators tends to be predominantly "quasiperiodic", or "regular", at low energies, the trajectories being confined to N-dimensional tori in phase space. At higher energies, the motion can become predominantly "chaotic" or "irregular",[15] displaying great sensitivity to small changes in inaccessible phase space. In a study of the quantum mechanical wavefunctions of a system with (2:1) Fermi resonance, we noted an analogous behavior; namely, in the quasiperiodic regime, the wavefunctions tended to be localized in well-defined regions of configuration space. On the other hand, in the chaotic regime, the wavefunctions spread over most of the allowed configuration space. The wavefunction for the trajectory in Fig. 1 is shown in Fig. 13.[4e,4g] The chaotic type wavefunction for the trajectory in Fig. 4 is shown in Fig. 14.[4e,4g]

Another well-known suggestion of quantum chaotic behavior is from Percival[16] and Pomphrey[17] and is related to the second differences. Second differences of eigenvalues were calculated from

$$\Delta^2 E_i = E_i(\lambda + \delta\lambda) - 2E(\lambda) + E_i(\lambda - \delta\lambda) \quad \text{(III-1)}$$

using a $\delta\lambda = 0.001$. The $\Delta^2 E_i$'s are divided by E_i for normalization. The results are plotted in Fig. 15. Apart from two distinctly large values of $\Delta^2 E_i$'s not in the figure and discussed below, the second differences seem to form distinct families. The family of smaller $\Delta^2 E_i$'s tended to belong to those states with high ℓ quantum numbers (internal angular momentum like quantum numbers[4d]), whereas the family with the larger $\Delta^2 E_i$'s belong to states with low ℓ. In this study, we found that only the cases of avoided crossing or crossing of eigenvalues in the $E(\lambda)$ vs. λ plots did extremely large second differences result, and so avoided crossings provide a mechanism for the origin of such large second differences.[6]

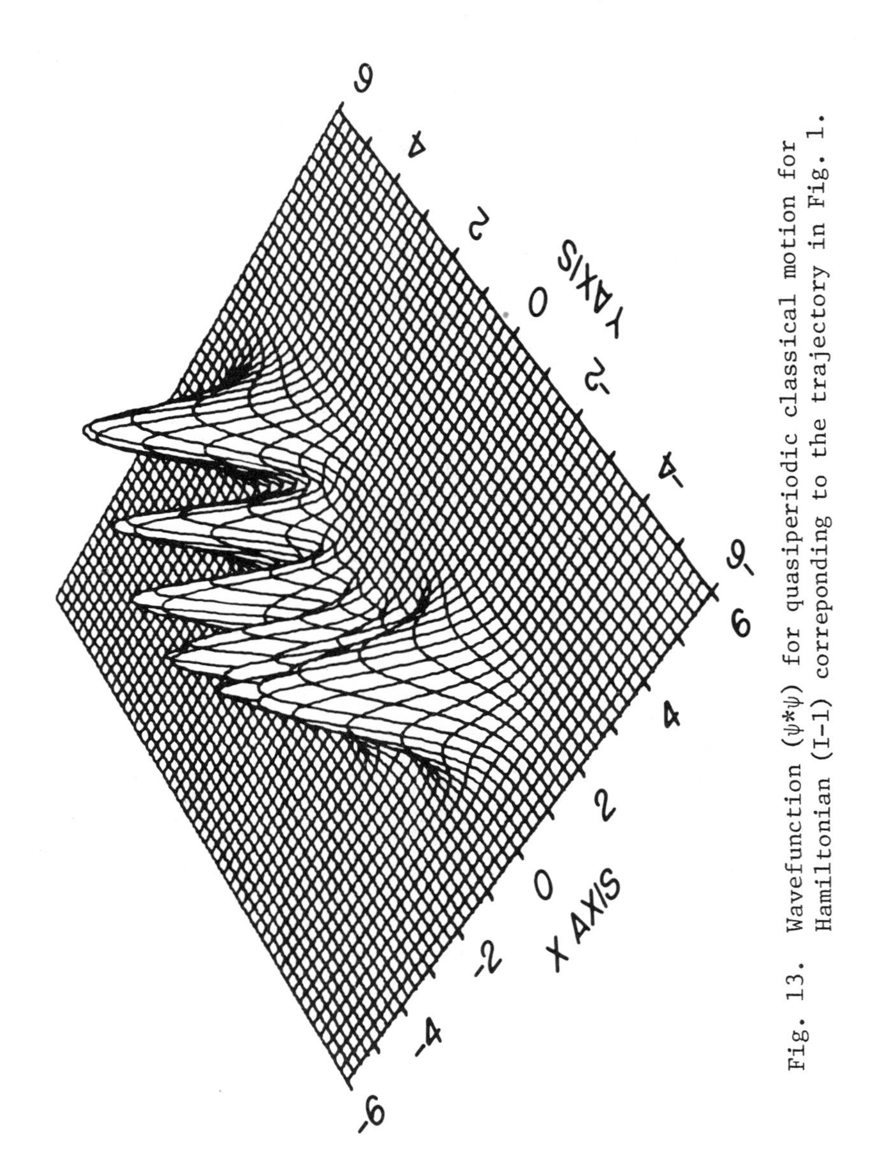

Fig. 13. Wavefunction $(\psi^*\psi)$ for quasiperiodic classical motion for Hamiltonian (I-1) correponding to the trajectory in Fig. 1.

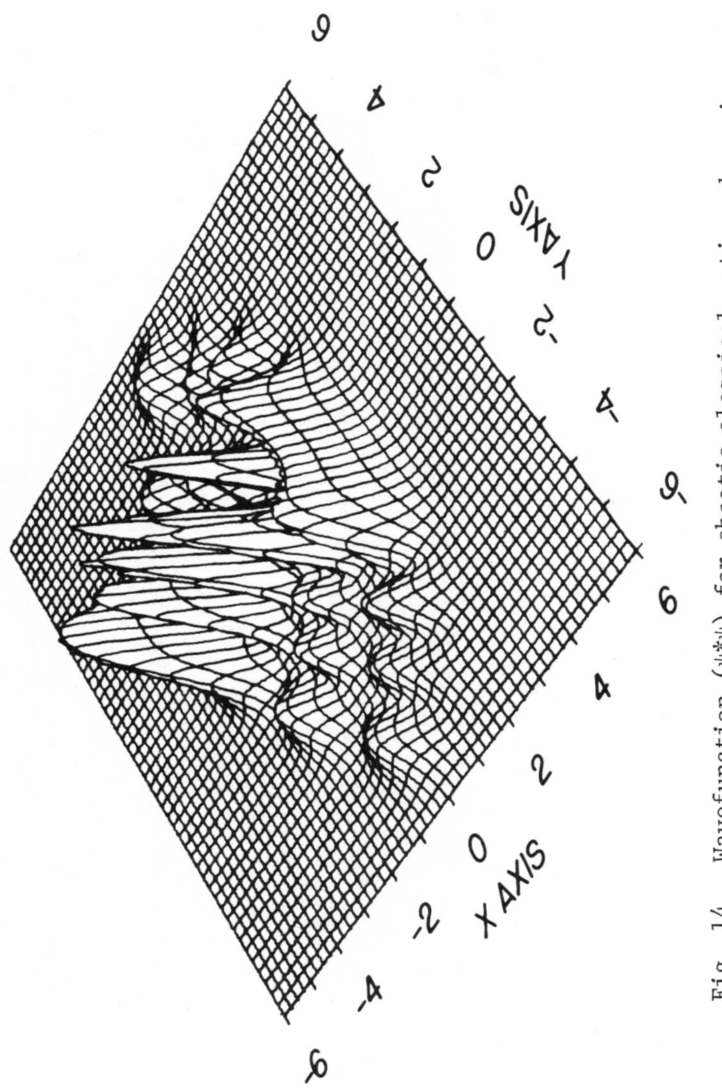

Fig. 14. Wavefunction ($\psi^*\psi$) for chaotic classical motion shown in Fig. 4 for Hamiltonian (I-1).

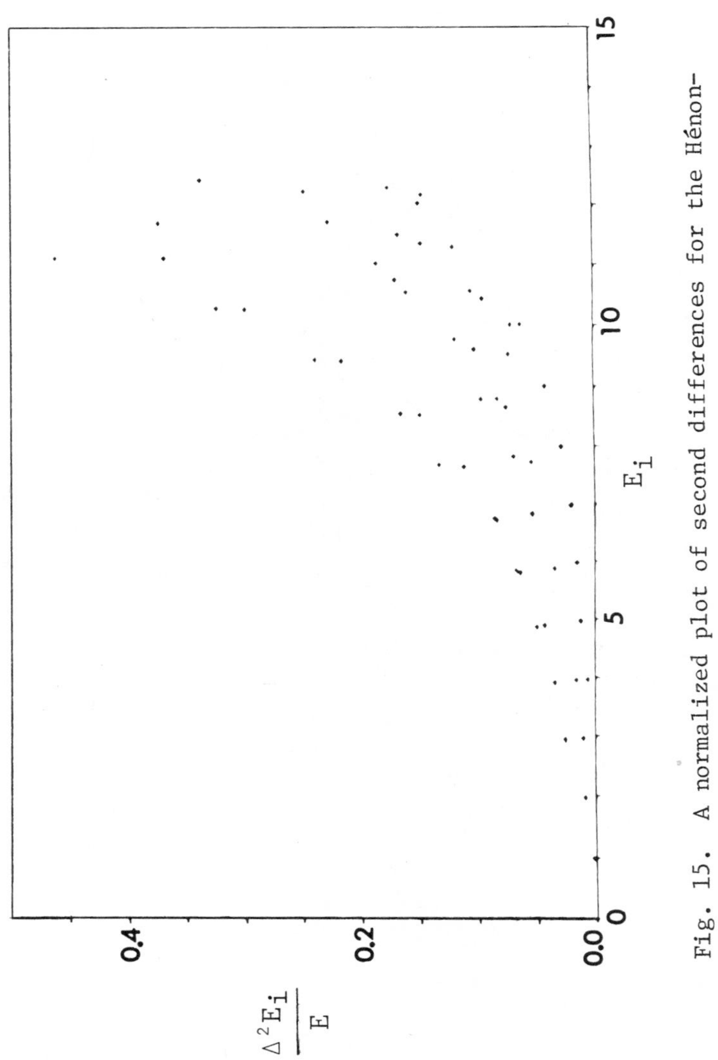

Fig. 15. A normalized plot of second differences for the Hénon-Heiles Hamiltonian.

In a recent manuscript,[18] we illustrated the avoided crossing behavior for a system of three coupled anharmonic oscillators with a Hamiltonian

$$H = \frac{1}{2}(p_x^{\,2} + p_y^{\,2} + p_z^{\,2} + \omega_x^{\,2}x^2 + \omega_y^{\,2}y^2 + \omega_z^{\,2}z^2)$$

$$+ K_{122}xy^2 + K_{133}xz^2 + K_{233}yz^2 + K_{111}x^3 + K_{222}y^3 \qquad \text{(III-2)}$$

$$+ K_{1122}x^2y^2 + K_{2233}y^2z^2 \; .$$

In Figs. 16 and 17, the eigenvalues E are plotted versus K_{122} for two ranges of energy. In the lower range, Fig. 16, no avoided crossings are observed. Above a certain energy, there are an increasing number of overlapping avoided crossings, as in Fig. 17. These figures were drawn through points at $-K_{122} = 0.10, 0.11, \ldots, 0.16$, and hidden symmetries were assumed absent. In that case, there are no curve crossings, since curve crossings would then correspond to a conical intersection in an energy versus multidimensional parameter space, and such crossings would be of "measure zero". Overlapping avoided crossings, we have suggested,[19] will produce a statistical behavior in the wavefunctions for the quantum mechanically chaotic regime.

Finally, we would like to comment on the relation of the idea of avoided crossings to some other work in the literature related to quantum chaotic behavior, and also to provide preliminary evidence for the expectation that as $h \to 0$, overlapping avoided crossings correspond to classical chaos. We have seen that avoided crossings produce large second differences in plots of eigenvalues versus perturbation parameter. Pomphrey studied second differences for the Hénon-Heiles potential, using a value of h smaller than ours, and found many more cases of large second differences. In quantum mechanical perturbation theory calculation,[20a] we also observed many more "zeroth order crossings" (which yield avoided crossings upon use of degenerate perturbation theory near the crossing) for Pomphrey's h than for ours. Similarly, in classical perturbation theory calculations for the Barbanis potential, we have found, using a grid of action variables, an onset of "zeroth order crossings" roughly at an energy where classical "chaos" begins.[20b]

Stratt, Handy and Miller[21] in their investigation of the Barbanis potential studied nodal patterns of quantum mechanical wavefunctions and found for some states major changes of nodal patterns. We believe, and are currently testing this possibility, that these changes are each associated with avoided crossings. In this study, it will be interesting to see whether the latter are isolated avoided crossings or overlapping ones. (Only the latter may prove to be associated with "chaotic" behavior.[19])

Nordholm and Rice[22] studied the wavefunctions ψ for the Hénon-Heiles potential and examined their projections on those of the unperturbed (harmonic oscillator) basis set. When a ψ was substantially distributed over a number of the latter states, it was termed "global" and was termed local otherwise. So defined, a state can appear to be global not only when its wavefunction is "statistical" (sometimes labeled 'stochastic', 'ergodic', 'chaotic'), but also when its shape is considerably distorted from the corresponding one of the unperturbed system. Indeed we were actually able to find classically the invariant tori, each of which corresponded individually semiclassically to many of the quantum states that had been termed "global", so that all of these particular global states certainly are of the second case, i.e. nonchaotic. Even though globality is, therefore, not necessarily, or even primarily, due to chaotic behavior, it is very interesting in its own right. It provides information on the nature of the wavefunction relative to that of some unperturbed system. Because of the basis set dependence of global vs. local states, similar calculations were made using natural orbitals as the basis set.[20] The comparison is still, to be sure, a coordinate dependent one.

IV. APPLICATIONS OF QUASIPERIODIC AND CHAOTIC BEHAVIOR

Scattering Processes

Scattering resonances receive much attention in the quantum dynamics literature and can contribute significantly to the total cross section of a scattering process. They are also an important source of disagreement between quasiclassical and fully quantum mechanical scattering calculations. A standard method of locating these resonances has been quantum mechanical scattering calculations.[23] However, when the resonances are very narrow (as in the following system reviewed in this paper), a quantum calculation faces a time-consuming and difficult task in locating each one. Alternatively, as the dimensionality of the problem increases, the quantum mechanical solution becomes correspondingly more difficult. Motivated by this difficulty, the variational approach (stabilization method)[24] was used by Eastes and Marcus[25] to locate resonances. They compared this variational calculation to an exact quantum mechanical calculation.

The compound state resonance studied arose from the coupling of a Morse interaction potential, which monotonically goes to the dissociation limit from the bottom of the well, to a harmonic oscillator A new semiclassical location of resonance energies[26] was recently made, which utilizes the Poincaré surface of section quantization of

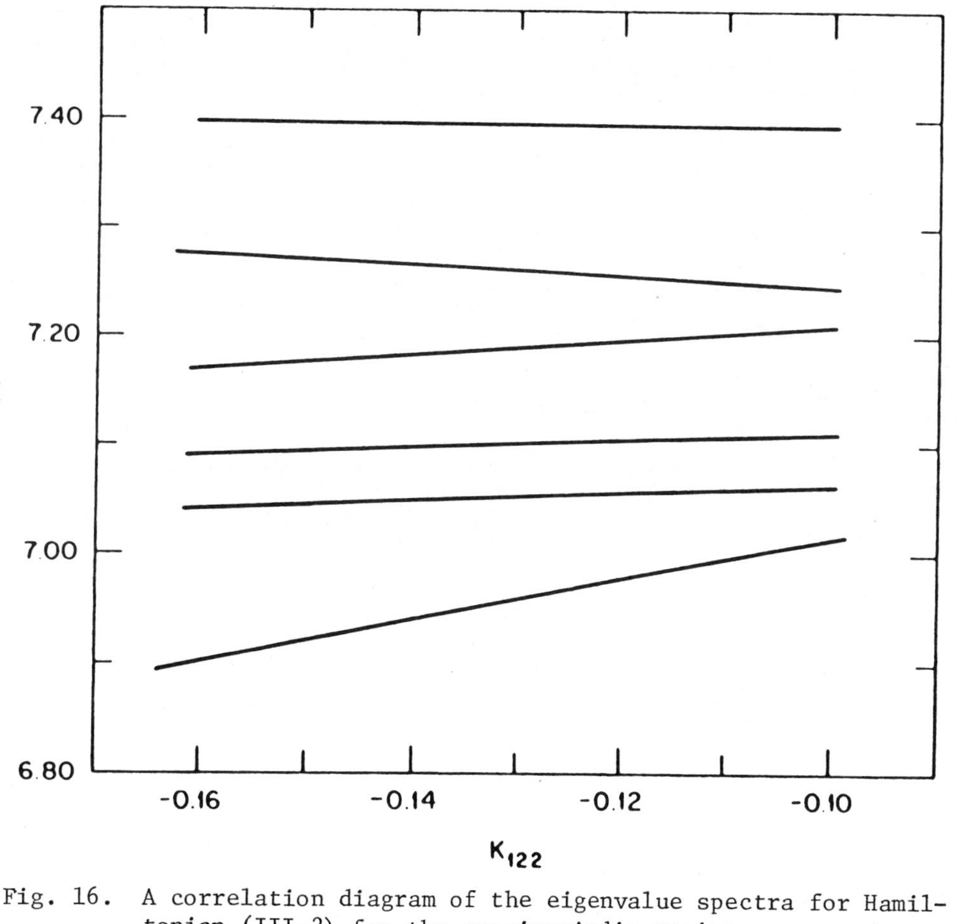

Fig. 16. A correlation diagram of the eigenvalue spectra for Hamiltonian (III-2) for the quasiperiodic regime.

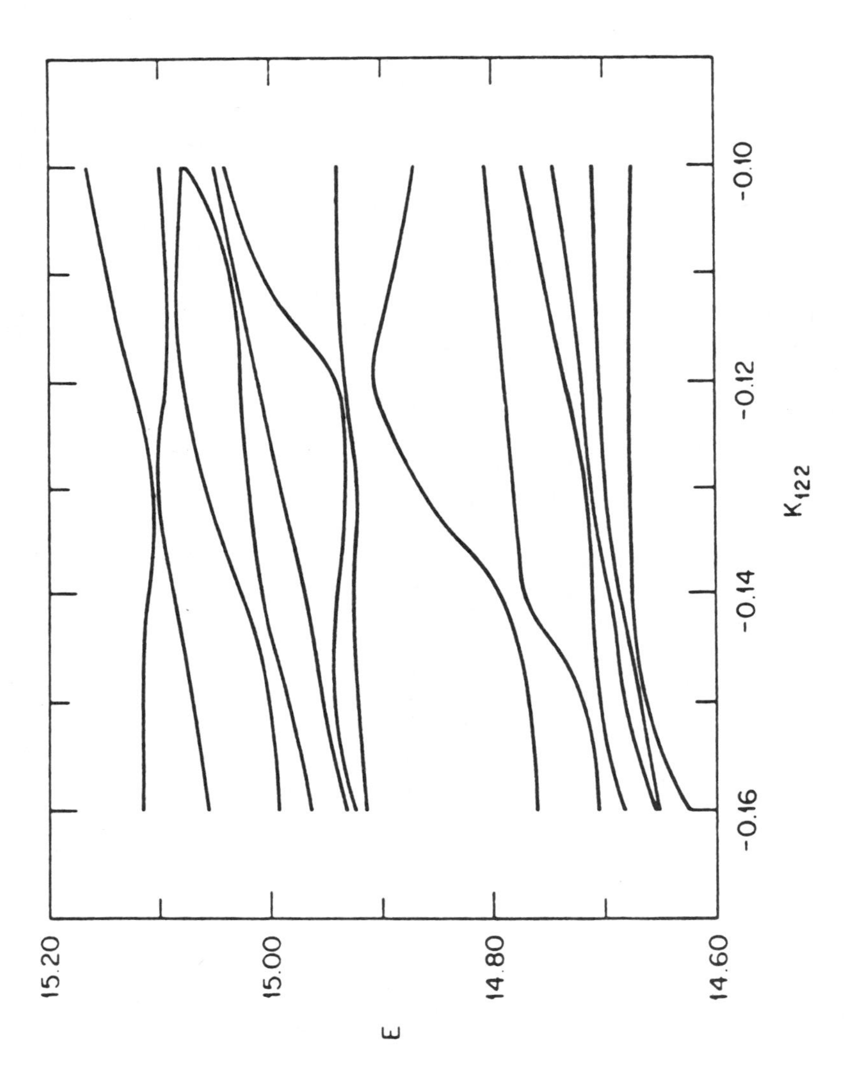

Fig. 17. As in Fig. 16, but for the chaotic regime.

the classical action integral. The eigentrajectories (trajectories with the correctly quantized actions) then correspond to quantum quasibound states; therefore, their energies correspond to the resonance energies.

The model Hamiltonian used was

$$H = \frac{p_x{}^2}{m} + p_y{}^2 + y^2 + De^{-\alpha(x-y)} - 2De^{-\alpha(x-y)/2} \qquad \text{(IV-1)}$$

which corresponds to a collision between a harmonic oscillator and a particle with a Morse interaction potential, where D is the depth, m is a mass ratio, and α is the steepness of the potential between the colliding partners. The parameters are the same as those in Ref. 25.

The Hamiltonian (1) will support no bound states along the exit channel (essentially the x coordinate) for energies greater than D, the dissociation limit of the Morse oscillator. However, even at such energies, some classically bound states of the entire system can be located, because of their quasiperiodic motion. An example is the trajectory in Fig. 18. They were then quantized using the Poincaré surface of section semiclassical method described earlier. Excellent agreement for the positions of the resonances was found between the semiclassical and full quantum results. Where no scattering resonances were found in Ref. 25, the trajectories did not produce quasiperiodic surfaces of section, but instead drifted, leading to escape of the particle. Such a chaotic type of unbound trajectory is shown in Fig. 19.

The existence of these two radically different types of motion (quasiperiodic and chaotic) led us to believe that there would be a change in the behavior of the inelastic cross section when the internal vibrational motion of one of the scattering partners reached the stochastic regime. One example with no observed effect has been reported.[27] We have also investigated this problem with a model consisting of a vibrator with two degrees of freedom undergoing a collinear collision with an atom.[28] The Hamiltonian for a system which exhibits more delocalized motion in the chaotic regime than that of Ref. 27 is

$$H = H_{vib} + H_p \qquad \text{(IV-2a)}$$

where

$$H_{vib} = \frac{p_x{}^2}{m} + p_y{}^2 + y^2 + D\left(\exp - 2\gamma(x-y) -2 \exp -\gamma(x-y)\right) \qquad \text{(IV-2b)}$$

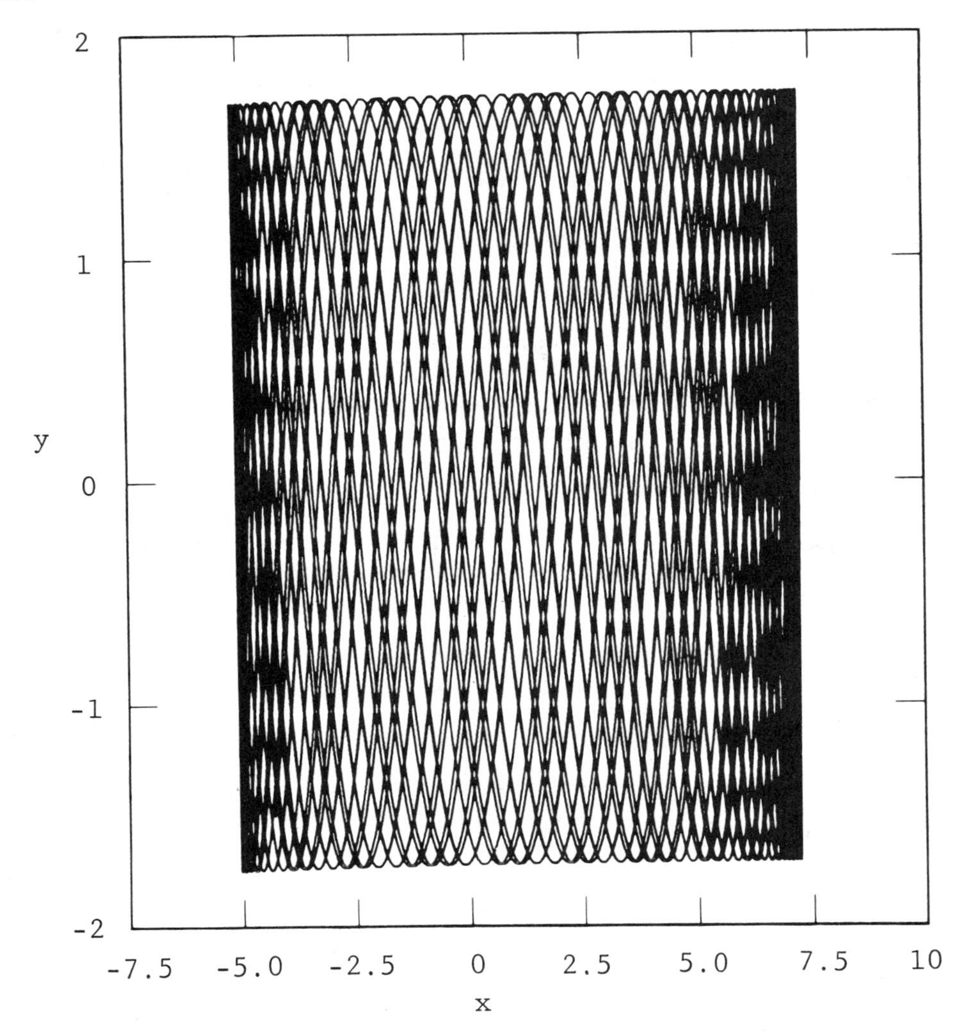

Fig. 18. A quasibound quasiperiodic trajectory for Hamiltonian (IV-1).

and

$$H_p = p_z^2 + 4\varepsilon\left(\left(\sigma/(z-x)\right)^{12} - \left(\sigma/(z-x)\right)^6\right) \tag{IV-2c}$$

with $D = 10.0$, $\gamma = 0.25$, $\varepsilon = 1.0$, and $\sigma = 1.0$. For large z, this Hamiltonian reduces to that of a free particle and a harmonic oscillator coupled to a Morse oscillator.

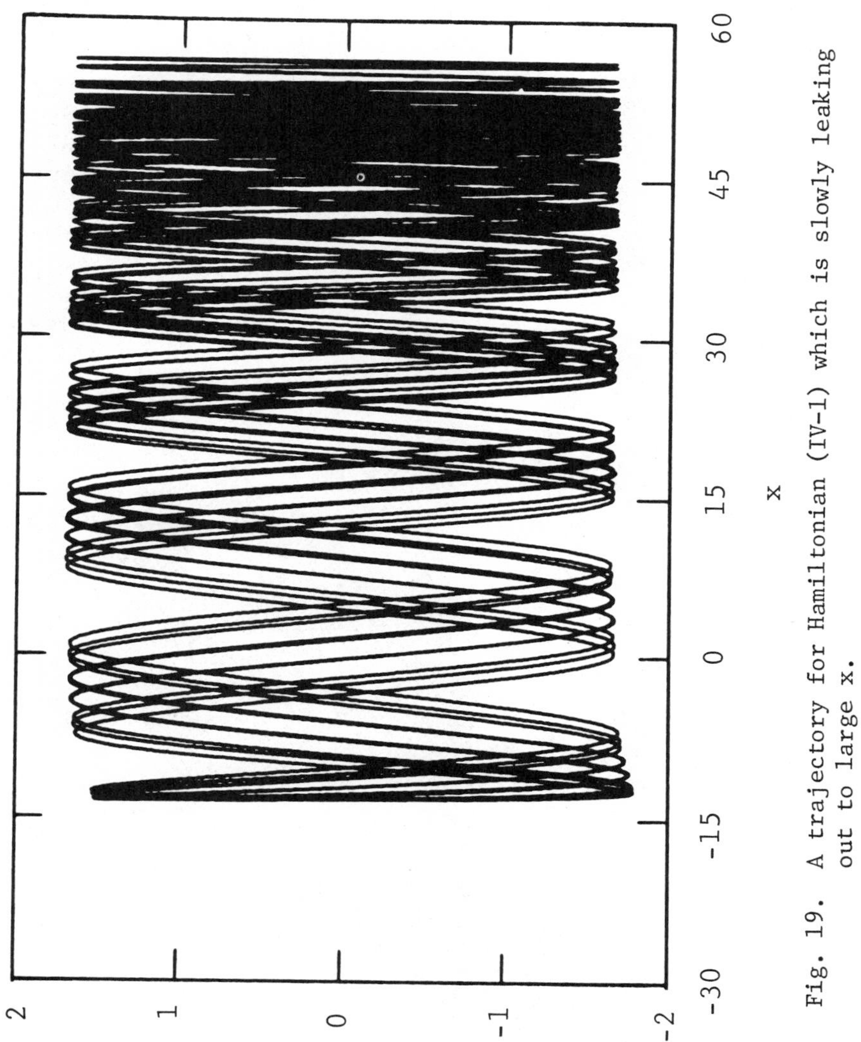

Fig. 19. A trajectory for Hamiltonian (IV-1) which is slowly leaking out to large x.

A trajectory for the molecular part of Hamiltonian (IV-2b), when the atom and molecule are far apart, is shown in Fig. 20 for the quasiperiodic regime and in the chaotic regime in Fig. 21. In order to investigate the scattering process, we calculated the average of the square of the energy transfer. The translational and "x-mode" vibrational energies were kept constant, and the average was calculated with 1000 trajectories over the vibrational phases. Fig. 22 shows the results as a function of total vibrational energy of the molecule E_M at two different translational energies. Fig. 22 shows that when the vibrational motion changes to chaotic (above $E \geqslant -2.5$), there is a difference in the behavior of the energy transfer probability. More detailed studies of this phenomenon are currently underway.

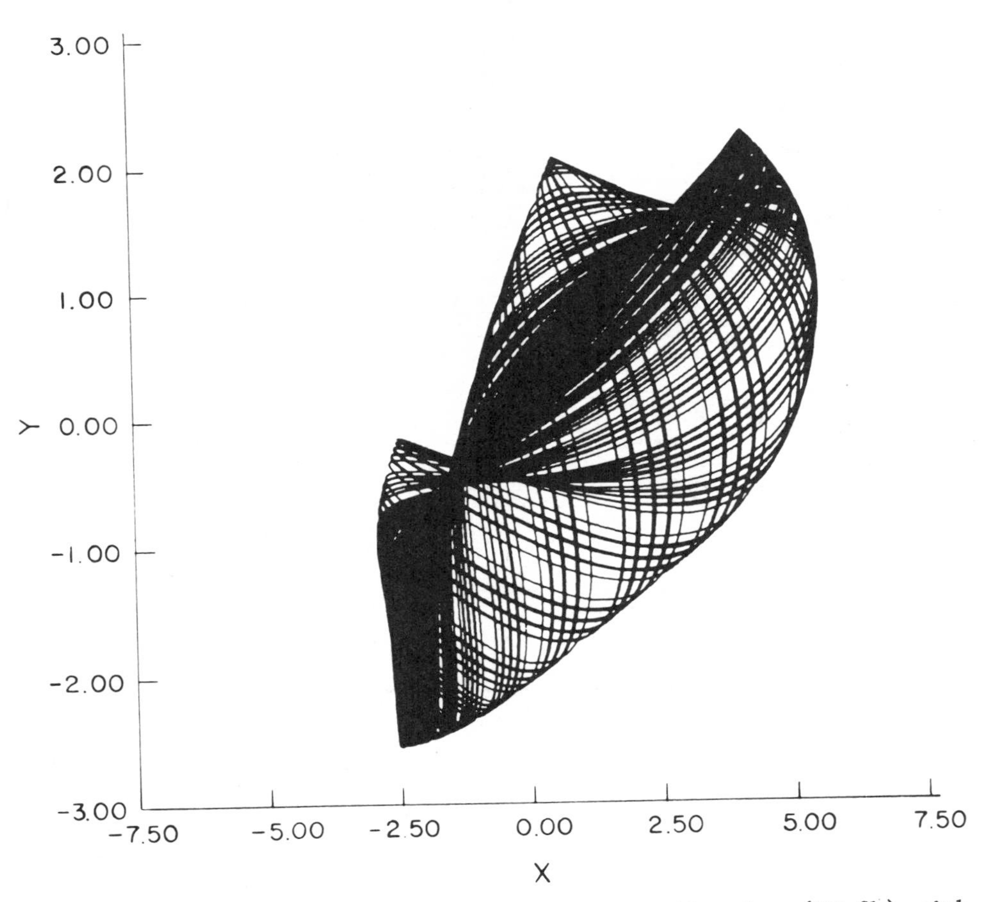

Fig. 20. Quasiperiodic trajectory for Hamiltonian (IV-2b) with $E = -3.5$.

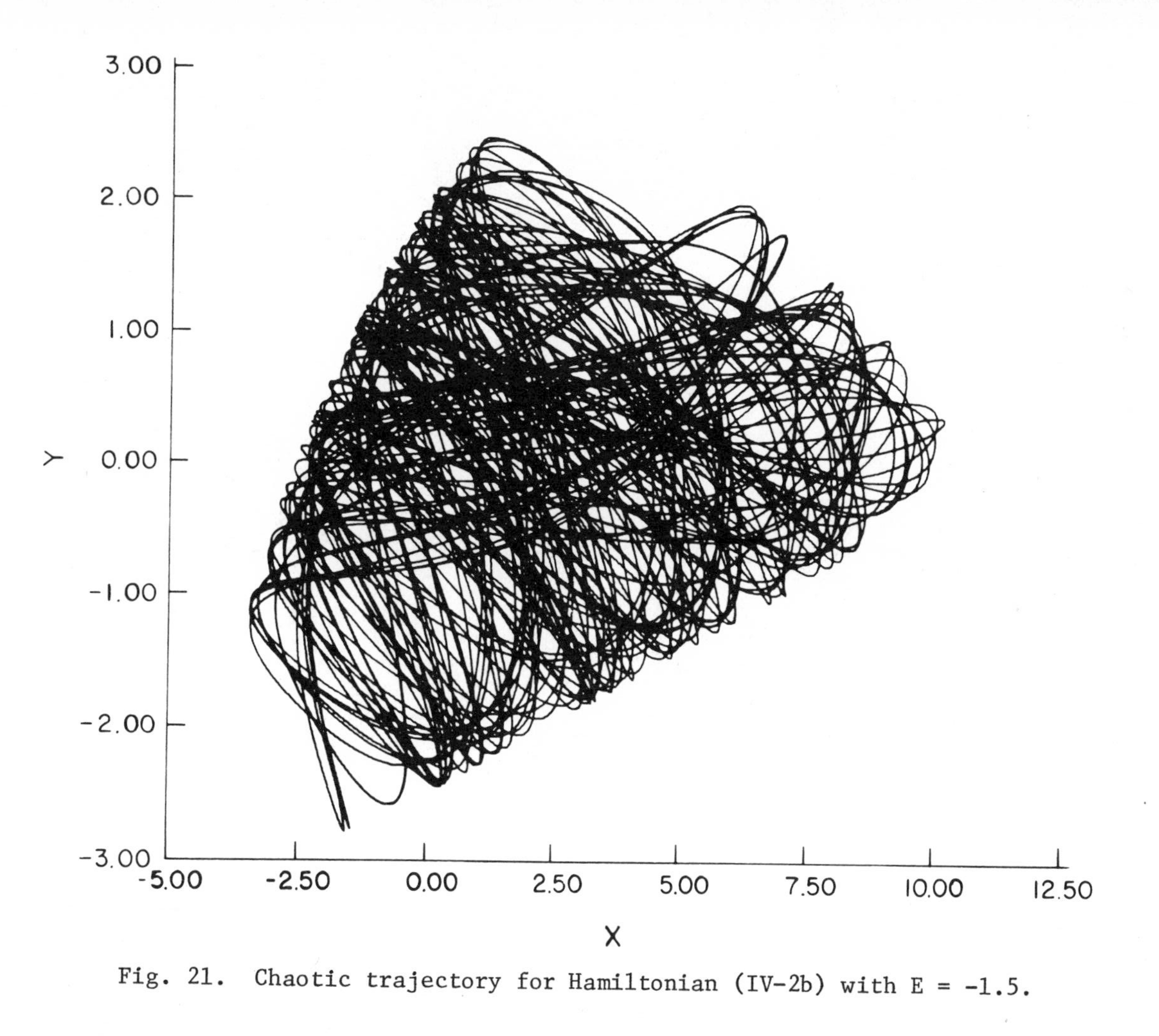

Fig. 21. Chaotic trajectory for Hamiltonian (IV-2b) with E = -1.5.

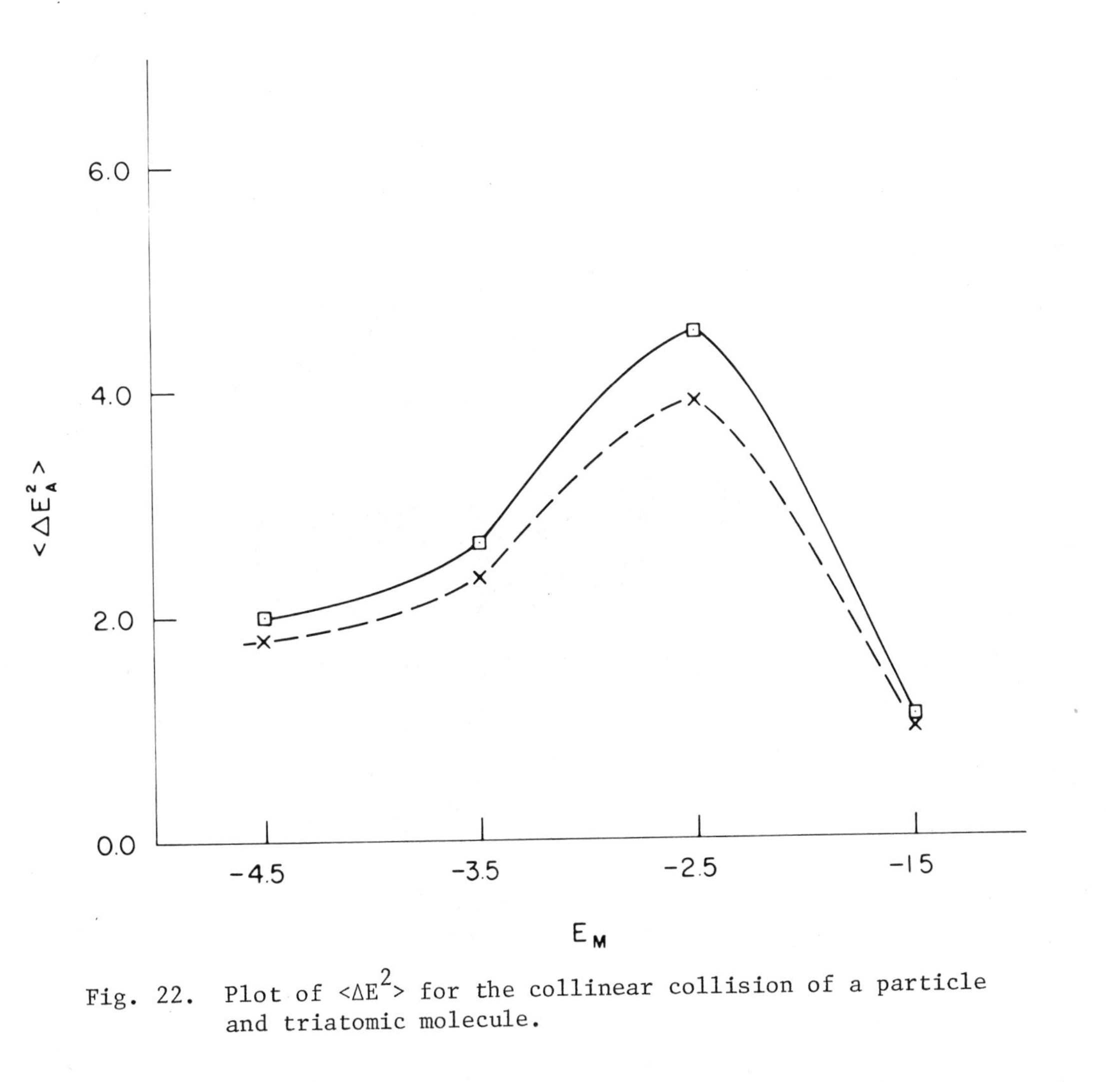

Fig. 22. Plot of $<\Delta E^2>$ for the collinear collision of a particle and triatomic molecule.

Infrared Multiphoton Dissociation

There has been considerable interest in infrared multiphoton dissociation processes during the past few years.[29-37] Particularly interesting experimental results have recently been obtained in which two infrared lasers of different powers and frequencies have been used to dissociate polyatomic molecules.[30] Their results indicate that polyatomic molecules may be dissociated much more easily (lower powers) by the use of two lasers than with just one.[30] These experiments also demonstrate that isotopic selectivity is significantly enhanced and that two-laser multiphoton dissociation may be more ideally suited for laser isotope separation.

We investigated the effect of one and two infrared lasers interacting with a diatomic molecule,[33,34] and we examined in detail the probability of dissociation for cases when the laser powers are equal and when one has a much higher power than the other. Although a model Hamiltonian was used to represent the diatomic molecule and its interaction with the lasers, the conclusions are analogous to those found experimentally for polyatomic molecules: the diatomic molecule was dissociated much more easily by using two laser frequencies than with just one. Recent results from nonlinear mechanics suggest that this two-laser effect may be due to a transition from quasiperiodic motion to a chaotic type motion.

Similar calculations were performed on a more realistic Hamiltonian where the parameters used correspond to the diatomic molecule being hydrogen fluoride.[34] It was found that in the two-laser case, a different type of dynamics occurs as opposed to the one-laser. An example of the one-laser energy adsorption as a function of time is shown in Fig. 23. The corresponding plot for the two-laser case is shown in Fig. 24. An explanation of this process may be the result of resonance overlap caused by the two-laser terms. The process has been postulated by Ford[15] and Chirikov[38] to be a source of stochastic instability in nonlinear mechanics. When the two resonant widths, $\Delta\omega$, associated with the two frequencies ω_1 and ω_2 are large, and when the ω_2 is red-shifted, chaotic behavior and dissociation occur.

In another study, we modeled the IR multiphoton process for a model of a triatomic molecule.[39] The maximum excitation energy is shown in Fig. 25 as a function of ω_I for $E_I = 0.05$, 0.10 and calculated both quantally and classically. In both cases, the agreement between the classical and quantal excitation is excellent. The only differences between the two calculations are (i) the quantal excitation falls more abruptly on the red side of the peak; (ii) the quantal threshold for dissociation is slightly higher than the classical. These effects may reflect the greater difficulty of exciting the discrete quantum levels, especially in the earlier states of excitation.

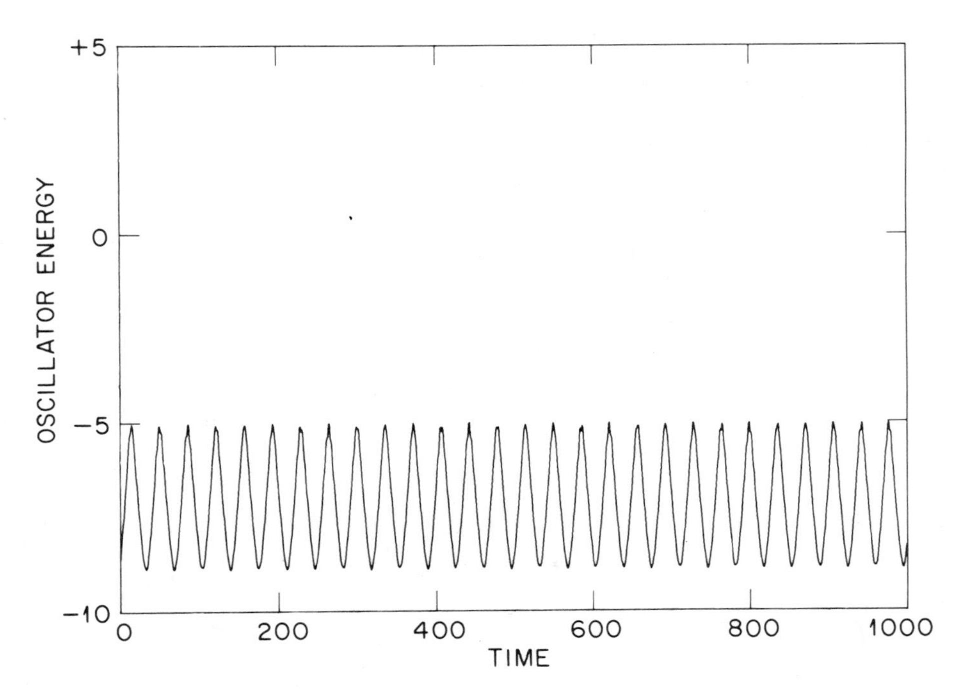

Fig. 23. Energy as a function of time for one-laser excitation.

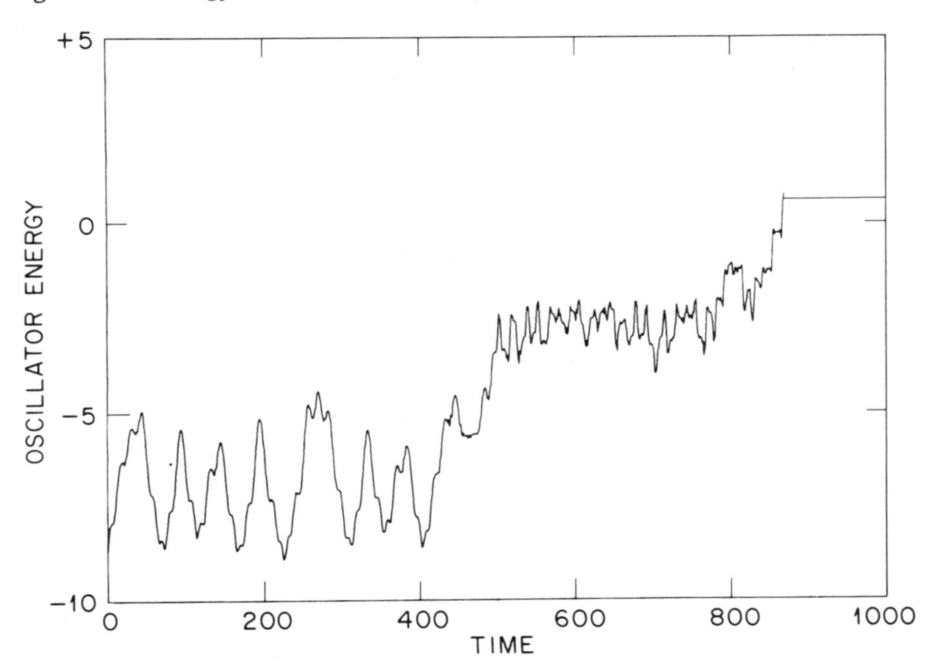

Fig. 24. Energy as a function of time for two-laser excitation.

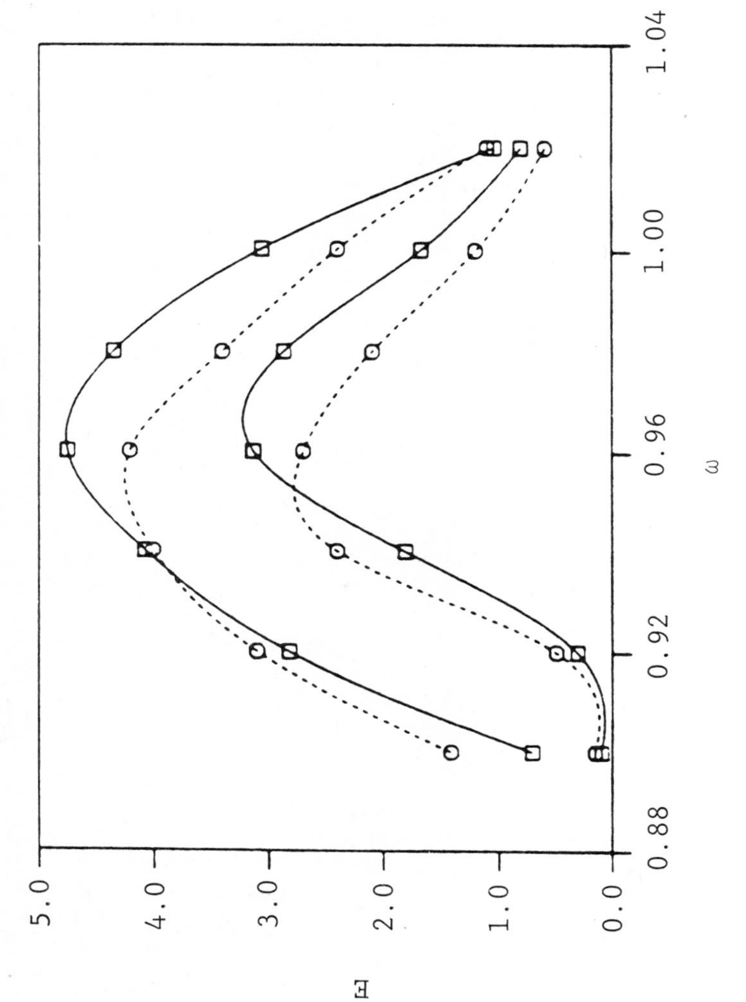

Fig. 25. A comparison of classical (circles) and quantum (squares) maximum excitation at two laser powers.

Franck-Condon Factors

The Franck-Condon factor[40] for transitions between different electronic states has recently been expressed in terms of the Wigner distribution function.[41] The various cases for quasiperiodic and stochastic type classical motion were considered.[42] It was found that the Wigner function is represented by very different types of functions in the quasiperiodic regimes.[43,44]

It was shown that in the classical limit, evaluation of the Franck-Condon factor requires solution of the dynamics if the vibrational state is quasiperiodic; in the chaotic regime, the Franck-Condon factor depends only on the global structure of the interaction surface.

Acknowledgment

We wish to thank Professor J. D. McDonald at the University of Illinois for many helpful discussions in the early stages of this work.

References

1. E.g. (a) J. B. Keller, Ann. Phys. (NY) 4:180 (1958).
   (b) M. V. Berry, Ann. Phys. (in press).
2. For a review of other methods, see I. C. Percival, Adv. Chem. Phys. 36:1 (1977).
3. M. Tabor, Adv. Chem. Phys. (in press).
4. (a) R. A. Marcus, Discuss. Faraday Soc. 55:34 (1973).
   (b) W. Eastes and R. A. Marcus, J. Chem. Phys. 61:4301 (1974).
   (c) D. W. Noid and R. A. Marcus, J. Chem. Phys. 62:2119 (1975).
   (d) D. W. Noid and R. A. Marcus, J. Chem. Phys. 67:559 (1977).
   (e) D. W. Noid, M. L. Koszykowski, and R. A. Marcus, J. Chem. Phys. 71:2864 (1979).
   (f) D. W. Noid, M. L. Koszykowski, and R. A. Marcus, J. Chem. Phys. 73:0000 (1980).
   (g) R. A. Marcus, D. W. Noid, and M. L. Koszykowski, "Semi-classical Studies of Bound States and Molecular Dynamics," in: Stochastic Behavior in Classical and Quantum Hamiltonian Systems, Volta Memorial Conference, Como, Italy, 1977, G. Casati and J. Ford, eds., Lecture Notes in Physics, Vol. 93, Springer Verlag, New York (1978/79) 283.
5. M. Toda, Phys. Lett. A 48:335 (1974).
   P. Brumer and J. W. Duff, J. Chem. Phys. 65:3566 (1976); ibid. 67:4898.
6. D. W. Noid, M. L. Koszykowski, M. Tabor, and R. A. Marcus, J. Chem. Phys. 72:6169 (1980).
7. D. W. Noid, M. L. Koszykowski, and R. A. Marcus, J. Chem. Phys. 67:404 (1977).

8. M. L. Koszykowski, D. W. Noid, and R. A. Marcus, J. Chem. Phys. (to be submitted).
9. K. Hansel, Chem. Phys. 33:35 (1978).
10. G. E. Powell and I. C. Percival, J. Phys. A 12:2053 (1979).
11. E. J. Heller, Chem. Phys. Lett. 60:338 (1979).
12. M. L. Koszykowski, D. W. Noid, M. Tabor, and R. A. Marcus, J. Chem. Phys. 73:0000 (1980).
13. K. C. Mo, Physica 57:455 (1972).
14. M. L. Koszykowski, D. W. Noid, and R. A. Marcus (to be submitted).
15. E.G., J. Ford, Adv. Chem. Phys. 24:155 (1973), and references cited therein.
16. I. C. Percival, J. Phys. B 6:559 (1973).
17. N. Pomphrey, J. Phys. B 7:1909 (1974).
18. D. W. Noid, M. L. Koszykowski, and R. A. Marcus, Chem. Phys. Lett. (in press).
19. R. A. Marcus, "Horizons in Quantum Chemistry," Proc., Third International Conference on Quantum Chemistry, Kyoto, Japan, Oct. 29-Nov. 3, 1979, eds. K. Fukui and B. Pullman, Reidel, Dordrecht, Holland (1980); R. A. Marcus, Ann. N. Y. Acad. Sci. (in press).
20. (a) R. Ramaswamy and R. A. Marcus, J. Chem. Phys. (submitted).
(b) R. Ramaswamy and R. A. Marcus, ibid. (submitted).
21. R. M. Stratt, N. C. Handy, and W. H. Miller, J. Chem. Phys. 71:3311 (1979).
22. S. Nordholm and S. A. Rice, J. Chem. Phys. 61:768 (1974); ibid. 62:157 (1975).
23. D. Secrest, Ann. Rev. Phys. Chem. (1973).
24. J. R. Taylor, Scattering Theory, John Wiley, New York (1972).
25. W. Eastes and R. A. Marcus, J. Chem. Phys. 59:4757 (1973).
26. D. W. Noid and M. L. Koszykowski, Chem. Phys. Lett. 73:114 (1980).
27. G. Schatz and T. Mulloney, J. Chem. Phys. 71:5257 (1979).
28. D. W. Noid and M. L. Koszykowski, Chem. Phys. Lett. (in press).
29. R. V. Ambartzumian and V. S. Letokhov, in: Chemical and Biochemical Applications of Lasers, Vol. III, ed. C. B. Moore, Academic Press, New York (1977).
30. D. K. Evans, R. D. McAlpine, and F. K. McClusky, Chem. Phys. Lett. 65:226 (1979).
31. D. W. Noid, M. L. Koszykowski, R. A. Marcus, and J. D. McDonald, Phys. Lett. 51:540 (1977).
32. R. A. Marcus, D. W. Noid, and M. L. Koszykowski, in: Advances in Laser Chemistry, ed. A. H. Zewail, Springer, New York (1978) 298.
33. D. W. Noid and J. R. Stine, Chem. Phys. Lett. 65:153 (1979).
34. J. R. Stine and D. W. Noid, Optics Comm. 31:161 (1979).
35. R. Walker and R. Preston, J. Chem. Phys. 67:2017 (1977).
36. K. D. Hansel, Chem. Phys. Lett. 57:619 (1978).
37. D. Poppe, Chem. Phys. 45:371 (1980).
38. B. V. Chirikov, Reports of Physics 52:263 (1979).

39. D. W. Noid, C. Bottcher, and M. L. Koszykowski, Chem. Phys. Lett. 72:397 (1980).
40. R. W. Nicholls and A. L. Stewart, in: Atomic and Molecular Processes, ed. D. R. Bates, Academic Press, New York (1962).
41. E. P. Wigner, Phys. Rev. 40:749 (1932).
42. W.-K. Liu and D. W. Noid, Chem. Phys. Lett. (in press).
43. M. V. Berry, Phil. Trans. Royal Soc. A 287:237 (1977).
44. M. V. Berry, J. Phys. A 10:2083 (1977).

PROBLEMS IN THE SEMICLASSICAL QUANTIZATION OF INTEGRABLE AND NONINTEGRABLE CLASSICAL DYNAMICAL SYSTEMS

William P. Reinhardt and Charles Jaffé†

Department of Chemistry, University of Colorado and Joint Institute for Laboratory Astrophysics, University of Colorado and National Bureau of Standards, Boulder, Colorado 80309

ABSTRACT

The problem of semiclassical quantization -- the determination of quantum energy levels and wave functions from classical dynamical input -- has been solved for a large class of integrable systems where classical trajectories are confined to Lagrangian manifolds which are topologically N-tori in the 2N dimensional phase space. In this paper we explore two cases where this quantization procedure breaks down. This occurs if the dynamics are nonintegrable, or if the classical tori have a complex structure on a scale small compared to Planck's constant, $\hbar$. In both cases the empirically correct procedure is to <u>smooth</u> the classical dynamics before applying a quantization procedure. The concepts are illustrated using the nonintegrable two-dimensional Henon-Heiles system, where it is shown that excellent results are obtained.

I. INTRODUCTION

The problem of semiclassical quantization of classical Hamiltonian systems continues to be of interest to workers in many areas of mathematics, physics, and chemistry.[1] After setting the stage for recent developments by a brief review of the Einstein-Brillouin-Keller (EBK) quantization procedure for nonseparable but

---

†Current address: Department of Chemistry, University of Toronto, Toronto, Canada.

integrable multidimensional classical system we outline recent progress in two areas: 1) adaptation of the EBK quantization to "weakly" nonintegrable systems; and 2) the need for multidimensional uniform approximations even in the integrable case, where the EBK primitive quantization procedure can fail to yield a correct zero order manifold of quantum states. In both of these areas the size of Planck's constant as compared to the scale of complexity of the classical dynamics plays an essential role, further illustrating the nonuniformity of the $\hbar \to 0$ limit in quantum mechanics as emphasized by Berry, among others.[2]

II. INTEGRABLE HAMILTONIAN DYNAMICS AND EBK QUANTIZATION

Conservative Hamiltonian dynamical systems are defined by the 2N equations of motion

$$\dot{p}_i = -\frac{\partial H}{\partial q_i} \quad ; \quad \dot{q}_i = \frac{\partial H}{\partial p_i} \qquad i = 1,2,\ldots N \tag{1}$$

where $\dot{p}_i = dp_i(t)/dt$, $\dot{q}_i = dq_i(t)/dt$, and where $H = H(\bar{p},\bar{q})$ is not an explicit function of the time, t. If such a system has N independent isolating integrals of the motion it is said to be <u>integrable</u>. The motion of an individual trajectory, defined by Eq. (1) and suitable initial conditions, is then confined to an N-dimensional hypersurface of the full 2N dimensional phase space of the p_i, q_i $(i = 1,2,\ldots N)$. All separable systems are integrable; but integrable systems are the exception for nonseparable systems. However, nonintegrable systems may show integrable behavior in some regions of phase space if they are "close" enough to an integrable system either in the sense of being a small perturbation of an exactly integrable system, <u>or</u>, if the energy, E ($E \equiv H(\bar{p},\bar{q})$ is always conserved for the systems under consideration) is "small." This is the content of the celebrated KAM Theorem. The concept of integrable-like motion is illustrated in Figs. 1 and 2 showing integrable-like dynamics for the two-dimensional Henon-Heiles system defined by

$$H^o = \frac{1}{2}p_1^2 + \frac{1}{2}q_1^2 + \frac{1}{2}p_2^2 + \frac{1}{2}q_2^2 \tag{2a}$$

$$H = H^o + \lambda\left(q_1^2 q_2 - \frac{1}{3}q_2^3\right) \quad . \tag{2b}$$

For sufficiently low energies or small extensions of q_1, q_2 from zero, the cubic terms of Eq. (2b) are small compared to the quadric, harmonic terms of the unperturbed Hamiltonian, H^o, of Eq. (2a), suggesting the possibility of integrable motion, as illustrated in Figs. 1 and 2. Figure 2 indicates the surface of section defined by $[q_2(t),\ p_2(t)]$ such that $q_1(t) = 0$, and is, in a sense, a slice through the four-dimensional phase space (q_1, q_2, p_1, p_2). The motion is apparently on a two-dimensional surface in

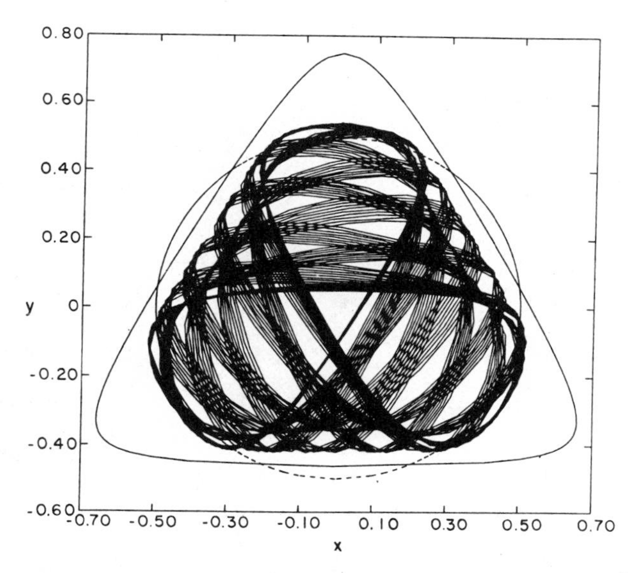

Fig. 1. Trajectories for the Henon-Heiles problem $H = \frac{1}{2}(p_x^2 + x^2 + p_y^2 + y^2) + x^2y - y^3/3$. Figure 1 is a quasi-periodic trajectory, lying (at least to an excellent approximation) on an invariant torus (see Figs. 2, 3).

the phase space, corresponding to existence of an invariant torus as illustrated in Fig. 3.

Einstein,[3,4] in 1917, suggested the semiclassical quantization condition

$$\frac{1}{2\pi}\int_{C_i} \bar{p}\cdot d\bar{q} = n_i\hbar \tag{3}$$

for motion defining (or defined by!) a torus of the type shown in Fig. 3. The 1-form $\bar{p}\cdot d\bar{q}$ is invariant to canonical transformations, and the integral of Eq. (3) invariant to homotopically equivalent distortions of the integration paths, C_i, on the surface of the torus, which follows from the fact that the tori are Lagrangian manifolds. The work of Keller and Maslov requires that the "$n_i\hbar$" on the right of Eq. (3) be replaced by "$[n_i+(\alpha_i/4)]\hbar$," where α_i is the Maslov index which represents a correction for differing possible behavior at classical turning points.[5] The quantization condition (3), with the inclusion of the appropriate Maslov index is commonly referred to as the EBK quantization condition.[5] There has been much recent work on semiclassical quantization of integrable motion of coupled oscillator systems following the EBK condition.

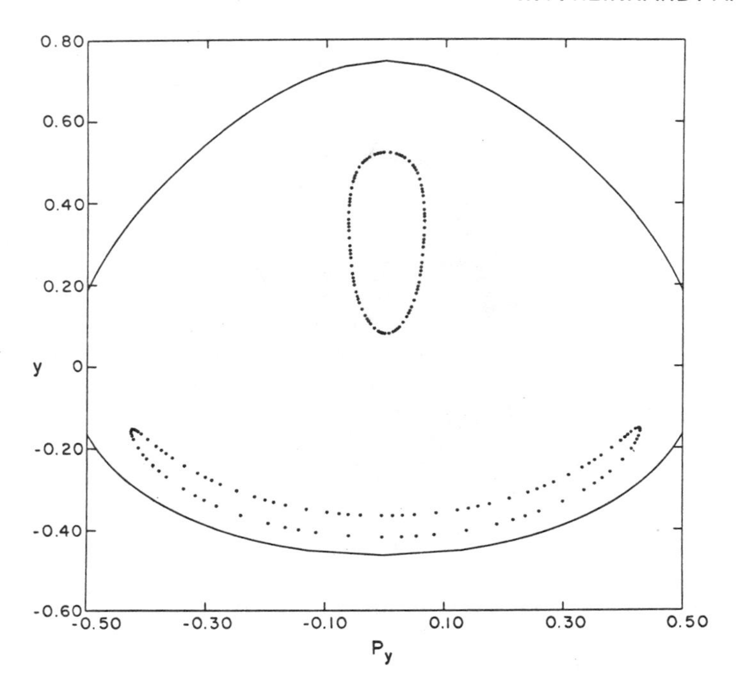

Fig. 2. A Poincaré surface of section for a "stable" trajectory run for the Henon-Heiles coupled oscillator problem. The actual coordinate space trajectory is that of Fig. 1. The surface of section is a slice through the four-dimensional phase space x, y, p_x, p_y, for the Hamiltonian system. More precisely it is defined as the set of points (y, p_y) generated by following a trajectory and keeping only those values of (y, p_y) which arise each time x=0 along the trajectory. In this case the collection of points thus generated appears to lie on two smooth curves, which are, in fact, cross sections of an invariant torus. This behavior is to be contrasted with that of Fig. 5.

However, the EBK condition is clearly not adequate for treatment of classical motion in two important cases: 1) nonintegrable motion where the motion is not on classical tori (manifolds), thus making determinations of the paths "C_i" of Eq. (3) problematic and 2) integrable dynamics of such complexity that the condition

$$\frac{1}{2\pi} \int_{C_i} \bar{p} \cdot d\bar{q} = \left(n_i + \frac{\alpha_i}{4}\right) \hbar \tag{4}$$

is <u>not</u> satisfied for small n_i, even though quantum states must exist. These inadequacies are, respectively, the subject of Secs. III and IV.

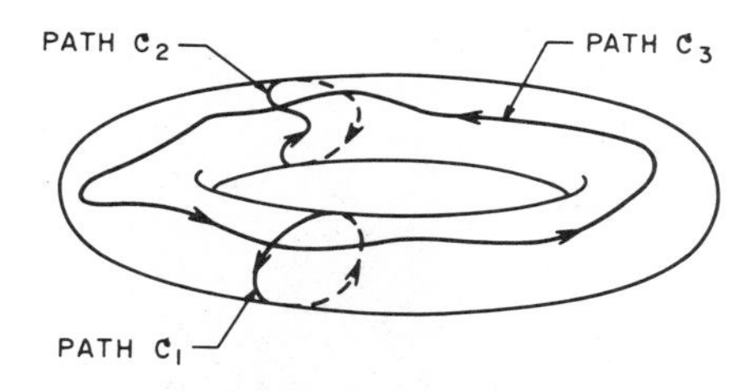

Fig. 3. Schematic illustration of integration paths on an invariant classical manifold for a system of two degrees of freedom. As noted in the text, the two different quantum numbers needed to describe quantization arise from two different topologically distinct paths on the torus, C_1 and C_3. The properties of integration of 1-forms on the Lagrangian manifold insure path independence for homotopically equivalent paths; that is, integration of $\bar{p}\cdot d\bar{q}$ over path C_1 gives an identical result to that over C_2. We note that the integration paths are not classical trajectories. The classical trajectories evolve on the surface of the two-dimensional torus, embedded in a four-dimensional phase space, x, y, p_x, p_y.

III. EBK QUANTIZATION ON VAGUE TORI

The problems raised in the consideration of semiclassical quantization of nonintegrable motion are dramatically illustrated in Figs. 4 and 5 where a chaotic trajectory (run at the same energy as the trajectory of Figs. 1, 2) is shown, first in the q_1, q_2 (x,y) configuration space, and then in surface of section space. The motion is visually of quite a different character. In particular, the chaotic trajectory apparently fills a volume in the phase space. This leads to the obvious questions: how are we to modify the EBK quantization condition of Eq. (3) when the invariant manifolds, underlying both integrable dynamics and the formulation leading to Eq. (3), no longer exist. Einstein, himself, recognized this as a fundamental problem, pointing out that the condition(s) of Eq. (3) only applied to what we nowadays refer to as integrable motion.

The problem of semiclassical quantization of nonintegrable classical dynamics is in its infancy. Gutzwiller,[6] in a combinatorial tour de force, has analytically quantized the problem of a free particle on a Riemannian surface of constant negative curvature, using a periodic orbit quantization procedure derived as the stationary phase approximation to the Feynman path integral. Free motion on a surface of constant negative curvature (first discussed by Hopf) is known to be both ergodic and mixing: it is of the C-system type, and has strong statistical properties. Quantization of even free motion in this multidimensional nonintegrable case must be regarded as a triumph; but, the techniques utilized by

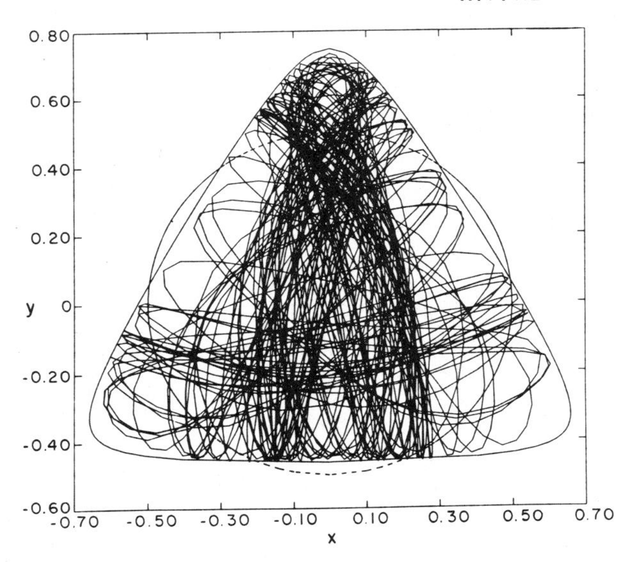

Fig. 4. A chaotic classical trajectory for the Henon-Heiles system. This trajectory was run at the identical energy as that of Fig. 1, but occupies a distinct part of phase space as is made clear by examination of the surface of section, shown in Fig. 5.

Gutzwiller do not immediately lend themselves to more general situations.

There is however an intermediate situation. We wish to claim that the surface of section of Fig. 5 is misleading in the sense that it overemphasizes the chaotic nature of the trajectories. Arguing qualitatively from the primitive stability analyses of Brumer-Duff,[7] and Cerjan-Reinhardt[8]; or from the determination of local constants of the motion by Padé approximants obtained from Birkhoff-Gustafson normal forms as determined by Shirts and Reinhardt,[9] the actual phase space geometry is deduced to be closer to that of Fig. 6, than of complete chaos. The implication of Fig. 6 is that motion in the chaotic regions of the classical phase space is not at all random. On the contrary, large pieces of the manifold structures (tori) are at least approximately intact as indicated by local convergence of the Padé summed Birkhoff-Gustafson normal form. However, the surfaces are no longer complete tori, there being patches which act as strong sources of chaos, in the sense that the Padé approximants do not converge. Actual classical trajectories run in the corresponding regions of phase space appear to be approximately confined to a torus for many classical periods, to pass through chaotic transition regions, and to reappear on new approximate tori -- we refer to these approximate tori as _vague tori_.

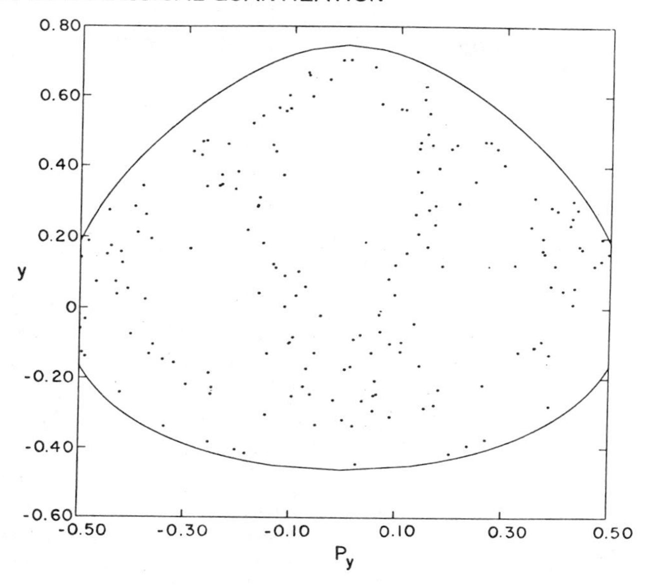

Fig. 5. A Poincaré surface of section for the chaotic trajectory shown in Fig. 4. (See the caption of Fig. 2 for a definition of surface of section.) Although it is not possible to make rigorous mathematical statements from such a computational result, it appears from the figure that: (1) the motion is no longer on an invariant torus, (2) the motion seems to fill a subvolume of phase space -- and may be described as chaotic in that the individual points which comprise the surface of section show extreme sensitivity to initial conditions, and apprently do not appear in any regular order as an individual trajectory evolves. See however, Sec. III.

The qualitative picture contained in Fig. 6 implies an empirical quantization procedure: use the approximate manifold surfaces of Fig. 6 to perform the EBK quantization. With the assumption that the destroyed patches on the toroidal surfaces are small compared to Planck's constant, $\hbar$, the quantum states should not be sensitive to the breakdown of integrability. Such systems are only weakly chaotic with respect to algorithms for semiclassical quantization. Swimm and Delos[10] and Jaffé and Reinhardt[11] have obtained, respectively, primitive, and uniform semiclassical EBK-type quantum levels implicitly based on these ideas. Typical results are indicated in Table I: all of these states correspond to a volume of phase space where individual trajectories are chaotic. The success of the EBK based procedure on the remnants of tori, as well as the Padé work, seems to justify the feeling that the classical dynamics is only weakly chaotic. Thus, simple patching up of the EBK theory, by quantization on vague tori, suffices to obtain reasonable approximations to quantum levels.

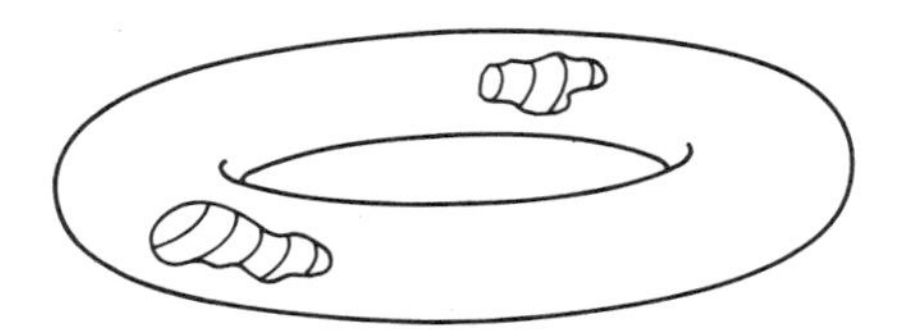

Fig. 6. Highly imaginative view of a "partially destroyed" torus corresponding to the type of nonintegrable behavior predicted by the Padé approximants obtained from the coefficients of the Birkhoff-Gustafson normal form by Shirts and Reinhardt (Ref. 9). The analysis suggests that it is not the case that invariant tori disappear in chaotic (or quasi-ergodic) volumes of phase space; rather, they simply grow localized "patches" where the formerly global constants of the motion disappear, leaving much of the manifold structure intact. If the patches (where tori interleave and mix on Cantor sets) are "small" compared to the larger scale of the remnant "vague" torus, and "small" on a scale of $\hbar$, a natural smoothing over the "patches" leads to an excellent semi-classical scheme, as clearly demonstrated in Table I.

Table I. Uniform semiclassical of Jaffé and Reinhardt (JR), Ref. 11, eigenvalues of some high-lying states for the Henon-Heiles problem of Eqs. (2a,b) with $m=1$, $\hbar=1$, $\lambda^2=1/80$. The results are compared with exact quantum results (QM) and primitive semiclassical results of Delos and Swimm, (DS), Ref. 10, the latter not splitting the degeneracies of the "A" states.

Quantum Numbers[a]	Symmetry[a]	QM	JR	DS
12 ± 0	A	11.966	12.011	11.864
12 ± 2	E	11.968	12.017	--
12 ± 4	E	12.206	12.217	--
12 ± 6	A	12.277 12.334	12.274 12.332	12.310
12 ± 8	E	12.480	12.490	12.491
12 ± 10	E	12.712	12.749	12.750
12 ± 12	A	13.077 13.087	13.0975 13.0976	13.097

[a]The quantum numbers and symmetry notation are explained in Noid and Marcus, Ref. 12.

IV. BREAKDOWN OF PRIMITIVE EBK QUANTIZATION FOR INTEGRABLE SYSTEMS

The quantization of the nonintegrable dynamics discussed in Sec. III followed from the basic concept that one can ignore classical chaos which occurs on a scale small compared to $\hbar$ (or to a de Broglie wavelength). However, integrable dynamics can also show complexity on a fine scale. If such integrable motion takes place on a scale small compared to h the resulting problems for semiclassical quantization are just as great as those for nonintegrable classical motion. It is the purpose of this section simply to indicate the nature of the problem, and to propose a conjecture as to its ultimate solution.

The problem is simply posed. Figure 7 shows part of a surface of section for dynamics on the Henon-Heiles potential surface as determined by direct integration of Hamilton's equations. The phenomenon of interest is the chain of seven islands lying between two larger scale tori. Such island chains are often symptomatic of nonintegrability, but it is easy to give examples of integrable dynamics which also show this phenomenon. The islands result from

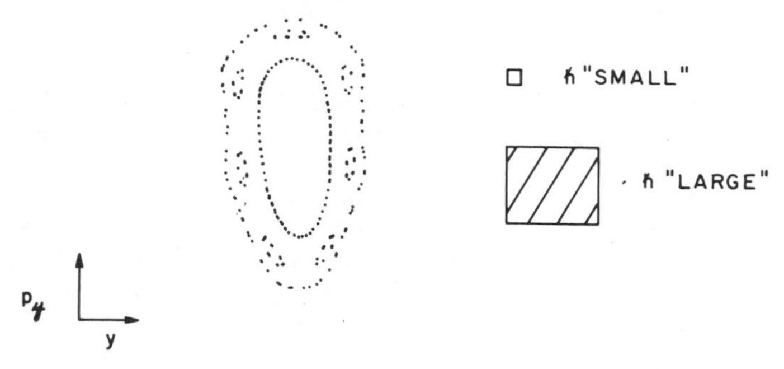

Fig. 7. Portion of composite classical surface of section for the Henon-Heiles surface. Tori corresponding to three distinct trajectories are shown. For the "small" value of $\hbar$ the detailed "island" structure between the larger scale tori must be taken into account, or quantum states will be omitted from a primitive semiclassical analysis. For the "large" value of $\hbar$, indicated, one does not expect the "islands" to play an important role, and thus we conjecture that an appropriate "smoothing" of the classical dynamics (as illustrated in Fig. 8) would be in order. This type of smoothing is actually implicit in the work of Refs. 10, 11, 12 for semiclassical quantization of almost integrable systems, although the classical fine structure is of <u>very</u> small measure for such systems, and not even easily observed in computer calculations. The successful quantization of the more strongly irregular Henon-Heiles motion (Refs. 10, 11) also assumes appropriateness of smoothing the classical complexities occurring on a scale "small" compared to $\hbar$.

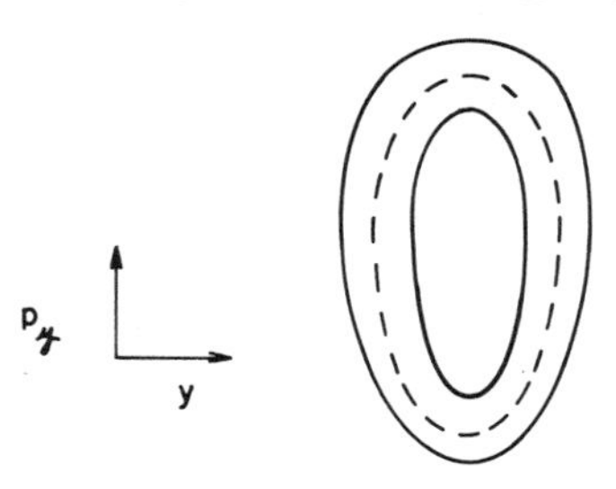

Fig. 8. Smoothed classical dynamics. If the Birkhoff-Gustafson normal form for the Henon-Heiles problem is simply truncated at tenth or twelfth order in λ, the level surfaces corresponding to the parts of phase space of Fig. 7 are smoothed as shown here. This smoothed classical dynamics offers no problems for the applications of the primitive EBK quantization condition -- which cannot be directly applied to the island regions of Fig. 7, for the "large" value of $\hbar$ shown.

the fact that the motion is on a torus which itself winds back and forth through the sectioning plane (x = 0) many times. The difficulty with the primitive EBK quantization scheme is now evident: for the "large" value of $\hbar$ shown in Fig. 7, the integral around the short shank of the torus producing the islands will always be substantially less than "$\hbar$," thus yielding no primitive quantum state. What is clearly needed is a uniform version of the EBK quantization formula. This is, of course, a difficult problem: as yet there is no generically multidimensional approach to the problem of uniform approximations. However, we conjecture that the essence of such a multidimensional uniform result will be to replace the underlying classical dynamics by smoothed classical dynamics (see Fig. 8), followed by primitive quantization of the smoothed motion. Thus, as is the case of quantization of the weakly nonintegrable motion discussed in Sec. III we replace actual classical dynamics with smoothed dynamics, with the necessary fine structure kept determined by the size of Planck's constant.

V. SUMMARY AND CONCLUSIONS

We have discussed quantization of two types of classical dynamics where problems arise because of classical complexity on a scale small compared to Planck's constant. In both cases the intuitively appropriate and empirically successful remedy was smoothing of the classical dynamics by truncation of the divergent Birkhoff normal expansion, followed by the usual EBK procedure. The problem of derivation of these types of results from, say, the Feynman path formulation of quantum mechanics is equivalent to solving the problem of making generically multidimensional uniform approximations and will be difficult. However, if the conjectured results of the present discussion are correct (even if only in the weak sense of being reasonable approximations to the "correct" approximation) the problem of finding multidimensional uniform approximations will be made simpler by the possibility of being able to work in both directions toward the solution.

ACKNOWLEDGMENTS

Helpful conversations with R. Shirts and C. Holt are gratefully acknowledged, as is the support of the National Science Foundation, through Grants PHY79-04928 and CHE80-11442 to the University of Colorado.

REFERENCES

1. See, for example, G. Casati and J. Ford (Eds.) Stochastic Behavior in Classical and Quantum Hamiltonian Systems - Volte Numerical Conference, Como 1977, Springer-Verlag, Berlin, Lecture Notes in Physics 93, 1979; V. I. Arnold, Mathematical Methods of Classical Mechanics, Springer-Verlag, Berlin, New York, Graduate Texts in Mathematics 60, 1978; S. Jorna (Ed.) Topics in Nonlinear Dynamics, AIP Conference Proceedings #46, AIP, New York 1978; see also Ref. 5, below.
2. M. V. Berry, J. Phys. A 10, 2083 (1977) and M. V. Berry and S. Jorna (Ed.), Ref. 1.
3. A. Einstein, Dent. Phys. Ges. Berlin Verh. Vol. #19, #9/10, May 30, 1917.
4. A translation of Ref. 3 by Charles Jaffé may be obtained as Joint Institute for Laboratory Astrophysics Technical Report #116, September 1980.
5. This work is reviewed by I. C. Percival, Adv. Chem. Phys. 36, 1 (1977), where a complete list of references may be found.
6. M. Gutzwiller, Phys. Rev. Lett. 45, 150 (1980), and references to earlier work therein.
7. P. Brummer and J. Duff, J. Chem. Phys. 65, 3566 (1976).
8. C. Cerjan and W. P. Reinhardt, J. Chem. Phys. 71, 1819 (1979).
9. R. B. Shirts and W. P. Reinhardt, (which contains references to the work of Birkhoff and Gustafson), this volume, and J. Chem. Phys. (to be submitted).
10. R. T. Swimm and J. B. Delos, J. Chem. Phys. 71, 1706 (1979).
11. C. Jaffé and W. P. Reinhardt, J. Chem. Phys. (to be submitted); C. Jaffé, Ph.D. Thesis, University of Colorado, 1979 (unpublished).
12. D. W. Noid and R. A. Marcus, J. Chem. Phys. 62, 2119 (1975); S. Chapman, B. C. Garrett and W. H. Miller, J. Chem. Phys. 64, 502 (1976); I. C. Percival and co-workers, see Ref. 5.

NONSELFADJOINT OPERATORS IN DIFFRACTION AND SCATTERING

C. L. Dolph and A. G. Ramm

Department of Mathematics
University of Michigan
Ann Arbor, MI 48109

1. In many problems of interest for physicists and engineers nonselfadjoint operators arise naturally. For example the EEM (eigenmode expansion method) can be described as follows. Let Ω be an exterior domain with a smooth closed compact boundary Γ,

$$(\Delta + k^2)u = 0 \quad \text{in } \Omega\,, \; u|_\Gamma = f, \; |x|\left(\frac{\partial u}{\partial |x|} - iku\right) \to 0 \quad \text{as } |x| \to \infty\,, \quad k^2 > 0\,. \tag{1}$$

Let us look for a solution of (1) in the form

$$u = \int_\Gamma \sigma(t)\, \exp(ikr_{xt})(4\pi r_{xt})^{-1}\, dt\,, \quad r_{xt} = |x - t|\,. \tag{2}$$

For the unknown σ we get equation

$$A\sigma \equiv \int_\Gamma \exp(ikr_{st})(4\,\pi\, r_{st})^{-1}\, \sigma(t)dt = f(x)\,. \tag{3}$$

Operator A is nonselfadjoint in $L_2(\Gamma)$. Suppose that its root system forms a basis of $H = L^2(\Gamma)$. Then we can look for a solution of (3) in the form of the series, expanding $\sigma(t)$ and $f(x)$ according to the root vectors of $A(k)$. This is called EEM for solution of (1). Thus we have question 1) Does the root system of $A(k)$ form a basis of H?

It is easy to prove that the Green function $G(x,y,k)$ of the exterior Dirichlet (or Neumann) problem can be meromorphically continued on the whole complex plane k and its poles k lie in the lower half-plane $\text{Im } k < 0$. If f is a smooth function with compact support, $v = \int G(x,y,k)f(y)dy$, $\int = \int_\Omega$ and (*) $|v| \leq C(1 + |k|^a)^{-1}$, $a > 0.5$, $C = \text{const.}$, $C = C(\text{Im } k)$, then the solution of the problem

$$u_{tt} = \Delta u \quad \text{in } \Omega, \quad t > 0, \; u|_\Gamma = 0, \; u\big|_{t=0} = 0, u_t\big|_{t=0} = f(x) \qquad (4)$$

can be represented in the form

$$u(x,t) = \Sigma_{j=1}^{n} \exp(-ik_j t)v_j(x,t) + o(\exp\{-|\text{Im } k_n|t\}), t \to +\infty \qquad (5)$$

where $v_j(x,t)$ grow not faster than t^m as $t \to +\infty$ and m is some integer. Expansion (5) is an example of SEM(singularity expansion method). This leads to questions: 2) What can be said about location of the poles k_j ? ; 3) When is (*) valid?; 4) to what extent does the set $\{k_j\}$ determine the obstacle?; 5)how can the k_j be calculated?; 6)whether the poles k_j are simple?

2. The answer to question 1) is given in [1] and is described below. In [2]-[4] some results about bases with brackets are given. Some answers to question 2) are given in [5]-[7].Answers to question 3) are given in [8]-[10]. Answer to question 4) is unknown. Answer to question 5) is given in [11],[12]. Answer to 6) is unknown, but some engineers (C.E. Baum e.g.) think that if Γ is convex then the Green function of the exterior Dirichlet problem has simple poles. Some particular cases when this is true are discussed in [13]. In [14], a survey of the SEM is given and [15] presents an engineering point of view. In [16] a survey of what is known about questions 1)-6) is given and in [17] some relevant results can be found. In [18] the relation between SEM and the matehmatical scattering theory is discussed. In [19],[20] variational principles are discussed for nonselfadjoint problems.

3. In this section we answer question 1). If $A = L + T$, where L is a selfadjoint operator on a Hilbert space H, the spectrum $\{\lambda_j\}$ of L is discrete, $\lambda_n = cn^p(1 + O(n^{-\delta}))$ where c,p,δ are

some positive constants, and $|Tf| \leq c_1 |L^a f|$, $c_1 > 0$, $a < 1$ for all $f \in D(L^a)$, $D(L^a) \subset D(T)$, then: 1) the root system of A forms a Riesz basis of H ($A \in R$) if $\delta > 1$ and $p(1-a) > 2$; 2) if $\delta > 0$, $p(1-a) > 1$ it forms a Riesz basis of H with brackets, ($A \in R_b$); 3) if $\lambda_n^{1+a}(\lambda_{n+1} - \lambda_n)^2 \to 0$ as $n \to \infty$, then $A \in R$. Let us give the definition of the Riesz basis with brackets. Let the system $\{f_j\}$ form an orthonormal basis of H, $m_1 < m_2 < \cdots m_n < \cdots \to \infty$ is a sequence of integers, $\{F_j\}$ is the sequence of the subspaces, where F_j is the linear span of $f_{m_j}, f_{m_j+1}, \ldots, f_{m_{j+1}-1}$. Let $\{h_j\}$ be a minimal and complete system in H, H_j is the linear span of $h_{m_j}, \ldots, h_{m_{j+1}-1}$. If there exists a map $B \in L(H)$, $BH_j = F_j$, $j = 1,2,\ldots$, then the system $\{h_j\}$ is called a Riesz basis of H with brackets and the numbers m_j define the bracketing. By $L(H)$ we denote the set of linear bounded invertible operators which map H onto H.

4. Problems.

1. Find an asymptotic formula for $\tilde{k}_j$ as $|\tilde{k}_j| \to \infty$, where $\{\tilde{k}_j\}$ are the poles with minimal imaginary parts. For purely imaginary poles some information about asymptotic distribution is given in [7].

2. To what extent does the set $\{k_j\}$ determine the obstacle?

3. Is it true that the complex poles of the Green function of the exterior Dirichlet problem are simple provided that Γ is convex?

REFERENCES

1. A.G. Ramm, On the basis property for the root vectors of some non-selfadjoint operators, Jour.Math.Anal.Appl.(1980)
2. A. Marcus, The root vector expansion of a weakly perturbed selfadjoint operator, Sov. math. Doklady, 3, (1962),104-108.
3. V. Kacnelson, Conditions for a system of root vectors of some classes of non-selfadjoint operators to form a basis, Funct. anal. and appl. 1, (1967), 122-132.

4. N. Voitovich, B. Kacenelenbaum, A. Sivov, Generalized method of eigenoscillations in diffraction theory, Nauk. Moscow, 1977. Appendix written by M. Agranovich (Russian).
5. A.G. Ramm,Domain where the resonances are absent in the three dimensional scattering problem. Doklady, 166, (1966), 1319-1322. 34 #3902
6. P.D. Lax, R.S. Phillips, A logarithmic bound on the location of the poles of the scattering matrix,Arch. Rat. Mech. Anal. 40, 1971, 268-280.
7. P.D. Lax, R.S. Phillips, Decaying modes for the wave equation in the exterior of an obstacle, CPAM 22, (1969), 737-787.
8. P.D. Lax, R.S. Phillips, C. Morawetz, Exponential decay of solutions of the wave equation in the exterior domain of a star-shaped obstacle, CPAM 16, (1963), 477-486.
9. C. Morawetz, Exponential decay of solutions of the wave equation, CPAM, 19, (1966), 439-444.
10. A.G. Ramm, Exponential decay of solution of hyperbolic equation, Ibid., 6, (1970), 2099-2100. 44 #631. Et 1598-1599.
11. A.G. Ramm, Calculation of the quasistationary states in quantum mechanics, Doklady, 204, (1972), 1071-1074. 56 #14326
12. A.G. Ramm, On exterior diffraction problems, Radiotech. i Electron. 7, (1972), 1362-1365. 51 #4864, E.t. 1064-1067.
13. A.G. Ramm, Eigenfunction expansion corresponding the discrete spectrum, Radiotech. i Electron., 18, (1973), 496-501.
14. C.L. Dolph, R.A. Scott, Recent developments in the use of complex singularities, Electromagnetic scattering, Ed. P. Uslenghi, Acad. Press, N.Y., (1978), 503-570.
15. C.E. Baum, Emerging technology for transient and broad analysis and synthesis of antennas and scaterers, Proc. IEEE, 64, (1976), 1598.
16. A.G. Ramm, Nonselfadjoint operators in diffraction and scattering, Math. methods in appl. sci., (1980)
17. A.G. Ramm, Theory and applications of some classes of integral equations, Springer-Verlag, N.Y., 1980 (to appear in 1980).
18. C.L. Dolph, S. Cho, On the relationship between the singularity expansion method and the mathematical theory of scattering, (to appear in Trans. IEEE Antennas and Propag.)
19. A.G. Ramm, A variational principle for resonances. J. Math. Phys. (1980)
20. A.G. Ramm, Variational principles for spectrum of compact nonselfadjoint operators, J. Math. Anal. Appl. (1980).

Supported by AFOSR 800204.

ONE-DIMENSIONAL CRYSTALS IN AN EXTERNAL FIELD

James S. Howland[1]

Department of Mathematics
University of Virginia
Charlottesville, Virginia 22903

ABSTRACT

For a certain class of analytic potentials $V(x)$, matrix elements of the resolvent of $H_F = -d^2/dx^2 - Fx + V(x)$ with entire vectors of the translation group have meromorphic continuations from $\mathrm{Im}\, z > 0$ to the whole complex plane. The poles of these continuations are restricted to a discrete set independent of the analytic vectors chosen. Certain random potentials corresponding to an infinite number of particles distributed on the points of a Poisson set lie in this class with probability one as do a large class of periodic potentials.

In this paper, we shall discuss resonances which arise when an infinite one-dimensional crystal, described by a Hamiltonian

$$p^2 + V(x)$$

is placed in a uniform electric field of strength F, so that the new Hamiltonian becomes

$$p^2 + V(x) - Fx.$$

The crystal may either be *pure*, which means that $V(x)$ is periodic with some period a, or it may contain *impurities* distributed *randomly* throughout the crystal. One may even consider the case in which $V(x)$ is completely random, and contains no periodic component.

1. The Stark Ladder. The case of a pure crystal leads to the Stark ladder problem [1]. Let us recall for a moment the ordinary

---

[1]Supported by NSF Grant MCS-79-02490

Stark effect. If $V(x)$ is a nice function vanishing at infinity, the operator

$$H_0 = -d^2/dx^2 + V(x)$$

has a spectrum consisting of the positive half-line $[0,\infty)$ and possibly one or more discrete bound states. When the field is turned on, the spectrum of

$$H_F = -d^2/dx^2 + V(x) - Fx$$

becomes the entire real line. For $F > 0$, the particle eventually move off to the right in an accelerated motion [2].

If $V(x)$ is periodic, with period a, then H_0 has a spectrum consisting of an infinite number of intervals, with (usually) non-zero gaps in between. The intervals of $\sigma(H_0)$ are called conduction bands; electrons can only propagate in the crystal at band energies. The problem is discussed mathematically by Reed and Simon [3] and physically by Kittel [4], who draws an analogy between conduction bands and bound states.

In the ordinary Stark effect, for a given bound state energy E of H_0, there is often, when F is small, a second sheet resonance pole $\lambda(F) - i\Gamma(F)$ of $(H_F - z)^{-1}$ near E. The real part $\lambda(F)$ is the shifted energy level computed by perturbation theory in textbooks; the imaginary part, the <u>width</u> $\Gamma(F)$ of the resonance, is essentially the decay rate of the corresponding quasi-bound state. (Typically, $\Gamma(F)$ is exponentially small, of order something like $e^{-1/F}$.)

Does anything of the sort occur in the periodic case? It has been suggested that the answer is "yes" -- that when F is positive, but small, then for each band there is an infinite sequence of resonances with constant imaginary part, each separated by a distance aF from the next:

$$\Lambda_n = \lambda + naF - i\Gamma \qquad n = 0, \pm 1, \pm 2, \ldots$$

Such a sequence of resonances is called a <u>Stark ladder</u> [1].

A heuristic argument may be given as follows. The potential $V(x) - Fx$ looks like $V(x)$, except that it slopes gently off to the right. Picture a quasi-bound state resembling a standing wave, distributed over many lattice sites, decaying rapidly to the left, but which eventually tunnels out on the right, just as in the ordinary Stark effect. This gives rise to one resonance $\lambda - i\Gamma$. Now shift the whole picture right one period, a. Physically, everything is exactly the same, except that the energy is decreased by aF. This gives the new resonance, $\lambda - i\Gamma - aF$. Additional translations generate the whole ladder.

At the present writing, there is apparently no proof of the existence of these ladders. Before such a proof could be given, however, a (mathematically) more basic question must be answered: what are the resonances poles of? Of an analytic continuation of certain matrix elements

$$<(H_F-z)^{-1}\phi,\phi>$$

across $\sigma(H_F)$, certainly; but this idea alone has no significance, since literally any point can be a pole if ϕ is chosen appropriately. One standard procedure is to take ϕ to be an analytic vector of a suitable unitary group; for example, the group of dilations or scale transformations (see the review article [5]). For the Stark ladder Hamiltonian, it is more appropriate to use the group of translations.

2. Translation Analyticity. Let us indicate formally a method for continuing the matrix element $<(H-z)^{-1}\phi,\phi>$, where

$$H = p^2 + V(x) - x$$

and $p^2 = -d^2/dx^2$. Let $U(\tau)$ be the unitary group of translations

$$U(\tau)f(x) = f(x-\tau)$$

and set

$$H(\tau) = U(\tau)HU(-\tau) = p^2 + V(x-\tau) - x + \tau$$

and

$$\phi(\tau) = U(\tau)\phi \quad .$$

We then have

$$\begin{aligned} <(H-z)^{-1}\phi,\phi> &= <U(\tau)(H-z)^{-1}U(-\tau)U(\tau)\phi,U(\tau)\phi> \\ &= <(H(\tau)-z)^{-1}\phi(\tau),\phi(\tau)> \ . \end{aligned}$$

This is valid for all real τ . Fix $\operatorname{Im} z > 0$, and continue the right side to complex values $\tau = ia$, $a > 0$. If ϕ is an entire vector for $U(\tau)$, this is possible for $\phi(\tau)$, while if $V(z)$ continues analyticly to a strip, then $V(x-ia)$ makes sense also, and so we obtain

$$<(H-z)^{-1}\phi,\phi> = <(p^2 + V(x-ia)-x+ia-z)^{-1}\phi(ia),\phi(-ia)> \ .$$

For $V \equiv 0$, the right side is analytic in $\operatorname{Im} z > a$, since p^2-x is self-adjoint. If, therefore, we know that

$$\sigma_{ess}(p^2 + V(x-ia) - x) = \sigma(p^2 - x) = \mathbb{R} \tag{1}$$

it would follow that $<(H-z)^{-1}\phi,\phi>$ continues meromorphically from $\operatorname{Im} z > 0$ to $\operatorname{Im} z > a$.

The difficulty in proving (1) is that the perturbation $V_a(x) = V(x-ia)$ is not relatively compact when $V(x)$ is periodic;

for example, we could have $V(x) \equiv 1$. On the other hand, the real part of V_a will not move the essential spectrum out of $\mathbb{R}$, while the imaginary part of V_a is small at least in the sense that its imaginary part must average to zero (by continuation). In any case, I.W. Herbst and the present author have recently succeeded in proving (1) for a rather general class of potentials analytic in a strip [6]. The assumptions on $V(x)$ are the following:

(i) $V(z)$ is analytic for $|\text{Im } z| < a$, for some $a > 0$

(ii) $|V(z)| \leq C(1+|z|)^{\frac{1}{2}-\varepsilon}$ for some $\varepsilon > 0$ and all z, $|\text{Im } z| < a$.

(iii) $V(x)$ is real-valued for real x.

We refer to [6] for a precise statement of the results. Note, however, that condition (ii) is surely satisfied for periodic $V(x)$, and even permits a relatively slow growth at infinity. This is important for the applications discussed in the final section.

The method of proof is to consider an operator

$$B(z,\tau) = -d^2/dx^2 - x + V(x-\tau) + Q(x,\tau,z) - z$$

for complex τ and z, where $Q(x,\tau,z)$ is a function of x vanishing at infinity, and to consider $H(\tau)-z$ as a relatively compact perturbation of $B(z,\tau)$. The point is that $B(z,\tau)$ can be inverted more-or-less explicitly because $Q(x,\tau,z)$ is chosen to make the solutions of

$$B(z,\tau)\phi = 0$$

the WKB approximations to the solutions of

$$H(\tau)\phi = z\phi \quad .$$

This procedure has much in common with Titchmarsh's treatment of the one-dimensional Stark problem and the work of Rejto and Sinha on Stark-like Hamiltonians [7].

3. Problems. We now wish to mention a few unanswered questions about the Stark ladder Hamiltonian.

(a) Existence of ladders. For small F, are there in fact ladders close to the real axis?

(b) Approximation. Can one obtain an asymptotic approximation for ladder resonances, valid for small F? The behavior of $\Gamma(F)$ is of particular interest because if ladders are to be observable, the width must vanish more rapidly than the ladder spacing aF; otherwise, the different resonances cannot be resolved.

Is $\Gamma(F) = o(F)$? For such calculations in the ordinary Stark effect and other cases, see, for example [8].

(c) Computation. Can the following idea be used to compute the positions of resonances numerically? The resonances are eigenvalues of the nonselfadjoint operator

$$H_F(i\tau) = -d^2/dx^2 + V(x+i\tau) - Fx - iF\tau.$$

Let $\phi_1, \phi_2, \ldots$ be an orthonormal basis, and compute numerically the eigenvalues of the finite matrix

$$\{\langle H_F(i\tau)\phi_i, \phi_j\rangle\}_{i,j=1,\ldots,N}$$

for some large N. Segments of one or more ladders should show up among them, if the basis set is well-chosen. Similar ideas have been used on atomic Hamiltonians, in connection with the dilation group [9].

(d) Spectral Concentration. Can any interesting result on spectral concentration be obtained here, in analogy with the work on the ordinary Stark effect? [10]. One very interesting result has just been obtained by R.B. Lavine [11] using his local spectral density which is reported on elsewhere in this volume. Let H be a Hamiltonian on the line, and Q the coordinate operator

$$Qu(x) = xu(x)$$

If f and g are nice functions, Lavine's Spectral Density is the measure ζ defined by

$$\iint |f(x)|^2 |g(E)|^2 d\zeta(x,E) = \mathrm{tr}\{\overline{g(H)}\,|f(Q)|^2 g(H)\} .$$

The measure ζ measures the spectral distribution of states of energy E. Lavine has shown that if ζ_F is the spectral density for the Stark ladder Hamiltonian

$$H_F = p^2 + V(x) - Fx$$

then as $F \to 0$, ζ_F is concentrated on the set

$$\{(x,E): E + Fx \in \sigma(H_0)\}$$

which consists of strips, based on the conduction bands of H_0, and sloping off to the right with slope $-F$.

4. Random Impurities. We wish to introduce a random potential which models the potential due to impurities distributed randomly throughout the crystal. We shall assume that the impurities are all particles of the same type, and are distributed at the points of a Poisson set.

Let $(P,\Omega,\mathcal{F})$ be a probability space for a Poisson ensemble of points on $\mathbb{R}$. For each ω, let $X_j(\omega)$, $j = 0,\pm1,\pm2,\ldots$ be the points of the set, numbered so that

$$\ldots < X_{-1} < 0 < X_0 < X_1 < \ldots$$

The number $N[S]$ of points in the Borel set S is Poisson, with mean $\alpha|S|$, where α is the density of points, and $|S|$ is the Lebesgue measure of S. If $u(x)$ is the potential of a single impurity stationed at the origin, then the potential due to the entire system of impurities is

$$V(x,\omega) = \sum_{j=-\infty}^{+\infty} u(x-X_j(\omega))$$

If $u(x)$ is bounded and integrable, then $V(x,\omega)$ is finite a.s. and, has finite mean

$$EV(x) = \alpha\int_{-\infty}^{+\infty} u(x)dx$$

and finite variance

$$E|V(x) - EV(x)|^2 = \alpha\int_{-\infty}^{+\infty}|u(x)|^2 dx$$

In the context of the shot effect in vacuum tubes, these formulas are known as Campbell's Theorem [12].

The result proved in [6] by Ira Herbst and the author is the following: Let $u(z)$ be analytic for $|\text{Im } z| < a_0$, and real for real z; if

$$\int_{-\infty}^{\infty} |u(x+ia)|dx < \infty$$

for $-a_0 < a < a_0$, then the potential $V(z,\omega)$ satisfies the conditions of section 2 for a.e. ω. Hence, meromorphic continuation of

$$\langle (p^2 + V(x,\omega) - Fx - z)^{-1}\phi,\phi\rangle$$

is possible for translation entire vectors ϕ for a.e. ω. In fact, a bit more is proved: for a.e. ω ,

(2) $V(x,\omega) = O(\log|x|)$

at infinity. (Of course, $V(x,\omega)$ does not really grow at infinity; it simply has large bumps near points where a large number of the points X_j are clustered. The estimate (2) says that one must go far out to find a big bump.)

The same continuation result must also hold for the operator

$$H_1 = p^2 + V_0(x) + V(x,\omega) - Fx$$

where $V_0(x)$ is a (non-random) periodic function satisfying the hypotheses of section 2. This represents a crystal with random impurities in an electric field, and the continuation result says that H_1 is absolutely continuous (except possibly for a set eigenvalues with no finite accumulation point).

For F = 0, the situation is quite different. For a class of random potentials very similar to ours - including the case $u(x) = \delta(x)$, the delta function - it has been shown [13] that, almost surely,

$$p^2 + V(x,\omega)$$

has <u>pure point spectrum</u>, <u>which is everywhere dense in</u> $[0,\infty)$! This sensational phenomenon is known as Anderson localization [14]. Our result shows that this behavior disappears when a field is applied.

There are some unanswered questions here,too. For one thing, the results of [13] do not appear to apply directly to our potential $V(x,\omega)$, so it would be desirable to have a proof that localization indeed takes place. If it does (as one certainly suspects), can anything be said asymptoticly about the location of resonances as $F \to 0$?

5. <u>Spectrum of</u> H_F. In this section, we give a rather expected result on the spectrum of H_F, which, however, was left implicit in [6]. The arguments are rather simple and general, and would apply in many similar contexts.

<u>Theorem</u>. <u>Under the hypotheses of section</u> 2, <u>the spectral measure of</u> H_F <u>is equivalent to Lebesgue measure</u>. <u>In particular</u>, H_F <u>is absolutely continuous</u>, <u>and</u> $\sigma_{ess}(H_F) = \mathbb{R}$.

For the proof, let $f_\pm(z) = \langle (H_F - z)^{-1}\phi, \phi\rangle$, for $\pm \mathrm{Im}\, z > 0$, with ϕ a translation entire vector. According to [6], $f_\pm(z)$ both continue meromorphicly to $|\mathrm{Im}\, z| < a$, and have poles contained in the fixed discrete set $\sigma_{disc}(H_F(i\tau))$, which is independent of ϕ. Excluding possible real poles, the spectral density for ϕ is

$$g(\lambda) = (2\pi i)^{-1}[f_+(\lambda)-f_-(\lambda)]$$

for λ real. If $g(\lambda)$ were to vanish on a set with accumulation point one would have $f_+(z) = f_-(z)$ for all z. This would mean that $<(H_F-z)^{-1}\phi,\phi>$ would be meromorphic in the plane, with only real poles, contained in a fixed discrete set S independent of ϕ. This would imply that $\sigma(H_F) \subset S$ and hence that H_F has discrete spectrum.

As R. Lavine has pointed out, this cannot occur. For the eigenvalues of H_F must have finite multiplicity. This can be seen by an abstract argument based on translation-analyticity (cf. [15]). Alternatively, the multiplicity cannot exceed two, because a second order linear ordinary differential equation has only two independent solutions. The operator $(H_F-z)^{-1}$ is therefore compact, since it can be written as a diagonal matrix with the diagonal elements $(\lambda_n-z)^{-1}$ tending to zero as n tends to infinity. However, this would imply that

$$(p^2 - Fx - z)^{-1} = (H_F-z)^{-1} - F(H_F-z)^{-1} x (p^2 - Fx - z)^{-1}$$

is also compact, and this is known to be false.

It remains to argue that there are, in fact, no eigenvalues of H_F. This follows from a result of J. Walter [16]. The potential

$$q(x) = V(x) - Fx$$

satisfies conditions (3) and (4) of [16] for $x > 0$ and any real λ. (This is clear when $V(x)$ is periodic; for the general case, Cauchy's estimate shows that $V'(z)$ and $V''(z)$ are also of the order $(1+|z|)^{\frac{1}{2}-\varepsilon}$ at infinity.) Walter proves, under this assumption, that no solution of

$$-u'' + q(x)u = \lambda u$$

with the boundary condition

$$u(0)\cos\alpha = u'(0)\sin\alpha$$

can be in $L_2[0,\infty)$. Since every solution satisfies such a condition for some α, there are no L_2 solutions of

$$H_F u = \lambda u \qquad \text{for real } \lambda .$$

REFERENCES

1. Avron, J.E., Phys. Rev. Lett. 37,1568 (1976); J. Phys. A. Math. Gen. 12, 2393-8 (1979).
Wannier, G.H., Phys. Rev. 117, 432 (1960); 181, 1364 (1969); Rev. Mod. Phys. 34, 645 (1962).
Zak, J., Phys. Rev. Letters 20, 1477 (1968); Phys. Rev. 181, 1366 (1969).
2. Avron, J.E. and Herbst, I.W., Comm. Math. Phys. 52, 239-254 (1977).
3. Reed, M. and Simon, B., "Methods of Modern Mathematical Physics" volume IV, Academic Press, New York 1978 (Section XIII.16).
4. Kittel, C., "Introduction to Solid State Physics", Wiley, New York, 2nd edition 1956 (Chapters 11 and 12).
5. Simon, B., Int'l J. Quantum Chem. 14, 529-542 (1978).
6. Herbst, I.W. and Howland, J.S., The Stark ladder and other external field problems, Comm. Math. Phys. (to appear).
7. Rejto, P.A. and Sinha, K., Helv. Phys. Acta 49, 389-413 (1976). See also reference 10.
8. Howland, J.S., J. Math. Anal. Appl. 50, 415-437 (1975).
Simon, B., Ann. Math. 97, 247-274 (1973).
Howland, J.S., Imaginary part of a resonance in barrier penetration, University of Virginia Preprint, 1979; J. Math. Anal. Appl. (to appear).
Simon, B. and E. Harrell, The mathematical theory of resonances whose widths are exponentially small, preprint 1980.
9. See: Junker, B.R., Int'l J. Quantum Chem. 14, 529-542 (1978); Cerjan, et al. ibid. 393-418; and other articles in the same issue.
10. Titchmarsh, E.C., Proc. Roy. Soc. A200, 34-46 (1949); A201, 473-479 (1950); A207, 321-328 (1951); A210, 30-47 (1951), J. Analyse Math. 4, 187-208 (1954/56); Proc. Lon. Math. Soc. 5, 1-21 (1955), Riddell, R.C., Pac. J. Math. 23, 377-401 (1967).
11. Lavine, R.B., The local spectral density and its classical limit, University of Virginia Preprint, 1980.
12. Feller, W., An Introduction to Probability Theory and its Applications, vol. II, 2nd edition, Wiley, New York, 1971.
13. Goldstein, I.Ja. and Molchanov, S.A., Dokl. Akad. Nauk SSSR 230, 761-764 (1976).
Goldstein, I.Ja., Molchanov, S.A. and Pastur, L.A., Funkts. Anal. Prilozhen. 11, 1-10 (1977).
Molchanov, S.A., Math. USSR Izvestia 12, 69-101 (1978).
14. Anderson, P.W., Phys. Rev. 109, 1492 (1958).
Thouless, D.J., Phys. Reports 13C, 95 (1974).
15. Aguilar, J. and J. Combes, Comm. Math. Phys. 22, 269-279 (1972).
16. Walter, J., Math. Zeit. 129, 83-94 (1972).

DYNAMICS OF FORCED COUPLED OSCILLATORS: CLASSICAL PHENOMENOLOGY OF INFRARED MULTIPHOTON ABSORPTION

Ramakrishna Ramaswamy and R. A. Marcus

Division of Chemistry and Chemical Engineering
California Institute of Technology
Pasadena, CA 91125

ABSTRACT

The classical mechanics of a system of two nonlinearly coupled oscillators of incommensurate frequencies driven by an oscillating electric field are studied. The presence of quasi-periodic and chaotic motion in the unforced system is shown to influence the nature of energy absorption. Two essentially different types of behavior are observed. In the first, energy is exchanged in a periodic manner between the system and the forcing field. The exact results are compared with perturbative analysis (based on Lie-transform techniques) employed in this regime. In the second regime, the energy exchange is erratic and a statistical analysis of a family of trajectories shows the role of the chaotic motion in the unforced system in the dissociation process. The results of the theory are compared with those obtained from an ensemble of exact classical trajectories and found to be in reasonable agreement.

INTRODUCTION

The phenomenon of multiphoton absorption in molecules has been intensively studied in the past few years.[1,2] Several theoretical models[3,4,5] have been proposed for the treatment of this problem. One prevalent qualitative scheme is based on the separation of molecular eigenstates into sets of discrete, quasi-continuous and continuous levels with coherent absorption of energy in the discrete set, and incoherent absorption by the quasi-continuum and continuum, followed by dissociation.

Of interest to such studies is the classical phenomenology of (forced) driven molecular systems and the nature of mode-mode energy transfer in facilitating dissociation. As an initial approach, we have studied[5] a system of nonlinearly coupled oscillators under the influence of an external field coupled to only one of the degrees of freedom. In essence, this resembles a simplified molecule interacting with a laser field.

In the first section we describe the classical Hamiltonian and some features of the autonomous and of the forced system dynamics, for a system without a zeroth order internal resonance. (For such nonlinear systems, exact analytic results are usually not possible; there is, however, the well-known KAM theorem[6] regarding the stability of motion under perturbations for both the autonomous and non-autonomous cases.) An approximate analytic and statistical theory is presented in the next section, followed by numerical results from trajectories and comparison with theory. A concluding discussion is given in the final section. The present symposium paper is an abbreviated version of a more detailed presentation given elsewhere.[5]

THE CLASSICAL HAMILTONIAN

Autonomous System

The Hamiltonian of the unforced coupled oscillator system investigated is

$$H' = \tfrac{1}{2}(p_x^2 + p_y^2 + \omega_x^2 x^2 + \omega_y^2 y^2) + \lambda x(y^2 + \eta x^2) \qquad (1.1)$$

where (x, p_x, ω_x) and (y, p_y, ω_y) denote the coordinate, momentum and zeroth-order frequency, respectively. The values of parameters chosen here are $\omega_x = 1.3$, $\omega_y = 0.7$, $\lambda = 0.1$, $\eta = -1$; this type of Hamiltonian has often been used in the nonlinear dynamics literature, and the parameters for the potential energy surface are similar to those used in Ref. 7, although the larger value of η here corresponds to a higher anharmonicity than that used previously.[7b] The three saddle points for the dissociation channels are located at $(x,y) = (5.63,0)$ and $(-2.45, \pm 7.71)$ with a minimum dissociation energy of $E = 6.54$ units (at the last two points).

It is well known that the dynamics of the Hamiltonian H' has a rich structure associated with it.[7a] The trajectories are either quasi-periodic in time or "chaotic". The basic difference between these two types of motion is that the former trajectories are confined to a torus in phase space while the latter are not. This difference is easily characterized by the Poincaré surfaces-of-section,[7] which for the former type are smooth curves, while for the latter they are a seemingly random set of points.

It has become convenient in the discussion of such systems to describe a critical energy[8] E_c above which most initial conditions lead to chaotic type trajectories. From the surfaces-of-section for the motion,[5] one can extract the relative fraction of phase space that leads to chaotic motion, by measuring the relative area not covered by smooth curves. This is shown in Fig. 1.

The Nonautonomous (Forced) System

The interaction with the driving term is chosen to occur through the y-degree of freedom, giving the total Hamiltonian,

$$H = H' - F\, y \cos \omega t \tag{1.2}$$

Here the driving frequency ω is equal to ω_y. Hamilton's equations obtained from (1.1) are

$$\begin{aligned} \dot{x} &= p_x, \quad \dot{y} = p_y \\ \dot{p}_x &= (\omega_x^{\,2}\, x + \lambda y^2 + 3\lambda\eta x^2) \\ \dot{p}_y &= -(\omega_y^{\,2} y + 2\lambda xy) + F \cos \omega t \end{aligned} \tag{1.3}$$

and can be integrated numerically.

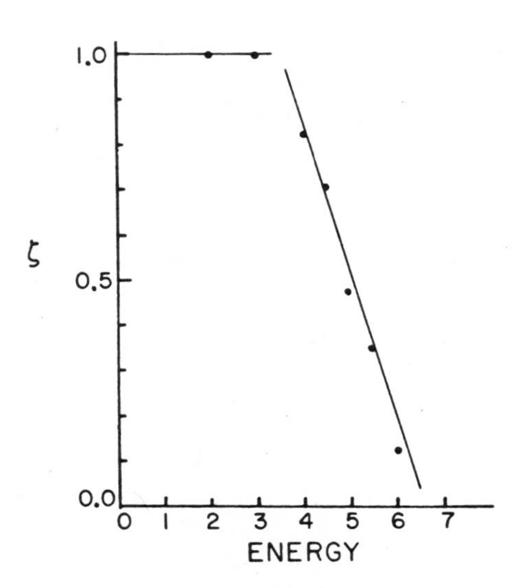

Fig. 1: Fraction of phase space covered by tori as a function of energy.[5]

We first consider some principal qualitative features associated with a typical trajectory. Shown in Fig. 2 is the total energy content of the oscillators, E (i.e., H'(t)) as a function of time. Two regions of behavior may be identified, separated by a vertical line in Fig. 2: Regular energy exchange between the system and the field, with a definite set of associated frequencies. Erratic energy exchange between the system and the field with several associated frequencies--in marked contrast to the previous region. Arrival at this region was, for all trajectories studied here, ultimately accompanied by dissociation, as in Fig. 2.

A related type of behavior was observed in an earlier classical trajectory study[9a] of multiphoton absorption in CD_3Cl (Fig. 4 of Ref. 9a).[5] (See also Ref. 9b for an examination of individual trajectories.) The actual behavior of individual trajectories can differ considerably in the extent to which they sample the two regimes. Before presenting the numerical results obtained by integrating Eq. (1.3), we first present an approximate analytic and statistical theory of the process.

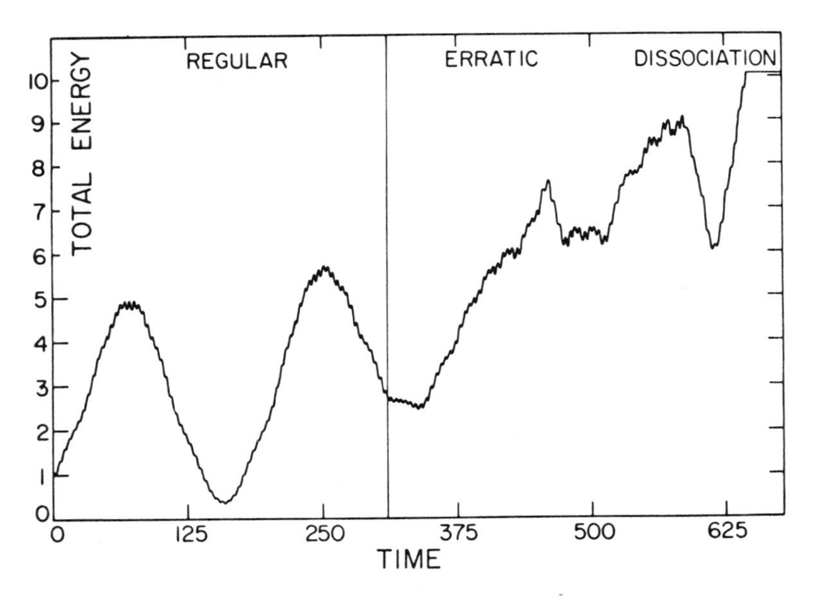

Fig. 2: Time dependence of the total energy for a given initial condition.[5]

THEORIES FOR THE TWO REGIMES

The Regular Regime

In this regime, we find[5] that the overall behavior of the total energy is well duplicated simply by treating the total system (with $F \neq 0$) as a single, y-type oscillator of altered frequency $\tilde{\omega}_y$, driven by the external field. Thus, we now get the equation of motion

$$\ddot{y} + \tilde{\omega}_y^{\,2}\, y = F \cos \omega t \tag{2.1}$$

where $\tilde{\omega}_y$ is not equal to ω_y and may be determined[5] by standard classical perturbation methods.[10]

Solution of Eq. (2.1) yields

$$y(t) = K \sin (\tilde{\omega} t + \theta) + \frac{F}{\tilde{\omega}y^2 - \omega^2} \cos \omega t \tag{2.2}$$

where K and θ are determined by the initial conditions. The total energy behavior, i.e., H'(t) determined mainly from (2.2), is shown in Fig. 3, and it can be seen by comparison with Fig. 2, that the overall features have the same behavior as the numerical results in the regular regime. In the event that mixing is widespread in the autonomous system (as in a 1:1 or 1:2 resonance case), replacement of the coupled system by a single oscillator may not be possible. The frequency

$$\beta = \tfrac{1}{2}(\tilde{\omega}_y - \omega) \tag{2.3}$$

determines the long periodicity in the energy behavior, while the shorter oscillations have period equal to π/ω_y.

The Erratic Regime

In this regime, the total energy of the oscillators fluctuates in time in an irregular manner. The difference between this and the previous regime is similar to that between quasi-periodic and chaotic motion in the autonomous system. The current absence of rigorous analytic methods that are applicable for treatment of chaotic behavior makes a statistical approach a useful first alternative.

We made[5] the following assumptions for our present system. 1) In the periodic regime for an individual trajectory, the total energy is approximately the same for all maxima. 2) The erratic regime sets in for an individual trajectory when the forcing term leads the system into a portion of phase space where the underlying motion is chaotic. 3) All systems in the erratic regime ultimately dissociate. 4) The probability of entering the erratic regime

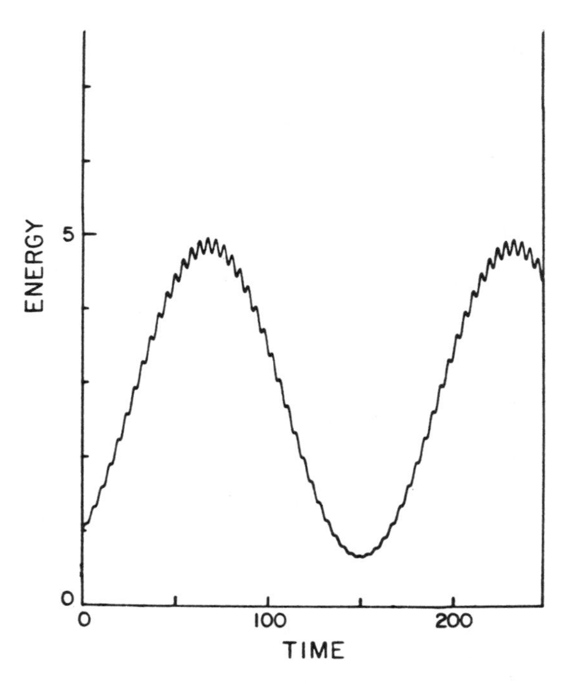

Fig. 3: Time dependence of total energy in the regular regime predicted by the Lie transform analysis.[5]

at a given peak is simply that fraction, $(1-\zeta)$, of phase space not covered by tori.

With the above assumptions, we deduced[5] that the fraction $f_s(t)$ of trajectories that are surviving at time t is given approximately by

$$f_s(t) = \zeta^{\beta t/\pi} \tag{2.4}$$

Here β depends on the properties of the regular regime, and ζ is dependent on those of the autonomous dynamical system in the regular-chaotic regime. β is given by (2.3).

NUMERICAL RESULTS

Regular Regime

In the quasi-periodic regime, a quantum state has its analogue in an eigen torus.[7b] We analyze the forced system in terms of a family of trajectories with initial conditions uniformly chosen over the torus corresponding to the ground state of the system (1.1).

Using the perturbation method of the Lie transform,[10] which is described in Ref. 5, the dependence of various dynamical quantities on system parameters is somewhat complicated. However, the long periodicity, which is determined by β, was 150 time units compared to the observed value of 160 time units. The amplitude of the first peak in Fig. 2, was 5.1, which agrees well with the value of 5.0 predicted by the Lie transform analysis[5] (cf Fig. 3).

Erratic Regime

The connection between the onset of erratic behavior in the forced system and chaotic behavior in the forced system may be examined by turning off the field along the trajectory and then allowing the system to evolve from this "initial condition". When the field was turned off in the regular regime, a quasi-periodic trajectory was obtained, while turning off the excitation in the erratic regime produced a typically chaotic trajectory.

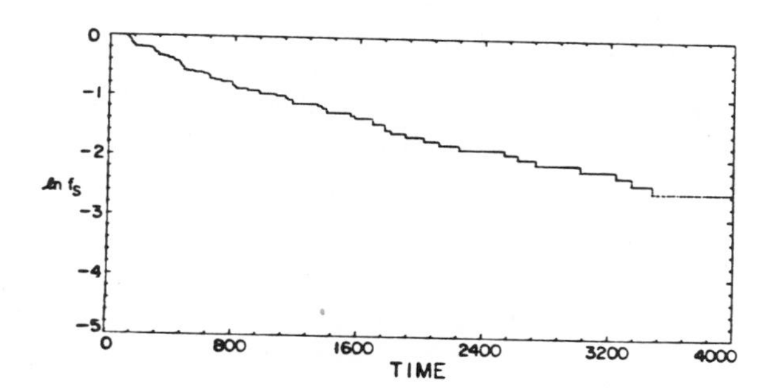

Fig. 4: Ln of fraction $f_s(t)$ of surviving trajectories as a function of time for F = 0.1, ω = 0.7.[5]

On a logarithmic scale, $f_s(t)$ is very roughly linear, (during the first few half lives) in accord with Eq. (3.4) and has a slope at half life of -0.002. In applying Eq. (3.4), we estimate $\zeta = 0.5$ at $E_{max} = 5.0$ (from Fig. 1). Thus $(\beta/\pi) \ln \zeta = -0.004$. This value agrees to a factor of two with the empirical value obtained from Fig. 4. It must be emphasized that the model is only a first approximation. The decrease in slope of the $\ln f_s(t)$ vs t curve with time in Fig. 4, may be due to the residual unreacted systems being "locked" into a regular part of phase space instead of sampling the latter more randomly.

CONCLUSION AND SUMMARY

In our study of this problem we have formulated a classical phenomenology of a driven coupled oscillator system. Two essentially different kinds of behavior can be distinguished: the exchange of energy between system and field occurs with a well-defined time scale deriving from several aspects of the motion in the forced and unforced system. Secondly, the energy behavior can become highly erratic, and the latter motion inevitably led to dissociation under the conditions studied. The contrasting regular and erratic regimes in the forced system have their immediate analogue in quasi-periodic and chaotic motion in the autonomous system.

A time lag is observed in the appearance of dissociation when an ensemble of trajectories is analyzed. The distribution of lifetimes shows a power-law decay, which is tied into the extent of stability in the motion of the autonomous system. This distribution can be understood semi-quantitatively, and it is demonstrated how the parameters governing the decay rate can be approximately related to the parameters of the system.

In larger systems that are more typically "molecular", the analysis will necessarily become more complicated although the essential physics of multidimensional forced systems is likely to be similar for systems without internal resonances. The presence of several coupled modes will probably reduce the overall periodicity,[9] and indications are that chaotic motion can occur at fairly low energies (compared to dissociation). In higher-dimensional systems, methods based on spectral characteristics[11] or on the mean rate of separation of nearby trajectories[12] are more suited for measuring the extent of chaos since surfaces of section become considerably more difficult to compute. At this time considerable effort is being devoted in the literature to the prediction of widespread chaos in such systems,[8] and it may ultimately be possible to obtain ζ without recorse to numerical experiments.

ACKNOWLEDGMENT

We are pleased to acknowledge the support of this research by a grant from the National Science Foundation.

REFERENCES

1. See, e.g.: E. Yablonovitch and N. Bloembergen, Physics Today 31, 23 (1978); M. F. Goodman, J. Stone and E. Thiele, in "Multiple Photon Excitation and Dissociation of Polyatomic Molecules, Topics in Current Physics," ed. C. D. Cantrell (Springer Verlag, NY 1980).
2. See, e.g.: a) R. V. Ambartzumian and V. S. Lekhotov, in "Chemical and Biochemical Applications of Lasers, Vol. 2," ed. C. B. Moore (Academic, NY, 1977); b) D. Bomse, R. L. Woodin and J. Beauchamp, in "Advances in Laser Chemistry," ed. A. H. Zewail (Springer, NY, 1978); c) Aa. Sudbø, P. A. Schulz, E. R. Grant, Y. R. Shen and Y. T. Lee, J. Chem. Phys. 70, 912 (1978); J. Chem. Phys. 69, 2312 (1978); d) P. A. Hackett, C. Willis and M. Gauthier, J. Chem. Phys. 71, 2682 (1979);

e) P. Kolodner, C. Winterfield and E. Yablonovitch, Optics Commun. 20, 119 (1977).

3. a) M. Quack, J. Chem. Phys. 68, 1281 (1978); b) N. Bloembergen, C. Cantrell and D. Larsen, in: "Tunable Lasers and Applications," eds. A. Mooradian, et al. (Springer Verlag, NY 1967); c) M. Goodman and J. F. Stone, Phys. Rev. A 18, 2618 (1978); d) S. Mukamel and J. Jortner, J. Chem. Phys. 65, 5204 (1976); e) E. R. Grant, P. A. Schulz, Aa. S. Sudbø, Y. R. Shen and Y. T. Lee, Phys. Rev. Letts. 40, 115 (1978); f) J. G. Black, E. Yablonovitch, N. Bloembergen and S. Mukamel, Phys. Rev. Letts. 38, 1131 (1977); g) E. Yablonovitch, Opt. Lett. 1, 87 (1977); h) J. Lyman, J. Chem. Phys. 67, 1868 (1977).
4. J. T. Lin, Phys. Lett. 70A, 195 (1979); also, W. E. Lamb, paper presented at Conference on Laser Chemistry, Steamboat Springs, Colorado, 1976.
5. R. Ramaswamy and R. A. Marcus, J. Chem. Phys. 73, 0000 (1980).
6. a) V. I. Arnold, "Mathematical Methods of Classical Mechanics," (Springer Verlag, NY 1978); b) J. Ford, in: "Fundamental Problems in Statistical Mechanics," ed. E. D. G. Cohen (Elsevier, 1975); c) see Appendix 8 in Ref. 6a.
7. a) M. Hénon and C. Heiles, Astron. J. 69, 73 (1964); b) D. W. Noid and R. A. Marcus, J. Chem. Phys. 62, 2119 (1975); D. W. Noid, M. L. Koszykowski and R. A. Marcus, J. Chem. Phys. 71, 2864 (1979); and references therein.
8. a) See P. Brumer, Adv. Chem. Phys. (in press) for a review. b) See M. Tabor, Adv. Chem. Phys. (in press) for a review. c) R. Ramaswamy and R. A. Marcus, to be published.
9. a) D. W. Noid, M. L. Koszykowski, R. A. Marcus and J. D. McDonald, Chem. Phys. Lett 51, 540 (1977); b) K. D. Hansel, ibid. 57, 619 (1978).
10. a) G. Hori, Publ. Astr. Soc. Japan 19, 229 (1967); b) G. E. O. Giacaglia, "Perturbation Methods in Nonlinear Systems," (Springer, NY, 1972).
11. R. A. Marcus, D. W. Noid and M. L. Koszykowski, in: "Advances in Laser Chemistry," ed. A. Zewail (Springer, NY, 1978); D. W. Noid, M. L. Koszykowski and R. A. Marcus, J. Chem. Phys. 67, 404 (1977); C. E. Powell and I. Percival, J. Phys. A12, 2053 (1979).
12. G. Benettin, L. Galgani and J. M. Strelcyn, Phys. Rev. A14, 2338 (1976).

IRREVERSIBILITY AND STOCHASTICITY OF CHEMICAL PROCESSES

K. Gustafson, R.K. Goodrich, and B. Misra

Dept. of Mathematics, University of Colorado, Boulder, Colorado 80309
Inst. de Physique et Chimie Solvay, ULB, Campus Plaine 1050 Brussels, Belgium

Abstract. Given a unitary process U_t on a state space describing statistically an underlying deterministic dynamics in a phase space Ω, can one make a change of representation Λ in which there is no information loss under either Λ or Λ^{-1} and yet such that the transformed process $W_t^* = \Lambda U_t \Lambda^{-1}$ is no longer deterministic? We show the answer is no. Lemma: Let Λ be a closed operator in $L^2(\Omega, s, \mu)$ with domain and range containing the characteristic functions of Borel sets in S, $\Lambda(1) = 1$, $\int \Lambda(\rho) = \int \rho$ for $\rho \in D(\Lambda)$, $\Lambda(\rho) \geq 0$ when $\rho \geq 0$. Then one cannot have $\Lambda^{-1}(\rho) \geq 0$ for $\rho \geq 0$ unless Λ is induced from a point transformation of the phase space. Results of this type establish the principle often used intuitively in chemistry that a forward moving (e.g., Markov) process that loses information cannot be reversed.

1. INTRODUCTION

The distinction between reversibility and irreversibility on the microscopic level is an important element in understanding the basic mechanisms of interacting particles. Let us cite two important instances.

In physics, as is well known, the difficulties of properly incorporating all retardation effects into the description of two interacting charged particles in an electromagnetic field are responsible for the renormalization rules used in quantum electrodynamics. More generally, similar considerations underly one of the main questions of modern physics, the dynamical definition of an elementary particle.

In chemistry, much recent work, e.g., as described in other papers at this conference, is based upon the view that regions of the phase plane in classical dynamics that exhibit ergodic behavior have their analogues in the quantum dynamics of molecules. Roughly, the idea is that a randomization of internal energy precedes bond breaking.

There are some recent mathematical investigations which bear upon these two questions just mentioned. We shall mention those bearing on the first only peripherally, and as concerns the second, rather than getting into the description of the phase plane analyses, some of which are exposed in other work reported on at this conference, we shall instead put forth recent work that looks at the problem statistically. These results do not pretend to answer those mentioned fundamental questions of physics and chemistry. They do, however, bring out and clarify the roles of two other important aspects of reversibility, namely, determinism and information loss in processes.

2. BACKGROUND

In George, Henin, Mayne, and Prigogine [1] and in other physics and chemistry papers, attention is focussed on reconciling a dynamical reversible evolution U_t and a related thermodynamic irreversible evolution W_t. Following [1], one may attempt to establish a change of representation Λ so that the given deterministic evolution is transformed into an evolution displaying thermodynamic properties. In particular, in [1] such Λ and corresponding Lyapunov entropy functionals are shown to exist in a Friedrichs field theoretical model.

Let us display this situation as follows, bearing in mind not only the Friedrichs model situation of [1] but also the more general possibility in chemical processes of including within a description more general effects incorporating dissipation, decay, dispersion, and possibly other nonconserved properties of the basic mechanism as it interacts with its environment. In terms of the infinitesimal generators we may therefore study transformations effecting the following:

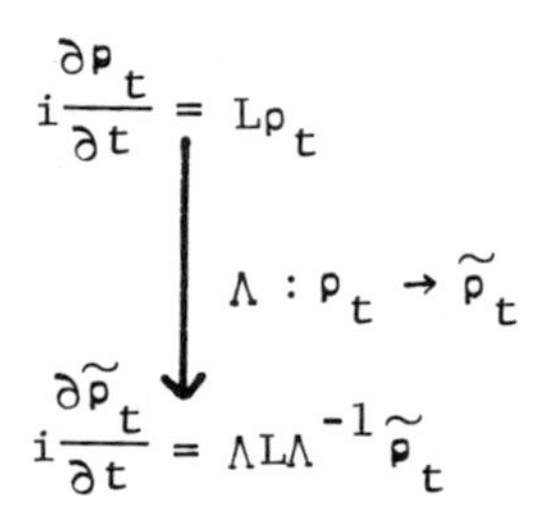

In Misra, Prigogine, and Courbage [2], the above problem is studied further, with emphasis on the change from deterministic systems to probabilistic descriptions. By requiring Λ to be invertible rather than, say, a projection, there is no drastic "contraction of description" or "coarse graining" involved in going to the stochastic description. In particular it is shown in [2] that such transformations Λ exist for all Bernoulli systems.

In [1] the Λ considered turned out to be star unitary: $\Lambda^*(L) \equiv \Lambda^\dagger(-L) \equiv \overline{\Lambda^*}(-L) = \Lambda^{-1}(L)$. This permitted the transformed Hamiltonian $i\Lambda L\Lambda^{-1}$ to have negative real part and hence dissipativity. The latter leads to an irreversibility in terms of an increasing entropy defined by an appropriate Lyapunov functional. In the more general setting of [2] Λ was to be (i) bounded selfadjoint on a measure space $L^2(\Omega, S, \mu)$, (ii) positivity preserving, (iii) trace preserving, (iv) information preserving, and (v) equilibrium preserving. In [2] it was conjectured that Λ^{-1}, even when unbounded, could not be of the same type. On a physical basis, this amounts to concluding that Λ^{-1} cannot be positivity preserving, for otherwise $W_t^* = \Lambda U_t \Lambda^{-1}$ would not be a truly probabilistic description to be arrived at from a deterministic evolution U_t.

3. IRREVERSIBILITY

In Goodrich, Gustafson, and Misra [3] the above conjecture is answered affirmatively, and for a larger class of transformations Λ.

Theorem [3]. Let (i) Λ be a closed not necessarily self-adjoint operator with domain $D(\Lambda)$ containing all characteristic functions of Borel sets E in S , (ii) $\Lambda(\rho) \geqq 0$ for $\rho \geqq 0$ and $\rho \in D(\Lambda)$, (iii) $\int \Lambda(\rho) d\mu = \int \rho d\mu$ for $\rho \in D(\Lambda)$, (iv) range $R(\Lambda)$ containing all characteristic functions E , (v) $\Lambda(1) = 1$, and (vi) $\Lambda^{-1}(\rho) \geqq 0$ for $\rho \geqq 0$ and $\rho \in R(\Lambda)$: then Λ is induced from a point transformation of the phase space.

In [2] the transformation Λ constructed effected a transformation from the dynamical group U_t to a strong Markov semigroup W_t^* . Thus the result of [3] establishes the following pringiple often used intuitively in chemistry.

<u>Theorem</u> (see [3]). A forward moving (e.g., Markov, chemical) process that loses information cannot be reversed.

For by the above result of [3] , given a reversible evolution U_t and change of representation Λ satisfying (i) - (vi) , the corresponding $W_t^* = \Lambda U_t \Lambda^{-1}$ must also be reversible. Thus by reversing this argument, beginning with a forward moving W_t^* that loses information, should there exist a way to convert it to a reversible U_t via a change of representation Λ within the class allowed above, then one could come back to W_t^* now no longer information losing.

4. PICTURE

Accepting the famous old adage that a good picture, even if well-known, is worth a thousand words, the following shows the relation between the evolution on states and the underlying dynamics:

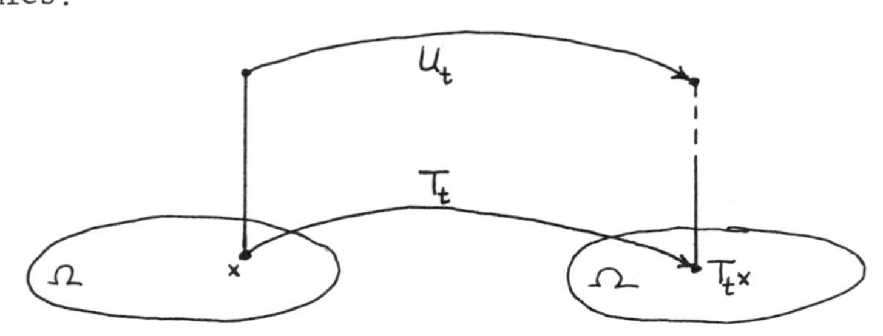

where

$$(U_t \rho)(x) = \rho(T_t x) \quad , \quad x \in \Omega \quad ,$$

and where

Ω = a constant energy surface in a phase space Γ ,

μ = an invariant finite measure with respect to an underlying point transformation family T_t on Ω ,

ρ = an $L^2(\mu)$ element of a space of states.

Also following the principle (e.g. as ennunciated by no less than Einstein, among others) that if a simple example cannot be supplied there is some lack of understanding on the part of the supplier, we offer the following simple example of a Λ satisfying (i) - (v) but not generated by an underlying point transformation of the phase space. Let $X = \Omega = \{0,1\}$, $\mu(0) = \mu(1) = 1/2$, and

$$\Lambda = \begin{bmatrix} 1/4 & 3/4 \\ 3/4 & 1/4 \end{bmatrix} .$$

Then we may see that (i) - (v) hold as follows.

(i) Λ is bounded selfadjoint invertible: $L^2 \to L^2$.

(ii) $\rho \geq 0 \Rightarrow \Lambda_\rho \geq 0$ (by positive entries).

(iii) $\int \Lambda \rho \, d\mu = \int \rho \, d\mu$ may be verified directly, or else indirectly by noting that $\Lambda^*(1) = 1$, $\Lambda = \Lambda^*$ here.

(iv) $D(\Lambda) = R(\Lambda) = L^2 \supset$ all characteristic functions.

(v) $\Lambda\binom{1}{1} = \binom{1}{1}$.

Suppose $(\Lambda\rho)(x) = \rho(Tx)$. Let $\rho = \binom{1}{0}$. Then $\rho(Tx)$ is either $\binom{0}{1}$ or $\binom{1}{0}$. But $(\Lambda\rho)(x) = \binom{1/4}{3/4}$.

5. DECAY

In Gustafson and Misra [4,5] a connection between certain stochastic processes of a "shift" type and quantum mechanical decay laws was investigated. There one ends up considering reduced evolutions of the form $Z_t = PU_tP$ in models for unstable particles such as mesons. In other words, "Λ" is a projection P , and if interpreted as in the present paper, Λ loses information by restricting the overall evolution to functions in a subspace or more visually to functions of restricted support in the case that P is given by a spatial characteristic function.

There is a similar loss of information in the "coarse graining" approaches to dissipative description in statistical mechanics. In coarse graining schemes one in essence projects (via a conditional expectation) onto a reduced system exhibiting dissipative behavior from a total system on which there is a reversible evolution. The complementary subspace is regarded as a space of heat bath states.

In unstable particle models the complementary subspace is the subspace into which the decay products disappear.

The same philosophy underlies the development of quantum dynamical semigroups for quantum measurement theory, see Davies [6]. Interactions with the external environment in this case occur in the measuring process.

Nearly always, (noncommuting) interactions with an external environment produce a semigroup rather than a group. Indeed, as Hille [7] remarked over thirty years ago, everywhere he looks, he seems to see a semigroup!

6. ENTROPY

In Misra [8] Lyapunov functionals of the form $\langle \Psi, U_{-t} MU_t \Psi \rangle$ are constructed for certain processes of mixing type. We leave a more full description of those results to [8] and [9]. Here we would like to point out that such Lyapunov variables are usually arrived at in one of two ways. Either M is constructed in terms of a change of representation Λ of the type described herein (e.g., $M = \Lambda^* \Lambda$), or M may be taken as a function $h(Q)$ where Q is a conjugate operator to the given Hamiltonian L, somewhat in the spirit of [4,5].

Let us recall some gradations of mixing processes.

K - flow $\Rightarrow$ Liouvillian L restricted to the orthogonal complement of the one dimensional subspace N_0 of constant functions on the energy surface is absolutely continuous with uniform spectral multiplicity and spectrum $(-\infty, \infty)$

$\Rightarrow$ (see Misra [6]) the existence of a Lyapunov variable describing a nonequilibrium entropy

$\Rightarrow$ L restricted to $N_0^{\perp}$ is absolutely continuous

$\Rightarrow$ mixing

$\Rightarrow$ ergodicity (N_0 = Null space of L)

In Misra [8] Lyapunov functionals of the form $\langle \Psi, U_{-t} MU_t \Psi \rangle$ which increase with t for all $\Psi \notin N_0$ are constructed for mixing processes under the conditions shown above.

There is no uniqueness of the entropy in [8] and as is well known there are many different entropies, the choice depending on the situation. See for example Marchand [10] .

As pointed out in [8] , to find ways to escape the necessity of requiring conditions of mixing in the justifications of non-equilibrium thermodynamics it was suggested in [1] and succeeding works that it would be desirable to allow M to be unbounded. As shown in [3] this can be done for M constructed in terms of Λ's if one does not demand too much on the latter.

7. BERNOULLI SYSTEMS

See [2] for an entropy construction and transformation from the deterministic Baker's transformation to a genuinely stochastic Markov process. As shown in [2] the same equivalence holds from general Bernoulli systems to stochastic Markov processes via an appropriately constructed change of representation Λ found as the square root of a Lyapunov variable M .

Combining the considerations of [2] and [3] , we would like to close with a strategy for providing an alternate proof of Ornstein's isomorphism theorem on Bernoulli systems. Consider two Bernoulli systems possessing the same Kolmolgorov-Sinaı entropy, and let U_1 and U_2 be their induced unitary operators. It is known that any two such U_1 and U_2 are equivalent to a unitary V which moreover can be chosen so that $V(1) = 1$. If V can be chosen positivity preserving, then by [3] the corresponding underlying point isomorphism T establishes the Ornstein isomorphism theorem.

References

1. C.I. George, F. Henin, F. Mayne, and I. Prigogine, New quantum rules for dissipative systems, Hadronic Journal 1 (1978), 520-573.
2. B. Misra, I. Prigogine, and M. Courbage, From deterministic dynamics to probabilistic descriptions, Physica, 98A, 1 , to appear.
3. K. Goodrich, K. Gustafson, and B. Misra, On a converse to Koopman's Lemma, Physica, to appear.
4. K. Gustafson and B. Misra, Correlations and evolution equations, Proc. III Mexico-United States Symp. on Diff. Eqns, Mexico City, January, 1975, Boletin de la Sociedad Matematica Mexicana, 1975.
5. K. Gustafson and B. Misra, Canonical commutation relations of quantum mechanics and stochastic regularity, Letters in Mathematical Physics 1 (1976), 275-280.
6. E.B. Davies, Quantum theory of open systems, Academic Press,

London, 1976.

7. E. Hille, Functional analysis and semigroups, Amer. Math. Soc. Colloq. Public. 31, Providence, Rh. I. (1948).
8. B. Misra, Nonequilibrium entropy, Lyapunov variables, and ergodic properties of classical systems, Proc. Nat. Acad. Sci. USA 75 (1978), 1627-1631.
9. B. Misra, On nonunitary equivalence between unitary groups of deterministic dynamics and contraction semi-groups of Markov processes, these proceedings.
10. J.P. Marchand, Statistical inference in quantum mechanics, these proceedings.

ON THE REPRESENTATIONS OF THE LOCAL CURRENT ALGEBRA AND THE GROUP OF DIFFEOMORPHISMS (I)

Daoxing Xia*

Recently, several physicists and mathematicians[1-8] have investigated the representations of the local current algebra and the group of diffeomorphisms, that are motivated by the theory of quantum physics and statistical physics. These representations are closely related to quasi-invariant measures[1-3],[6]. But the investigation on the measure, which is quasi-invariant with respect to the group of diffeomorphisms, began only a few years ago. In this paper, firstly we give the analytic expression of the Radon-Nikodym derivative of measure on the space of generalized functions, which is quasi-invariant with respect to the group of diffeomorphisms. Secondly, by means of this expression we give a method of finding out a class of representations of local current algebra in the theory of quantum physics.

Let φ be a diffeomorphism of n-dimensional Euclidean space R^n onto itself, which is identical mapping outside some compact set depending upon φ. The set of all such mappings is denoted by $\mathrm{Diff}(R^n)$. The set $\mathrm{Diff}(R^n)$ becomes a group by introducing the composition of mappings as group operation. The topology in $\mathrm{Diff}(R^n)$ is defined such as in [2]. Let K_n^m be the space of all test functions that are C^∞ mappings from R^n to R^m with compact supports. K_n^m becomes linear topological space, if it is endowed with Schwartz's topology[9]. In quantum physics, it is

---

*The author thanks Professor Karl Gustafson for his invitation to attend the meeting and kind hospitality. The author is also acknowledging Professor Mityagin and the Department of Mathematics, The Ohio State University for their help to complete this paper during his visit in Columbus, Ohio.

reasonable to consider the subgroup $D_0(R^n)$ of all $\varphi \in \mathrm{Diff}(R^n)$ such that there is a continuous function, $\varphi_t \in \mathrm{Diff}(R^n)$, $t \in [0,1]$, with piecewise continuous derivative φ_t', satisfying $\varphi_0 = I$ (the identical mapping) and $\varphi_1 = \varphi$. This connected subgroup of topological group $\mathrm{Diff}(R^n)$ is an infinite dimensional Lie group. In the linear space K_n^n, we introduce a non-associative operation as follows, $[g,h] = h^j\partial_j g - g^j\partial_j h$. Then K_n^n becomes the Lie algebra of the Lie group $D_0(R^n)$. The exponential mapping can be expressed as follows. For $g \in K_n^n$, the solution φ_t of the differential equation

$$\frac{d}{dt}\varphi_t = g \circ \varphi_t \tag{1.1}$$

with initial condition $\varphi_0 = I$ is denoted by φ_t^g. The exponential mapping is exp: $g \mapsto \varphi_t^g$, i.e. $\exp[tg] = \varphi_t^g$. In quantum physics, when the local current algebra[3-4],[6-8] is investigated, the following infinite dimensional Lie group $\mathfrak{G}$ must also be considered.

In the theory of local current algebra, the commutational relations of the continuous unitary representation U and V are considered:

$$U(f_1)U(F_2) = U(f_1 + f_2), \quad V(\varphi)U(f) = U(f \circ \varphi)V(\varphi),$$
$$V(\varphi_1)V(\varphi_2) = V(\varphi_1 \circ \varphi_2). \tag{1.2}$$

Let $U(tf) = e^{it\rho(f)}$ and $V(\exp[tg]) = e^{it\,J[g]}$ define the operator-valued generalized functions $\rho(x)$ and $J_j(x)$ by the equations

$$\rho(f) = \int \rho(x)f(x)dx, \quad J(g) = \int g^j(x)J_j(x)dx$$

respectively. The operators $\rho(f)$ and $J(g)$ satisfy the formal commutation relations

$$[\rho(f_1), \rho(f_2)] = 0, \quad [\rho(f), J(g)] = i\rho(g^j\partial_j f),$$
$$[J(g_1), J(g_2)] = iJ([g_1,g_2]). \tag{1.3}$$

The local current algebra $\{\rho(x), J(x)\}$ can be considered as an infinite dimensional Lie algebra with operations defined by (1.3).

Let K_n^* be dual space of K_n^1, i.e. the space of generalized functions, and $\mathfrak{B}$ be the smallest complete Borel field of subsets of K_n^* such that every linear functional (F,f), $F \in K_n^*$, $f \in K_n^1$ is measurable with respect to $\mathfrak{B}$. Let μ be the σ-finite non-negative measure on $(K_n^*,\mathfrak{B})$. For arbitrary $\varphi \in D_0(R^n)$, we define the mapping φ^* from K_n^* to K_n^* by the equations

$$(\varphi^* F, f) = (F, f \circ \varphi),\ F \in K_n^*,\ f \in K_n^1.$$

If for every $\varphi \in D_0(R^n)$, the measure $\mu(\varphi^*(\cdot))$ is equivalent to the measure $\mu(\cdot)$, then we say that μ is a quasi-invariant measure. In quantum theory of physics, the following unitary representation[3-4],[6-8] of group $\mathfrak{G}$ is considered. Let μ be a quasi-invariant measure, $\mathcal{H} = L^2(K_n^*, \mathfrak{B}, \mu)$. For any $(f, \varphi) \in \mathfrak{G}$, we construct the operators $U(f)$ and $V(\varphi)$ in the unitary representation of $\mathfrak{G}$ as follows:

$$(U(f)\Psi)(F) = e^{i(F,f)}\Psi(F), \tag{1.4}$$

$$(V(\varphi)\Psi)(F) = \Psi(\varphi^* F)(d\mu(\varphi^* F)/d\mu(F))^{1/2},\ \Psi \in \mathcal{H}, \tag{1.5}$$

where $d\mu(\varphi^* F)/d\mu(F)$ is the Radon-Nikodym derivative of measure $\mu(\varphi^*(\cdot))$ with respect to $\mu(\cdot)$. In order to investigate about unitary representation, we must investigate $d\mu(\varphi^* F)/d\mu(F)$.

In the representations (1.4-1.5), there is a vector $\Omega(F) \equiv 1$ in $\mathcal{H}$, which is called ground state and has a special physical significance. In physics, it is often needed that

$$\Omega \in D(J(g)),\quad g \in K_n^n, \tag{1.6}$$

where $D(J(g))$ denote the domain of self-adjoint operator $J(g)$. The condition (1.6) is equivalent to the existence of $\frac{d}{dt} V(\varphi_t^g)\Omega\Big|_{t=0}$. From (1.5), we know that it is also equivalent to the existence of function

$$S(g,F) = \frac{d}{dt}\left(\frac{d\mu(\varphi_t^{g*}F)}{d\mu(F)}\right)^{1/2}\Bigg|_{t=0} \tag{1.7}$$

and $S(g,\cdot) \in L^2(K_n^*, \mathfrak{B}, \mu)$ simultaneously. It is obvious that $S(g,\cdot) = iJ(g)\Omega$.

2. Now we shall investigate the relation between $S(g,\cdot)$ and the Radon-Nikodym derivative $d\mu(\varphi^* F)/d\mu(F)$ and its general form under some general assumption.

<u>Lemma 1</u>. Suppose that (1.6) holds and there is a subspace S^* of K_n^* with μ-outer-measure 1 such that $S(g,F)$ is a continuous functional on S^* for every $g \in K_n^n$. Then

$$\frac{d\mu(\varphi_t^{g*}F)}{d\mu(F)} = e^{2\int_0^t S(g,\varphi_\tau^{g*}F)\,d\tau},\quad g \in K_n^n,\ F \in S^*. \tag{2.1}$$

Lemma 2. Under the assumption of Lemma 1, if further $S(g,\varphi^*F)$ is a continuous functional of two variables g and φ for every $F \in S^*$, then for every $\varphi_t \in D_0(R^n)$, when φ_t is a continuous function of $t, t \in [0,1]$ and $g_t = \varphi_t' \circ \varphi_t^{-1\dagger}$ is a piecewise continuous function of t we have

$$\frac{d\mu(\varphi_t^*F)}{d\mu(F)} = \exp\left\{2\int_0^t S(g_\tau,\varphi_\tau^*F)d\tau\right\} , \quad F \in S^* . \tag{2.2}$$

3. In the following, we always suppose that the μ-outer measure of S^* is 1, S^* is an invariant subspace of K_n^* with respect to $D_0(R^n)$ and every non-empty open subset of S^* has positive μ-measure. We also suppose that there is a space S_* of test functions, i.e. continuous functions, $S_* \supset K_n^1$, and a topology T in S_* such that (S_*,T) is a linear topological space, the relative topology of T in K_n^1 is weaker than the topology of K_n^1, S_* is the completion of K_n^1 with respect to the topology T, K_n^1 is the space of multiplicator of S_* and S_* is invariant under the transformation group $D_0(R^n)$. Further, the functional in S^* can be continued from K_n^1 to S_* and becomes linearly continuous functional on S_*, which has such property that function $f(\cdot)$ locally belongs to S_*, if $\partial_j f(\cdot) \in S_*$ for all j.

Let $S_*^{\otimes m}, m = 1,2,\ldots$ be the symmetric tensor product of M copies of S_* and $S_*^{\otimes 0} = R^1$. We suppose that the pair of spaces S^* and S_* has the following property (U); for any sequence of functions $\{a_m\}$, $a_m \in S_*^{\otimes m}$, if

$$\sum_{m=0}^{\infty} \int a_m(x_1,\ldots,x_m)F(x_1)\ldots F(x_m)dx_1\ldots dx_m = 0 ,$$

(where the term corresponding to $m = 0$ is constant a_0) for every $F \in S^*$, then $a_m = 0$, $m = 0,1,2,\ldots$. For example, if S^* is a linear space and there are sufficient functionals in S^* such that for any $\varphi = 0$, then S^* and S_* have the property (U). The above assumptions about the structures of S^* and S_* are not very strong.

Now, we suppose that (1.6) holds and $S(g,F)$ has the following properties. For every $g \in K_n$ and every non-negative integer m, there is $S(g;x_1,\ldots,x_m) \in S_*^{\otimes m}$ such that

$$S(g,F) = \sum_{m=1}^{\infty} \int S(g;x_1,\ldots,x_m)F(x_1)\ldots F(x_m)dx_1\ldots dx_m, \tag{3.1}$$

where the term corresponding to $m = 0$ is $S(g)$. Hereinafter, we shall not give any explanation about the term corresponding to

---

† φ_t^{-1} is the inverse mapping of φ_t .

$m = 0$ again. Since $S(g,F)$ is linear with respect to g, we know from the property (U) of S^* and S_* and (3.1) that $S(g;x_1,\ldots,x_m)$ is linear with respect to g, for any m. We say that $S(g,F)$ is an analytic functional if $g\mapsto S(g;x_1,\ldots,x_m)$ is linearly continuous functional on K_n^n for every $x_1,\ldots,x_m$; $S(g_\tau;x_1,\ldots,x_m)$ and $\partial_{x_1 j}S(g_\tau;x_1,\ldots,x_m)$ are continuous function of $x_1,\ldots,x_m$ and τ when g_τ is a continuous function of τ; the mapping τ $X(g_\tau,\varphi_\tau,(x_1),\ldots,\varphi_\tau(x_m))$ from $[0,1]$ to $S_*^{\otimes m}$ is continuous for continuous function $\varphi_\tau \in D_0(R^n)$ of τ, and the series

$$S(g_\tau,\varphi_\tau^* F) = \sum_{m=0}^{\infty} \int S(g_\tau,\varphi_\tau(x_1),\ldots,\varphi_\tau(x_m))F(x_1)\ldots F(x_m)dx_1\ldots dx_m \tag{3.2}$$

is uniformly convergent for $\tau \in [0,1]$. Here the restriction about the analyticity of $S(g,F)$ is strong in mathematics but is reasonable and acceptable in physics. If we want to weaken the restriction, then we must establish a more complex theory, since it must use variation. Some other paper is prepared to discuss this problem.

We notice that for every $\varphi \in D_0(R^n)$, there are many continuous φ_t and g_t satisfying

$$\frac{d}{dt}\varphi_t = g_t \circ \varphi_t\,, \quad 0 < t < 1;\ \varphi_0 = I\,,\quad \varphi_1 = \varphi\,, \tag{2.4}$$

and corresponding to the same φ. But the left-side of equation (2.2) only depends upon φ at $t = 1$. Since we suppose that every non-empty open set in S^* has positive measure, we know from the continuity of $S(g,F)$ that for different φ_t satisfying (2.4), the value of the functional $\int_0^1 S(g_\tau,\varphi_\tau^* F)d\tau$ at $F \in S^*$ depends upon φ only. Hence

$$\int_0^1 S(g_\tau,\varphi_\tau(x_1),\ldots,\varphi_\tau(x_m))d\tau,\ m = 0,1,2,\ldots, \tag{3.3}$$

only depends upon φ and is independent of concrete φ_τ and g_τ because of (3.2) and the property (U) of spaces S^* and S_*.
Take the variation of φ_τ and keep $\delta\varphi_0 = \delta\varphi_1 = 0$. We notice that $\frac{d}{d\tau}\delta\varphi_\tau^j = \delta g_\tau^j \circ \varphi_\tau + \delta\varphi_\tau^k(\partial_k g_\tau \circ \varphi_\tau)$. Let $D_\tau(x) = \det(\partial_l \varphi_\tau^k(x))$. Thus we have $\frac{d}{d\tau}\ln D_\tau(x) = (\partial_j g_\tau^j) \circ \varphi_\tau(x)$ by Liouville's Theorem. If we denote

$$S(g;x_1,\ldots,x_m) = \int S_j(x,x_1,\ldots,x_m)g^j dx\,,$$

then we have

$$D_\tau(x)[g^k_\tau(\varphi_\tau(x))\partial_{\varphi^k_\tau(x)} S_j + g^k_\tau(\varphi_\tau(x_1))\partial_{\varphi^k_\tau(x_1)} S_j$$
$$+S_j \partial_{\varphi^k_\tau(x)} g^k_\tau(\varphi_\tau(x)) + S_k \partial_{\varphi^j_\tau(x)} g^k_\tau(\varphi_\tau(x))]$$
$$- \delta(x - x_1)\partial_{\varphi^j_\tau(x_1)} S(g_\tau, \varphi_\tau(x_1), \ldots, \varphi_\tau(x_m)) = 0, \quad (3.4)$$

where $S_j = S_j(\varphi_\tau(x), \varphi_\tau(x_1), \ldots, \varphi_\tau(x_m))$, since the variation of (3.3) is zero. Putting $\tau = 0$ in (3.4), multiplying $h^j(x)dx$ to (3.4), summing up j and integrating with x, we obtain

$$S([g,h], x_1, \ldots, x_m) + g^k(x_1)\partial_{x^k_1} S(h; x_1, \ldots, x_m)$$
$$- h^k(x_1)\partial_{x^k_1} S(g; x_1, \ldots, x_m) = 0, \quad (3.5)$$

where g^k is g^k_0, x^l_k is the l-th coordinate of $x_k, g, k \in K^n_n$ and $x_1, \ldots, x_m \in R^n$.

Now, we have to determine the concrete form of $S(g; x_1, \ldots, x_m)$. First of all, putting $m = 0$ in (3.5), we can prove that $S_j(x) \equiv 0$, i.e. $S(g) = 0$.

Next, consider the case $m = 1$. We shall prove that the support of the generalized function $S_j(x, x_1)$ as function of x is the set $\{x_1\}$ of single element x_1, where x_1 is considered as a parameter. But then (3.5) implies $S([g,h], x_1) = 0$.

$$\int [S_l(x; x_1)h^l(x) - S_j(x; x_1)(x^l - x^l_1)\partial_l h^j(x)]dx = 0.$$

Thus, when $x \neq x_1$, $S_l(x; x_1) = 0$, h^l and h^j are independent of $l \neq j$. Hence the form of $S_j(x; x_1)$ must be

$$S_j(x; x_1) = P_{jK}(x_1)D^K\delta(x - x_1), \quad (3.6)$$

where $K = (k_1, \ldots, k_n)$ and $D^K = \partial_1^{k_1} \cdots \partial_n^{k_n}$. Under the assumption of $S(g; x_1)$, we know that $P_{jK}(\cdot) \in S_*$ and $\partial_j P_{jK}(\cdot) \in S_*$. Now we have to determine P_{jK}. Denote $|K| = k_1 + \ldots + k_n$. Substituting (3.6) for $S_j(x; x_1)$ in (3.5), we have

$$\sum_K (-1)^{|K|}\{P_{jK}(x)D^K(h^i(x)\partial_i g^j(x) - g^i(x)\partial_i h^j(x))$$
$$+ g^i(x)\partial_i(P_{jK}(x)D^K h^j(x)) - h^i(x)\partial_i(P_{jK}(x)D^K g^j(x))\} = 0. \quad (3.7)$$

The coefficients of $h^j(x)$ in (3.7) must be zero. We denote P_{jK} by P_{j0} for $K = (0,0,\ldots,0)$. Thus we obtain

$$g^j(x)(\partial_j P_{i0}(x) - \partial_i P_{j0}(x)) - \sum_{|K|>1} (-1)^{|K|}(\partial_i P_{jK}(x))D^K g^j(x) = 0. \tag{3.8}$$

Since $g^j(x)$ is arbitrary, from (3.8) we obtain that (i) $\partial_j P_{i0}(x) - \partial_i P_{j0}(x) = 0$, i.e. there is a function $P(x)$ such that $P_{i0}(x) = \frac{1}{2}\partial_i P(x)$ and (ii) $P_{jK}(x) \equiv$ constant (we denote it by P_{jK}) for $|K| \geq 1$. Hence (3.7) reduces to

$$\sum_{|K|>1} (-1)^{|K|} P_{jK} \sum_{\substack{K_1+K_2=K \\ K_2 \neq K}} C^K_{K_1K_2}(D^{K_1}h^j)(D^{K_2}\partial_i g^j)$$
$$- \sum_{|K'|>1} (-1)^{|K'|} P_{jK'} \sum_{\substack{K'_1+K'_2=K' \\ K'_2 \neq K'}} C^{K'}_{K'_1K'_2}(D^{K'_1}g^i)(D^{K'_2}\partial_i h^j) = 0, \tag{3.9}$$

where $C^K_{K_1,K_2}$ is the coefficient in the expansion. In this case, we can prove that the terms in (3.9) for $|K| > 1$ and $|K'| > 1$ which can be cancelled, are the terms for which $D^{K_1} = K^{K'_2}\partial_j$, $D^{K_2}\partial_j = D^{K'_1}$ only, $P_{iK} = 0$ for $|K| > 1$ and $P_{jK} = 0$ except $D^K = \partial_m$ for $|K| = 1$. When $D^K = \partial_m$, we denote the corresponding P_{jK} by P_{jm}. From (3.9) we can prove $P_{jm} = -\frac{1}{2}Q_1\delta_{jm}$, where Q_1 is constant. Thus

$$S_j(x; x_1) = \frac{1}{2}\delta(x - x_1)\partial_j P(x_1) - \frac{1}{2}Q_1\partial_j\delta(x - x_1).$$

By the same method, we can prove in general that there are a function $P(x_1,\ldots,x_m)$ which is symmetric with respect to arguments $x_1,\ldots,x_m$, and a constant Q_m such that

$$S_j(x; x_1,\ldots,x_m) = \frac{1}{2}\delta(x - x_\ell)\partial_{x^j_\ell}P(x_1,\ldots,x_m)$$
$$- \frac{1}{2}Q_m\partial_j\sum_{\ell=1}^{m}\delta(x - x_1).$$

Thus, we obtain the following expansion

$$S(g,F) = \frac{1}{2} \sum_{m=1}^{\infty} \int [g^j(x_\ell)\partial_{x_\ell^j} P(x_1,\ldots,x_m) + Q_m \partial_j \sum_{\ell=1}^{m} g^j(x_\ell)]F(x_1) \ldots F(x_m)dx_1\ldots dx_m. \tag{3.10}$$

Let $C_N(\lambda)$ be the partial sum of the power series $C(\lambda) = \sum_{m=1}^{\infty} Q_m \lambda^{m-1}$ and $P_N(F)$ be the partial sum of the functional series

$$P(F) = \sum_{m=1}^{\infty} \int P(x_1,\ldots,x_m)F(x_1)\ldots F(x_m)dx_1\ldots dx_m. \tag{3.11}$$

From (3.5) we obtain

$$\frac{d\mu(\varphi^*F)}{d\mu(F)} = \lim_{N\to\infty} \exp\left\{C_N\left(\int F dx\right)(F, \ln D(x,\varphi)) + P_N(\varphi^*F) - P_N(F)\right\}, \tag{3.12}$$

where $D(x,\varphi) = \det(\partial_i \varphi^j)$. In (3.12), if $\int F dx$ does not exist, then the function $C_N(\cdot)$ must be a constant C. If the power series $C(\lambda)$ is convergent in the range of $\lambda = \int F dx$, then (3.11) and (3.12) can be rewritten as

$$\frac{d\mu(\varphi^*F)}{d\mu(F)} = \exp\left\{C\left(\int F dx\right)(F, \ln D(x,\varphi)) + P(\varphi^*F) - P(F)\right\}, \tag{3.13}$$

where

$$P(\varphi^*F) - P(F) = \sum_{m=1}^{\infty} \int (P(\varphi(x_1),\ldots,\varphi(x_m)) - P(x_1,\ldots,x_m))F(x_1) \ldots F(x_m)dx_1\ldots dx_m \tag{3.14}$$

is a convergent series.

We must notice that the integral $J_m = \int [P(\varphi(x_1),\ldots,\varphi(x_m)) - P(x_1,\ldots,x_m)]dx_1\ldots dx_m$ in (3.14) is convergent and the functional $P_N(\varphi^*F) - P_N(F) = \sum_{m=1}^{N} J_m$ is also well-defined. Thus we obtain the following Theorem.

Theorem. If $S(g,F)$ is an analytical functional, then the Radon-Nikodym derivative $d\mu(\varphi^*F)/d\mu(F)$ has the expression (3.12).

In particular, if the power series $C(\lambda)$ is convergent, then $d\mu(\varphi^*F)/d\mu(F)$ has the expression (3.13).

Example. Consider the Poisson measure[7]. Let $f(x)$ be a fixed smooth function on R^n, $0 < f(x) < \infty$ and m be a measure on R^n, $m(U) = \int_U f(x)dx$. If $\{x_n\}$ is a sequence of points in R^n without limit point and $x_m \neq x_{m'}$ for $m \neq m'$, then we construct $\sum_{m=1}^{\infty} \delta(x - x_m)$. Let S^* be the set of all such generalized functions. If μ_0 is the Poisson measure on S^*,

$$\mu_0\left(\left\{ \sum_{m=1}^{\infty} \delta(x - x_m) \in X^* \,\middle|\, \text{there are } n \text{ points of } \{x_m\} \text{ in } U \right\}\right)$$

$$= \frac{(\lambda m(U))^n}{n!} e^{-\lambda m(U)},$$

then we have

$$\frac{d\mu_0(\varphi^*F)}{d\mu_0(F)} = \exp\left(F, \ln\left(D(x,\varphi)\frac{f(\varphi(x))}{f(x)}\right)\right).$$

If we take some measurable functional $P(F)$ such that $0 < \exp(P(F) - (F, \ln f)) < \infty$ for almost all F and define a measure μ as $\mu(E) = \int_E \exp(P(F) - (F, \ln f))d\mu_0(F)$, then

$$\frac{d\mu(\varphi^*F)}{d\mu(F)} = \exp((F, \ln D(x,\varphi)) + P(\varphi^*F) - P(F)) \tag{3.15}$$

4. In [6], Menikoff considered the case when the Hamiltonian H for a system of particles is of the form $H = H_K + H_P$, where H_K and H_P are the kinetic energy term and potential energy term respectively,

$$H_K = \frac{1}{8} \int dx D(x)^{\dagger} \frac{1}{\rho(x)} K(x),$$

$$H_P = \frac{1}{2} \iint dxdy\, \rho(x)(\rho(y) - \delta(x - y))V(x - y). \tag{4.1}$$

In (4.1), $K(x) = \nabla\rho(x) + 2iJ(x)$ and the j-th component of ∇ is ∂_j. In the papers [6-8], Menikoff investigated the problem to find the representation space. Now we investigate this problem by means of the results in the above. We have to find out a quasi-invariant measure μ such that in the representation of the current algebra determined by the measure μ, we can give a definite interpretation of H such that H becomes a non-negative self-adjoint operator with ground state as eigenvector of H corresponding to eigenvalue zero. Here H_P is not necessary to be the concrete form of (4.1), but H_P may be a general functional operator of $\rho(\cdot)$.

Suppose that μ satisfies the condition in our Theorem. We use some notations in [3], but we do not use the boldfaced letter for vectors. We consider the case of R^n instead of R^3. We also notice that $\rho(x)$ is $F(x)$. Let e_j be the unit vector whose j-th component is 1 and other components are zero. Let

$$R_j(x,F) = S(e_j\delta(-x),F) + \frac{1}{2}\partial_j F(x). \tag{4.2}$$

From [6] the equation $H\Omega = 0$ is equivalent to

$$\int e^{i(F,j)}\left(\iint\left[\frac{1}{2}\frac{\sum_{j=1}^{n} R_j(s,F)^2}{F(x)} + \frac{i}{2}R_j(s,F)\partial_j f\right]dx + H_P(F)\right)d\mu(F) = 0, \tag{4.3}$$

where $H_P(F)$ is H_P. Now, we consider such case that $S(g,F)$ has the expression (3.14) but $C\left(\int F(x)dx\right) = 1$. Hence (4.3) becomes

$$\int e^{i(F,f)}\left(\iint\left[\frac{1}{8}F(x)\sum_{j=1}^{n}\left(\partial_j\frac{\delta P(F)}{\delta F(x)}\right)^2 + \frac{i}{4}F(x)\partial_j f(x)\cdot\partial_j\frac{\delta P(F)}{\delta F(x)}\right]dx \right.$$

$$\left. + H_P(F)\right)d\mu(F) = 0 \tag{4.4}$$

For a given $H_P(F)$, our aim is to find out the **functional** $P(F)$ satisfying (4.4).

We construct a one-parameter family of measurable transformations β_t such that $\beta_t S^* = S^*$, $\beta_0 F = F$ and

$$\frac{d}{dt}\beta_t F\Big|_{t=0} = -\partial(FR_j(\cdot,F)).$$

we have

$$\int e^{i(F,f)}\left[-i(\partial_j(FR_j(\cdot,F)),f) + \frac{d}{dt}\frac{d\mu(\beta_t F)}{d\mu(F)}\right]d\mu(F) = 0. \tag{4.5}$$

Hence from (3.13-14), we obtain

$$H_P(F) = \iint\left[\frac{1}{8}F(x)\sum_{j=1}^{n}\left(\partial_j\frac{\delta P(F)}{\delta F(x)}\right)^2 + \frac{1}{2}\frac{\delta P(F)}{\delta F(x)}\nabla^2 F\right]dx, \tag{4.6}$$

where $\nabla^2 = \sum_{j=1}^{n} \partial_j^2$. For a given $H_P(F)$, we must take $P(F)$ to satisfy the equation (4.6).

Suppose that

$$H_P(F) = V_0 + \sum_{m=1}^{\infty} \int V(x_1, \ldots, x_m) F(x_1) \ldots F(x_m) dx_1 \ldots dx_m, \quad (4.7)$$

where V_0 is a constant, and $V(x_1, \ldots, x_m)$ is a symmetric function. For example, the Hamiltonian in (4.1) is of the form

$$H_P(F) = -\frac{1}{2} V(0) \int F(x) dx + \frac{1}{2} \iint V(x_1 - x_2) F(x_1) F(x_2) dx_1 dx_2,$$

where the corresponding $V_0 = 0$, $V(x_1) = -\frac{1}{2} V(0)$, $V(x_1, x_2) = \frac{1}{4}(V(x_1 + x_2) + V(x_1 - x_2))$ and $V(x_1, \ldots, x_m) = 0$ for $m \geq 3$. From (3.11) and (4.6-4.7), we obtain a relation between $P(x_1, \ldots, x_m)$ and $V(x_1, \ldots, x_m)$. Denoting the symmetrization of the function with respect to the arguments by the symbol $\mathbf{S}$, we deduce from (4.6) the following equation,

$$\frac{1}{8} \mathbf{S} \sum_{m_1 + m_2 = m} \frac{m!}{(m_1 - 1)!(m_2 - 1)!} \partial_{x^j_{m_1}} P(x_1, \ldots, x_m) \partial_{x^j_{m_1}} P(x_{m_1}, x_{m_1} + 1, \ldots, x_m)$$

$$+ \frac{1}{2} \sum_{\ell=1}^{n} \nabla^2 x_\ell P(x_1, \ldots, x_m) = V(x_1, \ldots, x_m), \quad (4.8)$$

and $V_0 = 0$. Hence the solution of the problem of representation is determined by the equation (4.8). In general, the solution $\{P(x_1, \ldots, x_m)\}$ is not unique.

REFERENCES

[1] Xia Dao-xing (Shah Tao-Shing): Measure and Integration Theory on Infinite-Dimensional Spaces, (translated by E.J. Brody) Acad. Press, N.Y., (1972)

[2] Vershik, A.M., Gelfand, I.M., Graev, I.M.: Uspehi Matem. Nauk, 30: 6 (1975), 1.

[3] Goldin, G.A.: Jour. Math. Phys., 12 (1971), 462.

[4] Goldin, G.A., Grodnik, J., Powers, R. and Sharp, D.H.: Jour. Math. Phys. 15 (1974), 88.

[5] Ismagilov, R.S.: Funkt. Analyz i ego Priloz, 9: 2(1975), 71.
[6] Menikoff, R.: Jour. Math. Phys., 15 (1974), 1138.
[7] ___________: Jour. Math. Phys., 15 (1974), 1394
[8] ___________: Jour. Math. Phys., 16 (1975), 2341, 2353.
[9] Schwartz, L.: Théorie des Distributions, Tom I, II.
[10] Skorohod, A.V.: Integration in Hilbert Space, Springer-Verlag, (1974).
[11] Trotter, H.F.: Proc. Amer. Math. Soc., 10 (1959), 545.

REGULAR AND CHAOTIC REGIMES IN QUANTUM MECHANICS*

Michael Tabor

Center for Studies of Nonlinear Dynamics
La Jolla Institute
P. O. Box 1434
La Jolla, CA 92038

INTRODUCTION

The ability of small Hamiltonian systems to show a transition from regular, integrable motion to irregular chaotic motion is now becoming well known and to some extent better understood.[1] This paper is concerned with the possible implications that this type of classical behaviour may have for the semiclassical limit of quantum mechanics. The paper will fall into essentially two parts: (i) will describe some concrete results that have been obtained recently[2] concerning the evolution of nonstationary states in the regular and irregular regimes and (ii) will be of a more speculative nature concerning what sort of criteria may or may not be useful for characterizing "quantum chaos" (what ever that is!) and what sort of parameters might be useful for quantifying such "chaos".

SEMICLASSICAL WAVEFUNCTIONS AND THEIR EVOLUTION

To start with we consider the nature of the wavefunction, which in the limit $\hbar \to 0$, provides an approximate solution to Schrödingers equation. In the case of the classical motion being integrable the wavefunction can be specified rather completely. Its most general form, for a system of N - degrees of freedom, is

$$\psi(\underset{\sim}{q}) = \sum_r \det\left(\frac{\partial^2 S_r}{\partial q_i \partial I_j}\right)^{\frac{1}{2}} \exp\left\{iS_r(\underset{\sim}{q},\underset{\sim}{I})/\hbar + i\delta_r\right\} \qquad (2.1)$$

where S is the classical action function

$$S = \int_{\underset{\sim}{q}_o}^{\underset{\sim}{q}} \underset{\sim}{p}(\underset{\sim}{q}', \underset{\sim}{I})\, d\underset{\sim}{q}' \tag{2.2}$$

where the momentum vector $\underset{\sim}{p}$ is expressed in terms of the coordinates $\underset{\sim}{q} = (q_1 \ldots q_N)$ and the constant classical action variables $\underset{\sim}{I} = (I_1 \ldots I_N)$. Since $\underset{\sim}{p}$ is a multivalued function of $\underset{\sim}{q}$ we have to sum over the different branches of S; this is the sum over r in eqn. (2.1). In the case of an eigenstate we require that $\psi(q)$ be a single-valued function of q. It is a standard result to show that this single-valuedness condition is satisfied if the classical actions satisfy the Einstein - Brillouin - Keller - Maslov quantization condition[3]

$$\underset{\sim}{I} = \left(\underset{\sim}{n} + \frac{\underset{\sim}{\alpha}}{4}\right)\hbar, \qquad \underset{\sim}{n} = (n_1 \ldots n_N), \quad n_i = 0, 1 \ldots \qquad \underset{\sim}{\alpha} = (\alpha_1 \ldots \alpha_N) \tag{2.3}$$

where the α are the Maslov indices which take care of the $\frac{\pi}{2}$ loss of phase that occurs at the classcical turning points. The stationary states can thus be associated with the family of trajectories lying on the torus with the set of actions $\underset{\sim}{I}$ satisfying the EBKM conditions (2.3).

The amplitude of the wavefunction is a measure of the density of classical paths at the point $\underset{\sim}{q}$. To see this we form the semi-classical probability density

$$|\psi(q)|^2 = \sum_r \sum_s \det\left(\frac{\partial^2 S_r}{\partial \underset{\sim}{q} \partial \underset{\sim}{I}}\right)^{\frac{1}{2}} \det\left(\frac{\partial^2 S_r}{\partial \underset{\sim}{q} \partial \underset{\sim}{I}}\right)^{\frac{1}{2}} \exp \frac{i}{\hbar}\left\{S_r(\underset{\sim}{q}, \underset{\sim}{I}) - S_s(\underset{\sim}{q}, \underset{\sim}{I})\right\}. \tag{2.4}$$

Although this is the $\hbar \to 0$ limit for the probability density there are always quantum oscillations present due to the interference between the different branches of S. (Hence we can never write for such quantities an analytic power series in h of the form: quantum = classical + $O(\hbar)$ + $O(\hbar^2)$ + ...).

In order to see the nature of the actual classical limit, i.e., the h = 0 limit, as opposed to the $\hbar \to 0$ limit, we must perform some form of local averaging[4] i.e.,

$$\bar{f}(q) = \frac{1}{\Delta}\int_{q-\Delta/2}^{q+\Delta/2} f(q')\, dq' \tag{2.5}$$

where

$$\lim_{\hbar \to 0} \Delta = 0$$

$$\lim_{\hbar \to 0} \hbar/\Delta = 0$$

which wipes away the quantum interference leaving

$$|\overline{\psi(\underset{\sim}{q})}|^2 = \sum_r \det\left|\frac{\partial^2 S_r}{\partial \underset{\sim}{q} \partial \underset{\sim}{I}}\right| \quad . \tag{2.6}$$

Now since S is the generating function that effects the canonical transformation between (p, q) and $(\underset{\sim}{I},\underset{\sim}{\theta})$ variables ($\underset{\sim}{\theta}$ are the conjuate angle variables to $\underset{\sim}{I}$) we have

$$\underset{\sim}{\theta} = \nabla_{\underset{\sim}{I}} S(\underset{\sim}{I},\underset{\sim}{q}) \tag{2.7}$$

and hence

$$|\psi|^2 = \sum_r \left|\frac{d\underset{\sim}{\theta}_r}{d\underset{\sim}{q}}\right| \quad . \tag{2.8}$$

This has a nice geometrical interpretation. Since the trajectories are uniformly distributed in $\underset{\sim}{\theta}$ around the associated torus eqn. (2.8) represents the projection of the torus onto the coordinate plane. These ideas can be clearly seen by considering the one dimensional

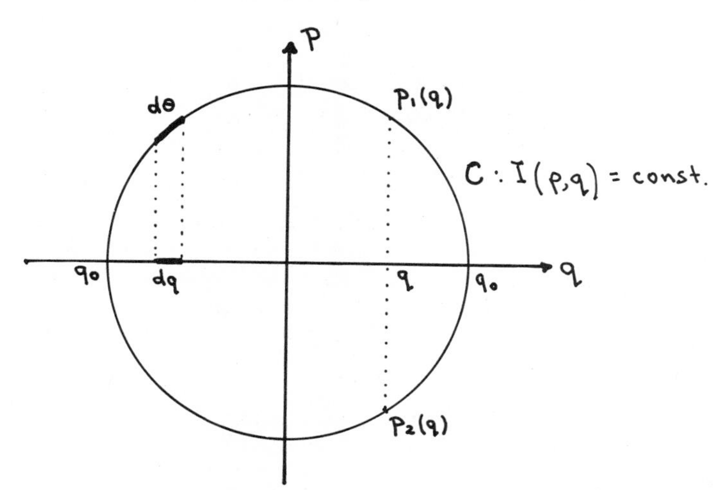

Fig. 1. Level curve C of one dimensional bounded system.

case. In the (p,q) phase plane we draw the curve C of constant action (also contant energy in this one-dimensional case) which represents the family of trajectories uniformly distributed about C in the angle variable θ.

We see that p is a two valued function of q - the two branches coalescing at the turning points q_o. This results in the projection $d\theta/dq$ being singular at these points.

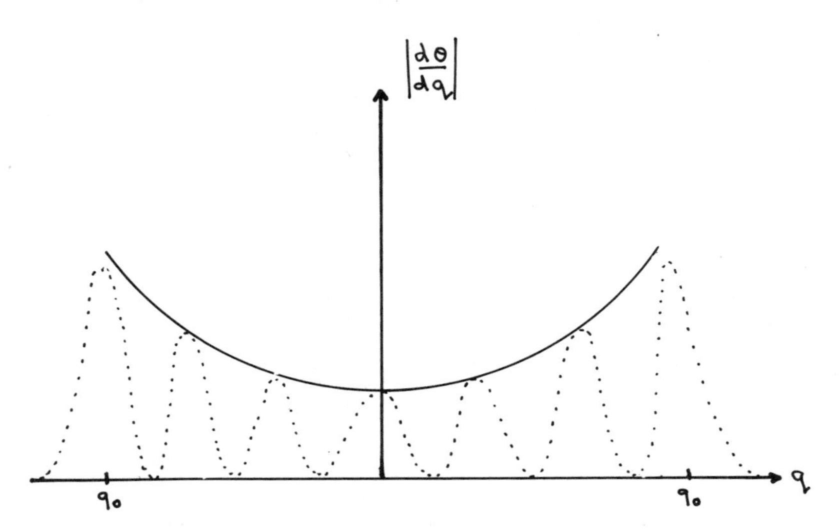

Fig. 2. Projection of C onto coordinate axis showing singularities at the turning points q_o. The projection provides the smooth envelope for the actual quantal probability density (dotted line).

These singularities are know as <u>caustics</u>. Since the projection of C onto the coordinate plane can be regarded as a gradient map the caustics can be regarded as the singularities of the gradient map and hence classified accordingto Thom's theorem[4].

The coarse grained probability density $|d\theta/dq|$ provides in the limit $\hbar \to 0$ the smooth envelope, in the classically allowed regions, of the oscillations of the actual quantal probability density. In effect the "rôle" of $\hbar$ is to add a regular oscillatory structure to a smooth classical background. Clearly the simple form of semiclassical wavefunction eqn. (2.2) breaks down at the caustics and has to be modified by some form of uniform approximation involving Airy Functions.[5] A simple geometrical criterion for estimating the vicinity in which the (nonuniformized) semiclassical wavefunction breaks down is that the area enclosed by C between two nearby branches is of $O(\hbar)$ or less

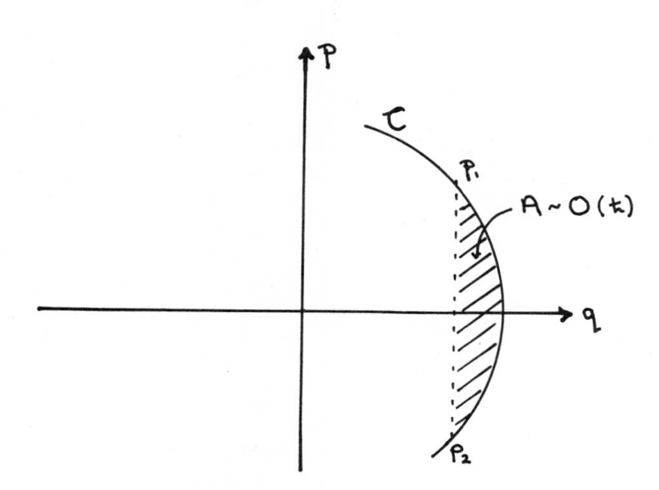

Fig. 3. Area of $\mathcal{O}(h)$ enclosed by C between the branches of p.

Overall we have seen that the semiclassical stationary states for integrable motion can be given a rather complete description in terms of the EBKM quantization conditions and the semiclassical wavefunction. What is nothing like as well understood is the nature of the semiclassical state for nonintegrable chaotic motion and the evolution of nonstationary states in both regimes. I shall be concerned here with the latter problem and we shall see, as has already been mentioned, that h plays a crucial rôle in these problems.

Returning to the one dimensional problem we have seen that the semiclassical wavefunction can be associated with a smooth family of trajectories lying on some curve C. In the case of a stationary state this curve is an invariant curve, i.e., under the Hamiltonian flow it remains unchanged, i.e.,

$$C_t(p(t, q(t)) = C_o(p(o), q(o)) \quad . \tag{2.9}$$

On the other hand in the case of a nonstationary state, C is no longer an invariant curve and will develop under the flow e.g.,

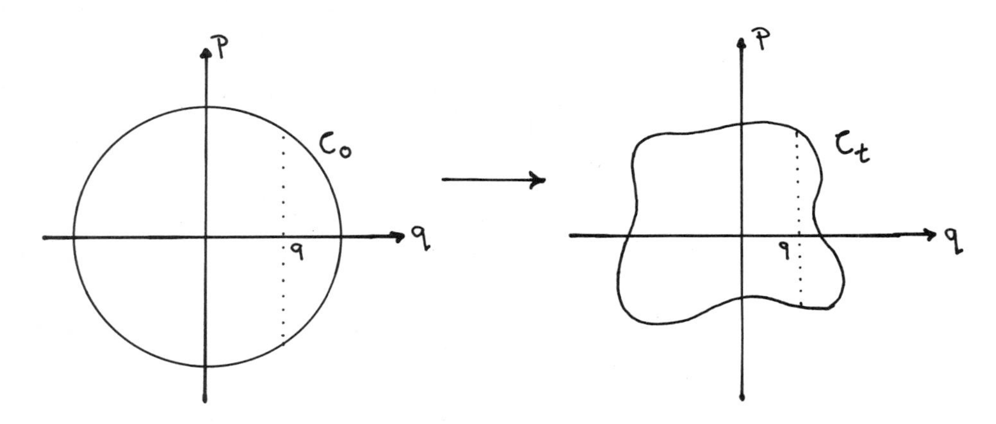

Fig. 4. Evolution of a noninvariant curve

The way in which C_t develops, i.e., the types of convolution it can go through, depends on whether the motion is regular or chaotic.

One can still define a semiclassical wavefunction, now time dependent, in terms of the evolving curve (essentially C_t is a Lagrangian manifold since the phase flow preserves symplectic structure.[6] That is, the wavefunction can still be expressed as a sum of contributions, of the form path density x phase, over the different braches of p(q). However although the wavefunction $\psi(t)$ is valid at fixed t for all $\hbar \to 0$, it is not always valid at fixed $\hbar$ for $t \to \infty$.[7] The problem is that as C_t evolves its convolutions become ever more complicated. Once these convolutions become more complicated than scales of $O(\hbar)$ the simple form of wavefunction, i.e., sum over branches, breaks down. But even worse, once the caustics start clustering on scales of $O(\hbar)$ uniformization procedures can no longer be implemented to patch up the singularities.

The nature of this breakdown of the semiclassical wavefunction is of great importance and depends crucially on whether the underlying motion is regular or chaotic.

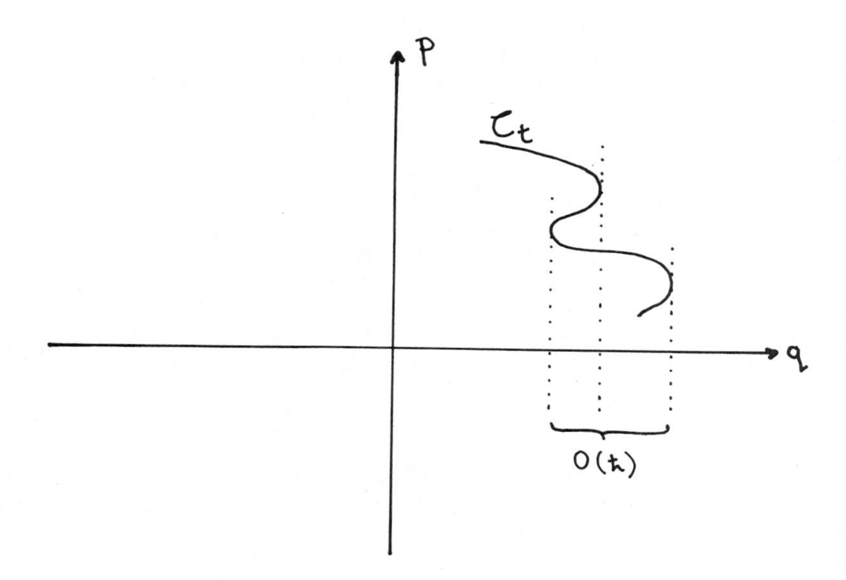

Fig. 5. Complicated convolutions of C_t leading to clusterings of caustics on scales of $\mathcal{O}(\hbar)$.

In the case of conservative one dimensional systems the classical motion is integrable. So although a noninvariant curve will still develop in time we cannot use these systems for determining the differences between regular and chaotic motion.

However if we have a time dependent Hamiltonian of the form

$$\begin{aligned} H &= p^2/2\gamma \qquad 0<t<\gamma T \qquad (0<\gamma>1) \\ &= V(q)/(1-\gamma) \qquad \gamma T<t<T \end{aligned} \tag{2.10}$$

the equations of motion when integrated from time nT to (n+1)T take the form

$$\begin{aligned} q_{n+1} &= q_n + p_n \\ p_{n+1} &= p_n - \left(\frac{\partial V}{\partial q}\right)_{q=q_{n+1}} \end{aligned} \tag{2.11}$$

which have the form of an algebraic area preserving mapping.[2] If V(q) is a polynomial of greater than order 2 the mapping is of the form of a Cremona Transformation which can be proven to be nonintegrable. We show here examples of orbits evolving under this mapping in the case $V(q)=q^4/4$.[2] We see the generic mixture of invariant curves, hyperbolic and elliptic fixed points and chaotic trajectories typical of nonintegrable mappings.[1,2]

Of course we don't just have to consider the evolution of individual orbits - we can also look at the evolution of whole families of orbits. We now show the evolution of the curve (C) lying in the outermost hyperbolic/escaping region of the map.[2] The small curls are due to segments of the curve wrapping themselves around the elliptic fixed points of the map. We term these convolutions whorls - they grow approximately linearly in time. The rapidly growing strands, which we term tendrils, grow more than exponentially fast and are due to the portions of C passing through the hyperbolic fixed points of the mapping.

We will shortly see how the projections of these evolving curves, corresponding to the coarse - grained evolving probability density can be compared with the corresponding, exact, quantal probability density.

The quantum mechanics of our mapping eqn. (2.11) is easily performed.[2] The operator $\hat{U}$ that transforms the state $|\psi_n\rangle$ at 'time' n to state $|\psi_{n+1}\rangle$ at 'time' n+1, i.e.,

$$|\psi_{n+1}\rangle = \hat{U}|\psi_n\rangle \ , \tag{2.12}$$

is simply

$$\hat{U} = e^{-i\hat{p}^2 T/2\hbar} \, e^{i\hat{V}(q)T/\hbar} \ . \tag{2.13}$$

In coordinate representation this becomes

$$\psi_{n+1}(q) = \int dq' \, \langle q|\hat{U}|q'\rangle \psi_n(q') \tag{2.14}$$

where the matrix elements $\langle q|\hat{U}|q'\rangle$ are easily evaluated.

The choice of initial state $\psi_o(q)$ and associated initial curve C_o is made quite simply by assuming that at t = 0 the Hamiltonian is just

$$H = 1/2 \ p^2 + V(q) \tag{2.15}$$

(which is just the time average of eqn. (2.11)) before the time dependent 'perturbation' is switched on. The initial state ψ_o is thus taken to be a stationary state, $\Psi_N(q)$, of eqn. (2.15) and the initial curve C_o is the corresponding - in the semiclassical sense - invariant curve, i.e., the curve C_o for which

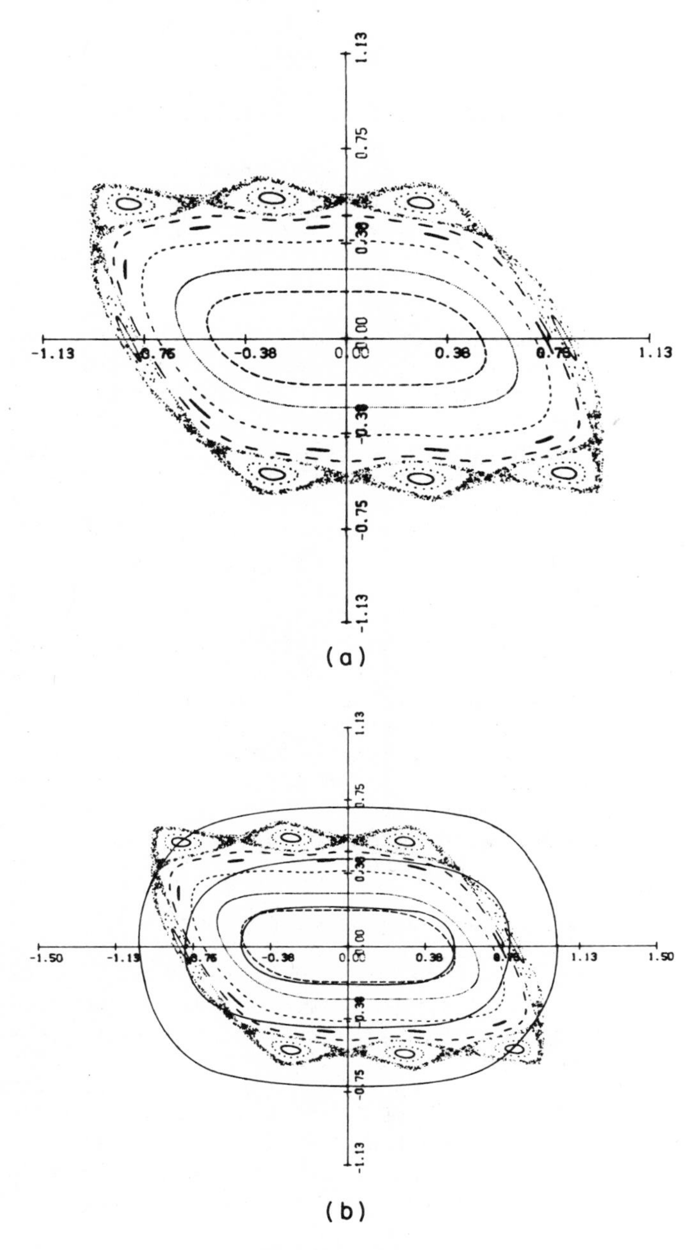

Fig. 6. Typical phase plane orbits of mapping (2.11) with $V(q)=q^4/4$. (b) same as (a) but with three invariant curves of (2.15) superimposed.[2]

$$\frac{1}{2\pi}\oint_{C_o} p\,dq = (N + 1/2)\hbar \tag{2.16}$$

where N is the quantum number of the stationary state $\Psi_N(q)$.

We now show the evolution of $|\psi_n|^2$ under the "quantum mapping" eqn.(2.14). In this case ψ_o is the eighteenth eigenstate of eqn. (2.15) (with $V(q) = q^4/4$) and $\hbar$ was chosen such that the associated invariant curve C_o is precisely the one whose evolution we saw previously.

We now look at the projections of C and compare them with $|\psi_n|^2$. For n=0 and n=1 we can see that the projected curve provides the envelope of the oscillations of $|\psi_n|^2$. The time dependent form of the semiclassical wavefunction discussed earlier is still applicable at n=1. But from n=2 onwards the wavefunction has made a transition to a new type of behaviour which is manifested by the rapid increase in the number of scales of oscillations. This is due to the development of the tendrils which rapidly introduce whole new families of branches of p(q). (There is also a loss of the original regular nodal pattern but at this stage we are not too sure what this means).

That there is no clear relationship between $|\psi_n|^2$ and the projected curve for $n \geq 2$ is due to the presence of clusters of caustics that cannot be resolved on scales of the order of a de Broglie wavelength. The quantum state can therefore no longer follow this fine detail which is 'wiped out' by quantum interference. Thus in order to make a comparison between $|\psi|^2$ and the projections we must smooth both over a width of the order of a de Broglie wavelength. When this is done we can see good agreement between the two probability densities.[2]

What these results show is that the presence of classical chaos[+] leads to a rapid (probably exponentially fast) proliferation in classical caustic structure. The semiclassical representation of the quantum states will break down once this structure can no longer be resolved on scales of $\mathcal{O}(\hbar)$, i.e., the fine structure is wiped out by quantum interference leaving a wavefunction with structure on all scales down to $\mathcal{O}(\hbar)$. (This contrasts with the regular case where the "rôle" of $\hbar$ is to add a regular oscillatory structure to a smooth classical background). Clearly the smaller $\hbar$ the longer the quantum state can follow classical chaos and hence develop ever greater "quantum complexity" reflecting the underlying

---

[+]In the case of non chaotic motion the development of 'whorls' will also introduce new caustic structure leading to eventual break down of the wavefunction. However, 'whorlification' tends to occur linearly in time whereas 'tendrilization' occurs exponentially fast.

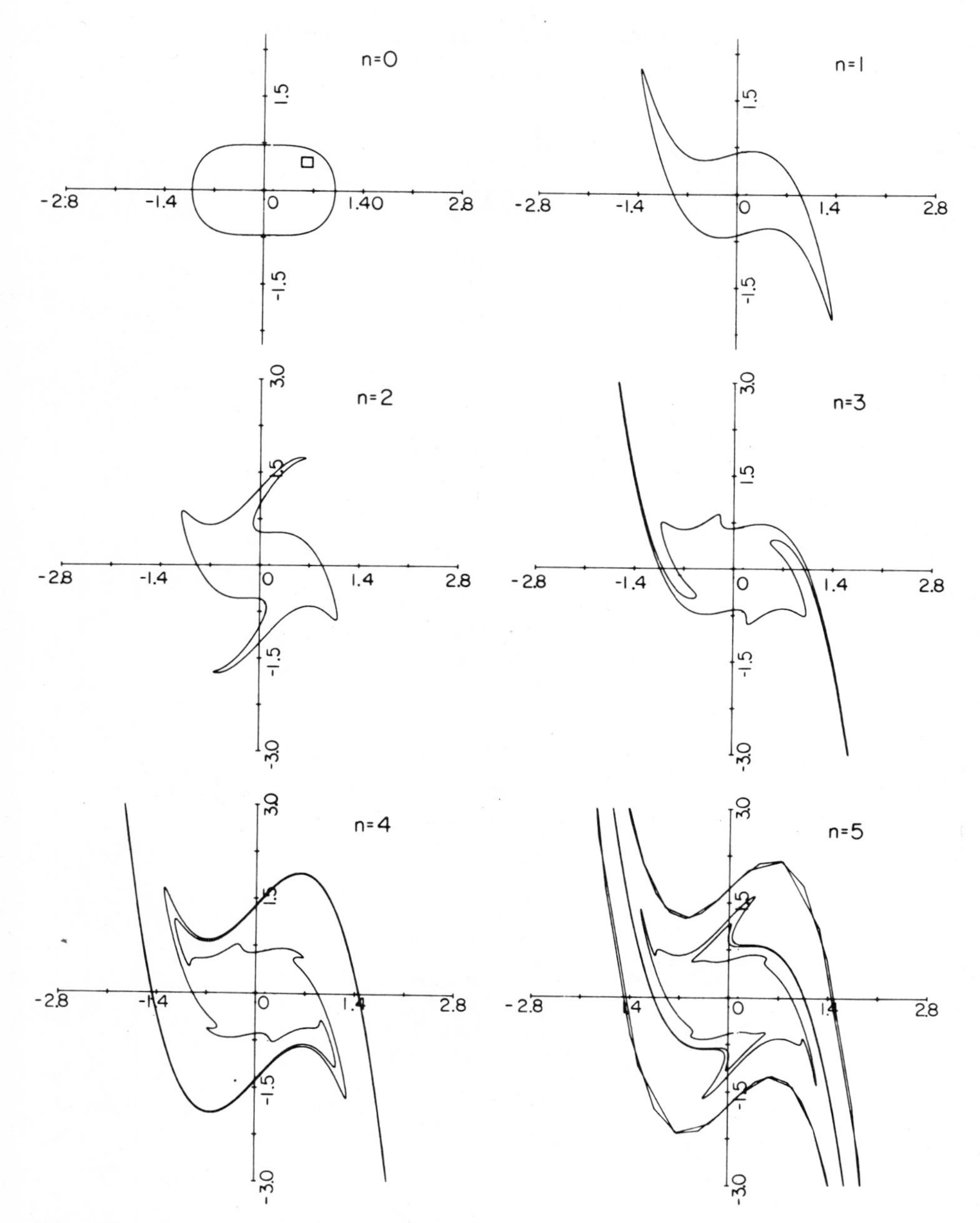

Fig. 7. Evolution of outermost curve in Fig. 6(b) under mapping.

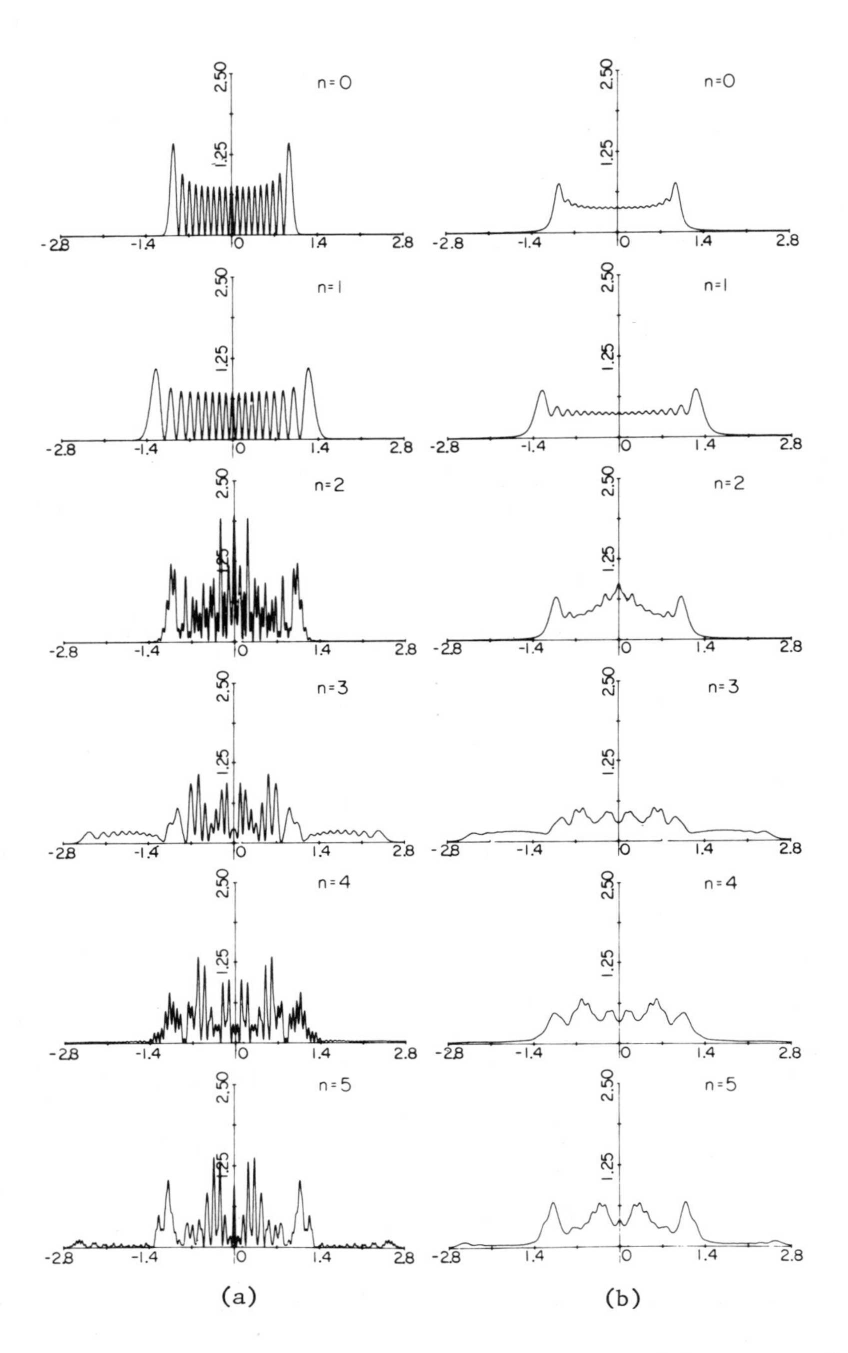

Fig. 8. Evolving quantal probability density: (a) exact and (b) smoothed over same width as in Fig. 8. Ref.[2]

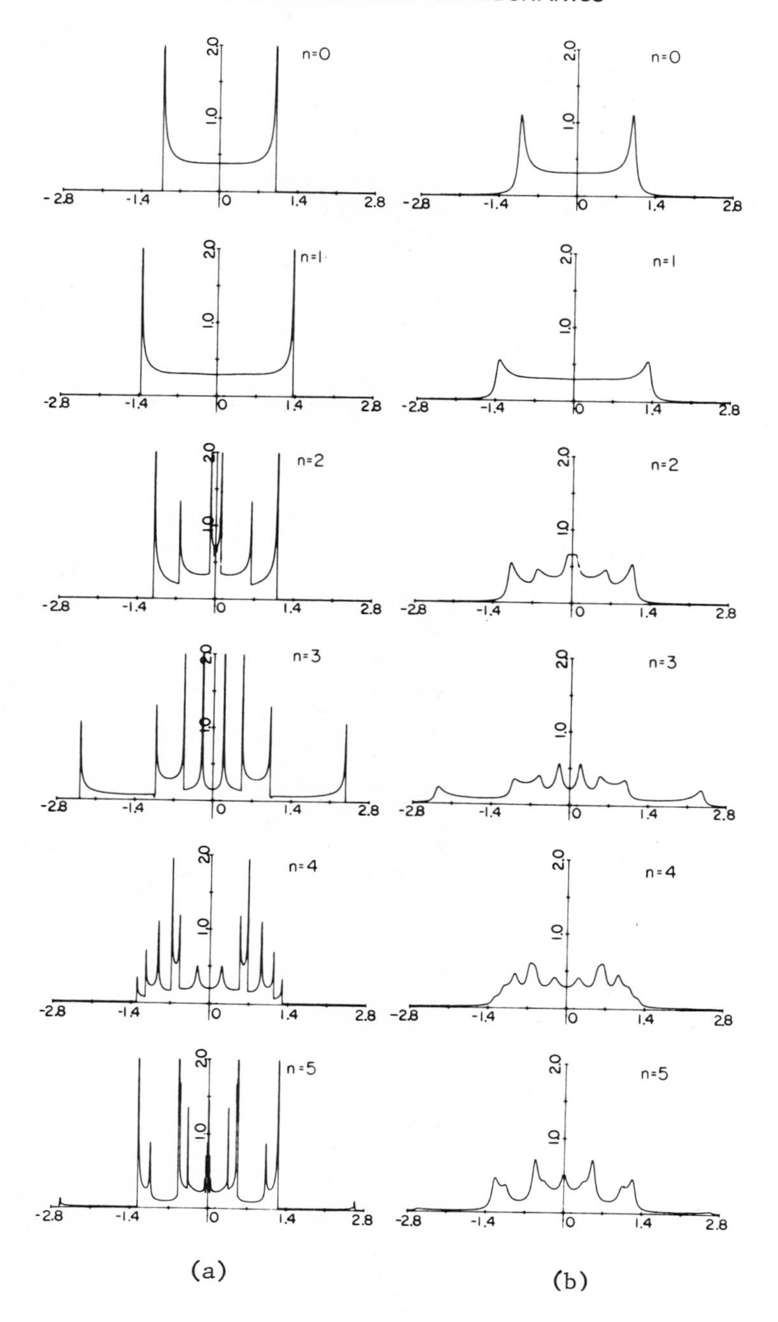

Fig. 9. Projections of evolving curves onto q-axis (a) and smoothed projections (b).[2]

chaos. It is this sort of consideration that I will now use to suggest possible criteria for "quantum chaoticity".

CRITERIA FOR QUANTUM CHAOS

First of all I think it is necessary to carefully define terminology. In classical mechanics the notions of regular and chaotic motion can be clearly distinguished. In the case of regular motion the trajectories are forever confined to tori around which they wrap themselves uniformly. (With the exception of closed orbits the motion is, in fact, ergodic on the tori). Chaotic trajectories are easily distinguished from regular ones by virtue of their great sensitivity to initial conditions and the way in which they separate from each other exponentially fast exploring large portions of the energy shell. The exponential separation of trajectories is characteristic of a _mixing_ process.[8]

In trying to define notions of 'quantum ergodicity' or 'quantum chaoticity' I will be concerned with those definitions that give some measure of the _extent_ to which the underlying classical behaviour is _realized_ by the quantum mechanics. This approach has the advantage that the ensuing definitions of 'quantum ergodicity', 'mixing', 'Bernoulli Shift' etc. are all based on the well defined classical concepts. This approach is, of course, of a semiclassical nature and guarantees that the quantum definitions approach the underlying classical definitions in the limit $\hbar \to 0$.

First of all let us consider the notion of 'quantum ergodicity' where we use the term ergodicity in the sense of a uniform covering of some manifold. In this case we will consider a quantum mechanical microcanonical distribution.[9] This is defined as a uniform superposition of states lying in some narrow band of energy, i.e.,

$$\Psi(q,t) = \frac{1}{\sqrt{N}} \sum_{n=K}^{K+N} \phi_n(q) e^{-iE_n t/\hbar} \qquad (3.1)$$

where the states $\phi_n(q)$ are the eigenstates of the system Hamiltonian and the set N lie in the narrow range of energy ΔE, i.e., $E_{K+N} - E_N = \Delta E$. Furthermore we will take the system to be classically integrable. The probability density $P(q,t)$ is of course

$$P(q,t) = |\psi(q,t)|^2 = \frac{1}{N} \sum_n \sum_m \phi_n(q)\, \phi_m^*(q)\, e^{-i(E_n - E_m)t/\hbar}. \qquad (3.2)$$

We now consider the time average of P, i.e.,

$$\overline{P} \equiv \overline{P(q,t)} = \frac{1}{N} \sum_n |\phi_n(q)|^2 \qquad (3.3)$$

and examine its behaviour in the limit $\hbar \to 0$. To do this correctly we must perform a proper 'semiclassicalization' of P, i.e.,

(i) The quantum number is replaced by the classical action, i.e.,

$$n = I/\hbar - 1/2 \qquad (3.4)$$

(ii) The $\phi_n(q)$ are replaced by their semiclassical form

$$\phi_n(q) = \frac{1}{(2\pi)^{\frac{1}{2}}} \sum_r \left(\frac{\partial^2 S}{\partial q \partial I}\right) \exp\{iS_r(q,I)/\hbar\} \qquad (3.5)$$

(iii) The sum over n is replaced by an integration over I using the Poisson Sum formula.[5]

Thus we obtain

$$\overline{P} = \frac{\hbar\omega}{\Delta E} \frac{1}{2\pi\hbar} \sum_M e^{i\pi M} \sum_r \sum_s \int_{E/\omega}^{(E+\Delta E)/\omega} dI \left(\frac{\partial^2 S_r}{\partial q \partial I}\right)^{\frac{1}{2}} \left(\frac{\partial^2 S_r}{\partial q \partial I}\right)^{\frac{1}{2}} \times$$

$$\exp\left\{\frac{i}{\hbar} S_r(q,I) - \frac{i}{\hbar} S_s(q,I) + 2\pi \frac{iIM}{\hbar}\right\} \quad . \qquad (3.6)$$

Where the factor $(\hbar\omega/\Delta E)$ is the number of states lying in the energy interval ΔE.

For (extreme) simplicity we will consider the system to be just a one dimensional harmonic oscillator, i.e.,

$$H = 1/2(p^2 + \omega^2 q^2) = I\omega \quad . \qquad (3.7)$$

However, for this analysis the fact that a one dimensional system is both integrable and ergodic on the energy shell does not affect the final conclusions.

In the Poisson Sum series the terms $M \neq 0$ give a contribution of a higher order in $\hbar$ than the $M = 0$ term. Since we are interested in the extreme limit $\hbar \to 0$ we will consider only this term, i.e.,

$$\overline{P}_o = \frac{1}{2\pi} \frac{\omega}{\Delta E} \sum_r \sum_s \int_{E/\omega}^{(E+ E)/\omega} dI \left(\frac{\partial^2 S_r}{\partial q \partial I}\right)^{\frac{1}{2}} \left(\frac{\partial^2 S_r}{\partial q \partial I}\right)^{\frac{1}{2}} \times$$

$$\exp\left\{\frac{i}{\hbar} S_r(q,I - \frac{i}{\hbar} S_s(q,I)\right\} \quad . \qquad (3.8)$$

Again we are stuck with the quantum interference between the different branches of S. To make the connection with the actual classical, i.e., $\hbar = 0$, limit we have to perform some form of local averaging. This yields

$$\overline{\overline{P}}_o = \frac{1}{2\pi}\ \frac{\omega}{\Delta E} \sum_r \int_{E/\omega}^{(E+\Delta E)/\omega} dI \left(\frac{\partial^2 S_r}{\partial q \partial I}\right) \quad . \tag{3.9}$$

In the case of the harmonic oscillator, for which p has just two branches that only differ by a sign,

$$\overline{\overline{P}}_o(q) = \frac{\omega}{\pi\Delta E}\int_{E/\omega}^{(E+\Delta E)/\omega} dI \ \frac{1}{\sqrt{2(I\omega-\frac{1}{2}q^2\omega^2}} = \frac{\omega}{\pi\Delta E}\left\{\sqrt{2(E+\Delta E-\frac{1}{2}\omega^2 q^2)} - \sqrt{2(E-\frac{1}{2}\omega^2 q^2)}\right\} . \tag{3.10}$$

Which in the limit of a narrow packet, i.e., $\Delta E \to 0$, becomes

$$\overline{\overline{P}}_o(q) = \frac{\omega}{\pi} \ \frac{1}{\sqrt{2(E-\frac{1}{2}\omega^2 q^2)}} \quad . \tag{3.11}$$

Which is exactly the normalized probability density for the classical microcanonical distribution, i.e.,

$$\overline{\overline{P}}_o(q) = \frac{\int dp\,\delta(E-H(p,q))}{\int dp \int dq\,\delta(E-H(p,q))} \quad . \tag{3.12}$$

What this analysis tells us to that a microcanonical wavepacket collapses, in the limit $\hbar \to 0$, to the classical microcanonical distribution - as of course it must do. Thus such a wavepacket is ergodic on the energy shell, but only in the coarse grained limit. Without the coarse graining there would always be quantum interference present and we would never be able to demonstrate erogdicity. Furthermore since the quantal microcanonical distribution must always collapse to the classical microcanonical distribution as $\hbar \to 0$ (with coarse graining) the ergodic properties of the wavepacket are <u>independent</u> of whether the underlying classical motion is integrable or chaotic.

What I have tried to demonstrate in the above is that a <u>static</u>, time averaged, property such as 'quantum ergodicity' is not the most interesting property to look for. Rather we want some notion of 'quantum chaoticity' that gives some indication of the actual <u>dynamics</u> of the evolution. The previous discussion on quantum mechanical mappings demonstrated how the wavefunction attempts to follow the exponentially growing strands (the tendrils) of the classical path distribution thereby introducting great complexity into the wavefunction. The extent of this complexity is limited by the presence of quantum interference which eventually (in fact rather quickly) prevents the wavefunction from following the ever finer growing classical detail. The smaller $\hbar$ the greater the complexity the wavefunction can acheive and the faster it can spread probability density through the system.

How can we quantify this 'quantum complexity'. There are two competing processes. (i) the classical process corresponding to the exponentially fast growing separation of neighbouring trajectories (ii) the quantum interference effects that wash this away. The classical process can be conveniently parameterized in terms of the Lyapunov characteristic number, λ, of the trajectories.[10] This, crudely speaking, is the mean rate of separation of trajectories. In the case of regular motion λ is zero. What we require is the ratio of λ, the measure of classical spreading, to some measure of the quantum interference, Since λ has the dimensions of inverse time the construction of some dimensionless parameter to measure this ratio requires a quantal quantity which also has the dimensions of inverse time. That is

$$R_Q = \frac{\text{classical spreading}}{\text{quantum interference}} = \frac{\lambda}{\text{quantal quantity with dim. } t^{-1}} \ .$$

A quantal quantity that is a measure of quantum interference effects, has the desired dimensions and is easy to evaluate is

$$\frac{\omega_{\text{quant}}}{2\pi} = \frac{1}{2\pi}\,\frac{\Delta E}{\hbar} = \frac{1}{2\pi} \cdot \frac{1}{\substack{\text{density of states} \\ \text{a la Thomas - Fermi}}} \cdot \frac{1}{\hbar}$$

$$= \frac{(2\pi h)^N}{2\pi\hbar \int d\underset{\sim}{p} \int d\underset{\sim}{p}\,\delta(E-H(\underset{\sim}{p},\underset{\sim}{q}))}$$

hence

$$R_Q = \frac{\lambda}{(2\pi\hbar)^{N-1}} \int d\underset{\sim}{p} \int d\underset{\sim}{p}\;\delta(E-H(\underset{\sim}{p},\underset{\sim}{q})) \tag{3.13}$$

Where we evaluate the Thomas Fermi density of states at, say, the energy corresponding to the mean energy of the wave packet and λ is taken as some mean value for the associated family of trajectories.

R_Q might be called a "quantum Reynolds number". More precisely it is a <u>measure of the realizability of classical chaos in the quantum state</u>.

R_Q seems to have the right sort of properties. For regular motion it is zero. For chaotic motion it is small if $\hbar$ is large, i.e., quantal interference effects are still strong. It will only become large if $\hbar$ is small.

What about some order of magnitude estimates of R_Q? For the well know Henon-Heiles system supporting a total of, say, 100 bound states, at energies well into the chaotic regime we have

$$\frac{1}{2\pi\hbar}\int d\underset{\sim}{p}\int d\underset{\sim}{p}\ (E-H(\underset{\sim}{p},\underset{\sim}{q})) \sim 0\ (10)$$

Typical λ values are

$$\lambda \sim 0\ (1/10)$$

hence

$$R_Q \sim 1$$

to within an order of magnitude. For this system with this number of bound states there seems to be little quantum chaoticity.[11] Thus we would suggest the condition

$$R_Q >> 1$$

for significant quantum chaoticity. Suppose we require $R_Q \simeq 10$. This would require $\hbar$ to be x10 smaller. Since for this 2 degree of freedom system the number of states goes as $\hbar^{-2}$ this would require x100 more states than before. This would therefore correspond to a system supporting of the order of 10^4 states. What this very crude estimate tells us is that we should not expect to see strong quantum chaoticity in the sense that I've defined it unless the system supports many thousands of bound states. This high density of states is typical of molecules undergoing rapid internal energy transfer but still well beyond the reach of investigation by current computers.

REFERENCES

1. For a recent review see M. Tabor, 'The Onset of Chaotic Motion in Dynamical Systems' Adv. Chem. Phys. (1980).
2. M. V. Berry, N. L. Balazs, M. Tabor and A. Voros, Ann. Phys. 122, 26, (1979).

*This work sponsored in part by Office of Naval Research, Contract No. N00014-79-C-0537.

3. I. C. Percival, Adv. Chem. Phys. 36, 1 (1977).
4. M. V. Berry, J. Phys. A 10, 2083, (1977).
5. For a general review of semiclassical methods see, M. V. Berry and K. E. Mount, Rep. Progr. Phys. 35, 315, (1972).
6. V. I. Arnold, 'Mathematical Methods of Classical Mechanics'. Springer-Verlag Graduate Texts in Mathematics No. 60 (1978).
7. M. V. Berry and N. L. Balazs, J. Phys. A 12, 625, (1979).
8. V. I. Arnold and A. Avez, 'Ergodic Problems of Classical Mechanics', Benjamin, New York, (1968).
9. R. Jancel, 'Foundations of Classical and Quantum statistical Mechanics, Pergamon, London, (1968).
10. G. Benettin, L. Galgani and J. M. Strelcyn, Phys. Rev. A 14, 2338, (1976).
11. D. W. Noid, M. L. Koszykowski, M. Tabor and R. A. Marcus, J. Chem, Phys. 72, xxxx, (1980).

BARRIER PENETRATION AND EXPONENTIAL DECAY IN THE STARK EFFECT

Ira W. Herbst*

Department of Mathematics
University of Virginia
Charlottesville, Virginia 22903

ABSTRACT

Let $H = -\Delta + V + Fx_1$ with $V(x) = -\int \frac{\rho(y)}{|x-y|} d^3y$ and ρ a Gaussian charge distribution. Let ψ be an eigenvector of $-\Delta+V$ with negative eigenvalue. Among our results we show that for $F \neq 0$, $(\psi, e^{-itH}\psi)$ decays exponentially at a rate governed by the positions of the resonances of H. This exponential decay is in marked contrast to "conventional" atomic resonances for which power law decay is the rule.

I. INTRODUCTION

Temporal exponential decay is a well known quantum mechanical phenomenon. It often arises as a problem in barrier penetration: A particle of energy E inside a potential well as in Figure 1 will

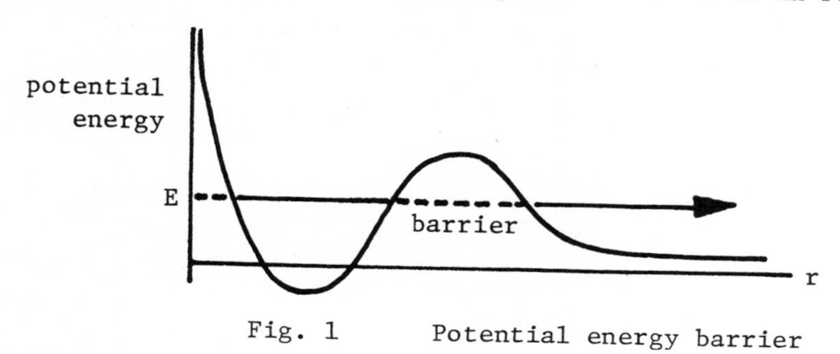

Fig. 1 Potential energy barrier

*Research supported by NSF Grant No. MCS 78-00101.

eventually tunnel through the classically forbidden region (barrier). The probability that the particle is still in the potential well at time t is given by $\exp(-t/\tau)$ to a good approximation under a wide variety of circumstances[4,13]. The quantity τ is called the lifetime of the state under consideration.

The problem of a hydrogen atom in an electric field is a classic example of such a barrier penetration problem. If we take the electric field in the x_1 direction with strength F and the electric charge $e = -1$, the total potential energy is

$$-r^{-1} + Fx_1.$$

This potential is shown in Figure 2 for a fixed value of $\vec{x}_\perp = (x_2, x_3)$ $(r = (x_1^2 + x_2^2 + x_3^2)^{-\frac{1}{2}})$. Note that as $F \downarrow 0$ the barrier width

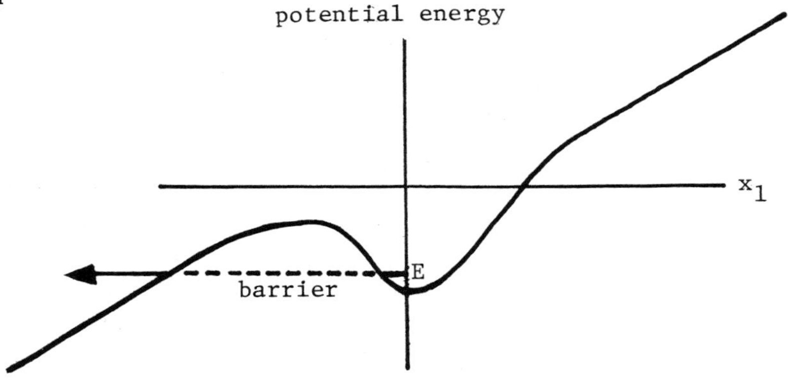

Fig. 2 Potential Energy barrier for Stark effect

goes to infinity. One therefore expects that for F small, an eigenstate of the hydrogenic Hamiltonian, $-\Delta - r^{-1}$, at energy E will remain localized around the origin for a long time. In fact, explicit WKB type calculations show[6,12,14] that this lifetime τ has an order of magnitude $\exp(c/F)$ for small $F > 0$ with $c > 0$ an explicit constant.

A way of looking at the barrier penetration problem which is very fruitful is to realize that even though an eigenstate of the Hamiltonian

$$H(0) = -\Delta - r^{-1}$$

of energy E_0 disappears into the continuum of

$$H(F) = -\Delta - r^{-1} + Fx_1$$

as soon as $F > 0$, it may become a second sheet pole of the resolvent of $H(F)$ (or rather, of certain matrix elements of this resolvent) which is very near E_0 for small F. The fact that this happens in hydrogen has been recently shown[5,7]. If this pole is at $E(F)$ with $\text{Im } E(F) < 0$ we expect that for small F

$$(\psi, e^{-itH(F)}\psi) \cong c(F,\psi)e^{-itE(F)} \tag{1.1}$$

where $H(0)\psi = E_0\psi$. Thus the probability that $e^{-itH(F)}\psi$ can be found in the state ψ at time $t > 0$ is approximately $|c(F,\psi)|^2 e^{-t/\tau}$ with $\tau = -1/(2 \text{ Im } E(F))$. One expects $|c(F,\psi)|$ to be close to 1 for F small if $||\psi|| = 1$. The quantity $c(F,\psi)$ should be related to the residue of the pole of $(\psi,(z-H(F))^{-1}\psi)$ at $z = E(F)$. Since presumably there are additional poles other than $E(F)$, one expects other terms to contribute to the right hand side of (1.1); but for small F the corresponding residues should be small so that at least for a resonable length of time (1.1) should be a good approximation.

If we go back to the potential V of Figure 1 which was drawn to indicate $V \geq -$ constant, it is known that exponential decay can not continue forever. In fact for such a potential one characteristically expects the behavior of $|(\psi,\exp(-it[-\Delta+V])\psi)|$ to be $(\text{constant})/t^{3/2}$ as $t \to \infty$ (in 3-dimensional space)[10,15]. However if the barrier is wide one should see exponential decay for quite some time. The mathematical reason why a potential such as that in Figure 1 cannot produce exponential decay is that if $|(\psi,e^{-itH}\psi)|^2 \leq \text{constant } \exp(-|t|/\tau)$ for $t > 0$, the same must be true for $t < 0$ by self-adjointness of H and thus the Fourier transform of $(\psi,e^{-itH}\psi)$ is analytic in the strip $\{\lambda \in \mathbb{C}: |\text{Im } \lambda| < (2\tau)^{-1}\}$. Thus the spectral measure $dE_H(\lambda)$ cannot vanish for any open set of energies λ and in particular H cannot be bounded below.

In the case of the Stark effect, the Hamiltonian is not bounded below and thus we are not prevented from obtaining exponential decay for all long times. In the next section we show that for a smoothed out version of the Coulomb potential one does indeed obtain exponential decay.

For a more detailed discussion of the problem considered here we refer the reader to [2,8] on which this work is based.

II. TRANSLATION ANALYTICITY AND EXPONENTIAL DECAY

Consider the Hamiltonian

$$H = H_0 + V, \qquad H_0 = -\Delta + Fx_1 . \tag{2.1}$$

Instead of $-\lambda r^{-1}$ we will take for technical reasons

$$V(\vec{x}) = \int \frac{\rho(\vec{y})}{|\vec{x}-\vec{y}|} d^3y = \frac{\rho(\vec{x}-\vec{y})}{|\vec{y}|} d^3y \tag{2.2}$$

where $\rho(\vec{x}) = -\lambda(2\pi\sigma^2)^{-3/2} \exp(-|\vec{x}|^2/2\sigma^2)$. This representa a nucleus of total charge λ with a smeared out nuclear charge distribution. The technical reason we do not take $\rho(\vec{x}) = -\lambda\delta(\vec{x})$ is that the resultant Coulomb potential, $-\lambda|\vec{x}|^{-1}$, is singular at the origin. As we will see our methods can be used to handle a large class of potentials which are analytic in the variable x_1. But this analyticity is crucial for our present methods.

In the following we will always assume $F > 0$. We will show:

THEOREM 2.1[8]: There exist complex numbers $\{E_n\}_{n=1}^{N}$ where $N \leq \infty$ obeying

a) $\text{Im } E_n < 0$

b) If $N = \infty$ then $\lim_{n\to\infty} \text{Im } E_n = -\infty$.

The E_n are called resonances of H. If ψ is an eigenvector of $-\Delta+V$ with negative eigenvalue then for any $\alpha>0$ we have the expansion:

$$(\psi, e^{-itH}\psi) = \sum_{\text{Im } E_n > -\alpha} c_n(\psi) e^{-itE_n} + O(e^{-\alpha t}); \qquad t > 0. \tag{2.3}$$

The remainder of this section will be devoted to proving Theorem 2.1. We first describe the resonances of H[2,7]: for $a \in \mathbb{C}$ let

$$H_a = -\Delta + Fx_1 + V_a + F_a \tag{2.4}$$

where here $V_a(\vec{x}) = V(x_1+a, \vec{x}_\perp)$, $\vec{x} = (x_1, \vec{x}_\perp)$, $\vec{x}_\perp = (x_2, x_3)$. We note that since V_z is a bounded analytic operator valued function of the complex variable z, the family of operators $\{H_z : z \in \mathbb{C}\}$ is an analytic family of type A in the sense of Kato[11].

LEMMA 2.2: $V_z(-\Delta+Fx_1+i)^{-1}$ is compact.

Proof: Suppose g is a bounded function of compact support. We claim $g(-\Delta+Fx_1+i)^{-1}$ is compact. To see this we compute

$$(-\Delta+Fx_1+i)^{-1} = (-\Delta+i)^{-1} - (-\Delta+i)^{-1}Fx_1(-\Delta+Fx_1+i)^{-1}$$
$$= (-\Delta+i)^{-1} - Fx_1(-\Delta+i)^{-1}(-\Delta+Fx_1+i)^{-1} - 2F\partial_1(-\Delta+i)^{-2}(-\Delta+Fx_1+i)^{-1}$$

where we have used $[x_1,f(-i\vec{\nabla})] = i(\partial_1 f)(-i\vec{\nabla})$. Since $g(-\Delta+i)^{-1}$, $gx_1(-\Delta+i)^{-1}$ and $g\partial_1(-\Delta+i)^{-2}$ are Hilbert-Schmidt operators it follows that $g(-\Delta+Fx_1+i)^{-1}$ is Hilbert-Schmidt. Since $\lim_{|\vec{x}|\to\infty} V_z(\vec{x}) = 0$, it follows that if X_A is the characteristic function of the set A,

$$X_{\{\vec{x}:|\vec{x}|\leq R\}} V_z(-\Delta+Fx_1+i)^{-1} \xrightarrow[R\to\infty]{||\cdot||} V_z(-\Delta+Fx_1+i)^{-1}$$

so that as the norm limit of compact operators, $V_z(-\Delta+Fx_1+i)^{-1}$ is compact.

If $\vec{p} = -i\vec{\nabla}$, explicit computation shows that $-\Delta+Fx_1 = U(p_\perp^2+Fx_1)U^{-1}$ where $U = \exp(ip_1^3/3F)$ and $\vec{p} = (p_1,\vec{p}_\perp)$. Thus the spectrum of $-\Delta+Fx_1$ is $(-\infty,\infty)$. Combining this information with Lemma 2.2 gives using standard results[16]

PROPOSITION 2.3: The essential spectrum of H_z is $\mathbb{R}+zF$.

We now discuss the discrete spectrum of H_z. For this purpose define the unitary translation operator

$$U(b)f(\vec{x}) = f(x_1+b,\vec{x}_\perp) \qquad b \in \mathbb{R}$$

and note that

$$U(b)H_zU^{-1}(b) = H_{z+b} \tag{2.5}$$

and thus $\sigma(H_z) = \sigma(H_{z+b})$ if $b \in \mathbb{R}$. Since the eigenvalues of H_z which are not in $\mathbb{R}+zF$ are branches of analytic functions of the variable z which do not depend on Re z, they are evidently constant as long as the essential spectrum, $\mathbb{R}+zF$, does not meet them. We want to show

PROPOSITION 2.4: If Im a < 0, the discrete spectrum of H_a is located in

$$\{ z:(\text{Im } a)F < \text{Im } z < 0\} .$$

Proof: If f and g are entire vectors for the translation group $\{U(b):b \in \mathbb{R}\}$ and $z \notin \sigma(H_a)$, Im $z > 0$, consider the function

$$M(a) = (f_{\bar{a}},(z-H_a)^{-1}g_a)$$

where g_a is the analtyic continuation of $U(b)g$ from $\mathbb{R}$ to a. M is analytic in $\{a:(\text{Im } a)F < \text{Im } z\}$. However for $b \in \mathbb{R}$

$$\begin{aligned} M(a) &= (U(b)f_{\bar{a}},U(b)(z-H_a)^{-1}g_a) \\ &= (f_{\overline{a+b}},(z-H_{a+b})^{-1}g_{a+b}) = M(a+b) \end{aligned}$$

and thus M is constant in $\{a:(\text{Im } a)F < \text{Im } z\}$. In particular in this region

$$(f_{\bar{a}},(z-H_a)^{-1}g_a) = (f,(z-H)^{-1}g) \quad . \tag{2.6}$$

Thus for any entire vectors f,g we have that for Im $a < 0$ fixed

$(f,(z-H_a)^{-1}g)$ is analytic in Im $z > 0$.

Similarly this function of z is analytic in Im $z < (\text{Im } a)F$. This shows that for Im $a < 0$, $\sigma_{disc}(H_a) \subseteq \{z:(\text{Im } a)F < \text{Im } z \leq 0\}$.

To show that H_a has no real eigenvalues for Im $a < 0$ we use the fact (announced in [1], proved in [2]) that H has no eigenvalues. The proof of Proposition 2.4 will thus be complete if we show that if H_a has an eigenvalue for Im $a < 0$ then H has an eigenvalue: If f,g are entire vectors and $(f,(z-H_a)^{-1}g)$ has a pole at $E_0 \in \mathbb{R}$, then by (2.6)

$$(f_{-\bar{a}},(z-H)^{-1}g_{-a}) = (f,(z-H_a)^{-1}g)$$

for z near E_0 and thus

$$\lim_{\varepsilon \downarrow 0} \varepsilon(f_{-\bar{a}},(E_0+i\varepsilon-H)^{-1}g_{-a}) \neq 0$$

so that H has an eigenvalue at E_0. This completes the proof of Proposition 2.4.

We define the set of resonances of H as

$$\mathcal{R} = \bigcup_{\text{Im } a<0} \sigma_{disc}(H_a) \quad .$$

It follows that $\mathcal{R}$ is at most countable and $E \in \mathcal{R} \Longrightarrow \text{Im } E < 0$. In addition $\mathcal{R}$ has no points of accumulation in the finite plane.

We must now show that the set R is bounded away from the real axis because otherwise one could not expect exponential decay of $(\psi, e^{-itH}\psi)$. This will follow from a bound of the form:

PROPOSITION 2.5: Fix $z \in \mathbb{C}$. Then

$$\lim_{\substack{E\in\mathbb{R} \\ |E|\to\infty}} ||V_z(-\Delta+Fx_1-E-i\lambda)^{-1}|| = 0$$

uniformly for λ in compact subsets of $(0,\infty) \cup (-\infty,0)$.

Proof: Suppose $K = (\delta,1/\delta) \cup (-1/\delta,-\delta)$ where $0 < \delta < 1$.

We need only show that

$$\lim_{|E|\to\infty} \sup_{\lambda\in K} ||g(-\Delta+Fx_1-E-i\lambda)^{-1}|| = 0 \tag{2.7}$$

for all $g \in C_0^\infty(\mathbb{R}^3)$ since if this is known then given $\varepsilon > 0$ we can find $g_\varepsilon \in C_0^\infty(\mathbb{R}^3)$ with $||g_\varepsilon - V_z||_\infty \leq \varepsilon\delta$ and using (2.7) we have

$$\overline{\lim_{|E|\to\infty}} \sup_{\lambda\in K} ||V_z(-\Delta+Fx_1-E-i\lambda)^{-1}|| \leq \varepsilon.$$

To prove (2.7) note that

$$||g(-\Delta+Fx_1-E-i\lambda)^{-1}||^2 = ||g(-\Delta+Fx_1-E-i\lambda)^{-1}(-\Delta+Fx_1-E+i\lambda)^{-1}\bar{g}||$$

$$\leq \frac{1}{2|\lambda|} (||g(-\Delta+Fx_1-E-i\lambda)^{-1}\bar{g}||+||g(-\Delta+Fx_1-E+i\lambda)^{-1}\bar{g}||)$$

so that we need only show

$$\lim_{|E|\to\infty} \sup_{\lambda\in K} ||g(-\Delta+Fx_1-E-i\lambda)^{-1}\bar{g}|| = 0.$$

Now write for $\lambda > 0$

$$(-\Delta+Fx_1-E-i\lambda)^{-1} = i\int_0^\infty \exp\{-it(-\Delta+Fx_1-E-i\lambda)\}dt$$

so that

$$g(-\Delta+Fx_1-E-i\lambda)^{-1}\bar{g} = i\int_0^\infty G_\lambda(t)e^{itE}dt$$

where

$$G_\lambda(t) = ge^{-it(-\Delta+Fx_1)}\bar{g}e^{-\lambda t} .$$

By a simple argument it is enough to show that for all $\varepsilon > 0$

$$\sup_{\delta^{-1}\geq\lambda\geq\delta} \int_{\varepsilon}^{\infty}||G_\lambda(t)e^{itE}dt|| \to 0$$

as $|E| \to \infty$. Doing an integration by parts, this will follow from

$$\sup_{\delta^{-1}\geq\lambda\geq\delta} \int_{\varepsilon}^{\infty}||\frac{d}{dt} G_\lambda(t)||dt < \infty \tag{2.8}$$

To prove (2.8) we use the formula[2]

$$e^{-it(-\Delta+Fx_1)} = e^{-itx_1F/2} e^{it\Delta} e^{-itx_1F/2} e^{-it^3F^2/12} \tag{2.9}$$

which can be derived as follows[9]: If $\psi_t = e^{-it(-\Delta+Fx_1)}\psi$, then performing the gauge transformation $\psi_t \to \phi_t \equiv e^{ix_1Ft}\psi_t$, we find

$$i\partial_t\phi_t = [(p_1-Ft)^2+p_\perp^2]\phi_t; \quad -i\vec{\nabla} = (p_1,\vec{p}_\perp)$$

and thus

$$\phi_t = e^{-i\int_0^t[(p_1-Fs)^2+p_\perp^2]ds}\psi \quad .$$

Using $e^{-ix_1a}p_1e^{ix_1a} = p_1+a$ to move $e^{-ix_1Ft/2}$ through to ψ we arrive at (2.9). For another simple derivation of (2.9) see [2]. (2.9) has a simple x-space kernel because $e^{it\Delta}$ has kernel $K_t(\vec{x},\vec{y}) = ct^{-3/2}\exp(i|\vec{x}-\vec{y}|^2/4t)$. Using this it is easy to see that the Hilbert-Schmidt norm of $\frac{d}{dt}G_\lambda(t)$ is bounded by

$$(\text{constant})\ t^{-7/2}\exp(-\delta t/2)$$

and this proves (2.8) and hence the proposition.

We use this bound to prove (b) of Theorem 2.1: Given $a < 0$ we show there are only finitely many resonances in the strip $S_a = \{z : 0 \geq \text{Im } z \geq a\}$. To see this consider the operator H_{ib} where $b = 2a/F$. The essential spectrum of H_{ib} is $\mathbb{R}+2ia$ and thus $S_a \cap \mathbb{R} \subseteq \sigma_{disc.}(H_{ib})$. Choose $E_0 > 0$ so that

$$||V_{ib}(-\Delta+Fx_1+2ia-z)^{-1}|| \leq 1/2$$

if $|\text{Re } z| \geq E_0$ and $\text{Im } z \in [a,1]$. Then

$$H_{ib}-z = \{1 + V_{ib}(-\Delta+Fx_1+2ia-z)^{-1}\}(-\Delta+Fx_1+2ia-z) \qquad (2.10)$$

so that $H_{ib}-z$ is invertible in $\{z:|\text{Re } z| \geq E_0\} \cap S_a$. Since there are at most finitely many points of R in $\{z:|\text{Re } z| < E_0\} \cap S_a$, $R \cap S_a$ is a finite set. This completes the demonstration of (b) of Theorem 2.1. We also see by inverting (2.10) that

LEMMA 2.6: Given $a < 0$ there is a number $E_0(a) > 0$ so that if $b = 2a/F$,

$$\sup\{||(z-H_{ib})^{-1}||:|\text{Re } z| \geq E_0(a),\ 0 \geq \text{Im } z \geq a\} < \infty$$

$$\sup\{||(z-H_{ib})^{-1}||:1 \geq \text{Im } z \geq 0\} < \infty \quad .$$

Before finishing the proof of Theorem 2.1, we need to know certain properties of the eigenvectors of $-\Delta+V$:

LEMMA 2.7: Suppose ψ is an eigenvector of $-\Delta+V$ with negative eigenvalue. Then ψ is an entire vector for the translation group $\{U(b):b \in \mathbb{R}\}$ and ψ_z is in the domain of H_z for all z.

<u>Proof</u>: $\{-\Delta+V_a:a \in \mathbb{C}\}$ is an analytic family of type A with $\sigma_{ess}(-\Delta+V_a) = [0,\infty)$ for all a. The latter follows from the compactness of $V_a(-\Delta+i)^{-1}$. Standard arguments[16] imply that ψ is thus an entire vector for $\{U(b):b \in \mathbb{R}\}$ and that ψ_a is an eigenvector of $-\Delta+V_a$. It remains only to show that ψ_a is in the domain of x_1. This follows from results of Combes and Thomas[3] who show in fact that $\psi_a \in \mathcal{D}(\exp(\delta|\vec{x}|))$ for some $\delta > 0$.

We now finish the proof of Theorem 2.1: If ψ is an eigenvector of $-\Delta+V$ with negative eigenvalue, we consider

$$(\psi,e^{-itH}\psi) = \int_{-\infty}^{\infty} d\lambda e^{-it\lambda}Q(\lambda) \quad . \qquad (2.11)$$

Here

$$Q(\lambda) = \lim_{\varepsilon\downarrow 0} (2\pi i)^{-1}\{(\psi,(H-\lambda-i\varepsilon)^{-1}\psi)-(\psi,(H-\lambda+i\varepsilon)^{-1}\psi)\}$$
$$= (2\pi i)^{-1}(K(\lambda)-G(\lambda)) \quad .$$

Note that for $\varepsilon > 0$ (we use $(H_z)^* = H_{\bar{z}}$)

$$(\psi,(H-\lambda-i\varepsilon)^{-1}\psi) = (\psi_{ib},(H_{-ib}-\lambda-i\varepsilon)^{-1}\psi_{-ib}) \quad b > 0$$

$$(\psi,(H-\lambda+i\varepsilon)^{-1}\psi) = (\psi_{-ib},(H_{ib}-\lambda+i\varepsilon)^{-1}\psi_{ib}) \quad b > 0$$

so that

$$K(\lambda) = (\psi_{ib}, (H_{-ib}-\lambda)^{-1}\psi_{-ib}) \qquad b > 0$$

$$G(\lambda) = (\psi_{-ib}, (H_{ib}-\lambda)^{-1}\psi_{ib}) \qquad b > 0$$

We note that K has a meromorphic continuation to $\mathbb{C}$ with poles possible only at points in $\mathcal{R}$. In fact denoting this continuation by K(z) we have for Im z > -bF, b > 0

$$K(z) = (\psi_{ib}, (H_{-ib}-z)^{-1}\psi_{-ib}).$$

Similarly G has a meromorphic continuation to $\mathbb{C}$ with poles possible only at points of $\overline{\mathcal{R}} = \{z : \bar{z} \in \mathcal{R}\}$ and for Im z < bF, b > 0 we have

$$G(z) = (\psi_{-ib}, (H_{ib}-z)^{-1}\psi_{ib}) .$$

Thus $Q(\lambda)$ has a meromorphic continuation to $\mathbb{C}$ given by $Q(z) = (2\pi i)^{-1}(K(z)-G(z))$.

Let $\alpha > 0$ be given and fix $b = 4\alpha/F$. Then by Lemma 2.6 there are no resonances in the region $W = \{z : |\text{Re } z| \geq E_0(-2\alpha),$ $0 \geq \text{Im } z \geq -2\alpha\}$ and $\sup_{z \in W} ||(z-H_{-ib})^{-1}|| < \infty$. Thus writing for $z \in W$

$$(z-H_{-ib})^{-1} = z^{-1} + (H_{-ib})z^{-2} + z^{-2}H_{-ib}(z-H_{-ib})^{-1}H_{-ib}$$

we see that using Lemma 2.7, for $z \in W$

$$\begin{aligned} K(z) &= -(\psi_{ib}, \psi_{-ib})z^{-1} + O(|z|^{-2}) \\ &= -||\psi||^2 z^{-1} + O(|z|^{-2}) . \end{aligned}$$

Similarly using Lemmas 2.6 and 2.7

$$G(z) = -||\psi||^2 z^{-1} + O(|z|^{-2})$$

so that $Q(z) = O(|z|^{-2})$ for $z \in W$. Using this bound we can shift the contour from the real axis in (2.11) to a line L parallel to $\mathbb{R}$ intersecting the imaginary axis at $(-\alpha-\varepsilon)i$ so that $\mathcal{R} \cap L = \emptyset$. We thus have

$$(\psi, e^{-itH}\psi) = \sum_{\substack{E_n \in \mathcal{R} \\ \text{Im } E_n > -\alpha-\varepsilon}} c_n(\psi)e^{-itE_n} + \int_L dz e^{-itz}Q(z)$$

where $c_n(\psi)$ is the residue of $-K(z)$ at $z = E_n$. Since

$$\left|\int_L dz e^{-itz} Q(z)\right| \leq e^{-(\alpha+\varepsilon)t} \int d\lambda \left|Q(\lambda-(\alpha+\varepsilon)i)\right|$$

and $Q(\lambda-(\alpha+\varepsilon)i) = O(|\lambda|^{-2})$, Theorem 2.1 is proved.

REFERENCES

1. S. Agmon: Proceedings of the Tokyo Int. Conf. on Functional Analysis and Related Topics, 1969.
2. J. Avron and I. Herbst: Commun. Math. Phys. 52, 239-254 (1977).
3. J.M. Combes and L. Thomas: Commun. Math. Phys. 34, 251-270 (1973).
4. M. Goldberger and K. Watson: "Collision Theory", Wiley, N.Y., 1964.
5. S. Graffi and V. Grecchi: Commun. Math. Phys. 62, 83-96 (1978).
6. E. Harrell and B. Simon: The Mathematical Theory of Resonances whose Widths are Exponentially Small, preprint.
7. I. Herbst: Commun. Math. Phys. 64, 279-298 (1979).
8. I. Herbst: Exponential Decay in the Stark Effect, to be published in Commun. Math. Phys.
9. W. Hunziker: Private communication via B. Simon.
10. A. Jensen and T. Kato: Duke Math. J. 46, 583-611 (1979).
11. T. Kato: "Perturbation Theory for Linear Operators", Berlin, Springer 1976.
12. L.D. Landau and E.M. Lifshitz: "Quantum Mechanics", New York, Pergamon Press, 1977.
13. E. Merzbacher: "Quantum Mechanics", New York, Wiley, 1961.
14. J.R. Oppenheimer: Phys. Rev. 31, 66-81 (1928).
15. J. Rauch: Commun. Math. Phys. 61, 149-168 (1978).
16. M. Reed and B. Simon: "Methods of Modern Mathematical Physics Vol.I: Functional Analysis", New York, Academic Press, 1972; Vol. IV, 1978.

A NONLINEAR SCHRÖDINGER EQUATION YIELDING THE "SHAPE OF MOLECULES" BY SPONTANEOUS SYMMETRY BREAKING

Peter Pfeifer

Laboratory of Physical Chemistry
ETH Zürich
8092 Zürich, Switzerland

ABSTRACT

It is reviewed how a molecule with an almost-degenerate ground state, if modelled as two-level system, is structurally unstable as follows: If the difference between the lowest two energy levels is below a certain critical value (determined by the free-molecule Coulomb Hamiltonian), then the coupling of the molecule to the quantized radiation field yields two symmetry-broken effective ground states of the molecule (one is the mirror image of the other) which are separated by a superselection rule orginating from the infrared singularity of the electromagnetic field. If this energy difference exceeds the critical value, then the ground state of the free molecule is not altered by the interaction with the field. It is shown how these results can be recovered from a Schrödinger equation for the molecule which, in addition to the free-molecule part, contains a nonlinear term incorporating the interaction with the radiation field.

§ 1. INTRODUCTION

Perhaps the simplest one of a series of problems recently revived under what has become known as question of "shape of molecules" (cf. e.g. Woolley, 1976, 1978; Claverie and Diner, 1980; Primas, 1980) is the so-called "paradox of optical isomers" (Hund, 1927; Rosenfeld, 1929; Born and Jordan, 1930, § 47). It is the question why so many molecules are never seen in an eigenstate of the associated (Coulomb) Hamiltonian of the free molecule, but rather in one of two classes of chiral states corresponding to left/right-handedness, where two such states that do not belong to the same class are separated by a superselection rule. In fact, this invalidation of the unrestricted superposition principle - or, equivalently, the fact that handedness is a classical observable for these molecules - is the major explanandum. On the other hand, a successful answer should also account for the absence of such a symmetry breaking in cases like ammonia.

To appreciate the problem and the starting point for its solution in some detail, consider the molecular Hamiltonian (in atomic units)

$$H = \sum_{\nu=1}^{N} \frac{1}{2m_\nu} |p_\nu|^2 + V(q_1, \ldots, q_N) , \tag{1a}$$

$$V(q_1, \ldots, q_N) = \sum_{\substack{\nu,\mu=1 \\ \nu<\mu}}^{N} \frac{z_\nu z_\mu}{|q_\nu - q_\mu|} \tag{1b}$$

with charge z_ν, mass m_ν, position operator q_ν, and momentum operator p_ν for the ν-th particle ($\nu=1, \ldots, N$). For simplicity we assume all particles to be distinguishable so that we can ignore spin and statistics. Suppose that the internal Hamiltonian $H - (2\sum_{\nu=1}^{N} m_\nu)^{-1} |\sum_{\nu=1}^{N} p_\nu|^2$ has at least two isolated eigenvalues $\varepsilon_1 < \varepsilon_2$ at the bottom of the spectrum, and corresponding (normalized) eigenfunctions $\Psi_1, \Psi_2 \in L^2(\mathbb{R}^{3N})$ which we may take to be real-valued and of product form, internal function $\times$ spherically symmetric center-of-mass function, where the latter is the same for Ψ_1 and Ψ_2. It follows that Ψ_1 is spherically symmetric , and in particular even; while Ψ_2 is expected to be odd. So Ψ_1 and Ψ_2 are not chiral and hence cannot represent states of a chiral molecule (a state $\Psi \in L^2(\mathbb{R}^{3N})$ is called chiral if Ψ is not eigenfunction of an improper-rotation operator).

But for a typical chiral molecule, $\varepsilon_2 - \varepsilon_1$ is practically zero. I.e. Ψ_1 and Ψ_2 are so sensitive to external perturbations that a physically significant description of such a molecule has to include essential parts of its surroundings. A minimal environment is given by the radiation field as generated by the molecule itself. With the standard dipole-velocity interaction, this leads to the following formal Hamiltonian for the molecule coupled to the transverse part of the electromagnetic field (quantized in a cube of length L):

$$H^{(L)} = H \otimes 1 + 1 \otimes \sum_{n=1}^{\infty} \omega_n^{(L)} b_n^* b_n + \sum_{n=1}^{\infty} (g_n^{(L)*} \otimes b_n + g_n^{(L)} \otimes b_n^*). \quad (2)$$

Here $n = 1, 2, \ldots$ refers to the n-th field oscillator with frequency $\omega_n^{(L)} > 0$, annihilation operator $b_n = 1 \otimes \ldots \otimes 1 \otimes b \otimes 1 \otimes \ldots$ (b in the n-th position acts in the one-oscillator Hilbert space $\mathcal{H}$ and satisfies $[b, b^*] \subset 1$), and coupling operator $g_n^{(L)}$ acting in $L^2(\mathbb{R}^{3N})$ ($g_n^{(L)}$ is related to the Fourier transform $\underline{g}$ of the electrical current density operator of the molecule,

$$\underline{g}(\underline{k}) = \frac{1}{2} \sum_{\nu=1}^{N} \frac{z_\nu}{m_\nu} (e^{-i\underline{k}\cdot\underline{q}_\nu} \underline{p}_\nu + \underline{p}_\nu e^{-i\underline{k}\cdot\underline{q}_\nu}), \quad \underline{k} \in \mathbb{R}^3). \quad (3)$$

Thus $H^{(L)}$ formally lives in the ("complete") infinite tensor-product Hilbert space $L^2(\mathbb{R}^{3N})\otimes(\bigotimes_{n=1}^{\infty}\mathcal{H})$. For future convenience, we also introduce the (normalized) coherent oscillator state $|z\rangle\in\mathcal{H}$ defined by $b|z\rangle=z|z\rangle$ $(z\in\mathbb{C})$.

Now focussing on the lowest molecular states Ψ_1 and Ψ_2 again, one may replace $H^{(L)}$ by a model Hamiltonian which treats the molecule as two-level system. The ensuing theory has been demonstrated to give a quantitative resolution of the paradox of optical isomers indeed (Pfeifer, 1980; Pfeifer and Primas, to be published). Two cornerstones thereof will be outlined in § 2 and will serve as background for the subsequent extension of the theory in § 3. The nonlinear Schrödinger equation so arising from $H^{(L)}$ promises to provide a framework wide enough for a proper understanding also of the shape of molecules other than optical isomers. (For nonlinear Schrödinger eqs. motivated by treating the charge-generated electromagnetic field classically, see Ulmer, 1980.)

§ 2. TWO-LEVEL MODEL FOR THE MOLECULE

When projecting $H^{(L)}$ onto the subspace $(\mathrm{span}\{\Psi_1,\Psi_2\})\otimes$ $\otimes(\bigotimes_{n=1}^{\infty}\mathcal{H})$, we obtain a model Hamiltonian which formally acts in $\mathbb{C}^2\otimes(\bigotimes_{n=1}^{\infty}\mathcal{H})$ and reads (apart from an irrelevant constant)

$$h^{(L)}=\tfrac{1}{2}(\varepsilon_1-\varepsilon_2)S_3\otimes 1+1\otimes\sum_{n=1}^{\infty}\omega_n^{(L)}b_n^*b_n+S_2\otimes\sum_{n=1}^{\infty}\lambda_n^{(L)}(b_n+b_n^*)$$

where

$$S_2=\begin{pmatrix}0&-i\\i&0\end{pmatrix},\quad S_3=\begin{pmatrix}1&0\\0&-1\end{pmatrix},$$

$$\lambda_n^{(L)}=i\langle\Psi_1|g_n^{(L)}\Psi_2\rangle=i\langle\Psi_1|g_n^{(L)*}\Psi_2\rangle\in\mathbb{R},$$

and where $\binom{1}{0}$, $\binom{0}{1}$ correspond to ψ_1, ψ_2, respectively. A rigorous analysis of the finite-mode truncations of $h^{(L)}$ produces strong evidence that the Hartree ground state(s), as given in the following result, and the exact ground state(s) of $h^{(L)}$ coincide asymptotically as we remove the infrared cutoff by letting $L \to \infty$:

Theorem 1 (Hartree ground state(s)). Put

$$\gamma^{(L)} = \min\{1, (\varepsilon_2 - \varepsilon_1)(4 \sum_{n=1}^{\infty} \lambda_n^{(L)2}/\omega_n^{(L)})^{-1}\}.$$

Then the energy $\langle \chi \otimes \Phi | h^{(L)}(\chi \otimes \Phi)\rangle$ (with normalized $\chi \in \mathbb{C}^2$ and $\Phi \in \bigotimes_{n=1}^{\infty} \mathcal{H}$) is minimized by taking

$$\chi = \frac{1}{\sqrt{2}} \begin{pmatrix} \sqrt{1+\gamma^{(L)}} \\ \pm i \sqrt{1-\gamma^{(L)}} \end{pmatrix},$$

$$\Phi = \bigotimes_{n=1}^{\infty} |\mp \sqrt{1-\gamma^{(L)2}}\, \lambda_n^{(L)}/\omega_n^{(L)}\rangle.$$

Both for $\lambda_1^{(L)} = \lambda_2^{(L)} = \dots = 0$ and for $\varepsilon_1 = \varepsilon_2$, $\chi \otimes \Phi$ is (are) the exact ground state(s) of $h^{(L)}$.

Although theorem 1 does yield two symmetry-broken ground states of the molecule for sufficiently small $\varepsilon_2 - \varepsilon_1$, it does not explain yet why superpositions of these ground states should not be physically realizable (as pure states). This is answered by the next result if we agree that only those operators acting in $\bigotimes_{n=1}^{\infty} \mathcal{H}$ can be considered as physical observables of the field which correspond essentially to finite-mode operations (for a general account of the underlying conceptual framework, cf. Primas, 1980, in particular § 4.8):

Theorem 2 (Superselection rule for the field states). Let $\mathcal{A}$ be the quasilocal algebra of field observables, defined as double commutant (relative to the bounded operators on $\bigotimes_{n=1}^{\infty}\mathcal{H}$) of the operator set $\{b_n | n=1,2,\ldots\}$. If $\lim_{L\to\infty}\gamma^{(L)}<1$ and condition (5) below holds, then the two field states

$$\Phi_{\pm 1}^{(L)} = \bigotimes_{n=1}^{\infty} |\pm\sqrt{1-\gamma^{(L)2}}\,\lambda_n^{(L)}/\omega_n^{(L)}\rangle$$

are asymptotically in two different superselection sectors (with respect to $\mathcal{A}$) of $\bigotimes_{n=1}^{\infty}\mathcal{H}$ as $L\to\infty$. I.e. every nontrivial linear combination $\sum_{s=\pm 1} c_s \Phi_s^{(L)}$ (with $c_{\pm 1}\in\mathbb{C}$) corresponds asymptotically (as $L\to\infty$) to a non-pure state on $\mathcal{A}$:

$$\lim_{L\to\infty}\{\langle\sum_{s=\pm 1} c_s\Phi_s^{(L)}|A\sum_{s=\pm 1} c_s\Phi_s^{(L)}\rangle - \sum_{s=\pm 1}|c_s|^2\langle\Phi_s^{(L)}|A\Phi_s^{(L)}\rangle\} = 0$$

for all $A\in\mathcal{A}$.

One arrives so at the following conclusion: Let

$$\gamma = \min\{1, (\varepsilon_2-\varepsilon_1)\pi^2 c^2/\int_{\mathbb{R}^3}|\langle\psi_1|\underline{g}(\underline{k})\psi_2\rangle\times\underline{k}|^2|\underline{k}|^{-4}d(\underline{k})\} \tag{4}$$

where c is the velocity of light and $\underline{g}$ is defined by (3) (all in a.u. still). If $\gamma<1$ and

$$\langle\psi_1|\sum_{\nu=1}^{N} z_\nu \underline{q}_\nu \psi_2\rangle \neq \underline{0}, \tag{5}$$

then the molecule has available two effective ground states $(\sqrt{1+\gamma}\,\psi_1 \pm i\sqrt{1-\gamma}\,\psi_2)/\sqrt{2}$ which are chiral and which are separated by a superselection rule because the associated coherent field states lie in different physical sectors as generated by the quasilocal field observables. Conversely, if $\gamma=1$ then the molecule has, as for effective ground state, only the achiral free-molecule ground state ψ_1, while the field is in the vacuum state. Numerical estimates of this sharp borderline between chiral and achiral molecules compare rather well with experimental data.

§ 3. NONLINEAR SCHRÖDINGER EQUATION

Guided by the remark preceding theorem 1 and by the fact that quantum-mechanical correlations between the molecule and the radiation field must be negligible in the ground state of the combined system (else we could not ever treat the two other than as holistic entity - which contradicts experience; cf. also Primas, 1980, ch. 5), we now go beyond the two-level scheme for the molecule by applying the Hartree method to the full Hamiltonian $H^{(L)}$ as given by eqs. (2) and (1a) (the specific form (1b) of the potential V is inessential for what follows). In spirit, this approach resembles recent work of Davies (1979). There are two major differences, however: (i) The coupling of the molecule to the radiation field (rather than to a little-specified phonon field) seems to cover a more fundamental situation and leads to a different type of nonlinearity. (ii) Here a symmetry breaking is typically accompanied by a field-induced superselection rule.

<u>Theorem 3</u> (Hartree ground state(s)). In the limit $L \to \infty$, the energy $\langle \varphi \otimes \Phi | H^{(L)}(\varphi \otimes \Phi) \rangle$ (with normalized $\varphi \in L^2(\mathbb{R}^{3N})$ and $\Phi \in \bigotimes_{n=1}^{\infty} \mathcal{H}$) is minimized by taking φ to minimize

$$E(\varphi) = \langle \varphi | H \varphi \rangle - (2\pi c)^{-2} \int_{\mathbb{R}^3} |\langle \varphi | g(\underline{k}) \varphi \rangle \times \underline{k}|^2 |\underline{k}|^{-4} d(\underline{k})$$

where c and g are as before, and by choosing Φ to be the coherent field state $\bigotimes_{n=1}^{\infty} | \omega_n^{(L)-1} \langle \varphi | g_n^{(L)} \varphi \rangle \rangle$.

Some properties of E:

a) E is well-defined on the (operator) domain $\mathcal{D}(H)$ of H. Indeed, $\varphi \in \mathcal{D}(H)$ implies that $|\langle \varphi | \underline{g}(\underline{k}) \varphi \rangle| = o(|\underline{k}|^{-1})$ as $|\underline{k}| \to \infty$.

b) E is invariant under all elements of the Euclidean group that leave $V(\underline{q}_1, \dots, \underline{q}_N)$ invariant.

c) E is invariant under time reversal, i.e. $\mathsf{E}(\varphi) = \mathsf{E}(\varphi^*)$ for $\varphi \in \mathcal{D}(H)$. If $\varphi \in \mathcal{D}(H)$ and $\varphi = z\varphi^*$ for some $z \in \mathbb{C}$, then $\mathsf{E}(\varphi) = \langle \varphi | H\varphi \rangle$.

d) If $\varphi \in \mathcal{D}(H)$ is sufficiently well-behaved, then

$$\mathsf{E}(\varphi) = \langle \varphi | H\varphi \rangle - (2c^2)^{-1} \int_{\mathbb{R}^3} \int_{\mathbb{R}^3} \frac{\underline{j}_\varphi^\perp(\underline{r}) \cdot \underline{j}_\varphi^\perp(\underline{r}')}{|\underline{r} - \underline{r}'|} d(\underline{r}) d(\underline{r}')$$

where the transverse part $\underline{j}_\varphi^\perp$ of the electrical current density $\underline{j}_\varphi(\underline{r}) = \langle \varphi | \frac{1}{2} \sum_{\nu=1}^{N} \frac{z_\nu}{m_\nu} \{ \delta(\underline{r} - \underline{q}_\nu) \underline{p}_\nu + \underline{p}_\nu \delta(\underline{r} - \underline{q}_\nu) \} \varphi \rangle$ $(\underline{r} \in \mathbb{R}^3)$ is given by

$$\begin{aligned} \underline{j}_\varphi^\perp(\underline{r}) &= (2\pi)^{-3} \int_{\mathbb{R}^3} e^{i\underline{r}\cdot\underline{k}} \, \underline{k} \times (\langle \varphi | \underline{g}(\underline{k}) \varphi \rangle \times \underline{k}) |\underline{k}|^{-2} d(\underline{k}) \\ &= \left(\underline{\partial} \times \int_{\mathbb{R}^3} \frac{(\underline{\partial} \times \underline{j}_\varphi)(\underline{r}')}{4\pi |\cdot - \underline{r}'|} d(\underline{r}') \right)(\underline{r}) \qquad (\underline{r} \in \mathbb{R}^3). \end{aligned}$$

The following definitions serve to express the effects of the radiation field on the molecule in purely molecular terms similar as in conclusion of § 2:

<u>Definition</u>.

1) φ is called ground state of E if $\varphi \in \mathcal{D}(H)$, $\|\varphi\| = 1$, and $\mathsf{E}(\varphi) = \inf \{ \mathsf{E}(\psi) | \psi \in \mathcal{D}(H); \|\psi\| = 1 \}$.

2) A ground state φ of E is called trivial if $\mathsf{E}(\varphi) = \langle \varphi | H\varphi \rangle$.

3) Two ground states φ, φ' of E are said to be separated by a superselection rule (inherited from the radiation field) if the associated field states

$$\bigotimes_{n=1}^{\infty} |\omega_n^{(L)-1} \langle \varphi | g_n^{(L)} \varphi \rangle\rangle \, , \quad \bigotimes_{n=1}^{\infty} |\omega_n^{(L)-1} \langle \varphi' | g_n^{(L)} \varphi' \rangle\rangle$$

are asymptotically in different superselection sectors (with respect to the quasilocal observables $\mathcal{A}$ on $\bigotimes_{n=1}^{\infty} \mathcal{H}$) as $L \to \infty$.

Physical arguments and the above properties b), c) of E suggest that E is bounded from below if and only if H is so, and that E has a ground state if and only if H does so (but note that functionals of type $\langle \cdot | H \cdot \rangle$ plus quartic form are known which have ground states without H doing so (Lieb, 1977; Davies, 1979)). So if the second conjecture holds then, for translation-invariant potentials V, the foregoing definitions and subsequent results should be read as referring to minimizing sequences rather than to single minimizing vectors. Note also that, by property c), nontrivial ground states of E are intrinsically complex-valued and always come in pairs, of which one is the time-reversed image of the other.

The functional E gives rise to a nonlinear Schrödinger equation in the sesquilinear-form sense as follows:

Theorem 4 (Nonlinear Schrödinger equation). If φ is ground state of E, then there is an $\varepsilon \in \mathbb{R}$ such that

$$\langle \psi | H \varphi \rangle - (2\pi^2 c^2)^{-1} \int_{\mathbb{R}^3} (\langle \varphi | \underline{g}(\underline{k}) \varphi \rangle^* \times \underline{k}) \cdot (\langle \psi | \underline{g}(\underline{k}) \varphi \rangle \times \underline{k}) |\underline{k}|^{-4} d(\underline{k}) = \varepsilon \langle \psi | \varphi \rangle \tag{6}$$

for all $\psi \in \mathcal{D}(H)$.

Apart from rather singular exceptions, the existence of symmetry-broken and/or superselection-separated ground states of E obviously requires E to have nontrivial ground states. To this end, we have:

Theorem 5 (Nontrivial ground states). If φ is ground state of E, and if H has at least two eigenvalues $\varepsilon_1 < \varepsilon_2$ (where $\varepsilon_1 = \inf\{\langle\psi|H\psi\rangle | \psi \in \mathcal{D}(H);\ \|\psi\| = 1\}$) with normalized real-valued eigenfunctions ψ_1, ψ_2 such that $\gamma < 1$ (where γ is given by (4)), then φ is nontrivial. In fact, these hypotheses imply $E((\sqrt{1+\gamma}\,\psi_1 \pm i\sqrt{1-\gamma}\,\psi_2)/\sqrt{2}) < \varepsilon_1$.

Application of theorem 5 to the special case where $N=1$ and $H = (2m_1)^{-1}|\underline{p}_1|^2 + zz_1/|\underline{q}_1|$ ($m_1 > 0$, $zz_1 < 0$) nourishes one's expectations that, among other factors, high particle charges will favor nontrivial ground states of E. Indeed, if ψ_1, ψ_2 are normalized real-valued 1s,2p-eigenfunctions with eigenvalues $\varepsilon_1, \varepsilon_2$ of H, then eq. (4) reads $\gamma = \min\{1, \frac{1}{5}(\frac{3}{2})^9 c^2 |zz_1^3|^{-1}\}$.

The final result gives a partial converse of the fact that a superselection rule can occur only if at least one of the involved ground states of E breaks both time-reversal and space-inversion symmetry (note that condition (7) below implies that φ is neither eigenfunction of the time-reversal operator nor of any improper-rotation operator other than a reflection-at-a-plane operator):

Theorem 6 (Superselection rule). If φ, φ' are ground states of E such that

$$\sum_{\nu=1}^{N} \frac{z_\nu}{m_\nu} (\langle\varphi|\underline{p}_\nu\varphi\rangle - \langle\varphi'|\underline{p}_\nu\varphi'\rangle) \neq \underline{0} ,$$

then φ and φ' are separated by a superselection rule. In particular, if φ is ground state of E with

$$\sum_{\nu=1}^{N} \frac{z_\nu}{m_\nu} \langle \varphi | p_\nu \varphi \rangle \neq 0 , \tag{7}$$

then φ and φ^* are separated by a superselection rule.

If the premises of theorem 5 are satisfied and if we assume that "the" resulting nontrivial ground state φ of E is given by $\varphi = (\sqrt{1+\gamma}\,\psi_1 + i\sqrt{1-\gamma}\,\psi_2)/\sqrt{2}$, then conditions (5) and (7) are equivalent. Thus the nonlinear Schrödinger equation (6), together with theorem 6, really covers as much of the two-level model in § 2 as it possibly can.

REFERENCES

Born, M., and Jordan, P., 1930, "Elementare Quantenmechanik," Springer, Berlin.

Claverie, P., and Diner, S., 1980, The concept of molecular structure in quantum theory: interpretation problems, Israel J. Chem., 19: to appear.

Davies, E.B., 1979, Symmetry breaking for a nonlinear Schrödinger equation, Commun. math. phys., 64: 191.

Hund, F., 1927, Zur Deutung von Molekelspektren III, Z. Phys., 43: 805.

Lieb, E.H., 1977, Existence and uniqueness of the minimizing solution of Choquard's nonlinear equation, Stud. Appl. Math., 57: 93.

Pfeifer, P., 1980, "Chiral molecules - a superselection rule induced by the radiation field," thesis, ETH Zürich.

Primas, H., 1980, Foundations of theoretical chemistry, in: "Proc. NATO Adv. Study Inst., 1979, Quantum dynamics of molecules," R.G. Woolley, ed., Plenum Press, New York.

Rosenfeld, L., 1929, Quantenmechanische Theorie der natürlichen optischen Aktivität von Flüssigkeiten und Gasen, Z. Phys., 52: 161.

Ulmer, W., 1980, On the representation of atoms and molecules as self-interacting fields with internal structure, Theor. Chim. Acta, 55: 179.

Woolley, R.G., 1976, Quantum theory and molecular structure, Adv. Phys., 25: 27.

Woolley, R.G., 1978, Must a molecule have a shape?, J. Amer. Chem. Soc., 100: 1073.

Note added in proof. It has been pointed out by E.B. Davies that the functional E, as defined in theorem 3, is not bounded below for a large class of potentials V of physical interest. It turns out that this is because the Hamiltonian (2) does not include the so-called A^2-term. Indeed, if $H'^{(L)} = H^{(L)} + A^2$-term then, as $L \to \infty$, minimization of $\langle \varphi \otimes \Phi | H'^{(L)} (\varphi \otimes \Phi) \rangle$ (with $\varphi \in L^2(\mathbb{R}^{3N})$ and coherent state $\Phi \in \bigotimes_{n=1}^{\infty} \mathcal{H}$, $\|\varphi\| = \|\Phi\| = 1$) leads to a functional E' which is bounded below at least for nonnegative V. But the crude estimate

$$0 \leq E'(\varphi) - E(\varphi) \leq c^{-2}(\langle \varphi | H \varphi \rangle - E(\varphi)) \frac{8}{3} \left(\frac{2}{\pi}\right)^{1/3} \cdot \left[\int_{\mathbb{R}^3} \left| \langle \varphi | \sum_{\nu=1}^{N} \frac{z_\nu^2}{2m_\nu} \{ \delta(\underline{r} - q_\nu) + \delta(\underline{r} + q_\nu) \} \varphi \rangle \right|^{3/2} d(\underline{r}) \right]^{2/3}$$

(for all $\varphi \in \mathcal{D}(H)$) suggests that in general $E'(\varphi)$ will differ only little from $E(\varphi)$ for φ minimizing E', and that a good approximation for this minimizing φ may be obtained from minimization of E restricted to a suitable subset (not invariant under dilations) of $\mathcal{D}(H)$. Details will be given elsewhere.

CLASSICAL LIMIT OF THE NUMBER OF QUANTUM STATES

Richard Lavine

Mathematics Department, University of Virginia
Charlottesville, Virginia 22903
(Permanent address: University of Rochester, NY. 14627)

ABSTRACT

A heuristic principle in quantum mechanics says that the number of quantum states supported by a subset S of classical phase space (of dimension 2d) is given by $h^{-d}|S|$, where h is Planck's constant. One precise version of this is that as $h \to 0$, h^d times the number of energy eigenvalues $\leq$ E approaches the volume of the set in phase space with energy less than E. We generalize this to apply to continuous spectrum as well, showing that h^d times the number of states with energy below E and position in $B \equiv h^d \mathrm{tr}(\chi_B(Q)\chi_{(-\infty,E)}(H))$ approaches the volume of the corresponding subset of phase space, for $H = -h^2\Delta + V(Q)$, with mild regularity conditions on V. (Q = multiplication by position.) The quantity we study has a statistical interpretation, and it exhibits the resonances of H.

INTRODUCTION

The behavior of a quantum mechanical particle in $\mathbb{R}^n$ is determined by its Hamiltonian operator, which typically takes the form $H = P^2 + V(Q)$, operating in $L^2(\mathbb{R}^n)$. (We write $P = -i\hbar\nabla$, where $\hbar = h/2\pi$ and h is Planck's constant, $Q\phi(x) = x\phi(x)$, and we take functions of self-adjoint operators in the sense of the functional calculus.) Let χ_I be the characteristic or indicator function, which takes the values 1 and 0 according to whether the variable is in I or its complement. The operator H is determined up to unitary equivalence by its spectral projection valued measure $I \to \chi_I(H)$.

But in some ways, spectral theory (the study of such measures) fails to reflect perfectly the physical situation, especially when there is continuous spectrum. For example,

(i) A small change in $V(x)$ for values of x too "far away" to be physically relevant can change the spectral measure from discrete to absolutely continuous.

(ii) If $V(x) \to 0$ suitably at infinity, then the spectral measure of H is absolutely continuous, and for all non-null measurable subsets I of $(0,\infty)$, $\dim \chi_I(H)L^2(\mathbb{R}^n) = \infty$. There is no way to distinguish resonant energies by properties of the spectral projections $\chi_I(H)$.

(iii) If H has purely discrete spectrum below E_0, then as $h \to 0$, the number of eigenvalues of H below E_0, $\dim(\chi_{(-\infty,E_0)}(H)L^2(\mathbb{R}^n)) = \mathrm{tr}(\chi_{(-\infty,E_0)}(H))$ is asymptotically equal to h^{-n} times the volume of the set $\{(x,p) \in \mathbb{R}^{2n}: p^2 + V(x) < E_0\}$[1] (the "classical limit"). But if $V(x) \to 0$ at infinity and $E_0 > 0$, then both the above trace and the above volume are infinite, and the statement loses most of its interest.

A remedy for such difficulties is to restrict the position as well as the energy, and consider, instead of $\mathrm{tr}(\chi_I(H))$ a quantity we call the local spectral density for H:

$$\rho_h(I\times B) \equiv h^n\, \mathrm{tr}(\chi_I(H)\chi_B(Q)\chi_I(H)) \tag{1}$$

It can be shown[2] that for a wide class of potentials V, (1) is finite and ρ_h extends uniquely to a measure on $\mathbb{R}^{n+1}$, which takes finite values on bounded measurable subsets of $\mathbb{R}^{n+1}$.

This measure can be described in various ways. For example, its density is essentially the E-derivative of the spectral function $e_E(x,x)$, where $e_E(\cdot,\cdot)$ is the integral kernel of the operator $\chi_{(-\infty,E)}(H)$. Its behavior as $E \to \infty$ has been investigated[3,4]. If H has a complete set $\{\phi_1,\phi_2,\ldots\}$ of orthonormal eigenfunctions with corresponding eigenvalues $\{E_1,E_2,\ldots\}$, then

$$\rho_h(I\times B) = h^n \sum_{E_k \in I} \int_B |\phi_k(x)|^2 d^n x\ .$$

In general,

$$\int |f(x)|^2 |g(E)|^2 \, d\rho_h(E,x) = h^n ||f(Q)g(H)||_2^2 = h^n ||g(H)f(Q)||_2^2 \qquad (2)$$

where $||\cdot||_2$ denotes the Schmidt norm. If $H = P^2$, these are explicit integral operators and we can calculate

$$\begin{aligned} \int |f(x)|^2 |g(E)|^2 d\rho_h(E,x) &= h^n ||f(Q)g(P^2)||_2^2 \qquad (3) \\ &= \int_{\mathbb{R}^n}\int_{\mathbb{R}^n} |f(x)|^2 |g(p^2)|^2 d^n p \, d^n x \\ &= \int_{\mathbb{R}^n} |f(x)|^2 \frac{S^{n-1}}{2} \int_0^\infty |g(E)|^2 E^{n/2-1} dE, \end{aligned}$$

where S^{n-1} is the area of the unit sphere in $\mathbb{R}^n$.

The local spectral density does not play any role in standard quantum mechanics, but it seems natural to interpret $\int \chi_B(x) f(E) d\rho_h(E,x)$ as the (expected) number of particles in $B \subset \mathbb{R}^n$, given a situation which would be described classically by a particle density $f(p^2 + V(x))$ on phase space[2].

Here we shall show how this notion can be used to solve problems (i) and (iii), assuming that the potential V is well behaved. More general potentials and problem (ii) are considered elsewhere[2].

THE CLASSICAL LIMIT

The quantity in classical mechanics analogous to ρ_h is the measure ρ_0 on $\mathbb{R}^{n+1}$ which assigns to a measurable product set $I \times B \subset \mathbb{R}^{n+1}$ the volume of $\{(x,p) \in \mathbb{R}^{2n}: p^2 + V(x) \in I, \ x \in B\}$. Thus

$$\begin{aligned} \int |f(x)|^2 |g(E)|^2 d\rho_0(E,x) &= \int \left(\int |f(x)|^2 |g(p^2+V(x))|^2 d^n p \right) d^n x \qquad (4) \\ &= \int |f(x)|^2 \left(\int_{V(x)}^\infty g(E) \frac{S^{n-1}}{2} (E-V(x))^{n/2-1} dE \right) dx, \end{aligned}$$

so ρ_0 has density $S^{n-1}[E-V(x)]^{n/2-1}\chi_{[V(x),\infty)}(x)/2$. Comparing with (3) we see that if $V \equiv 0$ we have $\rho_h = \rho_0$, but this is not true in general. The two measures are illustrated for a typical one-dimensional potential in Figure 1.

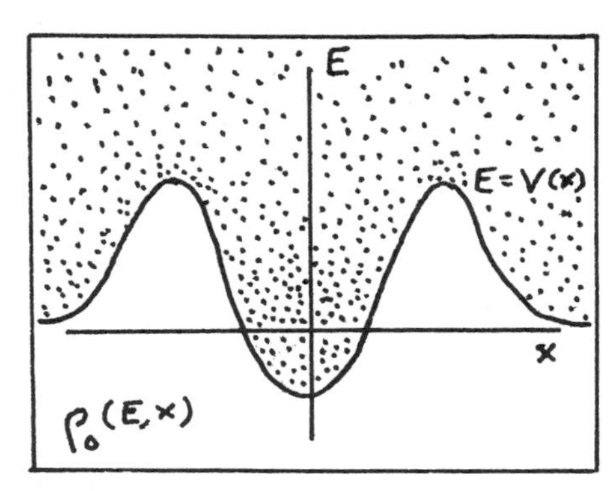

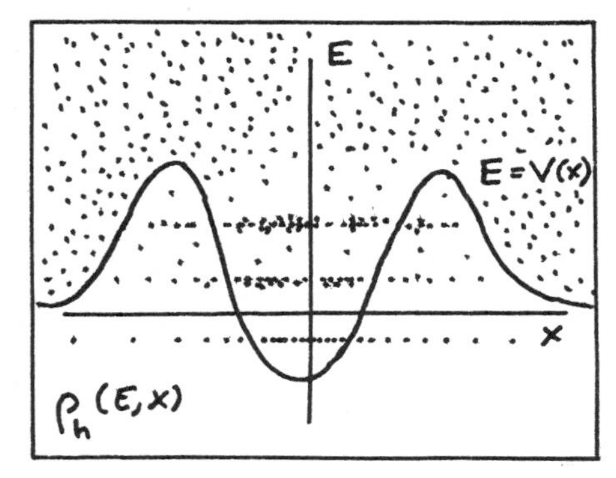

Fig. 1 The measure of a set is indicated by the number of dots it contains. Some features of ρ_h not shared by ρ_0 are

(a) subsets of the region $V(x) > E$ have nonzero measure
(b) some subsets of $V(x) < E < 0$ have zero measure
(c) ρ_h is singular for $E < 0$, supported on $\sigma(H)\times\mathbb{R}$
(d) ρ_h is highly concentrated inside the barrier at certain energies (resonances)[2].

We want to show that ρ_h approaches ρ_0 as $h \to 0$. Our strategy is straightforward. Integrals with respect to ρ_h are given by expressions of the form $||g(H)f(Q)||_2^2$, and we know how to compute Schmidt norms of the form $||g(P^2)f(Q)||_2$. If we take $g(E) = (E-z)^{-1}$ we can use the resolvent equation to relate what we want to what we know. Let us concentrate on a point $x_0 \in \mathbb{R}^n$, and approximate $(P^2 + V(Q)-z)^{-1} \equiv R$ by $(P^2 + V(x_0)-z)^{-1} \equiv R_0$. If we take the function f to be supported in a small neighborhood of x_0, this should be a good approximation. We have

$$\begin{aligned}[R-R_0]f(Q) &= R[V(x_0)-V(Q)]R_0 f(Q) \\ &= R[V(x_0)-V(Q)]\{f(Q) - R_0[P^2,f(Q)]\}R_0 \qquad (5) \\ &= R[V(x_0)-V(Q)]f(Q)R_0 + (R-R_0)[-\hbar^2\Delta f(Q)+2i\hbar P\cdot\nabla f(Q)]R_0.\end{aligned}$$

For the Schmidt norms to be finite, we need the dimension $n \leq 3$. Then we have

$$||(R-R_0)f(Q)||_2 \leq ||R||\ ||[V(x_0)-V(Q)]f(Q)R_0||_2$$
$$+ 2\hbar||(R-R_0)P||\ ||\nabla f(Q)R_0||_2 \qquad (6)$$
$$+ \hbar^2||R-R_0||\ ||\Delta f(Q)R_0||_2 \ .$$

In order to evaluate $||(R-R_0)P||$ we need

LEMMA 1. Suppose that $H = P^2 + V(Q)$, where V is bounded below by V_0 and $\mathcal{D}(H) \subset \mathcal{D}(P)$. Then for $E_0 \in \mathbb{R}$ and $\varepsilon \neq 0$,

$$||P(H-E_0-i\varepsilon)^{-1}|| \leq \frac{[(E_0-V_0)^2+\varepsilon^2]^{\frac{1}{4}}}{\varepsilon} \qquad (7)$$

Proof: Let $\phi \in \mathcal{D}(H)$. Then

$$||P\phi||^2 \leq \langle\phi,(H-E_0)\phi\rangle + (E_0-V_0)||\phi||^2.$$

For any $\gamma > 0$,

$$||P_\phi||^2 \leq \frac{\gamma}{2}\ ||(H-E_0)\phi||^2 + [\ \frac{1}{2\gamma}\ + (E-V_0)]||\phi||^2.$$

Choosing $\gamma = \{[(E_0-V_0)^2+\varepsilon^2]^{\frac{1}{2}} + E_0-V_0\}/2\varepsilon^2$ so that $\varepsilon^2 = (2\gamma^2)^{-1} + 2(E_0-V_0)/\gamma$, we have

$$||P\phi||^2 \leq \frac{\gamma}{2}(||(H-E_0)\phi||^2 + \varepsilon^2||\phi||^2) = \frac{\gamma}{2}\ ||(H-E_0-i\varepsilon)\phi||^2$$

which implies (7), since
$[(E_0-V_0)^2+\varepsilon^2]^{\frac{1}{2}} + E_0-V_0 \leq 2[(E_0-V_0)^2+\varepsilon^2]^{\frac{1}{2}}$. ■

Let us suppose that

$$|x-x_0| < \eta \Rightarrow |V(x)-V(x_0)| < \alpha \qquad (8)$$

and choose $f(x) = f_\eta(x) = \eta^{-n/2}u([x-x_0]/\eta)$, where $||u||_2 = 1$ and $u(x) = 0$ for $|x| > 1$.

Using (7) and (3) to estimate the right hand side of (6), with $z = E_0+i\varepsilon$, gives

$$||(R-R_0)f_\eta(Q)||_2 \leq \left\{\frac{\alpha}{\varepsilon} + 4\hbar\frac{[(E_0-V_0)^2+\varepsilon^2]^{\frac{1}{4}}}{\varepsilon\eta}\ ||\nabla u||_2 + \frac{2\hbar^2}{\varepsilon\eta^2}\ ||\Delta u||_2\right\}\left\{\frac{S^{n-1}}{2}\int_{V(x_0)}^{\infty}\frac{[E-V(x_0)]^{n/2-1}}{|E-z|^2}\ dE\right\}^{\frac{1}{2}} .$$

Thus we have an approximation of

$$\int \frac{|f_\eta(x)|^2}{|E-z|^2} \, d\rho_h(E,x) = ||Rf(Q)||_2^2$$

by

$$||R_0 f(Q)||_2^2 = \frac{S^{n-1}}{2} \int \frac{[E-V(x_0)]^{n/2-1}}{|E-z|^2} \, dE$$

$$= \frac{S^{n-1}}{2} \iint \frac{[E-V(x)]^{n/2-1}}{|E-z|^2} \, \delta(x-x_0) d^n x dE$$

$$= \int \frac{\delta(x-x_0)}{|E-z|^2} \, d\rho_0(E,x).$$

In fact, since

$$\left| ||Rf_\eta(Q)||_2^2 - ||R_0 f_\eta(Q)||_2^2 \right| \le$$

$$||(R-R_0)f_\eta(Q)||_2 (2||R_0 f_\eta(Q)||_2 + ||(R-R_0)f_\eta(Q)||_2)$$

we have

$$\left| \frac{\int \frac{|f_\eta(x)|^2}{|E-z|^2} \, d\rho_h(E,x)}{\int \frac{\delta(x-x_0)}{|E-z|^2} \, d\rho_0(E,x)} - 1 \right| \le 2\mu_h + \mu_h^2 \tag{9}$$

where

$$\mu_h(E_0,x_0) = \frac{\alpha}{\varepsilon} + \frac{c\hbar}{\varepsilon\eta} [(E_0-V_0)^2+\varepsilon^2]^{\frac{1}{4}} + \frac{c\hbar^2}{\varepsilon\,\eta^2} \tag{10}$$

and c depends only on our choice of u.

The integral of $f_\eta^2(x)|E-E_0-i\varepsilon|^{-2}$ is essentially the measure assigned by ρ_h to a set with energy within ε of E_0 and position within η of x_0. The corresponding set A in phase space is shown for n = 1 in Figure 2.

Using (8) and (9) we can show

THEOREM 1. <u>Suppose that</u> V <u>is bounded and continuous on</u> $\mathbb{R}^n$ <u>for</u>

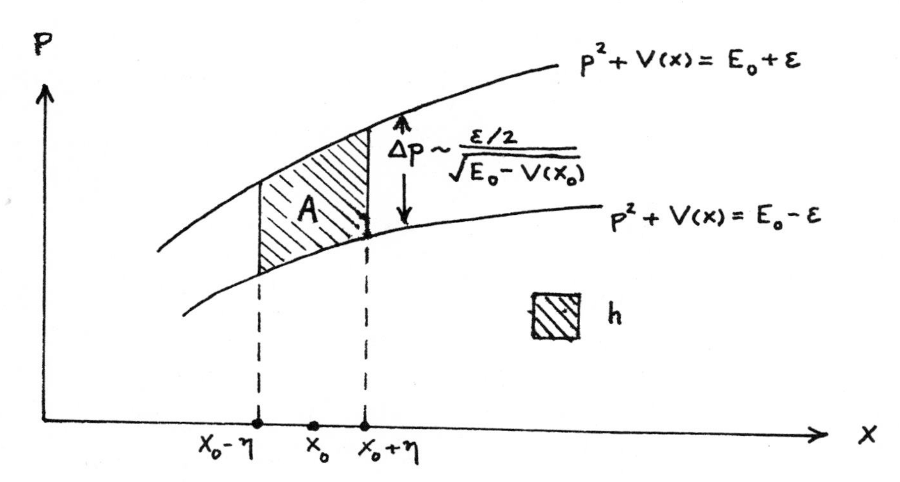

Fig. 2 If $\varepsilon \leq E - V(x_0)$, the error expression (10) is essentially $\alpha/\varepsilon + c\hbar/\Delta p\eta + c\hbar^2\varepsilon\eta^2$, which is small if α/ε is small, and $h \ll \eta\Delta p \sim |A|$, as shown.

$n \leq 3$, _and let_ $H = P^2 + V(Q)$. _Let_ ρ_h _be the local spectral density measure for_ H. _Then for any continuous function_ F _of compact support on_ $\mathbb{R}^{n+1}$,

$$\lim_{h \to 0} \int F d\rho_h = \int F d\rho_0 .$$

Proof: Write

$$F_{\varepsilon,\eta}(E,x) = \frac{\varepsilon}{\pi} \iint \frac{F(E_0,x_0)\eta^{-n}\left|u\left(\frac{x-x_0}{\eta}\right)\right|^2}{|E-E_0-i\varepsilon|^2} dE_0 dx_0 ,$$

$$F_\varepsilon(E,x) = \frac{\varepsilon}{\pi} \int \frac{F(E_0,x)}{|E-E_0-i\varepsilon|^2} dE_0 .$$

Given $\alpha > 0$ we can choose ε and η so that (8) holds for all $(E_0,x_0) \in \text{supp } F$ and $||F-F_{\varepsilon,\eta}||_\infty + ||F-F_\varepsilon||_\infty < \alpha$. Then

$$|\int F d\rho_h - \int F d\rho_0| \leq \int |F-F_{\varepsilon,\eta}| d\rho_h + |\int F_{\varepsilon\eta} d\rho_h - \int F_\varepsilon d\rho_0| + \int |F-F_\varepsilon| d\rho_0 .$$

The first and third terms are small if α is small enough, and

$$|\int F_{\varepsilon,\eta} d\rho_h - \int F_\varepsilon d\rho_0| \leq c\int\int |F(E_0,x_0)[2\mu_h(E_0,x_0)+\mu_h^2(E_0,x_0)]| dE_0 dx_0,$$

which converges to zero as $h \to 0$, by (10). ■

STABILITY

Suppose that $V_1(x) = V_2(x)$ for $|x| \leq X$. Then the physics in the region where $x \leq X_0 << X$ determined by $H_j = P^2 + V_j(Q)$ should be pretty much the same for $j = 1$ and 2. Here we estimate the difference between the local spectral densities.

THEOREM 2. Suppose that V_1 and V_2 are real bounded measurable functions on $\mathbb{R}^n$ for $n \leq 3$, bounded below by V_0, and $|V_1(x)-V_2(x)| < \alpha$ for $|x-x_0| \leq X$. Let ρ_j be the local spectral density for $H_j = P^2 + V_j(Q)$ for $j = 1,2$. If f is a continuous function on $\mathbb{R}^n$ which vanishes when $|x-x_0| \geq X_0$ with $X_0 < X$, then for $\text{Im } z \neq 0$,

$$\left| \iint \frac{|f(x)|^2}{|E-z|^2} d\rho_2(E,x) \Big/ \iint \frac{|f(x)|^2}{|E-z|^2} d\rho_1(E,x) - 1 \right| \leq 2\mu+\mu^2 \tag{11}$$

where

$$\mu = \frac{1}{|\text{Im } z|}\left\{\alpha + \frac{c\hbar|z+V_0|^{\frac{1}{2}}}{X-X_0} + \frac{c\hbar^2}{(X-X_0)^2}\right\} \tag{12}$$

Proof: Take $R_j = (H_j-z)^{-1}$ and let g be a smooth function on $\mathbb{R}^n$ such that $fg = f$. Then by (5) (with $R_2 = R$ and $R_1 = R_0$)

$$\begin{aligned}(R_2-R_1)f(Q) &= (R_2-R_1)g(Q)f(Q)\\ &= R_2[V_2(Q)-V_1(Q)]g(Q)R_1f(Q)\\ &+ (R_2-R_1)[-\hbar^2\Delta f(Q)+2i\hbar P\cdot\nabla f(Q)]R_1f(Q).\end{aligned}$$

Now let us be more specific in choosing g. Take $\gamma \in C_0^\infty(\mathbb{R})$ with $0 \leq \gamma(s) \leq 1$, and $\gamma(s) = 1$ for $s \leq 0$, $\gamma(s) = 0$ for $s \geq 1$. Set $g(x) = \gamma([|x-x_0|-X_0]/[X-X_0])$. Then $|\nabla g| \leq c_1(X-X_0)^{-1}$ and $|\Delta g| \leq c_2(X-X_0)^{-2}$, so we have

$$\begin{aligned}\mu &= ||(R_2-R_1)f(Q)||_2 \,/\, ||R_1 f(Q)||_2 \\ &\le ||R_2||\ ||(V_1-V_2)g||_\infty + \hbar^2||(R_2-R_1)\Delta g(Q)|| \\ &\quad + 2\hbar||(R_2-R_1)P\cdot\nabla g(Q)|| \\ &\le \frac{1}{|\mathrm{Im}\ z|}\left\{\alpha + \frac{2\hbar^2 c_2}{(X-X_0)^2} + \frac{4c_1\hbar|z-V_0|^{\frac{1}{2}}}{X-X_0}\right\}\end{aligned}$$

from which (11) follows as in (9). ■

Herbst and Howland consider[5] a Hamiltonian $H_0 = P^2 + V(Q)$ in one dimension, where V is periodic, and its perturbation by a constant electric field: $H_F = P^2 + V(Q) - FQ$. The spectrum of H_0 is bounded below and consists of a number of bands, while that of H_F extends to $-\infty$ and does not have the band structure of H_0. However, the local spectral densities ρ_0 and ρ_F are related for small F. Taking $V_1(x) = V(x) - Fx_0$ and $V_2(x) = V(x) - Fx$ in Theorem 2, we get for supp $f \subset \{|x-x_0| < X_0\}$

$$\left|\int \frac{|f(x)|^2}{|E-z|^2}\,d\rho_F(E,x) \Big/ \int\frac{|f(x)|^2}{|E-z|^2}\,d\rho_0(E+Fx_0,x) - 1\right| \le 2\mu+\mu^2$$

where

$$\mu = \frac{1}{|\mathrm{Im}\ z|}\left\{FX + \frac{c\hbar|z+Fx_0+V_0|^{\frac{1}{2}}}{X-X_0} + \frac{c\hbar^2}{(X-X_0)^2}\right\} .$$

This can be made small by choosing X large enough to make the last two terms small and then choosing F small. For example, if we take $X = \sqrt{A/F}$ we get $\mu = O(\sqrt{F})$ as $F \to 0$. Since $V(Q)-FQ$ is unbounded, it does not satisfy the hypotheses of Theorem 2. The main problem is to estimate $||(R_2-R_1)P\cdot\nabla g(Q)||$. This can be done by a generalization of Lemma 1.[2]

REFERENCES

1. Reed, M. and Simon, B., "Analysis of Operators", Academic Press, New York (1978).
2. Lavine, R., The local spectral density and its classical limit, preprint, University of Virginia, 1980.

3. Agmon, S. and Kannai, Y., On the asymptotic behavior of spectral functions and resolvent kernels of elliptic operators, Israel J. Math. 5:1 (1967).
4. Hörmander, L., The spectral function of an elliptic operator, Acta Math. 121:193 (1968).
5. Herbst, I. and Howland, J., The Stark ladder and other one dimensional external field problems, to appear Comm. Math. Phys. See also their article in this volume.

FORMAL INTEGRALS FOR A NONINTEGRABLE DYNAMICAL SYSTEM: PRELIMINARY REPORT

Randall B. Shirts and William P. Reinhardt

Department of Chemistry, University of Colorado and
Joint Institute for Laboratory Astrophysics
University of Colorado and National Bureau of Standards
Boulder, Colorado 80309

ABSTRACT

The Birkhoff-Gustavson normal form has been generated to a high order for the Henon-Heiles model Hamiltonian. This analysis provides approximate integrable classical dynamics for the well-studied stochastic behavior of this system. Accelerated convergence of the series expansion for a second integral of the motion shows divergence of the series only in small regions of phase space, indicating approximate local integrability of the dynamics. The regions of divergence correspond well to the regions where stochastic motion first appears. If the regions of divergence are much smaller than Planck's constant, one would expect such behavior would be irrelevant with respect to quantum mechanics. This result lends support to the semiclassical quantization methods of Swimm and Delos and of Jaffé and Reinhardt since most of the invariant manifold structure remains intact.

I. INTRODUCTION

A classical Hamiltonian dynamical system is described by a Hamiltonian function in 2N phase space variables

$$H(q_i, p_i) \quad ; \quad i = 1,2,\ldots N \quad , \tag{1}$$

in which the dynamics is given by Hamilton's equations[1]

$$\dot{p}_i = -\frac{\partial H}{\partial q_i} \quad ; \quad \dot{q}_i = \frac{\partial H}{\partial p_i} \quad ; \quad i = 1,2,\ldots N \quad . \tag{2}$$

A Hamiltonian is termed integrable if there exist N linearly independent, single valued constants of the motion, C_i, in involution. Phase space functions are in involution if

$$[C_i, C_j] = 0 \quad ; \quad i,j = 1,2,\ldots N \tag{3}$$

where [A,B] is the Poisson bracket of A and B. In quantum mechanics, the analogous object is the commutator, and the variables are termed commuting variables when the commutator vanishes. Motion in an integrable system is termed regular; more specifically, integrable motion is quasi-periodic and evolves on an N-dimensional manifold imbedded within the 2N-dimensional phase space. This manifold has the topology of an N-dimensional torus and is called the invariant torus since once motion begins on the torus, it remains on the torus. For integrable systems, phase space is continuously filled with these invariant tori.[2,3]

Integrable systems are a special case of the set of all dynamical systems -- possibly a set of zero measure.[3] In a non-integrable system, N constants of the motion and their accompanying tori do not exist globally. Because a trajectory is not constrained to travel on an N-dimensional surface, it tends to wander through a volume of larger dimensionality. This seemingly random motion is termed stochastic, chaotic, or irregular.[2,3]

We are concerned here with the nature and source of non-integrability and the relevance of this classical phenomenon to quantum mechanics. We approach this problem through the methods of semiclassical quantization in which one attempts to obtain quantum mechanical information, at least approximately, from the classical dynamics. Recent work by Delos and Swimm[4] and by Jaffé and Reinhardt[5] has shown that accurate quantum energy eigenvalues can be obtained in some systems exhibiting chaotic classical motion. We show below why this may be justified.

In this report, we will be interested in coupled oscillators with two degrees of freedom, i.e., N = 2. This, however, is chosen for simplicity, and we note that the theory is generalizable to arbitrary N. For conservative systems, the energy is one constant of the motion, E = H. A two dimensional system is integrable if there exists one other constant of the motion, I, which is independent of energy and such that [I,H] = 0. The search for such a constant of the motion, if it exists, is the object of normal form theory.

II. NORMAL FORMS

In 1927, Birkhoff[6] developed the theory of normal forms for nonresonant systems. In its most useful form, we assume a

Hamiltonian is expanded about an equilibrium point in the following manner:

$$H = H_o + \varepsilon H_1 \quad ; \quad H_o = \sum_i^N \frac{\alpha_i}{2} (p_i^2 + q_i^2) \quad . \tag{4}$$

Birkhoff's formulation is only applicable to nonresonant systems. A system is said to be resonant when the independent oscillator frequencies, $\{\alpha_i, i = 1,2,...N\}$, are rationally related. Frequencies are rationally related if there exists a relation of the form

$$\sum_{i=1}^N n_i \alpha_i = 0 \quad , \tag{5}$$

where $\{n_i, i = 1,2,...N\}$ are integers.

A canonical transformation,[7] S, is sought such that the transformed Hamiltonian is a function of new variables $\{\tilde{\pi}_i, i = 1,2,...N\}$. That is

$$H(q_i, p_i) \xrightarrow{S} \tilde{H}(\tilde{\pi}_i) \quad , \tag{6}$$

where

$$\tilde{\pi}_i = \tfrac{1}{2} (\tilde{p}_i^2 + \tilde{q}_i^2) \quad . \tag{7}$$

Here we denote transformed variables by the tilde (~). The usefulness of the new form of the Hamiltonian is that the time derivatives of $\tilde{\pi}_i$, $i = 1,2,...N$ vanish, i.e.,

$$\frac{d\tilde{\pi}_i}{dt} = [\tilde{\pi}_i, \tilde{H}] = 0 \quad . \tag{8}$$

Thus the N functions $\tilde{\pi}_i$ are constants of the motion. They can easily be shown to be independent and in involution, thus, if S exists, the system is integrable.

The algebra of normal forms is described by the normal operator, D;

$$D = [\ , H_o] = \sum_{i=1}^N \alpha_i \left(p_i \frac{\partial}{\partial q_i} - q_i \frac{\partial}{\partial p_i}\right) \quad . \tag{9}$$

The normal operator is a linear operator on the space of functions of 2N variables $\{p_i, q_i, i = 1,2...N\}$. The null space of D is of paramount importance. If the frequencies of the oscillators of H_o, $\{\alpha_i, i = 1,2,...N\}$, are not rationally related, then the functions of the set $\{\pi_i, i = 1,2,...N\}$ span the null space of D, and if H can be transformed to the form $\tilde{H}(\tilde{\pi}_i)$ then $[\tilde{H},\tilde{H}_o] = D\tilde{H} = 0$. The system has therefore been shown to be integrable with integrals $\tilde{\pi}_i$.

The generalization to resonant systems, where a relation of the form (5) exists, is due to Gustavson.[8] He showed that additional basis functions besides $\{\pi_i,\ i = 1,2,\ldots N\}$ are needed to span the null space of D. To be in normal form, H must be a function of this extended set of null space basis functions. Although the individual members of this basis are not necessarily independent nor in involution, Gustavson shows how to find the N independent constants which are in involution.

We pause now to exhibit an example which will be used later. For N = 2 and $\alpha_1 = \alpha_2 = 1$ in Eq. (4) (a 1:1 resonance), one null space basis of D in the space of quadratic polynomials is

$$\{ {}^1\!/\!{}_2\,(p_i^2 + q_i^2)\ ,\quad {}^1\!/\!{}_2\,(p_2^2 + q_2^2)\ ,\quad p_1q_2 - p_2q_1\ ,\quad p_1p_2 + q_1q_2\}\ . \tag{10}$$

The functions of these four quantities span the null space of D. Note that two extra basis functions are included which would not be required for a nonresonant system. The constant of the motion which is independent of the energy is $\tilde{H}_o = {}^1\!/\!{}_2\,(\tilde{p}_1^2 + \tilde{q}_1^2) + {}^1\!/\!{}_2\,(\tilde{p}_2^2 + \tilde{q}_2^2)$ except in the trivial case that $\varepsilon = 0$.

We now outline a method to obtain a normalized power series for $\tilde{H}$ given any Hamiltonian, H, of the form (4). We follow Gustavson[8] with slight changes in notation. We assume H is normalized to order ε^{s-1} (s = 1 to begin since H_o is already normal) and find $S^{(s)}$ such that $\tilde{H}$ is normalized to order ε^s. Because we do not want to alter the terms of order less than ε^s, the transformation must be the identity plus some part proportional to ε^s

$$S^{(s)} = \sum_{i=1}^{N} \left(q_i^{(s-1)} \cdot p_i^{(s)}\right) + \varepsilon^s W_{s+2}\left(q_i^{(s-1)}, p_i^{(s)}\right)\ . \tag{11}$$

The final transformation, S, is then the sequence of transformations $S^{(s)}$ and is canonical if each transformation in the sequence is canonical.

The sth transformation is found by straightforward application of the theory of canonical transformations. The old variables and the new variables are related implicitly by

$$q_i^{(s)} = q_i^{(s-1)} + \varepsilon^s \frac{\partial W_{s+2}}{\partial p_i^{(s)}}\ ;\quad p_i^{(s-1)} = p_i^{(s)} + \varepsilon^s \frac{\partial W_{s+2}}{\partial q^{(s-1)}}\ . \tag{12}$$

We also have the relation

$$H\left(q_i^{(s-1)}, p_i^{(s-1)}\right) = \tilde{H}\left(q_i^{(s)}, p_i^{(s)}\right) \quad . \tag{13}$$

Substituting (12) into (13) and equating terms of order ε^s gives

$$DW_{s+2} = \tilde{H}_{s+2}\left(q_i^{(s-1)}, p_i^{(s)}\right) - H_{s+2}\left(q_i^{(s-1)}, p_i^{(s)}\right) \quad , \tag{14}$$

where subscripts on W, H, and $\tilde{H}$ refer to the order of polynomials in q_i and p_i in the case of cubic H_1.

Equation (14) is a linear matrix equation with unknowns W_{s+2} and $\tilde{H}_{s+2}$. The transformed Hamiltonian $\tilde{H}_{s+2}$ must be a linear combination of the null space vectors of D in the subspace of polynomials of order s+2 (or other order contained in H_{s+2}). $\tilde{H}_{s+2}$ is uniquely determined by requiring that a solution exist for W_{s+2}. A well-known theorem in linear algebra[9] states that the right-hand side of (14) must be orthogonal to the null space vectors of $D^\dagger$. $\tilde{H}_{s+2}$ can be chosen equal to the component of H_{s+2} in the null space of $D^\dagger$. The requirement that W_{s+2} have no component in the null space of D is useful and standard, but will turn out to be unimportant for us.

We have reduced the problem of finding N-1 constants of the motion to solving infinitely many matrix equations for an infinite sequence of canonical transformations and resulting in an infinite order normalized Hamiltonian. The pointwise convergence of this process will be examined. The ubiquitous Henon-Heiles system is chosen as an example.

III. RESULTS FOR THE HENON-HEILES SYSTEM

The Henon-Heiles system is a two-dimensional coupled oscillator system with a 1:1 resonance and a specified cubic coupling:

$$H = \frac{p_1^2}{2} + \frac{p_2^2}{2} + \frac{q_1^2}{2} + \frac{q_2^2}{2} + \varepsilon\left(q_1^2 q_2 - q_2^3/3\right) \quad . \tag{15}$$

The method of Sec. II has been applied on this system to find 11 successive canonical transformations and the normalized Hamiltonian to order ε^{11}. This work was done analytically on the MACSYMA algebraic manipulator[10] in terms of large polynomials with rational coefficients. The terms of order ε^{12} and higher are truncated. This procedure gives an integrable approximation to the Hamiltonian (15) which is believed to be nonintegrable. The two independent constants of the motion are chosen to be $\tilde{H} = E$ and $\tilde{H} - \tilde{H}_o = \tilde{I}$. We then transformed the functional form of $\tilde{I}$ successively through the

11 canonical transformations in reverse order giving I expressed in terms of the original (Cartesian) coordinates. The form of the resultant expression is

$$I = \varepsilon^2 I_4 + \varepsilon^3 I_5 + \ldots + \varepsilon^{11} I_{13} \quad , \tag{16}$$

where the subscripts denote the order of the polynomials involved. Terms of order ε^{12} and higher are truncated at this stage as well.

For $\varepsilon = 1$ and $E = 0.10$, we examine the pointwise convergence of the partial sums of (16). First, it is useful to review a study of the classical trajectories of this system. The $x = 0$ Poincaré surface of section[11] for many trajectories at this energy are shown in Fig. 1. Trajectory intersections with the $x = 0$ plane for a single trajectory tend to fall on smooth curves (regular motion) in four regions centered at points A, B, C and D, while chaotic trajectories appear near points E, F and G.

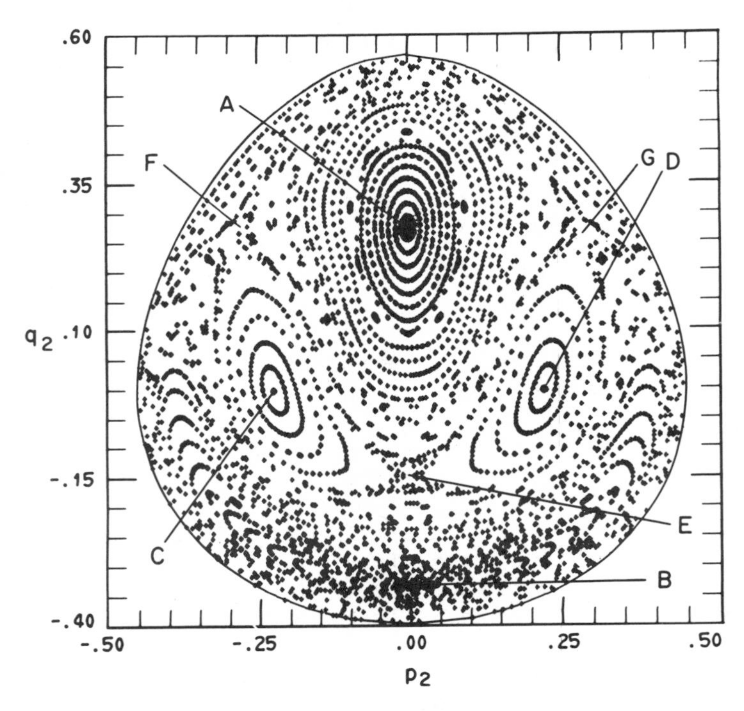

Fig. 1. Poincaré surface of section for 29 different trajectories in the Henon-Heiles system with energy 0.10. Points A through G are discussed in the text.

Figure 2 shows a contour map of the function I through order ε^{10} for the same energy. Note the excellent agreement in the qualitative shape of the level curves with the trajectory intersections in Fig. 1. This agreement suggests that I, which is an exact constant of the motion for the truncated normalized Hamiltonian, must be nearly constant on any given trajectory of the full Hamiltonian.

To investigate the convergence of the sequence of partial sums (16), we use the method of accelerated convergence.[12] This method, termed the epsilon algorithm, is numerically equivalent to making all possible Padé approximants of the series expansion (16) taken as partial sums. If the series is convergent for a given

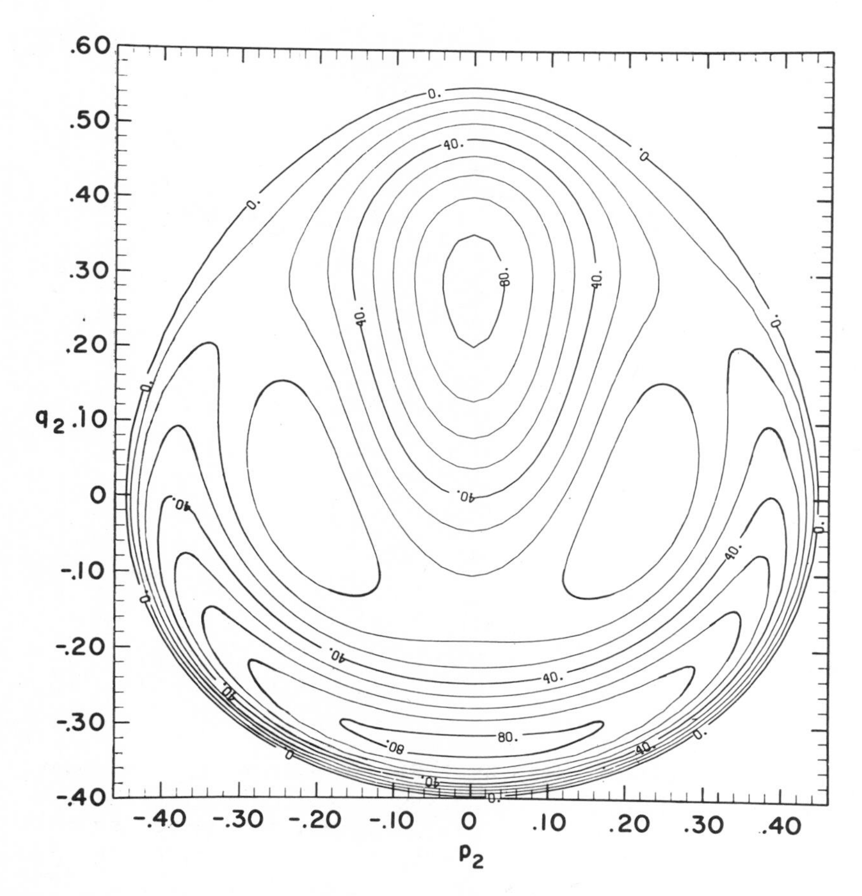

Fig. 2. Level curves of the formal constant of the motion in the Henon-Heiles system for energy 0.10 through tenth order in the coupling parameter.

phase space point, the high order Padé approximants should fall on a smooth surface near that point with a value near that of the higher partial sums. If the series is diverging at a given point in phase space, we expect the numerical procedure to give us values for I which differ widely at neighboring points and which differ substantially from the higher order partial sums of (16). The contours of the [5,4] Padé approximant for I are shown at low resolution in Fig. 3. The primary interest is that most of the

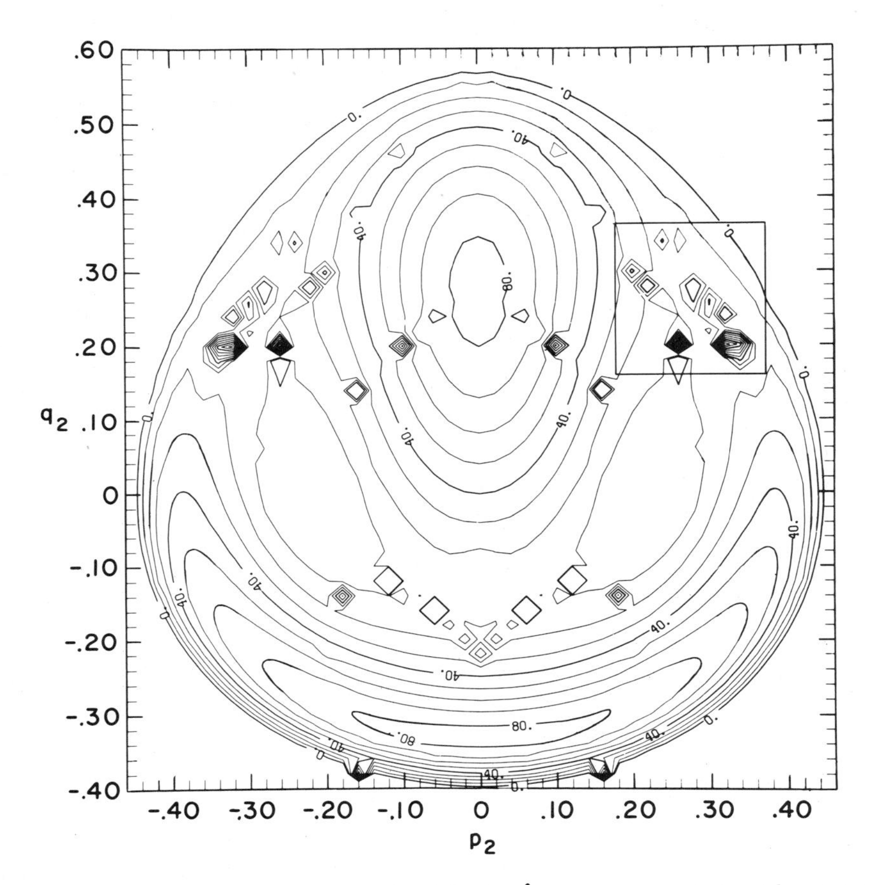

Fig. 3. Level curves of the [5,4] Padé approximant to the power series for the formal constant of the motion in the Henon-Heiles system for energy 0.10. The boxed region at upper right is shown at higher resolution in Fig. 4. The diamond-shaped complex structure in this figure and in Fig. 4 results from linear interpolation of data which are not smooth on a finite rectangular grid. The grid size is indicated along the figure borders.

contour lines remain largely unchanged. This agreement suggests that the series expansion is apparently convergent in most of the surface of section. At least we can say that any divergence in these regions must appear at higher order. Also of interest in Fig. 3 are the small diamond-shaped irregularities. These result from numerical values considerably removed from the smooth surface. These irregularities are due to a nearby zero of the denominator of the Padé approximant. Figure 4 shows a higher

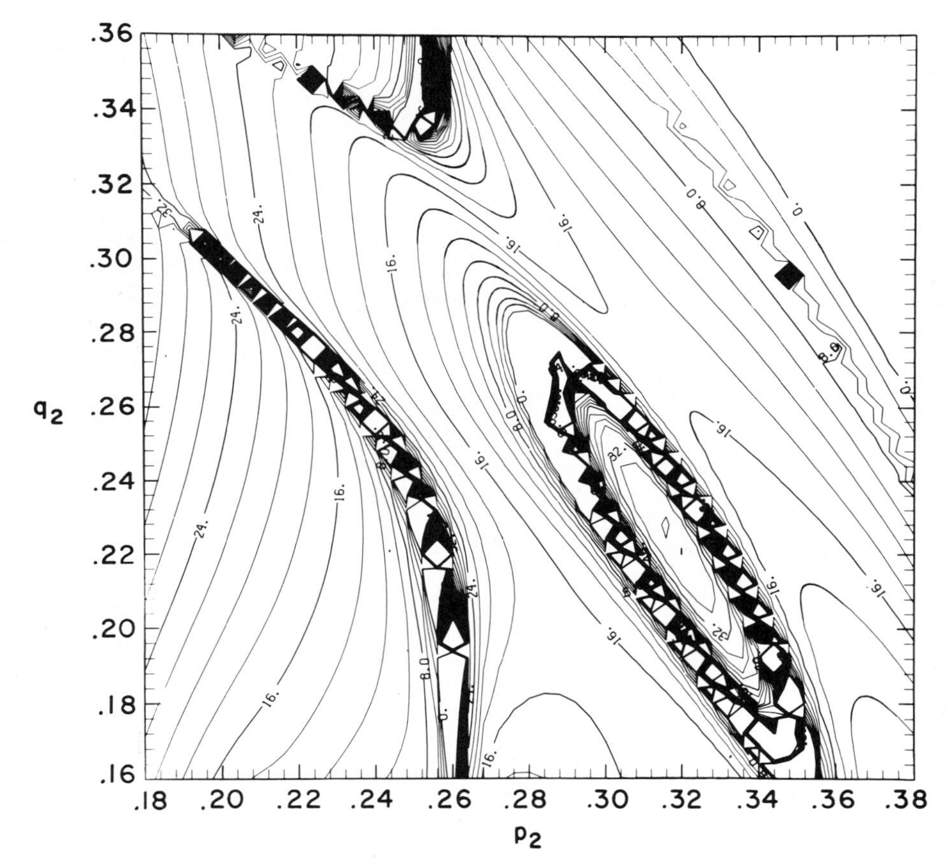

Fig. 4. High resolution level curves of the [5,4] Padé approximant to the power series for the formal constant of the motion in the Henon-Heiles system for energy 0.10 showing the region in the box in Fig. 3. This is the region where chaotic trajectories are first apparent as the energy is smoothly increased from E = 0 (see Fig. 1). Regions of complicated structure are regions of divergence of the formal constant and result from poles of the Padé approximant.

resolution contour map of the boxed area of Fig. 3. The locus of points for which the Padé approximant denominator has a zero consists of several smooth curves in this small region. The regions in the surface of section where these curves of zero denominator are most pronounced correspond precisely to the regions where chaotic motion appears in Fig. 1. These are the separatrix regions in Fig. 2. Similar results at lower and higher energies show that the extent of the irregularities increases rapidly with energy and is still present though imperceptible at low energy.

IV. CONCLUSIONS

We have used Birkhoff-Gustavson normal form to obtain approximate integrable dynamics for a Hamiltonian which shows chaotic motion typical of nonintegrable systems. We have found the series expansion in the coupling parameter for a second integral of the motion which is independent of the energy. Preliminary accelerated convergence studies of the partial sums of the series expansion suggest that for nonintegrable coupled oscillator systems, most of phase space is still composed of smooth surfaces, i.e., remnants of invariant tori. The regions where the surfaces are no longer smooth are small and grow with energy.

The Einstein-Brillouin-Keller (EBK) technique[13] of semiclassical quantization depends on the existence of smooth toroidal surfaces. The topologically distinct classical action integrals on these tori are then set equal to quantum values to determine quantum energy levels.[13] The truncation of high order nonintegrable terms in the normal form expansion allows one to implicitly smooth over destroyed regions of otherwise smooth tori. This preliminary analysis has demonstrated the localized nature of the destroyed regions. If the regions of phase space in which the tori are destroyed are much smaller than Planck's constant, it is clear that quantum mechanics will not sense their absence. In the work of Delos and Swimm[4] and of Jaffé and Reinhardt,[5] the size of Planck's constant is sufficiently large that the destroyed regions are negligible. Accurate quantum levels are obtained semiclassically in this case using the EBK technique.

It should be pointed out that the classical action integrals are not easily obtained from the normal form expansion as we have expressed it. Rather, one should normalize directly in action-angle variables.[4,5] The advantage of the present treatment is that it allows examination of the remnants of invariant tori in original untransformed coordinates, thus providing justification for the semiclassical quantization procedure performed in action-angle variables.

V. ACKNOWLEDGMENT

We would like to acknowledge the use of the MACSYMA algebraic manipulator maintained by the MIT Mathlab group (see Ref. 10) and supported, in part, by the U.S. Energy Research and Development Administration under contract number E(11-1)-3070 and NASA under grant NSG 1323. The support of the National Science Foundation through Grants PHY79-04928 and CHE77-16307 is gratefully acknowledged. We thank Dr. Alfred Maquet for his help in understanding the accelerated convergence procedure.

REFERENCES

1. H. Goldstein, "Classical Mechanics," Addison-Wesley, Reading, Mass. (1950) is one of many possible introductory treatments of classical dynamics.
2. See, for example, J. Ford, Adv. Chem. Phys. 24:155 (1973).
3. See, for example, M. V. Berry, in "AIP Conference Proceedings," No. 46, S. Jorna, ed., American Institute of Physics, New York (1978), p. 16.
4. R. T. Swimm and J. B. Delos, J. Chem. Phys. 71:1706 (1979).
5. C. Jaffé and W. P. Reinhardt, to be published; C. Jaffé, unpublished Ph.D. Thesis, Univ. of Colorado (1979).
6. G. D. Birkhoff, "Dynamical Systems," American Mathematical Society, New York (1927).
7. See Ref. 1, Ch. 8.
8. F. G. Gustavson, Astron. J. 71:670 (1966).
9. e.g. F. B. Hildebrand, "Methods of Applied Mathematics," Prentice-Hall, Englewood Cliffs, New Jersey (1956), p. 29.
10. MACSYMA is maintained by the Mathlab group, Laboratory for Computer Science, Massachussetts Institute of Technology, 545 Technology Square, Cambridge, Massachussetts 02139.
11. For a more complete description of the Poincaré surface of section as well as other figures relating to the Henon-Heiles Hamiltonian see W. P. Reinhardt and C. Jaffé in this volume, and Refs. 2, 3.
12. J. R. McDonald, J. Appl. Phys. 35:3034 (1964).
13. I. C. Percival, Adv. Chem. Phys. 36:1 (1977).

ON A GENERALIZED WEYL-VON NEUMANN CONVERSE THEOREM

M. Seddighin* and K. Gustafson

Department of Mathematics, University of Colorado,

Boulder, Colorado 80309 and Mashad University

Abstract. For two bounded selfadjoint operators A and $A+B$ with the same essential spectra, there exists a unitary operator U such that $B-(UAU^*-A)$ is compact (Weyl-Von Neumann). More generally, an operator B in $B(H)$ is compact iff $\sigma_e(A+B) = \sigma_e(A)$ for all $A \in B(H)$ (Gustafson-Weidmann), and in fact one needs only $\sigma(A+B) \cap \sigma(A)$ not empty for all $A \in B(H)$ (Dyer, Porcelli, Rosenfeld). Aiken (Is. Math. J., 1976) and Zemanek (Studia Math., to appear) have studied the question of when for an arbitrary Banach algebra with identity the last condition guarantees that B is in some proper two-sided ideal. We give new results for this question, including a number of examples.

*Partially supported by Mashad University.

1. Introduction

In perturbation theory, as is well known, compact operators B preserve the essential spectrum but not necessarily the continuous spectrum. For scattering theory one thus has potentials V such as the Coulomb potential $1/r$ that are relatively compact with respect to the Hamiltonians H_0 such as $H_0 = -\Delta$ preserving the essential spectrum $\sigma_e(H_0 + V) = \sigma_e(H_0) = [0,\infty)$ but not necessarily its continuity properties. Accepting this distinction between the essential spectrum σ_e and the continuous spectrum σ_c, nonetheless in practice in most cases the preservation of the essential spectrum by compact perturbations is tantamount to preserving σ_c and even σ_{ac}, the absolutely continuous spectrum, and in any case is the first step for so doing.

In the study of Banach algebras a different approach is usually taken. There the tendency is to begin not with Fredholm theory but rather with Calkin theory. Thus, for the case of $B(H)$, one defines the essential spectrum $\sigma_e(A)$ for an element A to be the spectrum $\sigma(\hat{A})$ of the image of A in the Calkin algebra $B(H)/K(H)$, $K(H)$ denoting the ideal of compact operators. In this way for arbitrary Banach algebras with identity one is led to a study of their ideal structures.

These two approaches have been for the most part pursued independently one of the other. For a partial survey of both approaches see Gustafson [1]. In this paper we concentrate on the latter approach.

2. Weyl's Theorem and Its Converse.

Definition (1). For a selfadjoint operator A on a complex Hilbert space H the essential spectrum $\sigma_e(A)$ is defined to be $\sigma_e(A) = \sigma(A) \sim \{\lambda \mid \lambda$ is an isolated eigenvalue of finite multiplicity$\}$, and the absolutely continuous spectrum $\sigma_{ac}(A)$ is defined to be

$$\sigma_{ac}(A) = \sigma(A|H^A_{ac}) , \quad H^A_{ac} = \{U \in H : \|E(S)U\|^2 = 0 \ \text{ if } \ m(S) = 0\}$$

where m is the Lebesgue measure and $\{E(S)\}$ is the spectral measure corresponding to A. Similarly $\sigma_c(A)$ is defined to be

$$\sigma_c(A) = \sigma(A/H^A_c) , \quad H^A_c = \{U: (E(-\infty,\lambda)U,U) \ \text{ is a continuous function of } \lambda\} .$$

For most selfadjoint operators we have $\sigma_{ac}(A) = \sigma_c(A)$. The following theorem is due to Weyl. (Weyl [2].)

<u>(Weyl's) Theorem (2)</u>. Let A be a selfadjoint operator on a complex Hilbert space and B be a selfadjoint and compact operator: then $\sigma_e(A+B) = \sigma_e(A)$.

It is not however true that $\sigma_c(A+B) = \sigma_c(A)$, in the general case. In fact there are selfadjoint operators A and B such that B is compact and perturbation by B sends the continuous spectrum of A into point spectrum. Hence compact operators are not sufficient for scattering theory, since in scattering theory the scattering operator $S = \Omega_+^*\Omega_-$ is not defined if $e_{ac}(H_0) \not\subset e_{ac}(H_0+B)$ where the selfadjoint operator H_0 is the Hamiltonian and $H_0 + B$ is the total Hamiltonian. This is why in scattering theory for preservation of σ_{ac} one has to consider perturbation by trace class operators.

The following three theorems can be considered as converses to Weyl's theorem, in increasing generality. The first is due to Von Neumann [3] , the second is due to Gustafson and Weidmann [4] , and the third is due to Gustafson [1] .

<u>Theorem (3A)</u>. If A_1 and A_2 are two bounded selfadjoint operators on a complex Hilbert space and $\sigma_e(A_1) = \sigma_e(A_2)$ then we have $A_2 - UA_1U^*$ compact for some unitary operator U .

<u>Theorem (3B)</u>. If $\sigma_e(A+B) = \sigma_e(A)$ for all A in $B(H)$ then B is compact.

<u>Theorem (3C)</u>. If $\sigma_e(A+B) = \sigma_e(A)$ for all A in $B(X)$, X an arbitrary Banach space, then B is inessential.

The inessential operators are a subclass of the Riesz operators ($\sigma_e(B) = \{0\}$) and forms the ideal of operators whose images in the quotient algebra of $B(X)$ modulo the uniform closure of the finite rank operators belong to the radical. In the case of the Banach algebra $B(H)$ for H separable as is well known the only closed two sided proper ideal is that of the compact operators, whereas for $B(X)$ and more generally for arbitrary Banach algebras the ideal structure can be much more complicated. Zemanek [5] observed the following extension of Weyl's Theorem and then raised an interesting converse question described below.

Theorem (4). If A is a complex Banach algebra with unit and r is an element in a proper two sided ideal of A then $\sigma(a+r) \cap \sigma(a) \neq \varphi$ for all elements a in A .

Zemànek also raised the question of whether in an arbitrary Banach algebra with identity the condition $\sigma(a+r) \cap \sigma(a) \neq \varphi$ for all a in the algebra implies that r belongs to some proper two-sided ideal? Aiken [6] proved (earlier than the question was independently asked by Zemànek) that this is not true for an arbitrary non-commutative Banach algebra by considering the C^*-algebra of B(H) generated by a unilateral shift. However Aiken [7] has positive results for W^*-algebras (Von Neumann algebras). Also it is known that the answer to the above question is positive for the case of algebras $B(\ell_p)$, $1 \leq p \leq \infty$. Earlier, Dyer, Parcelli and Rosenfeld [8] showed that $\sigma(A+B) \cap \sigma(A) \neq \varphi$ for all A in B(H) does indeed characterize the ideal K(H) of compact operators. The result of [8] thus may be regarded as an independent extension of the earlier result of [4] along the lines proposed by Zemanek [5] .

As pointed out by Zemanek [5] the converse question is easily resolved for commutative algebras. There the algebra is mapped into $C(X_0)$ by the Gelfand map, the ideals go to functions on closed subsets of X_0 , the condition $\sigma(A+B) \cap \sigma(A) \neq \varphi$ goes to the condition range $(a(x)+b(x)) \cap$ range $a(x) \neq 0$, so that the class $\{B\}$ of B specified by the condition is the ideal of elements B with 0 in $\sigma(B)$, i.e., $b(x_0) = 0$ for some x_0 .

The following two examples give an algebraic feeling for the class $\{B\}$ described by the condition $\sigma(A+B) \cap \sigma(A) \neq \varphi$ and show that it may be large or small.

Example (5A). If R is a two by two matrix on the field of complex numbers such that $\sigma(A+R) \cap \sigma(A) \neq \varphi$ for all two by two matrices A on the complex numbers then $R = 0$. This follows from the fact that $\{0\}$ is the only proper two sided ideal in the algebra of two by two matrices.

Example (5B). Consider the ring of polynomials. Only the nonzero constants have inverses in the algebra. Thus $\{B\}$ consists of 0 and all nonconstants and is large.

3. Weyl Algebras.

Definition (6). In a complex Banach algebra A an element r is called nonperturbing if $\sigma(r+a) \cap \sigma(a) \neq \varphi$ for all elements a in A . An element is called ideal if it belongs to a proper two sided ideal of A .

Definition (7). A Banach algebra is called a Weyl algebra if every nonperturbing element is ideal.

We have followed [8] in the choice of the term nonperturbing, we have used the term ideal to describe elements satisfying the (from Zemanek) spectral condition, and in general we will assume the algebra has an identity although (see Theorem (10) below) one need not do this. In an algebra with identity a proper two sided ideal has closure also proper so we may omit any specification of closed in the following.

Another suitable name for Weyl Algebras would be Nonperturbing Algebra.

Definition (8). An element x in a Banach algebra A is called quasi-regular if $x + y - xy = x + y - yx = 0$ for some $y \in A$. x is called quasi-singular if it is not quasi-regular.

Definition (9). For x in a Banach algebra the spectral radius $r(x)$ is defined by $r(x) = \lim_{n\to\infty} \|x^n\|^{1/n}$, and x is called quasi-nilpotent if $r(x) = 0$.

Theorem (10). Let A be a Banach algebra. Then the set $\bar{A} = \{(x,\alpha) : x \in A, \alpha \text{ complex}\}$ together with the operations $(x,\alpha) + (y,\beta) = (x+y,\alpha+\beta)$, $(x,\alpha)(y,\beta) = (xy+\beta x+\alpha y,\alpha\beta)$ and with the norm $\|(x,\alpha)\| = \|x\| + |\alpha|$ is a Banach algebra. $(0,1)$ is the identity for $\bar{A}$. $\sigma_{\bar{A}}((x,\alpha)) \subseteq \{\alpha\} \cup \{\sigma_A(x)+\alpha\}$ if A has already an identity and $\sigma_{\bar{A}}((x,\alpha)) \subseteq \{\alpha\} \cup \{\lambda \mid \lambda \neq \alpha, |\lambda-\alpha| \le r(x)\}$ if A does not have an identity. Moreover the set $A' = \{(x,0) \mid x \in A\}$ is a proper two-sided ideal in $\bar{A}$.

Proof. The only non-trivial part of the theorem is to show set inclusions. First suppose A has an identity. Suppose λ is a complex number. Then $(x,\alpha) - (0,\lambda) = (x,\alpha-\lambda)$. If $\lambda \neq \alpha$ then

$$(x,\alpha-\lambda)(y,\frac{1}{\alpha-\lambda}) = (xy+\frac{1}{\alpha-\lambda}x+(\alpha-\lambda)y,1).$$

If $\lambda \notin \sigma_A(x) + \alpha$ then the equation (*) $xy + \frac{1}{\alpha-\lambda}x + (\alpha-\lambda)y = 0$ has a solution. To see this we write (*) as $(\alpha-\lambda)xy + x + (\alpha-\lambda)^2 y = 0$ or $(\alpha-\lambda)[x+\alpha-\lambda] = -x$ or $y = \frac{1}{\alpha-\lambda}(x+\alpha-\lambda)^{-1}(-x)$. Now $(x+\alpha-\lambda)^{-1}$ exists since $\lambda \notin \sigma(x+\alpha) = \sigma(x) + \alpha$. This implies $\sim\{\alpha\} \cap \sim\{\sigma_A(x)+\alpha\} \subseteq \sim\sigma_{\bar{A}}((x,\alpha))$ and therefore we have

$\sigma_{\bar{A}}((x,\alpha)) \subseteq \{\alpha\} \cup \{\sigma_A(x)+\alpha\}$. Now suppose that A does not have an identity, and let $\lambda \neq \alpha$. If $\frac{1}{\alpha-\lambda}x$ is quasi-regular then there exists an element z in A such that $\frac{1}{\alpha-\lambda}xz + \frac{1}{\alpha-\lambda}x + z = 0$. If we take $y = \frac{1}{\alpha-\lambda}z$ then we have $xy + \frac{1}{\alpha-\lambda}x + (\alpha-\lambda)y = 0$. Hence $\sim\{\alpha\} \cap \sim\{\lambda \neq \alpha \mid \frac{1}{\lambda-\alpha}x$ is quasi-singular$\} \subseteq \sim\sigma_{\bar{A}}((x,\alpha))$, therefore $\sigma_{\bar{A}}((x,\alpha)) \subseteq \{\alpha\} \cup \{\lambda \neq \alpha : \frac{1}{\lambda-\alpha}x$ is quasi-singular$\}$.

We know for an element a in a Banach algebra the inequality $r(a) < 1$ implies that a is quasi-regular with quasi-inverse $a^0 = -\sum_{n=1}^{\infty} a^n$ (Rickart [9]). Hence for an element $\frac{1}{\lambda-\alpha}x$ to be quasi-singular it is necessary to have $r(\frac{1}{\lambda-\alpha}x) \geq 1$, that is, $r(x) \geq |\lambda-\alpha|$.

Theorem (11). If $y = (x,\alpha) \in \bar{A}$ is nonperturbing and $r(x) < |\alpha|$ then (x,α) is ideal.

Proof. Choose the element $z = (0,\beta)$ where β is any complex number. Suppose $|\alpha| > 0$ then $\sigma(z) = \{\beta\}$ and $\sigma(y+z) = \sigma((x,\alpha+\beta))$ is a subset of the circle with center at $\alpha+\beta$ and radius $r(x)$. This follows from Theorem (10). Now since $r(x) < |\alpha|$ and $|\alpha+\beta-\beta| = |\alpha|$, β is outside of the circle with center at $\alpha+\beta$ and radius $r(x)$. This shows that $\sigma(y+z) \cap \sigma(z) = \varphi$ a contradiction to the fact that y is nonperturbing. Therefore we must have $|\alpha| = 0$ i.e. $\alpha = 0$ (and hence $r(x) = 0$ too). Hence y has the form $y = (x,0)$ and therefore it belongs to the ideal A' of Theorem (10) and it is ideal.

Corollary (12). Let A be a Banach algebra such that each of its elements is quasi-nilpotent then $\bar{A}$ is a Weyl algebra.

Proof. We have $r(x) = 0$ for all x and hence $r(x) < |\alpha|$ for all x and all α. Thus by Theorem (11) every nonperturbing element is ideal.

Example (13). Let u and v be two symbols and let $\{\omega_n\}$ be the standard enumeration of words $u, v, u^2, uv, vu, v^2, u^3, u^2v, \cdots$. This enumeration can be formally performed as follows: $\omega_1 = u$, $\omega_2 = v$, hence ω_1 and ω_2 are the first and second elements of the ordered pair (u,v) . Now formally multiply (u,v) and (u,v) as follows: $(u,v)(u,v) = (u^2,uv,vu,v^2)$. Call the first element of the resulting fourth-tuple ω_3 , the second one ω_4 , the third one ω_5 and the fourth one ω_6 . Hence we have $\omega_3 = u^2$, $\omega_4 = uv$, $\omega_5 = vu$, $\omega_6 = v^2$. Now multiply (u,v) and (u^2,uv,vu,v^2) as follows $(u,v)(u^2,uv,vu,v^2) =$ $(u^3,u^2v,uvu,uv^2,vu^2,vuv,v^2u,v^3)$. Let $\omega_7 = u^3 \cdots$ and so forth. Let λ_n be the length of the word ω_n and let $A(u,v)$ be the algebra of all formal infinite series $f = \sum \alpha_n \omega_n$ where $\|f\| = \sum \frac{|\alpha_n|}{\lambda_n!} < \infty$ then each element in $A(u,v)$ is nilpotent and hence $\overline{A(u,v)}$ is a Weyl algebra. (See Bonsall and Duncan [10, p. 255].) In the following theorem by $\bar{\bar{A}}$ we mean $\overline{(\bar{A})}$.

Theorem (14). If each element in a Banach algebra A is quasi-nilpotent then the sets $I_1 = \{((a,\alpha),0)$, $\alpha \in A$, α complex$\}$ and $I_2 = \{((a,\alpha),\beta) \mid a \in A$, $\beta = -\alpha\}$ are proper two-sided ideals in $\bar{\bar{A}}$ and each nonperturbing element of $\bar{\bar{A}}$ is either in I_1 or I_2 .

Proof. Let $x = ((f,\alpha_1),\alpha_2)$ and $y = ((g,\beta_1),\beta_2)$. Since each element of A is quasi-nilpotent by Theorem (10) we have $\sigma_{\bar{A}}((f,\alpha_1)) = \alpha_1$ and $\sigma_{\bar{A}}((f,\beta_1)) = \beta_1$. Now since $\bar{A}$ has a unit, again by Theorem (10) we have $\sigma(x) \subseteq \{\alpha_2\} \cup \{\alpha_1+\alpha_2\}$, $\sigma(y) \subseteq \{\beta_2\} \cup \{\beta_1+\beta_2\}$, and $\sigma(x+y) \subseteq \{\alpha_2+\beta_2) \cup \{\alpha_1+\beta_1+\alpha_2+\beta_2\}$. Now $\sigma(x) \cap \sigma(x+y) \neq \varphi$ implies that $(\{\alpha_2+\beta_2\} \cup \{\alpha_1+\beta_1+\alpha_2+\beta_2\}) \cap (\{\beta_2\} \cup \{\beta_1+\beta_2\}) \neq \varphi$, which implies $\alpha_2 + \beta_2 = \beta_2$ or $\alpha_2 + \beta_2 = \beta_1 + \beta_2$ or $\alpha_1 + \beta_1 + \alpha_2 + \beta_2 = \beta_2$ or $\alpha_1 + \beta_1 + \alpha_2 + \beta_2 = \beta_1 + \beta_2$. Hence $\sigma(y) \cap \sigma(x+y) \neq \varphi$ implies $\alpha_2 = 0$ or $\beta_1 = \alpha_2$ or $\alpha_1 = -\beta_1 - \alpha_2$ or $\alpha_1 = -\alpha_2$. If $((f,\alpha_1),\alpha_2)$ is neither in I_1 nor I_2 then $\alpha_1 \neq -\alpha_2$ and $\alpha_2 \neq 0$. Therefore $\sigma(y) \cap \sigma(x+y) \neq \varphi$

for all $y = ((y,\beta_1),\beta_2)$, implies $\beta_1 = \alpha_2$ or $\beta_1 = -\alpha_1 - \alpha_2$ for all β, a contradiction, since for any element $y = ((y,\beta_1),\beta_2)$ in which $\beta_1 \neq \alpha_2$ and $\beta_1 \neq -\alpha_1 - \alpha_2$ we have $\sigma(y) \cap \sigma(x+y) = \varphi$. Hence $x \in I_1$ or $x \in I_2$. The final step is to show that I_1 and I_2 are ideals in $\bar{\bar{A}}$. That I_1 is an ideal follows from Theorem (10). To show I_2 is an ideal let $((f,\alpha_1),\alpha_2)$ be an element of I_2, noting that then $\alpha_1 = -\alpha_2$. Now let $((g,\beta_1),\beta_2)$ be an element of $\bar{\bar{A}}$. Then

$$((g,\beta_1),\beta_2)((f,\alpha_1),\alpha_2) = ((g,\beta_1)(f,\alpha_1) + \alpha_2(g,\beta_1) + \beta_2(f,\alpha_1),\alpha_2\beta_2)$$

$$= ((gf + \beta_1 f + \alpha_1 g + \alpha_2 g + \beta_2 f\,,\, \beta_1\alpha_1 + \alpha_2\beta_1 + \beta_2\alpha_1),\alpha_2\beta_2) \quad .$$

Now $\beta_1\alpha_1 + \alpha_2\beta_1 + \beta_2\alpha_1 = -\alpha_2\beta_1 + \alpha_2\beta_1 - \alpha_2\beta_2 = -\alpha_2\beta_2$, which shows that $((f,\alpha_1),\alpha_2)((g,\beta_1),\beta_2)$ is in I_2.

Remark 1. In most Weyl Banach algebras the set of all nonperturbing elements actually forms an ideal, i.e. the set $J = \{y : \sigma(x+y) \cap \sigma(x) \neq \varphi \ \forall x \in A\}$ is an ideal. However Theorem (14) shows that there are Weyl Banach algebras in which the above set J is not an ideal. In fact in the proof of Theorem (14) we showed that each nonperturbing element of $\bar{\bar{A}}$ is either in I_1 or I_2, but that is not enough to show that J is not an ideal (because the sum of the two ideals I_1 and I_2 may be a proper two sided ideal). Hence to show that J is not a proper two sided ideal we argue as follows. Let $x \in A$, then $u = ((x,1),0) \in J$ by Theorem (4) since $U \in I_1$, and $v = ((x,1),-1) \in J$ since $v \in I_2$. But $u - v = ((0,0),1)$, therefore elements of J do not form a proper two sided ideal in $\bar{\bar{A}}$.

Remark 2. In a Banach algebra an element a is ideal if and only if it belongs to a proper two-sided ideal, if and only if the ideal generated by a is a proper ideal, if and only if

$e \neq \sum_{k=1}^{n} s_k a r_k$ for any positive integer n and any choice of $s_k \in A$, and $r_k \in A$. This relation shows that the concept of idealness is a generalization of the concept of singularity. Also notice that the relation $\sigma(x+y) \cap \sigma(x) \neq \varphi$ for all x clearly implies that y is singular.

Remark 3. The (Jacobson) radical may be characterized algebraically as the intersection of all good ideals, or along Fredholm lines as the class $\{B \mid I + cB$ is invertible $\forall$ invertible $C\}$, equivalently as the class $\{B \mid A + B$ is invertible $\forall$ invertible $A\}$. One may roughly describe what we are doing in considering all non-perturbing elements as a complementary procedure to the construction of the Jacobson radical, in that we are characterizing a "super radical" as a union of all good ideals.

4. Convolution Algebras of Measures, Group Algebras

Let G be any locally compact topological group and let $M(G)$ denote the set of all complex-valued regular measures on Borel subsets of G. Let $\|m\|$ = total variation of m. For any Borel set E, set $(m * n)(E) = \int_G m(E\omega^{-1})\, d(\omega)$, then with this operation as multiplication for measures and regular addition of set functions and regular multiplication of a scaler with a set function, $M(G)$ is a Banach algebra. It is commutative if and only if G is commutative, and it has an identity if and only if G is discrete. On the group G we can also consider the space $L^1(G)$, the set of all complex-valued measurable functions f on G for which $\int_G |f| d\mu < \infty$, where μ is the Haar measure on the group G. On $L^1(G)$ we define the product of two elements f and g to be the convolution of f and g, i.e. $(f * g)(x) = \int f(xy)g(y^{-1})d\mu(y)$, where again the integral on the right is taken with respect to the Haar measure on G. Note that $L^1(G)$ is commutative if and only if G is commutative. If G is discrete with identity e, then the function $I(x) = \{1$ if $x = e$, 0 if $x \neq e\}$ is the identity for the algebra $L^1(G)$. In fact if G is discrete the Haar measure gives measure 1 to each point and we have $(f * g)(x) = \int f(xy)g(y^{-1})dy = \sum_y f(xy)g(y^{-1})$ $= \sum_{uv=x} f(u)g(v)$. Hence $(f * I)(x) = \int f(y)I(y^{-1}x)dy = \sum_{uv=x} f(u)I(v)$ $= f(x)$. It is easy to see that $L^1(G)$ is a subalgebra of $M(G)$ (actually a right ideal of $M(G)$). $L^1(G)$ is mapped onto $M(G)$ by the mapping $f \to m_f$, where for each Borel set E, $m_f(E)$ is defined to be $m_f(E) = \int_E f(\omega)d\omega$. If G is discrete one has $L^1(G) = M(G)$.

It is interesting to ask whether group algebras $L^1(G)$ are Weyl. We will show that $L^1(G)$ is a Weyl algebra for the first non-commutative group G. Note that S_3, the group of permutations on three objects, is the first non-commutative group.

Theorem (15). The group algebra $L^1(S_3)$ is Weyl.

Proof. First we discuss some group theoretical facts about S_3. This group is of order 6 with two generators $a = \begin{pmatrix} 1 & 2 & 3 \\ 2 & 3 & 1 \end{pmatrix}$ and and $b = \begin{pmatrix} 1 & 2 & 3 \\ 1 & 3 & 2 \end{pmatrix}$ whose generators a and b satisfy $a^3 = (1)$ ((1) means the identity permutation), $b^2 = (1)$, and $ba = a^2 b$. Hence as a free group generated by a and b, S_3 consists of elements $e, a, a^2, b, ab, a^2 b$, and the products of elements are defined according to the following table.

If $f \in L^1(S_3)$ and $g \in L^1(S_3)$ then $(f * g)(x) = \sum_{uv=x} f(u)g(u)$ so it follows from Table 1 that

$$(f*g)(e) = f(e)g(e) + f(a)g(a^2) + f(a^2)g(a) + f(b)g(b) + f(ab)g(ab) + f(a^2b)g(a^2b) ,$$

$$(f*g)(a) = f(e)g(a) + f(a)g(e) + f(a^2)g(a^2) + f(b)g(a^2b) + f(ab)g(b) + f(a^2b)g(ab) ,$$

$$(f*g)(a^2) = f(e)g(a^2) + f(a)g(a) + f(a^2)g(e) + f(b)g(ab) + f(ab)g(a^2b) + f(a^2b)g(b) ,$$

$$(f*g)(b) = f(e)g(b) + f(a)g(a^2b) + f(a^2)g(ab) + f(b)g(e) + f(ab)g(a) + f(a^2b)g(a^2) ,$$

$$(f*g)(ab) = f(e)g(ab) + f(a)g(b) + f(a^2)g(a^2b) + f(b)g(a^2) + f(ab)g(e) ,$$

$$(f*g)(a^2b) = f(e)g(a^2b) + f(a)g(ab) + f(a^2)g(b) + f(b)g(a) + f(ab)g(a^2) + f(a^2b)g(b) .$$

Let us call $x_1 = e$, $x_2 = a$, $x_3 = a^2$, $x_4 = b$, $x_5 = ab$, $x_6 = a^2b$ and $f_1 = f(x_1)$, $f_2 = f(x_2)$, $f_3 = f(x_3)$,

Table 1

	e	a	a^2	b	ab	a^2b
e	e	a	a^2	b	ab	a^2b
a	a	a^2	e	ab	a^2b	b
a^2	a^2	e	a	a^2b	b	ab
b	b	a^2b	ab	e	a^2	a
ab	ab	b	a^2b	a	e	a^2
a^2b	a^2b	ab	b	a^2	a	e

$f_4 = f(x_4)$, $f_5 = f(x_5)$, $f_6 = f(x_6)$. The above relations can be written as follows (after changing the order of terms):

$$
\begin{aligned}
(f*g)_1 &= f_1g_1 + f_3g_2 + f_2g_3 + f_4g_4 + f_5g_5 + f_6g_6 \\
(f*g)_2 &= f_2g_1 + f_1g_2 + f_3g_3 + f_5g_5 + f_6g_5 + f_4g_6 \\
(f*g)_3 &= f_3g_1 + f_2g_2 + f_1g_3 + f_6g_4 + f_4g_5 + f_5g_6 \\
(f*g)_4 &= f_4g_1 + f_5g_2 + f_6g_3 + f_1g_4 + f_3g_5 + f_2g_6 \\
(f*g)_5 &= f_5g_1 + f_6g_2 + f_4g_3 + f_2g_4 + f_1g_5 + f_3g_6 \\
(f*g)_6 &= f_6g_1 + f_4g_2 + f_5g_3 + f_3g_4 + f_2g_5 + f_1g_6 \; .
\end{aligned}
\qquad (*)
$$

Notice by the form that the identity element on $L^1(S^3)$ has, finding an inverse for $(f-\lambda I)$ is equivalent to solving the system

$$
\begin{bmatrix}
f_1-\lambda & f_3 & f_2 & f_4 & f_5 & f_6 \\
f_2 & f_1-\lambda & f_3 & f_5 & f_6 & f_4 \\
f_3 & f_2 & f_1-\lambda & f_6 & f_4 & f_5 \\
f_4 & f_5 & f_6 & f_1-\lambda & f_3 & f_2 \\
f_5 & f_6 & f_4 & f_2 & f_1-\lambda & f_3 \\
f_6 & f_4 & f_5 & f_3 & f_2 & f_1-\lambda
\end{bmatrix}
\begin{bmatrix} g_1 \\ g_2 \\ g_3 \\ g_4 \\ g_5 \\ g_6 \end{bmatrix}
=
\begin{bmatrix} 1 \\ 0 \\ 0 \\ 0 \\ 0 \\ 0 \end{bmatrix} .
$$

Hence $6(f)$ is equal to the spectrum of the matrix

$$
\begin{bmatrix}
f_1 & f_3 & f_2 & f_4 & f_5 & f_6 \\
f_2 & f_1 & f_3 & f_5 & f_6 & f_4 \\
f_3 & f_2 & f_1 & f_6 & f_4 & f_5 \\
f_4 & f_5 & f_6 & f_1 & f_3 & f_2 \\
f_5 & f_6 & f_4 & f_2 & f_1 & f_3 \\
f_6 & f_4 & f_5 & f_3 & f_2 & f_1
\end{bmatrix} .
$$

This suggests defining a mapping K from $L^1(S_3)$ into M_6 , the algebra of 6 by 6 matrices, by

$$
K(f) = \begin{bmatrix}
f_1 & f_3 & f_2 & f_4 & f_5 & f_6 \\
f_2 & f_1 & f_3 & f_5 & f_6 & f_4 \\
f_3 & f_2 & f_1 & f_6 & f_4 & f_5 \\
f_4 & f_5 & f_6 & f_1 & f_3 & f_2 \\
f_5 & f_6 & f_4 & f_2 & f_1 & f_3 \\
f_6 & f_4 & f_5 & f_3 & f_2 & f_1
\end{bmatrix}
$$

As we saw above this map preserves the spectrum, so it is natural to look at $K(L^1(S_3))$ the range of K in M_6 to see whether it is an algebra and it moreover is a Weyl algebra. If $K(f)$ and $K(g)$ are the images of f and g then the direct computation of the product of the six by six matrices $K(f)$ and $K(g)$ shows (we omit the lengthy multiplication of these matrices)

$$K(f)K(g) = \begin{bmatrix} (f*g)_1 & (f*g)_3 & (f*g)_2 & (f*g)_4 & (f*g)_5 & (f*g)_6 \\ (f*g)_2 & (f*g)_1 & (f*g)_3 & (f*g)_5 & (f*g)_6 & (f*g)_4 \\ (f*g)_3 & (f*g)_2 & (f*g)_1 & (f*g)_6 & (f*g)_4 & (f*g)_5 \\ (f*g)_4 & (f*g)_5 & (f*g)_6 & (f*g)_1 & (f*g)_3 & (f*g)_2 \\ (f*g)_5 & (f*g)_6 & (f*g)_4 & (f*g)_2 & (f*g)_1 & (f*g)_3 \\ (f*g)_6 & (f*g)_4 & (f*g)_5 & (f*g)_3 & (f*g)_2 & (f*g)_1 \end{bmatrix}$$

where $(f*g)_i$ are given by equations $(*)$. Hence $K(f)K(g) = K(f*g)$.

For any compact group G one can define an involution on $L^1(G)$ by $f^*(x) = \overline{f(x^{-1})}$. Hence for example for the case of $G = S_3$ we have from Table 1 that: $f_1^* = f^*(x_1) = f^*(e) = \overline{f(e)} = \overline{f_1}$, $f_2^* = f^*(x_2) = f^*(a) = f(a^2) = \overline{f(x_3)} = \overline{f_3}$, $f_3^* = f^*(x_3) = f^*(a^2) = \overline{f(a)} = \overline{f(x_2)} = \overline{f_2}$, $f_4^* = f^*(x_4) = f^*(b) = \overline{f(b)} = \overline{f_4}$, $f_5^* = f^*(x_5) = f^*(ab) = \overline{f(ab)} = \overline{f(x_5)} = \overline{f_5}$, $f_6^* = f^*(a^2b) = \overline{f(a^2b)} = \overline{f(x_6)} = \overline{f_6}$. Hence we have (using the Table):

$$K(f^*) = \begin{bmatrix} \overline{f_1} & \overline{f_2} & \overline{f_3} & \overline{f_4} & \overline{f_5} & \overline{f_6} \\ \overline{f_3} & \overline{f_1} & \overline{f_2} & \overline{f_5} & \overline{f_6} & \overline{f_4} \\ \overline{f_2} & \overline{f_3} & \overline{f_1} & \overline{f_6} & \overline{f_4} & \overline{f_5} \\ \overline{f_4} & \overline{f_5} & \overline{f_6} & \overline{f_1} & \overline{f_2} & \overline{f_3} \\ \overline{f_5} & \overline{f_6} & \overline{f_4} & \overline{f_3} & \overline{f_1} & \overline{f_2} \\ \overline{f_6} & \overline{f_4} & \overline{f_5} & \overline{f_2} & \overline{f_3} & \overline{f_1} \end{bmatrix} = \begin{bmatrix} f_1 & f_3 & f_2 & f_4 & f_5 & f_6 \\ f_2 & f_1 & f_3 & f_5 & f_6 & f_4 \\ f_3 & f_2 & f_1 & f_6 & f_4 & f_5 \\ f_4 & f_5 & f_6 & f_1 & f_3 & f_2 \\ f_5 & f_5 & f_4 & f_2 & f_1 & f_3 \\ f_6 & f_4 & f_5 & f_3 & f_2 & f_1 \end{bmatrix}^* = K(f)]^*$$

This shows that K is an algebraic isomorphism which preserves the spectrum and $*$. The fact that $K(f*g) = K(f)K(g)$ shows that K also preserves the ideals. So in order to show that $L^1(S_3)$ is a Weyl algebra it is enough to show that $K(L^1(S_3))$ is a Weyl algebra. Now $K(L^1(S_3))$ is a $*$ algebra (i.e. $x \in K(L^1(S_3))$

implies $x^* \in K(L^1(S_3))$), it contains the identity since the identity matrix $= K(I)$, and as a finite dimensional subspace of M_6 it is norm closed in M_6 . Since M_6 is finite dimensional $K(L^1(S_3))$ is strongly closed in M_6 . Thus by the Von Neumann multiplicity theorem (see, e.g. Naimark [11]) $K(L^1(S_3))$ is a Von Neumann algebra and therefore is a Weyl algebra. Hence $L^1(S_3)$ is a Weyl algebra.

One may use Similar reasoning to prove $L^1(G)$ is a Weyl algebra for any finite group.

5. Some Properties of Weyl Algebras

As we mentioned before, in an arbitrary Banach algebra an element is ideal if and only if the ideal generated by that element is a proper ideal. Hence an element a in a Banach algebra A is non-ideal if and only if for each element b in A there exist **a** positive integer n and $2n$ elements $s_1, s_2, \dots, s_n$ and $r_1, r_2, \dots, r_n$ such that $b = \sum_{k=1}^{n} s_k a r_k$ (see Remark 2). In other words each nonideal element generates the whole algebra. When the algebra A is a Weyl algebra we will show that for each nonideal element a of A any element b in A has a representation of the form $b = ad + cd - dc$ where d and c are elements of A .

Let A be any Banach algebra and let a and b be in A , then the operator T defined by $T(x) = bx - xa$ is a bounded operator from A into A . Many interesting properties of this operator have been established by Rosenblum [12] , such as the following theorem.

Theorem (16). If A is a Banach algebra and if the elements a and b are in A , then $\sigma(T) \subset \sigma(a) - \sigma(b)$, where T is the operator defined by $T(x) = ax - xb$, and where $\sigma(a) - \sigma(b) = \{\lambda_1 - \lambda_2 : \lambda_1 \in \sigma(a) , \lambda_2 \in \sigma(b)\}$.

Theorem (17). Let A be a Weyl algebra and let a be a nonideal element of A . Then for any element b in A there exist elements c and d in A such that $b = ad + cd - dc$

Proof. Since a is nonideal and A is a Weyl algebra there exists an element c of A such that $\sigma(a+c) \cap \sigma(c) = \varphi$. Therefore $0 \notin \sigma(a+c) - \sigma(c)$ and hence by Theorem (16) $0 \notin \sigma(T)$, where T is defined by $T(x) = (a+c)x - xc$. Therefore T is invertible and onto. Thus for any element b in A there exists an element d in A such that $b = T(d) = (a+c)d - dc = ad + cd - dc$.

As another property for Weyl algebras we generalize some properties of two by two matrices on the complex numbers to the space of two by two matrices over any Weyl Banach algebra. If b is a nonzero complex number one can easily verify that the two matrices $\begin{bmatrix} c & d \\ 0 & b+c \end{bmatrix}$ and $\begin{bmatrix} c & 0 \\ 0 & b+c \end{bmatrix}$ are similar. In fact we can easily check that $\begin{bmatrix} 1 & d/b \\ 0 & 1 \end{bmatrix}\begin{bmatrix} c & 0 \\ 0 & b+c \end{bmatrix} = \begin{bmatrix} c & d \\ 0 & b+c \end{bmatrix}\begin{bmatrix} 1 & d/b \\ 0 & 1 \end{bmatrix}$ where $\begin{bmatrix} 1 & d/b \\ 0 & 1 \end{bmatrix}$ is invertable with inverse $\begin{bmatrix} 1 & d/b \\ 0 & 1 \end{bmatrix}^{-1} = \begin{bmatrix} 1 & -d/b \\ 0 & 1 \end{bmatrix}$.

Theorem (18). If A is a Weyl algebra and b does not belong to any proper two sided ideal of A then there exists an element c in A such that the matrices $\begin{bmatrix} c & 0 \\ 0 & b+c \end{bmatrix}$ and $\begin{bmatrix} c & d \\ 0 & b+c \end{bmatrix}$ are similar for any element d in A.

Proof. First notice that any matrix of the form $\begin{bmatrix} e & x \\ 0 & e \end{bmatrix}$ is invertable with inverse $\begin{bmatrix} e & -x \\ 0 & e \end{bmatrix}$. Now since b is not in a proper two sided ideal there exists an element c such that $\sigma(b+c) \cap \sigma(c) = \varphi$. To show that $\begin{bmatrix} c & 0 \\ 0 & b+c \end{bmatrix}$ and $\begin{bmatrix} c & d \\ 0 & b+c \end{bmatrix}$ are similar for any d it is enough to find an element x in A such that $\begin{bmatrix} e & x \\ 0 & e \end{bmatrix}\begin{bmatrix} c & d \\ 0 & b+c \end{bmatrix} = \begin{bmatrix} c & 0 \\ 0 & b+c \end{bmatrix}\begin{bmatrix} e & x \\ 0 & e \end{bmatrix}$. The left hand side of the above equation is $\begin{bmatrix} c & d+(b+c)x \\ 0 & b+c \end{bmatrix}$ and the right hand side is $\begin{bmatrix} c & cx \\ 0 & b+c \end{bmatrix}$. Hence in order that equality holds we must have $(b+c)x + d = cx$ or $-d = (b+c)x - cx$. By Theorem (17) there exists an element x which satisfies the last equation.

Further Remarks

1. Some of these results constitute a part of Seddighin [13].

2. Both authors appreciate a discussion and correspondence with J. Zemanek concerning these problems.

3. The quasi-regularity in the second section is well-known in terms of the Jacobson circle operation. There are close connections between the nonperturbing elements and the inessential operators if one takes the spectrum in quotient or essential spectral sense. See [13].

4. When G is only a local compact group rather than a finite group the proof in Theorem (15) does not go through and one obtains only a C^* representation of the group.

5. In Theorem (15) one may obviate the multiplication of the matrices $K(f)$ and $K(g)$ by noting that K is (an integration of, by abuse of notation) the regular representation the group $(L^1(G))$ there. Hence one knows (e.g., using the machinery, see Naimark [11]) that $K(f^*) = K(f)^*$ and that $K(f) \cdot K(g) = K(f * g)$ without explicit calculation. On the other hand for larger groups one cannot always explicitly know this.

6. The property just mentioned (star representation) plus two others, isometricality and K (unit) $= I$, lead to an interesting analogy with the considerations of Gustafson, Goodrich, and Misra [14] . As shown there, in the Koopman picture under certain conditions one has a one-to-one correspondnece between representations of a group and an underlying dynamics taking place in the group. In particular, given a unitary equilibrium-preserving representing family for the group satisfying a positivity condition $U_s f \geq 0$ for all $f \geq 0$, a measure preserving underlying dynamics is assured. In the present paper one may say that we started with the dynamics (permutations) and generated the representations. The product rule of 5. above substitutes for the positivity.

7. The mapping K in Theorem (15) is a special case of a general mapping theorem by Segal [15] . Let G be a compact group, let $R(G)$ be $L^1(G)$ if $L^1(G)$ has an identity, $\overline{L^1(G)}$ if $L^1(G)$ does not have an identity, let $J(G) = L^1(G)$ if $L^1(G)$ has an identity, and let $J(G) = \{(f,\alpha) \in \overline{L^1(G)}$ with $\alpha = 0\}$ if $L^1(G)$ does not have an identity. Then [15, Theorem 3], if G is discrete, $J(G) = R(G)$, and if G is not discrete $J(G)$ is a maximal ideal in $R(G)$. Let M be any maximal ideal in $R(G)$ such that M is different from $J(G)$. Then: (i) there is a bounded continuous irreducible representation $g \to m(g)$ of G as matrices of some finite order n, such that the mapping $f \to \int_G m(g)f(g)d\mu$ if $L^1(G)$ has an identity, and $(f,\alpha) \to \alpha I_n + \int_G m(g)f(g)d\mu$ if $L^1(G)$ does not have an identity, on $R(G)$ to the ring M_n of all complex matrices of order n, is an exhaustive homomorphism. Here I_n is the unit matrix of order n, and the integrand is a matrix; (ii) the mapping given in (i) is continuous on $R(G)$ to M_n with the topology on $R(G)$ that is given by the norm $\|(f,\alpha)\| = \|f\|_1 + |\alpha|$ and the topology on M_n

the usual one; (iii) $f \in M$ if and only if $\int_G m(g)f(g)d\mu = 0$ when $L^1(G)$ has an identity, and $(f,\alpha) \in M$ if and only if $\alpha I_n + \int_G m(g)f(g)d\mu = 0$ when $L^1(G)$ does not have an identity. Furthermore, there corresponds to any bounded continuous irreducible representation of G as matrices of finite order a maximal ideal M in $R(G)$ such that (i) (ii) and (iii) hold. Finally, the correspondence is such that to different maximal ideals there correspond inequivalent representations, and conversely.

It is easy to check that the mapping K in Theorem (15) is obtained by the following representation of G as matrices of order σ.

$$m(x_1) = \begin{bmatrix} 1 & 0 & 0 & 0 & 0 & 0 \\ 0 & 0 & 0 & 0 & 0 & 0 \\ 0 & 0 & 0 & 0 & 0 & 0 \\ 0 & 0 & 0 & 0 & 0 & 0 \\ 0 & 0 & 0 & 0 & 0 & 0 \\ 0 & 0 & 0 & 0 & 0 & 0 \end{bmatrix}, \quad m(x_2) = \begin{bmatrix} 0 & 0 & 1 & 0 & 0 & 0 \\ 1 & 0 & 0 & 0 & 0 & 0 \\ 0 & 1 & 0 & 0 & 0 & 0 \\ 0 & 0 & 0 & 0 & 1 & 0 \\ 0 & 0 & 0 & 0 & 0 & 1 \\ 0 & 0 & 0 & 1 & 0 & 0 \end{bmatrix}$$

$$m(x_3) = \begin{bmatrix} 0 & 1 & 0 & 0 & 0 & 0 \\ 0 & 0 & 1 & 0 & 0 & 0 \\ 1 & 0 & 0 & 0 & 0 & 0 \\ 0 & 0 & 0 & 0 & 0 & 1 \\ 0 & 0 & 0 & 1 & 0 & 0 \\ 0 & 0 & 0 & 0 & 1 & 0 \end{bmatrix}, \quad m(x_4) = \begin{bmatrix} 0 & 0 & 0 & 1 & 0 & 0 \\ 0 & 0 & 0 & 0 & 1 & 0 \\ 0 & 0 & 0 & 0 & 0 & 1 \\ 1 & 0 & 0 & 0 & 0 & 0 \\ 0 & 1 & 0 & 0 & 0 & 0 \\ 0 & 0 & 1 & 0 & 0 & 0 \end{bmatrix}$$

$$m(x_5) = \begin{bmatrix} 0 & 0 & 0 & 0 & 1 & 0 \\ 0 & 0 & 0 & 0 & 0 & 1 \\ 0 & 0 & 0 & 1 & 0 & 0 \\ 0 & 0 & 1 & 0 & 0 & 0 \\ 1 & 0 & 0 & 0 & 0 & 0 \\ 0 & 1 & 0 & 0 & 0 & 0 \end{bmatrix}, \quad m(x_6) = \begin{bmatrix} 1 & 0 & 0 & 0 & 0 & 0 \\ 0 & 0 & 0 & 1 & 0 & 0 \\ 0 & 0 & 0 & 0 & 1 & 0 \\ 0 & 1 & 0 & 0 & 0 & 0 \\ 0 & 0 & 1 & 0 & 0 & 0 \\ 1 & 0 & 0 & 0 & 0 & 0 \end{bmatrix}$$

The following fact (which has been observed by Segal [15]) could be helpful in answering the question of whether $R(G)$ is a Weyl algebra for a general compact group G. Let on $R(G)$ $*$ be defined by $(f,\alpha)^* = (f^*,\bar{\alpha})$, where $f^*(x) = \overline{f(x^{-1})}$, then if M is a maximal ideal in $R(G)$ and if $z_i \in R(G)$ and $\sum_{i=1}^{n} z_i z_i^* \in M$, then each $z_i \in M$. $(i = 1,2,\ldots,n)$.

References

[1] K. Gustafson, Weyl's Theorems, Proc. Oberwolfach Conf. on Linear Operators and Approximation, 1971, International Series of Numerical Mathematics, 20, Birkhauser-Verlag (1972), 80-93.

[2] H. Weyl, Über beschränkte quadratische Formen, deren Differenz Vollsteting ist, Rend. cric. Math Palermo (1909), 373-392.

[3] J. Von Neumann, Charakterissierung des spektrums eienes integral operators, Actualites Sci. Indust. 229 (1935), 1-20.

[4] K. Gustafson and J. Weidmann, On the essential spectrum, J. Math. Anal. Applic. 25 (1969), 121-127.

[5] J. Zèmannek, Spectral characterization of two sided ideals in Banach algebras (to appeär).

[6] G. Aiken, Ph.D. dissertation, Louisiana State University, Baton Rouge, 1972.

[7] G. Aiken, A problem of Dyer, Porcelli, and Rosenfeld, Israel J. Math., vol. 25, Nos. 3-4 (1976), 191-197.

[8] J. Dyer, P. Porcelli, and M. Rosenfeld, Spectral characterization of two sided ideals in B(H) , Israel J. Math. 10 (1971), 26-31.

[9] C.E. Rickart, General Theory of Banach Algebras, D. Von Nostrand, Princeton (1960).

[10] E.F. Bonsall and J. Duncan, Complete Normed Algebras, Springer-Verlag, Heidelberg 1973.

[11] M. Naimark, Normed Rings, Nordhoff, Groningen, Netherlands (1959).

[12] M. Rosenbloom, On the operator equation $BX - XA = Q$, Duke Math. J. 23, (1956), 263-270.

[13] M. Seddighin, Ph.D. dissertation, University of Colorado at Boulder, to appear.

[14] K. Gustafson, R.D. Goodrich, B. Misra, Irreversibility and Stochasticity of Chemical Processes, these proceedings.

[15] I. Segal, The group ring of a locally compact group I , Proc. Nat. Acad. Sci. U.S.A. 27 (1941), 348-352.

SCATTERING THEORY IN MANY-BODY QUANTUM SYSTEMS. ANALYTICITY OF THE SCATTERED MATRIX*)+)

I. M. Sigal

Department of Mathematics
Princeton University
Princeton, N.J. 08544

Introduction. The aim of this paper is to formulate the result on and discuss the proof of the analytic properties of the S-matrix for the many-body quantum scattering. On a way to it we review the basic mathematical results and important methods of the many-body scattering theory. The statements are formulated and explained carefully and the proofs are outlined in sufficient detail to give a complete picture of the methods used. Complete proofs together with a rigorous discussion of related questions can be found in a paper of the author [21].

The many-body systems under consideration are short-range. This means that the potentials between particles vanish at

---

*) Talk presented at the 774th AMS meeting at Boulder, Colorado, March 27-29, 1980

+) Research partially supported by USNSF Grant MCS-78-01885.

infinity faster than $|x|^{-2}$. For notational simplicity we consider only single-channel systems (the definition will be given below). Our main theorem asserts that the scattering matrix has a meromorphic continuation in the energy parameter into a certain sector of the complex plane. The poles of this continuation occur only at eigenvalues of the dilation analytic family $H(\zeta)$ associated with the Hamiltonian H. By Balslev-Combes' theorem the latter eigenvalues lead to the poles of the meromorphic continuation of the matrix elements $\langle\phi,(H-z)^{-1}F\rangle$ on the dilation analytic vectors across the continuous spectrum into the second Riemann sheet. Moreover, if the potentials are nice enough so that the negative axis belongs to the meromorphic domain, then the poles on this axis occur at the negative eigenvalues of H.

This theorem settles in certain degree the problem of resonances in the many-body systems. Remember that the classical, one-particle definition of the resonances regards them as the poles of an analytic continuation of the S-matrix, while the poles of the meromorphic continuation of $\langle\phi,(H-z)^{-1}F\rangle$ are taken for the definition of resonances in the many-body systems [22].

1. Hamiltonian. Consider a system of N particles in $\mathbb{R}^\nu$ with masses m_i and interacting via pair potentials $V_\ell(x^\ell)$. Here ℓ labels pairs of indices and $x^\ell = x_i - x_j$ for $\ell = (ij)$. The configuration space of the system in the center-of-mass frame is defined as $R = \{x \in \mathbb{R}^{\nu N}, \sum m_i x_i = 0\}$ with the inner product

$(x,\tilde{x}) = \sum m_i x_i \cdot \tilde{x}_i$. Denote by V^ℓ and V_ℓ the multiplication operators on $L^2(\mathbb{R}^\nu)$ and $L^2(R)$ by the (real-valued) functions $V_\ell(y)$ and $V_\ell(x^\ell)$, respectively.

We assume that V^ℓ are Δ-compact, i.e. compact as operators from the Sobolev space $H_2(\mathbb{R}^\nu)$ to $L^2(\mathbb{R}^\nu)$. Then the operator

$$H = T + V,\ T = -1/2\ (\text{Laplacian on } L^2(R)),\ V = \sum V_\ell, \qquad (1)$$

is defined on $L^2(R)$ and is self-adjoint there.

Proposition 1. The potentials of the class $L^p(\mathbb{R}^\nu) + (L^\infty(\mathbb{R}^\nu))_\varepsilon$, where $p > \max(\frac{\nu}{2},2)$, if $\nu \neq 4$ and $p > 2$ if $\nu = 4$, and subindex ε indicates that L^∞-component can be taken arbitrarily small, are Δ-compact.

The proof of this proposition is a simple exercise [16] on embedding theorems for Sobolev spaces and on compact integral operators. We omit it here.

2. Dilation analyticity. Let $U(\rho)f(x) = \rho^{-(N-1)/2} f(\rho^{-1}x)$. Then $U(\rho)\ T\ U(\rho)^{-1} = \rho^2 T$ and $V_\ell(\rho) \equiv U(\rho)\ V_\ell\ U(\rho)^{-1}$ is the multiplication operator with $V_\ell(\rho^{-1}x^\ell)$. A T-bounded operator V_ℓ is said to be dilation analytic in $\mathcal{O} \subset \mathbb{C}$, $\mathcal{O} \cap \mathbb{R} \neq \emptyset$ iff $V_\ell(\rho)$, considered as an operator from $H_2(R^\ell) = D(T^\ell)$ to $L^2(R^\ell)$, has an analytic continuation into $\mathcal{O}$. In this case the family $H(\rho) \equiv U(\rho)\ H\ U(\rho)^{-1}$ has an analytic continuation from $\mathbb{R}$ to $\mathcal{O}$ with the common domain $D(T)$. Note also that if V_ℓ is dilation analytic in $\mathcal{O}$ it is dilation analytic in the sector $A = \{z \in \mathbb{C},\ |\arg z| < \alpha\}$, where $\alpha = \sup\{|\arg z|\ ,\ z \in \mathcal{O}\}$.

3. <u>Wave Operators and S-matrix.</u> In this section H and T are self-adjoint operators on a Hilbert space $\mathcal{H}$, E_p is the projection on the subspace of the point spectrum, $\sigma_p(H)$, of H and $E(\Delta)$ is the spectral projection for H. The strong limits

$$W^{\pm} = \text{s-lim}_{t \to \pm\infty} e^{iHt} e^{-iTt} , \tag{1}$$

whenever they exist on the absolute continuous subspace of T, are called the wave operators for the pair (H,T). In order not to carry an extra symbol we assume henceforth that T is absolute continuous. The scattering operator is defined on $\mathcal{H}$ by

$$S = W^{+*} W^{-} \tag{2}$$

It commutes with T, $[S,T] = 0$, and therefore is decomposable on a representation of $\mathcal{H}$ as a fiber direct integral $\int^{\oplus} \mathcal{H}_\lambda d\lambda$ with respect to T: $\pi S \pi^* = \int^{\oplus} S(\lambda) d\lambda$. Here $S(\lambda)$ is an operator on $\mathcal{H}_\lambda$, it is called the scattering matrix, and $\pi = \int^{\oplus} \pi_\lambda d\lambda$ is a unitary operator from $\mathcal{H}$ to $\int^{\oplus} \mathcal{H}_\lambda d\lambda$.

The definition of $W^{\pm}$ implies: (1) $W^{\pm}$ are isometric, $W^{\pm *} W^{\pm} = \mathbb{1}$, (2) $W^{\pm}$ are intertwining operators for (H,T), $HW^{\pm} = W^{\pm}T$, and (3) $R(W^{\pm}) \subset R(\mathbb{1} - E_p)$ or $E_p W^{\pm} = 0$. Property (2) yields that $W^{\pm} W^{\pm *}$ are projections from $\mathcal{H}$ into $R(W^{\pm})$. If $R(W^{+}) = R(W^{-})$, then S is unitary. We Call $W^{\pm}$ complete iff $W^{\pm} W^{\pm *} = \mathbb{1} - E_p$. In this case $\sigma_{s.c.}(H) = \emptyset$.

4. <u>Spectral Decomposition for the Free Hamiltonian.</u> Before proceeding to the theorem on a structure of the S-matrix we describe a fiber direct integral with respect to T. We define

$\int^{\oplus} H_\lambda d\lambda$ as $L^2(\mathbb{R}^+, L^2(\Omega))$, i.e. $H_\lambda = L^2(\Omega)$. Here Ω is the unit sphere in R (w.r. to the inner product defined in section 1). The unitary operator $\pi = \int^{+} \pi_\lambda d\lambda$ from H to $\int^{+} H_\lambda d\lambda$ is defined as $(\pi_\lambda f)(\omega) = C_N \lambda^\gamma \int e^{-i\sqrt{\lambda}\,\omega\cdot x} f(x)dx$, $\gamma = \frac{3(N-1)-2}{4}$, $C_N = \frac{1}{2}(2\pi)^{\frac{-3(N-1)}{2}}$. Obviously, $\pi_\lambda T = \lambda \pi_\lambda$ on $\mathcal{D}(T)$ and

$$\pi_\lambda = \lambda^{-1/2}\pi_1 U(\sqrt{\lambda}), \tag{2}$$

where $U(\rho)$ is the dilation group defined above in section 2.

5. Results. Let $R(z) = (H-z)^{-1}$.

Theorem 2. Let H be the Hamiltonian of a many-body, single channel system with real, dilation analytic potentials satisfying $V_\ell \in L^p \cap L^q(\mathbb{R}^\nu)$, $p > \frac{\nu}{2} > q$. Then

(1) $W^{\pm}$ and $W^{\pm *}$ exist as strong Abel limits, $W^{\pm}$ are isometric;

(2) $W^{\pm}$ are complete, i.e. $W^{\pm} W^{\pm *} = \mathbb{1} - E_p$

(3) The scattering operator and scattering matrix are unitary

(4) $T_\lambda = \text{s-lim}_{\varepsilon \downarrow 0} \pi_\lambda T(\lambda + i\varepsilon)\pi_\lambda^*$, where $T(z) = V - VR(z)V$, exists as a strongly continuous $B(L^2(\Omega))$-valued function uniformly bounded in $\lambda \in \mathbb{R}$ and $S(\lambda) = \mathbb{1}_\lambda + T_\lambda$.

Theorem 3. (I.M. Sigal) Let all the conditions of theorem 2 be satisfied and the explicit restriction on the potentials be obeyed also by $V_\ell(\zeta)$ for each $\zeta \in A$. Then the scattering matrix $S(\lambda)$ has a meromorphic continuation into the sector $\lambda \in 2A\backslash[1,\infty] \bigcup_{a \neq a_{max}} \sigma_p(H^a(e^{-i\alpha}))$. This continuation has poles in $A \cap \mathbb{C}^-$ only at eigenvalues of $H(e^{-i\alpha})$, i.e. at the points where the continuation of $(u, R(z), v)$ (on dilation analytic vectors u,v) from $\mathbb{C}^+$ across $\sigma(H)$ into the second Riemann sheet has its poles. If $\alpha > \frac{\pi}{2}$, then the poles of $S(\lambda)$ on the negative semiaxis occur at the eigenvalues of H.

The proof of (1) and (2) of theorem 2 is done in two steps. On the first step we develop a stationary scattering theory which connects the existence and asymptotic completeness of $W^{\pm}$ with the limiting absorption principle (boundary values of the resolvent). On the second step we prove the limiting absorption principle. (3) follows from (1) and (2). To show (4) we need in addition certain properties of the trace-operator families π_λ. Eqn. (2) and a slight modification of the limiting absorption principle are main ingredients of the proof of theorem 3.

6. Stationary Scattering Theory. If we replace the t-limit in the definition of $W^{\pm}$ by the weaker Abel-limit, then the resulting operators are called the stationary wave operators. They may exist even if $W^{\pm}$ do not. However, if $W^{\pm}$ do exist, then the stationary wave operators exist and equal $W^{\pm}$. It is convenient to define $Z^{\pm} = \text{s-Abel-lim}_{t \to \pm\infty} e^{iTt} e^{-iHt}$. If $W^{\pm}$ and $Z^{\pm}$ exist then $W^{\pm *} = Z^{\pm}$. The (stationary) wave operators and scattering matrix can be expressed in terms of the resolvents $R(z) = (H-z)^{-1}$ and $R_o(z) = (T-z)^{-1}$ (see 15]), e.g.

$$Z^+ = \text{s-lim } Z^{(\varepsilon)}, \quad Z^{(\varepsilon)} = \frac{\varepsilon}{\pi} \int R_o(\lambda - i\varepsilon) R(\lambda + i\varepsilon) d\lambda \qquad (3)$$

In the stationary case, where the integration over the spectral parameter is involved, it is useful to consider a local version of (3). We define for any Borel $\Delta \subset \mathbb{R}$

$$Z^{+}(\Delta) = \underset{\varepsilon \to \pm 0}{\text{s-lim}} \; Z^{(\varepsilon)}(\Delta), Z^{(\varepsilon)}(\Delta) = \frac{\varepsilon}{\pi}\int_{\Delta} R_{o}(\lambda-i\varepsilon)R(\lambda+i\varepsilon)d\lambda \quad (4)$$

These operators are the central objects in the stationary theory and $W^{\pm}$ are recovered through $W^{\pm}(\Delta) = Z^{\pm}(\Delta)$ by (see [20]):

Lemma 4. Let $Z^{\pm}(\Delta)$ exist for all Δ's from a directed sequence $\phi = \{\Delta\}$ of Borel subsets of $\mathbb{R}$ and satisfy $Z^{\pm}(\Delta)^{*}\, Z^{\pm}(\Delta') = E(\Delta \cap \Delta')$ for any Δ, $\Delta' \in \phi$. Assume that the Lebesgue measure of $\mathbb{R}\setminus \bigcup_{\Delta\in\phi} \Delta$ is zero. Then $\underset{\Delta \to \cup\Delta}{\text{s-lim}} \; Z^{\pm}(\Delta)$ and $Z^{\pm}$ exist and are equal.

Note that if $W^{\pm}$ are defined in a stationary way, then their isometry does not follow from their existence and has to be proven using an additional information about H and T.

Let $\delta_{\varepsilon}(A-\lambda) = \frac{1}{\pi}\, \text{Im}(A-\lambda-i\varepsilon)^{-1}$ and $T(z) = V - VR(z)V$, where $V = H-T$.

Theorem 5. Let there exist a Bannach space X with $X \cap H$ dense in X such that (i) for any compact interval $\Delta \subset \mathbb{R}\setminus\sigma_{p}(H)$, $\delta_{\varepsilon}(T-\lambda)u(\lambda)$ can be extended to a family of uniformly (in $\varepsilon \in \mathbb{R}^{+}$) bounded operators from $L^{2}(\Delta,X)$ to $L^{2}(\Delta,X')$, where X' is dual to X, (ii) $(H-T)R(\cdot + i\varepsilon)f \in L^{2}(\Delta,X)$ and has strong limits in $L^{2}(\Delta,X)$ as $\varepsilon \longrightarrow \pm 0$ for any f from a dense set $Y \subset H$. Then (a) $\sigma_{s.c.}(H) = \emptyset$, (b) $Z^{\pm}(\Delta)$ exist (as strong limits), (c) $W^{\pm}(\Delta)\, Z^{\pm}(\Delta') = E(\Delta \cap \Delta')$.

Before proceeding to the proof of the theorem we mention a few auxiliary relations, which follow from the density of $X \cap H$

in X and some facts about operator-valued functions [20]. Let T be a self-adjoint operator obeying [1] and $x, y \in L^2(\Delta, X)$ then

$$\int_\Delta \|(\delta_{\varepsilon'}(T-\lambda)^{1/2} - \delta_\varepsilon(T-\lambda)^{1/2})x(\lambda)\|^2 d\lambda \to 0(\varepsilon, \varepsilon' \to 0), \tag{5}$$

$$\int_{\Delta \times \Delta'} (\delta_\varepsilon(T-\lambda)\delta_\varepsilon(T-\nu)\, x(\nu), y(\lambda))d\lambda d\nu -$$

$$- \int_{\Delta \cap \Delta'} (\delta_\varepsilon(T-\lambda)\, x(\lambda), y(\lambda))\, d\lambda \to 0(\varepsilon \to 0) \tag{6}$$

Proof of theorem 5. (a) In virtue of Stone's theorem it suffices to prove that $\frac{\varepsilon}{\pi}\|R(\lambda+i\varepsilon)\, x\|^2$ for all $x \in Y$ is a Cauchy sequence in ε in $L^1(\Delta)$. We have

$\frac{\varepsilon}{\pi} \|R(\lambda+i\varepsilon)\, x\|^2 = \frac{\varepsilon}{\pi} \|R_o(\lambda+i\varepsilon)\, Q(\lambda+ i\varepsilon)x\|^2$, where $Q(z) = (T-z)(H-z)^{-1}$. The term on the right hand side is Cauchy in ε in $L^1(\Delta)$ by (5).

(b) $W^{(\varepsilon)}(\Delta)^*$ is bounded as follows from

$$\|W^{(\varepsilon)}(\Delta)^*u\| \leq \{\frac{|\varepsilon|}{\pi} \int_\Delta \|R_o(\lambda+i\varepsilon)Q(\lambda+i\varepsilon)u\|^2 d\lambda\}^{1/2} \leq M\|Q(\cdot+i\varepsilon)u\|_{L^2(\Delta,X)}$$

$W^{(\varepsilon)}(\Delta)^*$ converges strongly as $\varepsilon \to \pm 0$:

$$\|(W^{(\varepsilon')}(\Delta)^* - W^{(\varepsilon)}(\Delta)^*)u\| \leq 2\{\int \|(\delta_{\varepsilon'}(T-\lambda)^{1/2} - \delta_\varepsilon(T-\lambda)^{1/2}) \times Q(\lambda+i\varepsilon)u\|^2\, d\lambda\}^{1/2} + \{\int_\Delta \|\delta_\varepsilon(T-\lambda)(Q(\lambda+i\varepsilon') - Q(\lambda+i\varepsilon)u\|^2 d\lambda\}^{1/2}.$$

The right-hand side converges to 0 as ε, $\varepsilon' \longrightarrow \pm 0$ by (i), (ii) and (5).

(c) Using (6),

$$W^{(\varepsilon)}(\Delta)\, W^{(\varepsilon)}(\Delta')^* = \int_{\Delta \cap \Delta'} Q(\lambda+i\varepsilon)^*\, \delta_\varepsilon(T-\lambda)Q(\lambda+i\varepsilon)d\lambda \; + o\,(1)$$

weakly as $\varepsilon \longrightarrow \pm 0$. Using the definition of $Q(z)$,

$$\int_\Delta \delta_\varepsilon(H-\lambda)d\lambda = \int_\Delta Q(\lambda+i\varepsilon)^*\, \delta_\varepsilon(T-\lambda)Q(\lambda+i\varepsilon)d\lambda$$

as a sesquilinear form on Y. Comparing these two equations we arrive to the desired relation. □

Remark. A statement close to theorem 5 can be proven using Kato's relative smoothness method [12]. We choose the stationary method above since it can be generalized to the multichannel case (see [9, 14, 24, 20]).

Corollary 6. Under the conditions of theorem 5 the global have operators exist and $Z^{\pm} = s - \lim_{\Delta \to \mathbb{R}} Z^{\pm}(\Delta)$.

Theorem 7. Let there exist a Bannach space X with $X \cap H$ dense in X and such that (α) H-T is bounded from the dual X' to X, (β) $\delta_\varepsilon(T-\lambda)$ can be extended to a family of operators from X to X', uniformly bounded in $\varepsilon \in \mathbb{R}^+$ and $\lambda \in \mathbb{R}$, (γ) $(H-T)R(z)$, $z \in \mathbb{C}\backslash\mathbb{R}$, can be extended to a family of bounded operators on X which have strong boundary values on $\mathbb{R}\backslash\sigma_p(H)$. Then $T_\lambda \equiv s - \lim_{\varepsilon \downarrow o} \pi_\lambda T(\lambda+i\varepsilon)\pi_\lambda^*$ exists as a measurable $B(H_\lambda)$-valued function uniformly bounded in $\lambda \in \mathbb{R}$ and $S(\lambda) = \mathbb{1}_\lambda + 2\pi i\, T_\lambda$.

Proof. We begin with

Lemma 8. Let for some Bannach space X with $X \cap H$ dense in X, $\delta_\varepsilon(T-\lambda)$ be bounded from X to X' uniformly in $\varepsilon \in \mathbb{R}^+$ and $\lambda \in \mathbb{R}$. Then π_λ, as a family of operators from H to H_λ, is uniformly bounded in $\lambda \in \mathbb{R}$. If, in addition $\delta_\varepsilon(T-\lambda)$ has weakly continuous boundary values as $\varepsilon \downarrow o$, then $\|\pi_\lambda u\|_{H_\lambda}$ is continuous in λ.

Proof. First of all for any $u \in H$, $\|\pi_\lambda u\|_{H_\lambda} \in L^2(\mathbb{R})$. We can write $\langle\delta_\varepsilon(T-\lambda)u, u\rangle = \int \delta_\varepsilon(s-\lambda)\|\pi_s u\|^2_{H_s}\, ds$, a Poisson integral of $\|\pi_s u\|^2_{H_s}$. On the other hand, since $<\delta_\varepsilon(T-\lambda)u,u>$ is a harmonic, positive, uniformly bounded in $\lambda \in \mathbb{R}$ and $\varepsilon \in \mathbb{R}^+$, it is a Poisson integral of a uniformly bounded function, to which it converges as $\varepsilon \downarrow 0$ in the * - weak topology of L^∞ [23]. The proof now follows from the uniqueness of the Poisson integral. □

Lemma 9. If $W^{\pm *}$ exist then

$$S - \mathbb{1} = 2\pi i \lim_{\varepsilon \downarrow 0} \int \delta_\varepsilon(T-\lambda)\, W^{+ *} V\, \delta_\varepsilon(T-\lambda) d\lambda \qquad (5)$$

Proof. Using the definition of S and $W^{\pm} = \lim_{\varepsilon \to \pm 0} W^{(\varepsilon)}$,

$W^{(\varepsilon)} = \mathbb{1} - \int R(\lambda+i\varepsilon)V\ \delta_\varepsilon(T-\lambda)d\lambda$, we obtain

$$S - \mathbb{1} = \lim_{\varepsilon \downarrow 0} W^{+ *}(W^{(-\varepsilon)} - W^{(\varepsilon)}) \qquad (6)$$

substituting into here the expression for $W^{(\varepsilon)}$ and taking into account the intertwining property of $W^{+ *}$ one finds (5). □

Now we proceed to the proof of theorem 7. The properties of T_λ follow directly from lemma 8 and conditions (α) and (β). To obtain $S(\lambda) = \mathbb{1}_\lambda + 2\pi i\ T_\lambda$, we replace $W^{+ *}$ in (5) by $\mathbb{1} - \lim_{\varepsilon' \downarrow 0} \int \delta_{\varepsilon'}(T-\nu)VR(\nu+ i\varepsilon')d\nu$ take the diagonal $\varepsilon' = \varepsilon$ in the limit, apply to the obtained equation π_λ from the left and $\pi^*_{\lambda'}$ from the right and integrate out the variable ν (to justify this integration one uses conditions (α) and (β)).

Examples of the spaces X.

Definition 10. (T. Kato) An operator A is relatively smooth with respect to a self-adjoint operator T (or T-smooth) iff $D(T) \subset D(A)$ and $\int_{-\infty}^{\infty} \|A\, e^{-iTt}\, u\|\, dt \le M\|u\|^2$, $M < \infty$, for any $u \in D(T)$.

Denote by $A\mathcal{H}$ the completion of the range of A in the norm $\|x\|_{A\mathcal{H}} = \inf_{h,\ Ah = x} \|h\|_{\mathcal{H}}$. Note that $A\mathcal{H} \simeq \mathcal{H}/\mathrm{Ker}\, A$. In this notation, $\sum A_i\mathcal{H} = j(\oplus A_i\mathcal{H})$, where $j(\oplus \sum x_i) = \sum x_i$.

Lemma 11. Let operators A_i on $\mathcal{H}$ be T-smooth and define $X = \sum A_i\mathcal{H}$. Then $X \cap \mathcal{H}$ is dense in X and the $\delta_\varepsilon(T-\lambda)$ is bounded from X to X' uniformly in $\varepsilon \in \mathbb{R}^+$ and $\lambda \in \mathbb{R}$. If, in addition, $A_i^* \delta_\varepsilon(T-\lambda)A_j$ is weakly continuous on $\mathcal{H}$ as $\varepsilon \downarrow 0$ then so is $\delta_\varepsilon(T-\lambda)$ on X.

Proof. $X \cap \mathcal{H}$ dense in X by the definition of X as the completion of $\sum R(A_i)$ in the norm of X. Recall now that A_i is T-smooth if and only if $A_i^* \delta_\varepsilon(T-\lambda)A_i$ is uniformly in $\varepsilon \in \mathbb{R}^+$ and $\lambda \in \mathbb{R}$ bounded on $\mathcal{H}$. This implies that

$$\|A_j^* \delta_\varepsilon(T-\lambda)A_i\|^2 \le \|A_j^* \delta_\varepsilon(T-\lambda)A_j\|\ \|A_i^*\delta_\varepsilon(T-\lambda)A_i\|$$

is uniformly bounded on $\mathcal{H}$ for any i and j. The latter implies the first statement. The second one is obvious. □

Remark 12. Another proof of the lemma would be to use the original Kato definition of a relative smooth operator, which implies readily that $\sum A_i$ is T-smooth. Then one uses the equivalence of the norms in $\sum A_i\mathcal{H}$ and $(\sum A_i)\mathcal{H}$.

Lemma 13 (T. Kato). The potentials of the class $L^p \cap L^q(\mathbb{R}^\nu)$, $p > \nu > q$, are Δ-smooth.

Proof. See [12,16]. It is also a special case of lemma A.3 below.

7. Limiting Absorption Principle (Boundary Values of Resolvent)

Theorem 14. Let H be the Hamiltonian of an N-body, single-channel system with real, dilation-analytic potentials V_ℓ such that $V_\ell \in L^p \cap L^q(\mathbb{R}^\nu)$, $p > \frac{\nu}{2} > q$. Then for any pairs ℓ and s, $|V_\ell|^{\frac{1}{2}} R(z) |V_s|^{\frac{1}{2}}$ is an analytic in $z \in \mathbb{C}/\overline{\mathbb{R}^+}$ family of bounded operators on $L^2(R)$ with strong boundary values on $\overline{\mathbb{R}^+}$

A proof of this theorem is outlined in the appendix.

To prove a meromorphic continuation of the S-matrix we need a more elaborate result than theorem 14 (the notations are explained in the appendix).

Theorem 15. Let H be the Hamiltonian of an N-body, single-channel system with real, dilation analytic potentials V_ℓ such that $V_\ell(\zeta) \in L^p \cap L^q(\mathbb{R}^\nu)$, $p > \frac{\nu}{2} > q$, for each $\zeta \in A$. Then for all ℓ and s and $\zeta \in A$, $|V_\ell(\zeta)|^{1/2} R(z\zeta^2, \zeta) |V_s(\zeta)|^{1/2}$ is an analytic in $z \notin \bigcup_{a \neq a_{max}} \sigma_d(H^a(\zeta)) + \overline{\mathbb{R}^+}$ family of bounded operators on $L^2(R)$ with strong boundary values on $\overline{\mathbb{R}^+}$ if

$\operatorname{Im} z \cdot \operatorname{Im} \zeta < 0$ and on $\overline{\mathbb{R}^+} \cap [\mathbb{C}/(\zeta^{-2} \bigcup_{a \neq a_{max}} \sigma_d(H^a(\zeta)) + \overline{\mathbb{R}^+})]$ if

$\operatorname{Im} z \cdot \operatorname{Im} \zeta < 0$. Moreover, z is allowed to approach $\overline{\mathbb{R}^+}$ with angles other than $\frac{\pi}{2}$, e.g. $z = \lambda + i\varepsilon\, \zeta^{-2}$, $\varepsilon \longrightarrow \pm 0$. In both cases ($z = \lambda + i\varepsilon$ and $z = \lambda + i\,\varepsilon\, \zeta^{-2}$) the convergence is uniform in ζ from any compact subset of $(\mathrm{Re}\,z)^{-1}[2A/[1,\infty) \cup_{a \neq a_{max}} \sigma_d(H^a(e^{\mp i\alpha}))]$ for $\mathrm{Im}\, z > 0$.

For a proof of this theorem see the appendix.

8. Proof of Theorem 2. Theorems 5,7 and 14 imply all the statements except for the strong continuity of T_λ.

Lemma 16. Let ℓ be a pair of indices and let M be the multiplication operator by $f(x^\ell)$, $f \in L^p \cap L^q(\mathbb{R}^\nu)$, $p > \nu > q$, or $f \in L^2(\mathbb{R}^\nu)$. Then $\pi_s \dot{M}$ is a uniformly bounded family of operators: $L^2(R) \longrightarrow L^2(\Omega)$, strongly continuous in s. A similar statement is true also for $M^* \pi_s$.

Proof. Let $f \in L^p \cap L^q$, $p > \nu > q$. Then the statement follows from lemmas 8 and 11, prop. 13, eqn. (2) and strong continuity at $\rho = 1$ of $U(\rho)$ on L^p (different underlying spaces!). If $f \in L^2$, then the statement is obtained by the application of Cauchy-Schwartz inequality to $M\pi_s^* \phi$, $\phi \in L^2(\Omega)$, or $\pi_s Mu$, $u \in L^2(R)$, written explicitly as an integral. □

This lemma implies readily the strong continuity of T_λ. □

9. Proof of Meromorphic Continuation of S-matrix. We consider $T_\lambda \equiv \pi_\lambda V \pi_\lambda^* - \pi_\lambda VR(\lambda + io)V\pi_\lambda^*$. Let $A^\pm = \mathbb{C}^\pm \cap A$. Using (2), we find: $\pi_\lambda V\pi_\lambda^* = \lambda^{-1}\pi_1\, V(\sqrt{\lambda})\pi_1^*$. Therefore $\pi_\lambda V\pi_\lambda^*$ is analytic in $\lambda \in 2A$, where, remember, A is the sector of dilation analyticity of V_ℓ. Next we obtain

$$\pi_\lambda VR(\lambda + i\varepsilon)V\pi_\lambda^* = \lambda^{-1}\pi_1 V(\sqrt{\lambda})R(\lambda + i\varepsilon,\ \sqrt{\lambda})V(\sqrt{\lambda})\pi_1^* \quad (7)$$

The r.h.s. is a $B(L^2(\Omega))$-valued function, meromorphic in $\lambda \in 2A$ as long as $\lambda + i\varepsilon \notin \sigma_{ess}(H(\sqrt{\lambda})) = \bigcup_{a \neq a_{max}} [\sigma_d(H^a(\sqrt{\lambda})) + \lambda \overline{\mathbb{R}^+}]$. It follows from theorem 15 and lemma 14 that (5) converge in the $L^2(\Omega)$-operator norm as $\varepsilon \downarrow 0$, uniformly in λ from any compact subset of $B \equiv 2A/[1,\infty] \bigcup_{a \neq a_{max}} \sigma_d(H^a(e^{-i\alpha}))$ (take $z = 1 + i\varepsilon\lambda^{-1}$ and $\zeta = \sqrt{\lambda}$ in theorem 15). Then by the theorem on uniform convergence of analytic functions, the boundary value of (7) as $\varepsilon \downarrow 0$ is a meromorphic function in B, which can have poles only where $R(\lambda + io, \sqrt{\lambda})$ does. All the poles of the latter family are in A^- and at the eigenvalues of $H(e^{-i\alpha})$. The converse is also true. Indeed, for ϕ, F, two dilation vectors, and $\lambda \in 2A$ we have $\langle \phi, R(\lambda + io, \sqrt{\lambda})F \rangle = \langle \phi(\overline{\zeta}), R(\lambda, \sqrt{\lambda}\ \zeta)F(\zeta) \rangle$, where $\zeta \in A^-$ and $\phi(\zeta)$ is an analytic continuation of $U(\rho)\phi$. If $\lambda \in 2A^+$, then picking $\zeta = \lambda^{-1/2} \in A^-$ we can take $\zeta = \lambda^{-1/2}e^{-i\alpha} \in A^-$ to convince ourselves that it has poles exactly at eigenvalues of $H(e^{-i\alpha})$. □

Appendix. Proof of Limiting Absorption Principle.

Partitions. Let $a = \{C_i\}$ be a decomposition of the set $\{1,\ldots,N\}$ into nonempty, disjoint subsets C_i, called clusters. Denote by A the set of all such decompositions. A can be given the structure of a lattice; namely, if b is a partition obtained by breaking up certain subsystems of a, writing $b \subset a$. The smallest partition containing two partitions a and b will be denoted by $a \cup b$, i.e., $a \cup b = \sup(a,b)$. The largest partition contained in both a and b will be denoted by $a \cap b$: $a \cap b = \inf(a,b)$. We denote by a_{max} and a_{min} the maximal and minimal elements in A, respectively. A pair ℓ will be identified with the decomposition on N - 1 clusters, one of which is the pair ℓ itself and the others are free particles.

Define the configuration space of a system of A particles with the centers-of-mass of subsystems $C_i \in a$ fixed

$$R^a = \{x \in R, \sum_{j \in C_i} m_j x_j = 0 \ \forall C_i \in a\}$$

and the configuration space of the relative motion of the centers-of-mass of the clusters $C_i \in a$

$$R_a = \{x \in R, x_i = x_j \text{ if } i \text{ and } j \text{ belong to the same } C_k \in a\}$$

Then $R^a \perp R_a$ and

$$R^a \oplus R_a = R, \ L^2(R^a) \otimes L^2(R_a) = L^2(R).$$

Note that $R^b \subset R^a$ and $R_a \subset R_b$ for $b \subset a$, so we can define $R^a_b = R^a \ominus R^b = R_b \ominus R_a$. Then $R^b_c \oplus R^a_b = R^a_c$ and $H^a_b = L^2(R^a_b)$

satisfies

$$\mathcal{H}_c^a = \mathcal{H}_c^b \otimes \mathcal{H}_b^a \quad \text{for} \quad c \subseteq b \subseteq a .$$

We abbreviate if $\mathcal{H}_b = \mathcal{H}_b^a$ if $\#(a) = 1$ and $\mathcal{H}^a = \mathcal{H}_b^a$ for $\#(b) = N$.

Here $\#(a)$ is the number of clusters in the partition a.

The operator T_b^a on $\mathcal{H}_b^a$ is defined as the self-adjoint extension of $-1/2\, \Delta_b^a$, where Δ_b^a is the Laplacian on R_b^a. Then $T_c^a = T_c^b \otimes \mathbb{1}_b + \mathbb{1}_c^b \otimes T_b^a$. We define $H_b^a = T_b^a + \sum_{\ell \subset b} V_\ell$ on $\mathcal{H}^a$. and write $H^a = H_b^a$ for the $b = a_{min}$.

Regularizers for H-z. Let $\mathcal{H}$ be a Bannach space and T, an operator on $\mathcal{H}$ with a domain D(T). We call a bounded operator F from $\mathcal{H}$ to D(T) a (right,exact) regularizer for T iff (i) F is invertible and (ii) $T F - \mathbb{1}$, raised to some power, is compact on $\mathcal{H}$. In this section we construct regularizers for the family H-z and use them to study the spectral properties of H.

Definition A.1. Let A be a finite lattice and $\{H_a, a \in A\}$, a collection of operators on $\mathcal{H}$ with $\mathcal{D}(H_o) \subset D(H_a)$, where $H_o = H_{a_{min}}$ (as above, a_{min} and a_{max} are minimal and maximal elements in A). We define by induction on $a \in A$ the following families of bounded operator on $\mathcal{H}$:

$$A_a(z) = (H_a - z)(H_o - z)^{-1} \overrightarrow{\prod_{b \subset a}} A_b(z)^{-1} \tag{A.1}$$

where the arrow over the top of the product sign indicates the following order of the A^{-1}'s: if A_c^{-1} stands on the right of A_d^{-1} then $c \not\subset d$.

We set the family of bounded operator from $\mathcal{H}$ to $D(H_o)$:

$$F_a(z) = (H_o - z)^{-1} \overrightarrow{\prod_{b \subset a}} A_b(z)^{-1} \tag{A.2}$$

(A.1) and (A.2) imply that

$$(H_a - z)F_a(z) = A_a(z) \tag{A.3}$$

The obvious properties of A_a and F_a are listed for the reference convenience in the following two lemmas:

Lemma A.2. For any $a \in A$ and all $z \in \bigcap_{b \subset a} \rho(H_b)$, the operator $F_a(z)$ is bounded from H to $D(H_o)$ and has the bounded inverse (from $D(H_o)$ to H). Both operators are analytic in $z \in \bigcap_{b \subset a} \rho(H_b)$.

Lemma A.3. For any $a \in A$ and all $z \in \bigcap_{b \subset a} \rho(H_b)$, the operator $A_a(z)$ is bounded on H and is analytic in $z \in \bigcap_{b \subset a} \rho(H_b)$. It has the bounded inverse for $z \in \bigcap_{b \subseteq a} \rho(H_b)$ and the following statements are equivalent:

1. $0 \, \varepsilon \varepsilon \sigma(A_a(z))$ and $\phi \in \mathrm{Ker}\, A_a(z)$
2. $z \in \sigma_p(H_a)$ and $F(z)\phi \in \mathrm{Ker}(H-z)$.

In case when the H_a's are constructed as above the operators $F_a(z)$ and $A_a(z)$ have an additional structure ($H_o = T$):

Lemma A.4. The operators $F_a(z)$ and $A_a(z) - \mathbb{1}$ are finite, linear combinations of monomials of the form

$$R_o \, \pi[V_\ell R_b], \ \ell, \ b \subset a, \text{ and } \pi[V_\ell R_b] \ , \ b \subset a, \ \cup\ell = a \ , \qquad (A.4)$$

respectively. Here $R_o(z) = (T-z)^{-1}$.

The statement can easily be derived by induction. The details can be found in [18]. Note here only that since V_ℓ have T-bound 0, they are H_b-bounded as well. Therefore monomials of form (A.4) are bounded and analytic in $z \in \cap \rho(H_b)$.

Lemma A.5. For z with dist $(z,\sigma(T))$ sufficiently large, $A_a(z) - \mathbb{1}$ is a norm convergent series of monomials,

$$\pi_{\cup\ell = a} [V_\ell (T-z)^{-1}]$$

Proof. The statement follows from lemma A.4 and the fact that for $\mathrm{dist}(z,\sigma(T))$ large enough the following series are norm convergent

$$R_b(z) = (T-z)^{-1} \sum_{n=0}^{\infty} [\sum_{\ell \subset b} V_\ell (T-z)^{-1}]^n \qquad (A.5)$$

Indeed, $\|A(T-z)^{-1}\| \longrightarrow 0$ as $\mathrm{dist}(z,\sigma(T)) \longrightarrow \infty$ for any T-bounded operator A. □

Boundary values on $\mathbb{R}$. We introduce the parameter $g = (g_\ell)$, coupling constant, into the formulas replacing everywhere V_ℓ by $g_\ell V_\ell$ for all ℓ's. We define the domains (it suffices for us to keep g real. However, we do not use this restriction and therefore omit mentioning it explicitly): $G^a = \{g: \sigma_p(H^b(g)) = \phi$ for all $b \subset a\}$. In this section we prove only theorem 14. The proof of theorem 15 requires some simple, technical modifications which we leave to the reader as an exercise.

Theorem A.6. Let the conditions of the theorem 15 be obeyed. Then for each $a \in A$ and any pairs $c,d \subset a$, the family $|V_c|^{\frac{1}{2}} R^a(z,g)|V_d|^{\frac{1}{2}}$ is uniformly bounded on H^a, analytic in $z \in \mathbb{C}\backslash\sigma(T)$ and $g \in G$ and strongly continuous as $z \longrightarrow \sigma(T)$, uniformly in $g \in G^a$.

Proof. We conduct the proof by induction on $a \in A$. For $a = a_{min}$ we do not have H^a. Let the statement be true for all $b, b \subset a$, and prove it for a. In the sequel we suppress the superindex a.

Consider the operator $L(z,g) = A(z,g) = 1\!\!1$. Writing

$L(z,g) = \sum_f |V_f|^{\frac{1}{2}} L_f(z,g)$ and defining $L_{fg} = L_f|V_g|^{\frac{1}{2}}$ and

$F_{fg} = |V_f|^{\frac{1}{2}}$ we obtain

$$V_c^{\frac{1}{2}} R|V_g|^{\frac{1}{2}} + \sum_f V_c^{\frac{1}{2}} R|V_f|^{\frac{1}{2}} L_{fg} = F_{cg} \tag{A.6}$$

Proposition A.7. The operators $L_{fd}(z,g)$ and $F_{fd}(z,g)$ are bounded on H, analytic in $z \in \mathbb{C}\backslash\sigma(T)$ and $g \in G$ and strongly continuous as $z \longrightarrow \sigma(T)$ uniformly in $g \in G$.

Proof. The operators L_{cd} and F_{cd} are linear combinations of monomials of the form:

$$\overset{k}{\underset{i=1}{\overrightarrow{\pi}}} [V_{f_i}^{\frac{1}{2}} R_{b_i} |V_{f_{i+1}}|^{\frac{1}{2}}], \; b_i \neq a_{max} , \tag{A.7}$$

$f_1 = c$, $f_{k+1} = d$ with the conditions $\cup f_i = a_{max}$ and $f_i \neq a_{max}$, respectively.

We transform (A.7) so that each factor satisfies $f_i, f_{i+1} \subseteq b_i$ if $\#(b_i) \leq N-1$. To this end we use the equations $R_b = F_b - R_b L_b$ and $R_b = F'_b - L'_b R_b$, where $F'_b(z) = F_b(\overline{z})^*$ and $L'_b(z) = L_b(\overline{z})^*$, to surround each R_b in (A.7) with V_f, next on its left, and $|V_g|^{\frac{1}{2}}$, next on its right, satisfying $f,g \subset b$.

Lemma A.8. The operators $|V_c|^{1/2} R_b(z,g)|V_d|^{1/2}$, $c, d \subset b \subset a$, considered on H^a, are bounded, analytic in $z \in C\backslash\sigma(T)$ and $g \in G^b$ and strongly continuous as $z \longrightarrow \sigma(T)$ uniformly in $g \in G^b$.

Proof. Let $S_b = \mathbb{1}^b \otimes s_b$, where s_b is a unitary operator from

H_b to the direct integral $\int^{\oplus} H_\lambda d\lambda$ with respect to T_b: $(s_b T_b^a u)(\lambda) = \lambda(s_b u)(\lambda)$. Then (g is omitted)

$$S_b|V_c|^{1/2}R_b(z)|V_d|^{1/2}u = |V_c|^{1/2}R^b(z-\lambda)|V_d|^{1/2}(S_b u)(\lambda).$$

This equation together with the induction statement about $|V_c^b|^{1/2}R^b(z)|V_d^b|^{1/2}$ implies the lemma. □

If b_i has $\#(b_i) = N$, then we use

Lemma A.9. (Kato, Iorio-O'Connell, Combesqure-Ginibre, Hagedorn) Let U and W be the multiplication operators by functions $\phi(x^\ell)$ and $\psi(x^s)$, where $\phi, \psi \in L^p \cap L^q(\mathbb{R}^\nu)$, $p > \nu > q$ and ℓ and s are arbitrary pairs of indices. The family $W(-\Delta^a-z)^{-1}U$ is bounded on $L^2(R^a)$, analytic in $z \in \mathbb{C}\backslash\bar{\mathbb{R}}^+$, has strong boundary values on $\bar{\mathbb{R}}^+$ and is bounded in norm by

$$||W(-\Delta^a-z)^{-1}U|| \leq \text{const } ||\phi||_{L^p \cap L^q} ||\psi||_{L^p \cap L^q} \tag{A.8}$$

Moreover, if $\ell \cap s \neq \emptyset$, then the family is norm continuous as $\text{Im } z \longrightarrow \pm 0$.

Proof. See [7] (which is a close extension of [9,10,5]). □

Lemmas A.8 and A.9, the remark about L_{cd} and F_{cd} made at the beginning of the proof (up to lemma A.8) imply the statement of proposition A.7. □

Proposition A.10. The matrix $[L_{fd}(z,g)]^S$, $g \in G$, is compact for all $z \in \mathbb{C}\backslash\mathbb{R}$ up to the real axis.

Proof. Since the matrix is analytic in $g \in G$, it suffices to prove the proposition for a neighborhood of $g = 0$. It follows from lemma A.9 that the series (A.5) with $V_c \longrightarrow g_c V_c$ converges in the norm for all $z \in \mathbb{C}\backslash\mathbb{R}$ up to the real axis, as long as g is confined to a neighborhood of 0. Substituting such series for R_b's in (A.7) we conclude that $L_{fd}(z,g)$ for g in a neighborhood, V, of zero is a norm convergent series of terms of the form

$$g^k \prod_{\cup f_i = a_{max}}^{\rightarrow} [V_{f_i}^{\frac{1}{2}} R_o |V_{f_{i+1}}|^{\frac{1}{2}}] .$$

Lemma A.11 (I.M. Sigal) Let U_ℓ and W_ℓ be the multiplication operators by functions $\phi_\ell(x^\ell)$ and $\psi_\ell(x^\ell)$, respectively, where $\phi_\ell, \psi_\ell \in L^p \cap L^q(\mathbb{R}^\nu)$, $p > \nu > q$. Then a product of three operator-functions $\mathbb{C}\backslash\overline{\mathbb{R}}^+ \longrightarrow B(L^2(R^a))$ of the form $\pi[W_{\ell_i}(-\Delta^a - z)U_{\ell_{i+1}}]$, $\cup\ell_i = a$, (called the a-connected graphs) has norm-continuous boundary values on $\overline{\mathbb{R}^+}$. These boundary values are compact.

Proof. We begin with

Lemma A.12. Graphs $\pi[W_{\ell_i}(T^a - z)^{-1} U_{\ell_{i+1}}]]$ are norm continuous in $\phi_\ell, \psi_\ell \in L^p \cap L^q(\mathbb{R}^\nu)$, $p > \nu > q$, uniformly in $z \in \mathbb{C}\backslash\overline{\mathbb{R}}^+$.

Proof. The statement follows from the basic estimate (A.8).□

Lemma A.13. The product of three a-connected graphs with $\phi_\ell, \psi_\ell \in S(\mathbb{R}^\nu)$ is norm continuous on $L^2(R^a)$ as $\mathrm{Im}\, z \longrightarrow \pm 0$.

We will prove this lemma at the end of the appendix. Now we deduce the proof of lemma A.11 from lemmas A.12 and A.13. Indeed, since S is dense in L^p, there exist sequences $\phi_\ell^{(n)}$ and $\psi_\ell^{(n)}$ converging in $L^p \cap L^q(\mathbb{R}^\nu)$, $p > \nu > q$, to ϕ_ℓ and ψ_ℓ, respectively. Given a graph G we construct the new graphs, $G^{(n)}(z)$, by replacing in $G(z)$ all U_ℓ and W_ℓ by the operators $U_\ell^{(n)}$ and $W_\ell^{(n)}$ of multiplication by $\phi_\ell^{(n)}(x^\ell)$ and $\psi_\ell^{(n)}(x^\ell)$, respectively. By lemma A.12, $G^{(n)}(z) \longrightarrow G(z)$ is norm, uniformly in $z \in \mathbb{C}\backslash\bar{\mathbb{R}}^+$, as $n \longrightarrow \infty$. Now consider the product of three a-connected graphs and the norm approximation to this product constructed as above. By lemma A.2, this approximation is norm continuous as $\operatorname{Im} z \longrightarrow \pm 0$. Hence the product itself is norm continuous as $\operatorname{Im} z \longrightarrow \pm 0$. This completes the proof of lemma A.11

Now we return to the proof of proposition A.10 By lemma A.11 the matrix $[L_{fd}(z,g)]^s$ for $g \in V$ is a norm convergent series of terms each of which is, in virtue of lemma A.11, a compact operator on $\oplus H$ for all $z \in \mathbb{C}\backslash\mathbb{R}$ up to the real axis. By the theorem on the closedness of the set of compact operator in the uniform topology, $[L_{fd}(z,g)]^s$ is compact as well (for $g \in V$).

Proposition A.14. $-1 \in \sigma[L_{fg}(z,g)] \Leftrightarrow z \in \sigma_p(H(g))$.

Proof. (the parameter g is omitted henceforth). We begin with a few auxiliary statements:

Lemma A.15. The following two statements are equivalent:

(α) $-1 \in \sigma([L_{fd}(z)]$

(β) $(H-z)\phi = 0$ has a nontrivial weak solution in $R_0(z) \sum|V_\ell|^{\frac{1}{2}} H$.

Proof. If $f + [L_{cd}(z)]f = 0$, $f = \oplus f_\ell \in \oplus H$, then $\chi = \sum|V_\ell|^{\frac{1}{2}} f_\ell$ satisfies $A(z)\chi = 0$ and therefore, in virtue of (A.3), $\phi = F(z)\chi$ obeys formally (in the weak sense) $(H-z) = 0$. Since $F(z)$ has a bounded inverse we can go backward as well. $\square$

Corollary A.16. $(H-z)\phi = 0$ has no solutions in

$R_o(z) \sum|V_\ell|^{\frac{1}{2}}H$ for $z \notin \sigma(T)$.

Proof. Use that $\sigma(T) = \sigma(H_a)$ and properties of $[L_{fg}(z,g)]$. □

Lemma A.16. Let $\lambda \in \sigma(T)\backslash\sigma(T(\zeta))$, $\text{Im}\ \zeta \neq 0$, and $\phi \in R_o(\lambda \pm io)\sum|V_\ell|^{\frac{1}{2}} H$ be a weak solution to $(H-\lambda)\phi = 0$. Then either $\phi = 0$ or $\lambda \in \sigma_p(H)$.

Proof. We define $L_{cd}(z,\zeta)$ for $H(\zeta) = T(\zeta) + \sum V_\ell(\zeta)$. L_{cd} is analytic in ζ as long as $z \notin \sigma_{ess}(H(\zeta))$. We use

Lemma A.17. Let $G, \mathcal{D} \subset \mathbb{C}$ and $K(z,\zeta)$ be a family of compact operators, jointly norm continuous on $\overline{G} \times \mathcal{D}$ and analytic in $\zeta \in \overline{\mathcal{D}}/D$ as $z \in G/\overline{G}$. Let moreover $K(z,\zeta) = U(|\zeta|)K(z,e^{i\arg\zeta})$ $U(|\zeta|)^{-1}$, where $U(\zeta)$ is a unitary operator for $\rho \in \mathbb{R}^+$. Then for $z \in \overline{G}$ and $\nu \neq 0$

$$\nu \in \sigma(K(z,\rho)),\ \rho \in \partial\mathcal{D}\ ,\ \Leftrightarrow\ \nu \in \sigma(K(z,\zeta)),\ \zeta \in \mathcal{D}$$

Proof. See [21].

By lemma A.17

$$-1 \in \sigma([L_{cd}(\lambda \pm io)]) \Leftrightarrow -1 \in \sigma([L_{cd}(\lambda,\zeta)]),\ \text{Im}\ \zeta \neq 0.$$

Furthermore as in corollary A.16. □

$$-1 \in \sigma(L_{cd}(\lambda,\zeta) \Leftrightarrow \lambda \in \sigma_d(H(\zeta)) \text{ for } \lambda \notin \sigma_{ess}(H(\zeta)).$$

By the Balslev-Combes theorem [16]; $\sigma_d(H(\zeta)) \cap \mathbb{R} = \sigma_p(H) \setminus \{0\}$. This completes the proof of lemma A.16. □

To complete the proof of proposition A.14 we should consider the point $\lambda \in \sigma(T) \cap \sigma(T(\zeta)) = \{0\}$ for $\text{Im}\ \zeta \neq 0$.

Lemma A.18. Let $V_\ell \in L^p \cap L^q(\mathbb{R}^\nu)$, $p > \frac{\nu}{2}$ q. Then for all internal points of $G^{\ell} = \{g: \sigma_p(H^{\ell}(g)) = \emptyset\}$ the equation $H^{\ell}(g)\psi = 0$ has no nontrivial solutions in $(T^{\ell})^{-1}|V^{\ell}|^{\frac{1}{2}}L^2(R^{\ell})$.

Proof. Let, on the contrary, $H^{\ell}(g)\psi = 0$ with $\psi \in (T^{\ell})^{-1}V^{\ell}L^2(R^{\ell})$ and $\psi \neq 0$. Then $-g^{-1} \in \sigma(V^{\ell}(T^{\ell})^{-1})$ on $|V_{\ell}|^{\frac{1}{2}} L^2(R^{\ell})$. By the perturbation theory (we use here the fact that $V^{\ell}(T^{\ell}-\lambda)^{-1}$ is norm continuous as $\lambda \uparrow 0$) for any sufficiently small $\lambda < 0$ there exists g' such that $-g'^{-1} \in \sigma(V^{\ell}(T^{\ell}-\lambda)^{-1})$ and $g' \longrightarrow g$ as $\lambda \longrightarrow 0$. The latter implies that $\lambda \in \sigma_d(H^{\ell}(g'))$ for g' as close to g as we wish. However, this is impossible since g is an internal point of G.

Lemma A.19. $H^a \psi = 0$, $\psi \in (T^a)^{-1} \sum|V_{\ell}|^{1/2}L^2(R^a)$, implies that either $\psi = 0$ or $0 \in \sigma_p(H^a)$ for all a with $\nu(N-\#(a)) > 4$.

Proof. In the proof below we omit the superindex a. Let $\psi \in T^{-1} \sum|V_{\ell}|^{1/2} L^2(R)$, then $F(0)^{-1}\psi \in \sum|V_{\ell}|^{1/2} L^2(R)$. If, moreover, ϕ is a solution to $H\psi = 0$, then $\phi = F(0)^{-1}\psi$ is a solution to $\psi + L(0)\phi = 0$. Using methods of [20] one can show that if $\phi \in \sum|V_{\ell}|^{1/2} L^2(R)$ obeys the latter equation then in fact $\phi \in L^p(R)$, $3/2 - \varepsilon \leq p \leq 2$, and $\psi = F(0)\phi \in T^{-1} L^p(R)$. Hence by Sobolev's potential theorem $\psi \in L^2(R)$. Let now $\alpha \in C_0^{\infty}(\mathbb{R})$ and $\alpha(\lambda) = 1$ for $|\lambda| \leq 1$. Then $\alpha(T)\psi \in D(T)$. Furthermore, since $(1-\alpha(T))T^{-1}$ is bounded and maps H into $D(T)$, $(1-\alpha(T))T^{-1}T\psi \in D(T)$. □

Proposition A.19 is proven. □

Propositions A.7, A.10 and A.14 imply the statement of

theorem A.6 for H^a. This completes our inductive proof. □

Proof of Lemma A.13. Below we prove a statement stronger than lemma A.13. Namely, we show that the kernel of the product in lemma A.13 is a Hölder continuous, fast decreasing at infinity function, Hölder continuous in the parameter $z \in \mathbb{C} \backslash \overline{\mathbb{R}}^+$ up to $\overline{\mathbb{R}}^+$.

We set $\Delta_x^\nu(h)f(x,y) = |h|^{-\nu}(f(x+h,y)-f(x,y))$ if $0 < \nu \leq 1$ and $\Delta_x^\nu(h)f(x,y) = f(x,y)$ if $\nu = 0$.

Lemma A.20. (K. Hepp - I.M. Sigal) Let $G(z)$ be a product of three a-connected graphs with $\phi_\ell, \psi_\ell \in \mathcal{S}(\mathbb{R}^\nu)$. Then the Fourier transform $G(p,q,z)$, of its kernel satisfies the estimate

$$|\Delta_{p,q}^\nu(h)\Delta_z^\mu(w)G(p,q,z)| \leq \text{const.}\ (1+|p-q|)^{-r},\ r \in \mathbb{R}^+ \qquad (A.8)$$

Here p and q are two sets of independent variables in the space dual to $\mathbb{R}^a$ (i.e. in the corresponding momentum space).

Proof. The expression for the kernel of $G(z)$ in the momentum representation (i.e. the F. transform of the kernel) can be easily computed, since the kernels of U_ℓ, W_ℓ and $(T^a-z)^{-1}$ in this representation are known. It has the following form

$$G(p,q,z) = \int \frac{\phi(p,q,k)\ d^m k}{\prod_1^s [P_i(p,q,k)-z]}, \qquad (A.9)$$

where $\phi(p,q,k) \in C^\infty$ comes from the potential part (U_ℓ and W_ℓ)of $G(z)$ and $P_i(p,q,k)$ is the symbol of T^a expressed in the variables p,q,k, using an i-dependent linear function. The estimate of the decay of $G(p,q,z)$ at infinity can be easily obtained if we note that those of the P_i's which large enough p_k or q_k

(say $p_k^2 > 10$ Rez + 1) are not singular in the sense that P_i - Rez $\geq \delta > 0$. An estimation of the decay of $G(p,q,z)$ in such a p_k or q_k is a rather simple but, unfortunately, boring and longsome exercise. Since, moreover, the precise form of the estimating function is not important (what is important is its L^1-property) we omit here the derivation of the infinity-decay estimate.

To obtain the smoothness estimates for those variables p_j and q_j which stay in the bounded region of $\mathbb{R}^\nu$ and the smoothness estimates in z, we join those P_i, which contains variables (counting also the k-variables) from the vicinity of infinity specified above, to ϕ.

The resulting integral is of the form

$$J(u,z) = \int \frac{\phi(k)d^m k}{\overset{s}{\underset{1}{\pi}}[(x,R^i x)-z]}, \quad x = (k,u), \ k \in \mathbb{R}^{\nu m}, \ u \times \mathbb{R}^{\nu n}, \quad (A.10)$$

where u varies in a compact region of $\mathbb{R}^{\nu n}$, $\phi \in C_0^\infty(\mathbb{R}^{\nu m})$ and R^i are real, nonnegative, $(m+n) \times (m+n)$-matrices. R^i act on the space $\mathbb{R}^{\nu(m+n)}$, of which vectors are written as $p = (p_1 \ldots p_{m+n})$, $p_i \in \mathbb{R}^\nu$, according to the equation $(Rp)_i = \sum_{j=1}^{m+n} (R)_{ij} p_j$.

It is shown in lemma III.1 of [18] that such integrals are Holder continuous in u and z (including $z \longrightarrow \mathbb{R}$). The conditions of this lemma are satisfies by (A.10). For reader's convenience we reproduce the lemma under more restrictive conditions, which are still obeyed in our case, in a separate preprint which can be obtained from the author.

Here we mention only that to obtain the desired estimates on (A.10) we use first Feynmann identity,

$$\prod_{i=1}^{s} A_i^{-1} = \int_{[o,1]^s} s\,(\sum \alpha_i A_i)^{-s}\,\delta(1-\sum \alpha_i)\,d^s\alpha,$$

to transform the product of s polynomials (of the second degree) in the denominator into one polynomial (also of the second degree) but taken to the s-th power. Then we integrate by parts in k.

Literature Comments.

Section 2. A discussion of dilation analyticity as well as the Balslev-Combes theorems can be found in [16,1,2,22].

Section 5. Detailed references about results on asymptotic completeness can be found in [20].

Theorem 3 is our main result. The analytic extension of the N-body S-matrix was proven for the first time by G. Hagedorn [6] for (two-clusters) ⟶ (two-cluster) scattering. Our proof is completely different from that of G. Hagedorn. As in the two-particle scattering, he represents the integral kernels of the corresponding matrix elements of the S-matrix as sesquelinear forms and applies the Combes analytic continuation to those forms.

The analyticity of the diagonal elements of the S-matrix in the three body scattering was proven by E. Balslev [3].

Section 6. Lemma 4 was proven in [20]. Theorems 5 and 6 are special cases of the corresponding theorems from [20] (see also [14, 24]). An earlier result of this genre was established in [12, 15]. Lemma 8 was proven in [20]. A related result was obtained earlier in [13], The spaces $\sum A_i H$ of lemma 11 were introduced in [20] and in an earlier form [18].

Section 7. Theorems 14 and 15 were proven in [21]. Earlier versions of the limiting absorbtion principle are discussed in the literature notes of [20].

Section 8. A slightly weaker statement than lemma 16 was

proven first (in a different way) by M. Combesqure and J. Ginibre [5].

Appendix. Lemma A.9 was proven by G. Hagedorn [17] who extended close results of R.J. Iorio and M. O'Carroll [11] and M. Combesqure and J. Ginibre [5] which in turn are based on results and technique of T. Kato [12]. Lemma A.11 was proven by I.M. Sigal [18, 21]. Lemma A.17 goes back to G. Hagedorn [7], the present form with a detailed proof is given in [21]. Lemma A.20 was proven by I.M. Sigal [18], a close result was obtained earlier in a different way by K. Hepp [8].

References

1. D. Babbitt and E. Balslev, Dilation-Analyticity and Decay Properties of Interactions, Comm. math. Phys. 35 173-179 (1974).
2. D. Babbitt and E. Balslev, A Characterization of Dilation-Analytic Potentials and Vectors, J. of Funct. Anal. 18, 1-14 (1974).
3. E. Balslev, Aarhus, Preprint.
4. F.A. Berezin, Asymptotic Behavior of Eigenfunctions in Schrodinger's Equation for Many Particles, Dokl. Acad. Nauk SSSR, 163, 795-798 (1965).
5. M. Combesqure and J. Ginibre, Hilbert Space Approach to the Quantum Mechanical Three-Body Problem, Ann. Inst. H. Poincare, 21, 97-145 (1974).
6. G. Hagedorn, A Link Between Scattering Resonances and Dilation Analytic Resonances in Few Body Quantum Mechanics, Comm. Math. Phys., 65, 181-188 (1979).
7. G. Hagedorn, Asymptotic Completeness for Classes of Two, Three and Four Particle Schrodinger Operators, Trans. Amer. Math. Soc., 258, 1-75 (1980).
8. K. Hepp, On the Quantum Mechanical N-Body Problem. Helv. Phys., Acta., 42, 425-458 (1969).
9. J.S. Howland, Abstract Stationary Theory of Multichannel Scattering, J. Funct. Anal. 22 (1976) 250-282.
10. W. Hunziker, Time-Dependent Scattering Theory for Singular Potentials, Helv. Phys. Acta, 1967, 40 (1967), 1052-1062.

11. R.J. Iorio and M. O'Carroll, Asymptotic Completeness for Multiparticle Schrodinger Operators with Weak Potentials, Comm. Math. Phys. 27, 137-145 (1972).
12. T. Kato, Wave Operators and Similarity for Some Nonself-adjoint Operators, Math. Annalen 162, 258-279 (1966).
13. T. Kato, Smooth Operators and Commutators, Studio Mathematica XXXI, 535-546 (1968).
14. T. Kato, Two-Space Scattering Theory with Applications to Many-Body Problems, J. Fac. Sci. Univ. Tokyo, see IA, 24, 503-514 (1977).
15. T. Kato and S.T. Kuroda, The Abstract Theory of Scattering, Rocky Mountain J. Math. 1, 127-171 (1971).
16. M. Reed and B. Simon, Methods of Modern Mathematical Physics IV, Academic Press, 1978.
17. I.M. Sigal, On the Discrete Spectrum of the Schrodinger Operators of Multiparticle Systems, Commun. Math. Phys. 48, 137-154 (1976).
18. I.M. Sigal, Mathematical Foundations of Quantum Scattering Theory for Multiparticle Systems, a Memoir of AMS, n209 (1978).
19. I.M. Sigal, On Quantum Mechanics of Many-Body Systems with Dilation-Analytic Potentials, Bull AMS 84, 152-154 (1978).
20. I.M. Sigal, Scattering Theory for Multiparticle Systems I, II, Preprint ETH-Zurich (1977-1978) (an expanded version will appear in the Springer Lecture Notes in Mathematics).
21. I.M. Sigal, Mathematical Theory of Single-Channel Systems. Analyticity of Scattering Matrix, Princeton University, Preprint.
22. B. Simon, The Theory of Resonances for Dilation-Analytic Potentials and the Foundations of Time Dependent Perturbation Theory, Ann. Math. 97, 274-274 (1973).
23. E. Stein and G. Weiss, Introduction to Fourier Analysis on Euclidean Spaces, Princeton University Press, 1971.
24. K. Yajima, An Abstract Stationary Approach to Three-Body Scattering, J. Fac. Sci. Univ. Tokyo, Ser. IA, 25, 109-132 (1978).

EXISTENCE, UNIQUENESS, STABILITY AND CALCULATION OF THE STATIONARY REGIMES IN SOME NONLINEAR SYSTEMS

A.G. Ramm

University of Michigan
Department of Mathematics
Ann Arbor, MI 48109

1. INTRODUCTION

Consider a passive one-loop system consisting of a linear two-port L, a nonlinear two-port N and an electromotive force $e(t)$ which are sequentially joined. Let i be the current in the system, then $u_L = Zi$ is the voltage on L, $i = F(u)$ where $u = u_N$ is the voltage on N, Z is a linear operator, F is a nonlinear operator. The governing equation is

$$u + ZFu = e, \tag{1}$$

or

$$Au + Fu = J, \quad A = Z^{-1}, \quad J = Ae. \tag{2}$$

We assume that Z is invertible.(This is always the case in network theory: Z is the impedance of L and A is the admittance of L). The following questions will be discussed: 1) Suppose that $e(t) = e(t + T)$. Whether eq.(2) has a periodic solution (periodic here and below means T-periodic), whether this solution is unique? This solution is called a stationary regime in the system. 2) Whether the solution is stable under small (in some norm) periodic perturbations of $e(t)$? Whether it is stable with respect to arbitrary initial data on L (stability in the large)? 3) How to calculate the stationary regime? 4) Whether the system (the network) is convergent? The network is called convergent if for any initial data on L the transient regime in the network goes to the stationary regime as $t \to +\infty$. 5) The same questions in the case when $e(t)$ is almost periodic. 6) The same questions in the case when $e(t)$ is just bounded and measurable (e.g. a uniformly bounded sequence of random impulses).

Integral equation (1) can be written as

$$u(t) + \int_{-\infty}^{t} g(t,\tau)f(\tau, u(\tau))d\tau = e(t), \tag{3}$$

if the linear two-port is described as the linear operator with the kernel $g(t,\tau)$, $g(t,\tau) = 0$ if $t < \tau$ (causality principle) and the nonlinear two-port is described by the equality $Fu = f(t, u(t))$. The transient regime is described by the equation

$$u + \int_{0}^{t} g(t,\tau)f(\tau, u(\tau))d\tau = e(t) + m(t), \tag{4}$$

where $m(t)$ is the reaction of L on the initial data. The stationary regime is the solution U of equation (3), which is uniformly bounded on $I = (-\infty,\infty)$. The network is convergent if for more or less arbitrary $m(t)$ we have $|U(t) - u(t)| \to 0$ as $t \to +\infty$, where $u(t)$ is the solution of (4). For periodic solutions instead of equation (3) on non-compact domain we use the equation

$$u(t) + \int_{0}^{T} \Phi(t,\tau)\, f(\tau, u(\tau))d\tau = e(t) \tag{5}$$

with

$$\Phi(t,\tau) = T^{-1} \sum_{m=-\infty}^{\infty} \exp\{im\,\omega(t-\tau)\}\, Z(im\,\omega;\, t), \tag{6}$$

where

$$Z(im\,\omega;\, t + T) = Z(im\,\omega;\, t) = \int_{0}^{\infty} g(t, t-s)\exp(-i\lambda s)ds. \tag{7}$$

2. MAIN ASSUMPTIONS

Our physical assumption is the following (a natural one if we want uniqueness of the stationary regime): the two-port L - N is passive. Mathematical assumptions are essentially equivalent to this physical assumption and in addition we assume that the nonlinearity satisfies the inequality (*) $|Fu| \leq \varepsilon |u| + c(\varepsilon)$ for any $\varepsilon > 0$. In electronics many of the volt-amper characteristics of the electronic lamps and transistors are just bounded at infinity, so that (*) is valid. In the last section we consider the problem without restrictions on the growth of the nonlinearity at infinity, but this problem will be of interest mostly in quantum field theory and not in network theory.

The literature of the subject is vast (see [1]-[5] and references in these books). The new features of our results are: 1) the linear two-port is not time invariant (i.e. $g(t,\tau)$ is not a difference kernel); 2) no restrictions on the slope of the nonlinearity are made (i.e. $f'(u)$ can be arbitrarily large); 3) neither L nor N are assumed passive, only N - L is assumed passive, 4) an iterative process to calculate the stationary regime is given; 5) the degree arguments are not applicable since when $e(t)$ is almost periodic the operator in (3) is noncompact; 6) we treat the case when (2) is a semilinear elliptic equation on a non-compact domain, A is a positive definite elliptic operator of the second order (e.g. $A = -\Delta + 1$) and $f(u) \in C_{loc}$, $uf(u) \geq 0$ for $|u| \geq R$, where R is an arbitrarily large fixed number. In this case, of course, there is no uniqueness and we prove existence of solutions in L^∞ for $J \in L^\infty$. A new phenomenon here is that there is an example which shows that if $J \in L^\infty \cap L^2$ the solution $u \notin L^2$: in fact $u(\infty) = \text{const} \neq 0$ for some $J \in L^\infty \cap L^2$. This is impossible if $uf(u) \geq 0$ for all u (and not only near infinity as in our case).

3. MAIN RESULTS

We omit theorems and proofs and give only the references [6]-[13] where all of the above mentioned questions are answered (and some additional questions as well). In [13] there is an example which shows that the results obtained are in some sense final in passive network theory (if in one of the conditions we substitute $>$ by $\geq$ there will be no periodic regime in the network).

4. PRACTICAL PROBLEMS.

We consider (as an example of the application of this theory) stability and calculation of the stationary regime in a real scheme of a multicascade amplifier with a nonlinear final cascade and with a linear feedback. This is of course not a one-loop system but it can be reduced to such a system by applying the Thevenine theorem.

5. METHODS.

To answer questions 1)-3), 5) we prove some abstract theorems about equation (2) in a Hilbert space(or reflexive Banach space), where A is a closed, densely defined linear operator and F is a bounded hemicontinuous nonlinear operator. The main new assumption on A is the following: there exists a sequence of bounded linear operators A_n such that $A_n u \to Au, \forall u \in D(A)$ and $A_n^* u \to A^* u$, $\forall u \in D(A^*)$. Methods of proofs are quite simple. To answer questions 4), 6) we use function-theoretic arguments, which are also simple. Abstract assumptions on the operators A and F are interpreted in terms of network theory. They are easy to apply to verify.

REFERENCES

[1] J. Aggarval, M. Vidyasagar, Nonlinear systems. Dowden, Pennsylvania, 1977.

[2] J. Hsu, A. Meyer, Modern control principles and applications, McGraw-Hill, N.Y. 1968.

[3] J. Hale, Oscillations in nonlinear systems, McGraw Hill, N.Y., 1963.

[4] N. Bogoljubov, Tu. Mitropolsky, Asymptotic methods in non-linear oscialIations theory, Gordon and Breach, N.Y.,1967.

[5] J. Lions, Quelques methodes de resolution des problems aux limites nonlineares, Dunod, Paris, 1959.

[6] A.G. Ramm, Existence of periodic solutions to some nonlinear problems, Diff. eq. 13, (1977), 1186-1191.

[7] A.G. Ramm, Stability of control systems, ibid., 14, (1978), 1188-1193

[8] A.G. Ramm, An iterative process for calculation of periodic and almost periodic oscillations in some nonlinear systems, Rad. Engr. Electr. Phys., 21, (1976), 137-140; 24, (1979), 190-191.

[9] A.G. Ramm, Investigation of some classes of integral equations and their applications in collection "Abel inversion

and its generalizations", ed. N. Preobrazhensky, Sib. dep. of Acad. Sci. USSR, Novosibirsk, 1978, 120-179.

[10] A.G. Ramm, Existence uniqueness and stability of solutions to some nonlinear problems, Proc. Intern. Congr. on appl. math. in engineering, Weimar, 1978, 345-351.

[11] A.G. Ramm, Existence uniqueness and stability of periodic regimes in nonlinear networks, Proc. 3rd intern. symp. on network theory, Split, 1975, 623-628.

[12] A.G. Ramm, Theory and applications of some classes of integral equations. Springer, N.Y., 1980.

[13] A.G. Ramm, Stationary regimes in passive nonlinear networks, in "Nonlinear electromagnetics" ed. P.L.E. Uslenghi, Acad. Press, N.Y. ,(1980).

Supported by AFOSR 800204.

WEIGHTED TRIGONOMETRIC APPROXIMATIONS IN $\mathcal{L}_2(\mathbb{R}^n)$

R.K. Goodrich and K. Gustafson

Dept. of Mathematics, University of Colorado

Boulder, Colorado 80309

Abstract. One may describe the general problem as that of finding the best approximation to an arbitrary function from the closed subspace generated by functions of the form $e^{i(v,w)}f(w)$, where f is a fixed function and where the arguments v are taken to be from a given set S. By use of functional analytic and group theoretic methods we are able to treat a class of weighted trigonometric approximation problems for $\mathcal{L}_2(\mathbb{R}^n)$ with S either a half-space or a quadrant. In so doing we also arrive at a beginning of a new theory for inner and outer functions for $\mathcal{L}(\mathbb{R}^n)$, $n \geqq 2$. As a biproduct we obtain an abstract characterization of the regular representation of $\mathbb{R}^n$, answering a question of Chatterji (Springer Lec. Notes in Math. 645, 1978).

1. INTRODUCTION.

Let R_v be the regular representation of $\mathcal{L}_2(\mathbb{R}^n)$, i.e., $R_v(f)(w) = f(w - v)$ for v,w in $\mathbb{R}^n$ and f in $\mathcal{L}_2(\mathbb{R}^n)$. Then given a fixed cyclic vector φ in $\mathcal{L}_2(\mathbb{R}^n)$ we would like to find the best $\mathcal{L}_2(\mathbb{R}^n)$ approximation to an arbitrary function f from $\overline{\operatorname{span}}\{R_v(\varphi) \mid v \in S\}$, i.e., the closed linear subspace generated by the functions $\{R_v(\varphi) \mid v \in S\}$ where S is a fixed subset of $\mathbb{R}^n$. That is, we would like to find the projection of $\mathcal{L}_2(\mathbb{R}^n)$ onto the closed subspace $\overline{\operatorname{span}}\{R_v(\varphi) \mid v \in S\}$.

Let us illustrate with a simple example.

Example 1.1. Let $\varphi = \chi_{[-1,0]}$, the characteristic function of $[-1,0]$.

If we let $\mathcal{H}_s = \overline{\text{span}}\{R_t(\varphi) \mid t \leq s\}$, then one can show that $\mathcal{H}_s = \mathcal{L}_2(-\infty,s]$, i.e., $\mathcal{H}_s$ is the set of all $\mathcal{L}_2(\mathbb{R})$ functions with support in $[-\infty,s]$. Then the approximation problem is easy to solve; in fact the projection of $\mathcal{L}_2(\mathbb{R}^n)$ onto $\mathcal{H}_s$ is simply given by multiplication by $\chi_{(-\infty,s]}$.

Using Example 1.1 as a guide we may consider a slightly more general example.

Example 1.2. Let φ be any cyclic vector for the regular representation, i.e., $\overline{\text{span}}\{R_t(\varphi) \mid t \in \mathbb{R}\} = \mathcal{L}_2(\mathbb{R})$, and suppose further that

$$\overline{\text{span}}\{R_t(\varphi) \mid t \leq s\} = \mathcal{L}_2(-\infty,s] \quad . \tag{1.1}$$

As we have seen above the approximation problem is easy to solve for any such function for the case in which the set S is $(-\infty,s]$

If we take the Fourier transform then we obtain an equivalent approximation problem, i.e., for the case of one dimension, find the projection of $\mathcal{L}_2(\mathbb{R})$ onto $\overline{\text{span}}\{e^{itx}\hat{\varphi}(x) \mid t \leq s\}$, where $\hat{\varphi}$ denotes the Fourier transform. This is a weighted trigonometric approximation problem. This type of problem goes far back, for example, see the papers of Beurling [1, 2] and the books of Helson [3] and Dym and McKean [4] .

Note that if φ satisfies (1.1) then this weighted trigonometric approximation problem can be solved by first taking the inverse Fourier transform, multiplying by $\chi_{(-\infty,s]}$ and then taking the Fourier transform.

These examples may be used to illustrate and approach a more general problem. Let $\{U_t\}$ be a continuous unitary representation of the real line on a Hilbert space $\mathcal{H}$. We will assume $\{U_t\}$ has a cyclic vector φ . Let $\mathcal{H}_s = \overline{\text{span}}\{U_t(\varphi) \mid t \leq s\}$. The process $t \to U_t(\varphi)$ is called regular provided

$$\bigcap_s \mathcal{H}_s = \{0\} \quad . \tag{1.2}$$

Condition (1.2) is sometimes described as the emptyness of the infinitely remote past.

An example of such a process is of course given by Example 1.2. A basic result, given in Gustafson and Misra [5], is that any regular process is unitarily equivalent to Example 1.2, i.e., there is a unitary equivalence that takes U_t to R_t, the regular representation, and the cyclic vector for U_t is mapped to a function that satisfies condition (1.1).

Let us apply this last result to a special case. Suppose φ is a function in $\mathcal{L}_2(-\infty,s]$ such that $\hat{\varphi} \neq 0$ almost everywhere. (The condition that $\hat{\varphi} \neq 0$ almost everywhere is a necessary and sufficient condition to assure φ is a cyclic vector for R_t.) Then obviously $\mathcal{H}_s \subseteq \mathcal{L}_2(-\infty,s]$, and the process $t \to R_t(\varphi)$ is regular. By the above mentioned theorem of Gustafson and Misra [5] we know there exists a unitary equivalence taking R_t to R_t and φ into a function satisfying (1.1).

Since the just mentioned unitary equivalence takes R_t to R_t it must be of a very special form. In fact it is generated by an $\mathcal{L}_\infty(\mathbb{R})$ function in a sense made more precise by the next corollary.

<u>Corollary 1.1</u>. If ψ is any $\mathcal{L}_2(\mathbb{R})$ function such that $\hat{\psi} \neq 0$ almost everywhere and the support of ψ is contained in $(-\infty,0]$ then there exists a measurable function g such that $|g| = 1$ almost everywhere and such that $\varphi = {}^\vee(g\hat{\psi})$ is a cyclic vector for the regular representation satisfying condition (1.1).

Here ${}^\vee f$ denotes the inverse Fourier transform of a function f.

The actual solution to the approximation problem for such functions ψ is a bit complicated and requires the calculation of the $\mathcal{L}_\infty(\mathbb{R})$ function g mentioned in the corollary. This calculation involves the theory of inner and outer functions as found in [4]. From this point of view g^{-1} is the boundary value of an inner function and $h = g\hat{\psi}$ is the boundary value of an outer function. Thus $\hat{\psi} = g^{-1}h$ is factored into the product of an inner and outer function, and these factors are unique up to a scalar constant. Note that a function h is an outer function if and only if ${}^\vee h$ satisfies condition (1.1).

On the real line the weighted trigonometric approximation problem can be solved for functions generating regular processes. That is, functions generating regular processes can be decomposed into inner and outer functions and this allows us to "solve" the weighted trigonometric approximation problem for such functions.

This confluence of ideas breaks down in higher dimensions $(n \geq 2)$. The reason for this is that the theory of several complex variables in higher dimensions does not contain a theory of inner and outer functions that is sufficiently comprehensive for the approximation problem. This situation is related to the complicated structure of sets of zeros of analytic functions in several variables, see Zygmund [6, Chapter XVII, p. 316].

By use of functional analytic and group theoretic methods we are nonetheless able to treat this weighted trigonometric approximation problem for $\mathcal{L}_2(\mathbb{R}^n)$ with S either a half-space or a quadrant. We analyze these problems via several-parameter regular processes, without appeal to several complex variables. The method we employ makes strong use of cyclic vectors and is similar to that of Gustafson and Misra [5].

As a biproduct we obtain an abstract characterization of the regular representation. This answers a question of Chatterji [7], at least for $\mathbb{R}^n$.

For aspects of the prediction problem in the case of compact abelian groups with ordered dual, see Helson and Lowdenslager [8,9], and the book of Rudin [10]. For an interesting recent connection between the compact case and several complex variables see Rubel [11] where "external" and "internal" functions are studied.

2. TWO DIMENSIONAL REGULAR PROCESSES.

For simplicity we will state all of our results for $\mathbb{R}^2$. Analogous results hold in every case for $\mathbb{R}^n$ under suitable modifications. For the proofs of these results see Goodrich and Gustafson [12, 13].

Let $U_{(x,y)}$ be a continuous unitary representation of $\mathbb{R}^2$ defined on a Hilbert space $\mathcal{H}$ with cyclic vector φ. Let $E_{(s,t)}$, E_s and F_t be the projections of $\mathcal{H}$ onto respectively $\overline{\text{span}}\{U_{(x,y)}(\varphi) \mid x \leq s, y \leq t\}$; $\overline{\text{span}}\{U_{(x,y)}(\varphi) \mid x \leq s, -\infty < y < \infty\}$; $\overline{\text{span}}\{U_{(x,y)}(\varphi) \mid y \leq t, -\infty < x < \infty\}$. Denote the range of any projection P by $\mathcal{R}(P)$.

Definition 2.1. Let $U_{(x,y)}$ be a continuous unitary representation of $\mathbb{R}^2$ on a Hilbert space $\mathcal{H}$ with cyclic vector φ. We say $v \to U_v(\varphi)$ is a regular process if

$$\bigcap_s \mathcal{R}(E_s) = \{0\} = \bigcap_t \mathcal{R}(F_t) \tag{2.1}$$

and

$$E_s F_t = E_{(s,t)} \ . \tag{2.2}$$

We remark that there are many possible definitions for regularity here. See for example Chatterji [7] where $v \to U_v(\varphi)$ is defined to be regular provided the U is unitarily equivalent to the regular representation. Chatterji then poses the question of characterizing such functions. We shall say more about this question in the next section.

We mention here that our main motiviation for condition (2.2) is that assuming this condition allows us to show our projections $E_{(s,t)}$ are unitarily equivalent to the projections given by multiplication by $\chi_{(-\infty,s]\times(-\infty,t]}$. Thus if one could compute this unitary equivalence one could find the best $\mathcal{L}_2(\mathbb{R}^2)$ approximation to any function in $\mathcal{L}_2(\mathbb{R}^2)$ by functions in $\overline{\text{span}}\{U_v(\varphi) \mid v = (x,y) \ , \ x \le s \ , \ y \le t\}$, where φ is a given cyclic vector generating a regular process. Then, just as in the one dimensional case, one could find the projection of $\mathcal{L}_2(\mathbb{R}^2)$ onto $\overline{\text{span}}\{e^{i(v,w)}\psi(w) \mid v = (x,y) \ , \ x \le s \ , \ y \le t\}$ where ψ is a given vector generating a regular process. This is of course the weighted trigonometric approximation problem mentioned earlier for S equal to a quadrant.

Motivated by our earlier discussion of the one-dimensional case we define outer functions for $n = 2$.

Definition 2.2. A function $\psi \in \mathcal{L}_2(\mathbb{R}^2)$ is an outer function if $\overline{\text{span}}\{v \psi(v - w) \mid w = (x,y) \ , \ x \le 0 \ , \ y \le 0\} = \mathcal{L}_2(-\infty,0] \times (-\infty,0]$.

Let us now give an abstract characterization of regular processes.

Theorem 2.1. Let $v \to U_v(\varphi)$ be a regular process on $\mathcal{H}$ with cyclic vector φ . Then there exists a unitary mapping V of $\mathcal{H}$ onto $\mathcal{L}_2(\mathbb{R}^2)$ such that $VU_vV^{-1} = R_v$ is the regular representation and such that $\hat{\psi}$ is an outer function, where $\psi = V(\varphi)$.

The idea of the proof of Theorem 2.1 is to use the Stone-von Neumann theorem or the imprimitivity theorem, see Mackey [14, pgs. 174-181] . The details are similar to those found in [5] .

We may now obtain the existence and uniqueness of outer functions.

Corollary 2.1. Let ψ be any cyclic vector for the regular representation such that $v \to R_v(\varphi)$ is a regular process. Then there exists a g such that $|g| = 1$ almost everywhere, and such that $g\hat{\psi}$ is an outer function. The function g is unique up to a scalar multiple of absolute value one.

Corollary 2.1 may be viewed as a factorization theorem, i.e., if $v \to R_v(\psi)$ is a regular process then $\hat{\psi} = g^{-1}\psi$, where $|g^{-1}| = 1$ a.e. and ψ_1 is an outer function. This factorization is unique up to scalar constants of absolute value one.

It would be useful to have a complex variable integral formula for the outer factor as we do on the real line, see [4] .

We next give a less stringent condition than (2.2) .

Definition 2.3. Let $U_{(x,y)}$ be a continuous unitary representation with cyclic vector φ . We say $v \to U_v(\varphi)$ is a weak regular process provided condition (2.1) is true and

$$E_s F_t = F_t E_s \quad . \tag{2.3}$$

Definition 2.4. A function $\psi \in \mathcal{L}_2(\mathbb{R}^2)$ is a weak outer function if

$$\overline{\text{span}}\{\ \psi(v-w)\ |\ w = (x,y)\ ,\ x \le 0\ ,\ -\infty < y < \infty\} = \mathcal{L}_2(-\infty,0] \times (-\infty,\infty),$$

and

$$\overline{\text{span}}\{\ \psi(v-w)\ |\ w = (x,y)\ ,\ -\infty < x < \infty\ ,\ y \le 0\} = \mathcal{L}_2(-\infty,\infty) \times (-\infty,0)\ .$$

Results analogous to Theorem 2.1 and Corollary 2.1 hold for weak regular processes and weak outer functions. See [12] and [13] .

An example of a weak regular process that is not a regular process is given by $v \to R_v(\varphi)$ where $\varphi = \chi_A$, where A is the tilted (45 degree) square in the third quadrant with two vertices at $(-1,0)$, $(0,-1)$ Also χ_A is an example of a weak outer function that is not an outer function.

Weak regular processes allow us to solve the approximation problems mentioned in the introduction for half-planes, see [12, 13]

3. THE REGULAR REPRESENTATION.

It would be of interest to know in what generality condition (2.3) holds. If (2.3) always held then for any function φ with support in the third quadrant and $\hat{\varphi} \neq 0$ a.e. we would have $\hat{\varphi} = gh$ where $|g| = 1$ a.e. and where h is a weak outer function. In any case even if condition (2.3) isn't always true (and we don't believe it is) we are able to give an abstract characterization of the regular representation using condition (2.1) alone.

Theorem 3.1. A continuous unitary representation $U_{(x,y)}$ of $\mathbb{R}^2$ on a nontrivial Hilbert space $\mathcal{H}$ is unitarily equivalent to the regular representation if and only if there exists a cyclic vector φ for U such that

$$\bigcap_s \mathcal{R}(E_s) = \{0\} = \bigcap_t \mathcal{R}(F_t) \quad , \quad \text{i.e.,}$$

if and only if condition (2.1) holds.

This theorem also gives an answer to Chatterji's [7] question in the case of $\mathbb{R}^n$.

References.

[1] Beurling, A., On the spectral synthesis of bounded functions, Acta Math., 81 (1949), 225-238.
[2] Beurling, A., On two problems concerning linear transformations in Hilbert space, Acta Math., 81 (1949), 239-255.
[3] Helson, H., Lectures in Invariant Subspaces, Academic Press, New York (1967).
[4] Dym, H., McKean, H.P., Gaussian Processes, Function Theory, and the Inverse Spectral Problem, Academic Press, New York (1964).
[5] Gustafson, K., Misra, B., Canonical commutation relations of quantum mechanics and stochastic regularity, Letters in Math Physics, 1 (1976).

[6] Zygmund, A., Trigonometric Series, Cambridge Press, Cambridge (1968).

[7] Chatterji, S.D., Stochastic processes and commutation relationships, Proceedings of the Dublin Conference on Vector Space Measures and Applications, (1977), II, Springer Lec. Notes in Math. 645, Berlin (1978), 16-26.

[8] Helson, H., Lowdenslager, D., Prediction theory and Fourier series in several variables, Acta Math., 99 (1958), 165-202.

[9] Helson, H., Lowdenslager, D., Prediction theory and Fourier series in several variables II, Acta Math., 106 (1961), 176-213.

[10] Rudin, W., Fourier Analysis on Groups, Interscience, New York (1962).

[11] Rubel, L., Internal and external analytic functions of several complex variables, to appear.

[12] Goodrich, R.K., Gustafson, K., Weighted trigonometric approximation and inner and outer functions on higher dimensional Euclidean spaces, to appear.

[13] Goodrich, R.K., Gustafson, K., Regular representation and approximation, Proc. International Conference on Functions, Series, Operators, Colloquia Mathematica Societatis János Bolyai, North Holland Publishing Company (1980), to appear.

[14] Mackey, G.W., The Theory of Unitary Group Representations, The University of Chicago Press, Chicago (1976)

THE PHENOMENON OF MAGNETIC PAIRING AND EXACTLY SOLUBLE MODELS OF MAGNETIC RESONANCES

A. O. Barut

Department of Physics
University of Colorado
Boulder, Colorado 80309

ABSTRACT

The quantum mechanical charge-dipole and dipole-dipole interactions between elementary particles are considered. The incorrectness and difficulties of the standard perturbative treatment is shown. The nonperturbative treatment not only resolves these difficulties, but shows the existence of new type of resonance states which have far-reaching applications in nuclear and particle physics.

1. INTRODUCTION

Two opposite electric charges attract each other and can form a bound state. We shall refer to this as electric pairing, or the Coulomb-bound state regime. Two magnetic dipoles can either repel or attract each other depending on the total spin state and the sign of the magnetic moments. This problem alone is not well-defined in quantum mechanics due to the attractive $1/r^3$ - potential. However, if in addition one or both of the dipoles have an electric charge there is a charge-dipole interaction, and the sum of both the spin-orbit and the spin-spin interactions is, as we shall see, well-defined,. and the two-dipoles can form, in certain states, new resonance states, which we call the magnetic pairing, a new regime of dipole (quasi -) bound states, distinct from the Coulomb regime with its characteristic energy and size parameters.

The physical and mathematical significance of Coulomb bound states in quantum theory has been known and studied for over fifty years. Equally important is the dual case of magnetic bound states. This comes at first as a surprise, but the far-reaching significance of magnetic pairing is becoming to be recognized, albeit overlooked for a long time. In this work we elaborate the mathematical problems associated with the Hamiltonian of magnetically interacting particles, with emphasis on exactly soluble Hamiltonians, and indicate also the physical significance of the results.

2. NONRELATIVISTIC HAMILTONIANS WITH ELECTRIC AND MAGNETIC INTERACTIONS.

We consider a system of two particles each carrying both an electric charge and a magnetic dipole moment. The vector potential $\vec{A}$ produced by a dipole moment μ is given by

$$\vec{A} = \mu \frac{\vec{\sigma} \times \vec{r}}{r^3}$$

where $\vec{\sigma}$ are the Pauli matrices. Consequently the Hamiltonian of the system can be written as

$$H = \frac{1}{2m_1}\left[\vec{p}_1 - e_1\mu_2 \frac{\vec{\sigma}_2 \times (\vec{r}_1 - \vec{r}_2)}{|\vec{r}_1 - \vec{r}_2|^3}\right]^2 + \frac{1}{2m_2}\left[\vec{p}_2 - e_2\mu_1 \frac{\vec{\sigma}_1 \times (\vec{r}_2 - \vec{r}_1)}{|\vec{r}_2 - \vec{r}_1|^3}\right]^2$$

$$+ \frac{e_1 e_2}{|\vec{r}_1 - \vec{r}_2|} + \mu_1\mu_2 \frac{S_{12}}{|\vec{r}_1 - \vec{r}_2|^3} , \tag{1}$$

where

$$S_{12} = \frac{3\vec{\sigma}_1 \cdot (\vec{r}_1 - \vec{r}_2)\, \vec{\sigma}_2 \cdot (\vec{r}_1 - \vec{r}_2)}{|\vec{r}_1 - \vec{r}_2|^2} - \vec{\sigma}_1 \cdot \vec{\sigma}_2 \tag{2}$$

is the so-called spin-spin tensor operator. The last term in (1) represents the spin-spin interaction, the energy of a dipole in the magnetic field of the other dipole. This Hamiltonian can easily be generalized to N particles.

The system is translationally invariant, hence we can separate the center of mass variables. Introducing

$$\vec{P} = \vec{p}_1 + \vec{p}_2 \quad , \quad \vec{p} = \frac{m_2}{M}\vec{p}_1 - \frac{m_1}{M}\vec{p}_2 \quad , \quad \text{and } \vec{R} = \frac{m_1}{M}\vec{r}_1 + \frac{m_2}{M}\vec{r}_2 \quad ,$$

$$\vec{r} = \vec{r}_1 - \vec{r}_2 \quad , \quad M = m_1 + m_2$$

we obtain after some calculations, with $\frac{1}{\mu} \equiv \frac{1}{m_1} + \frac{1}{m_2}$,

$$H = \frac{1}{2M}\underline{P}^2 + \frac{1}{M}\underline{P}\cdot\left(e_2\mu_1\vec{\sigma}_1 - e_1\mu_2\vec{\sigma}_2\right)\times\frac{\vec{r}}{r^3} + \frac{1}{2\mu}p^2 - \vec{p}\cdot\left(\frac{e_1\mu_2}{m_1}\vec{\sigma}_2\right.$$

$$\left. + \frac{e_1\mu_2}{m_2}\vec{\sigma}_1\right)\times\frac{\vec{r}}{r^3} + \left[\frac{e_1^2\mu_2^2}{m_1} + \frac{e_2^2\mu_1^2}{m_2}\right]\frac{1}{r^4} + \frac{e_1 e_2}{r} + \mu_1\mu_2\frac{S_{12}}{r^3} . \tag{3}$$

In the center of mass frame, P = 0, the reduced Hamiltonian becomes with

$$\mu_i = \frac{e_i}{4m_i} g_i \quad ,$$

where g is the so-called Landé magnetic moment factor,

$$H_{rel} = \frac{1}{2\mu} p^2 - \lambda \vec{p} \cdot \vec{\Sigma} \times \frac{\vec{r}}{r^3} + \lambda^2 g^2 \frac{1}{r^4} + \frac{e_1 e_2}{r} + \lambda^2 m_1 m_2 \frac{S_{12}}{r^3}$$

$$\lambda = \frac{e_1 e_2}{4m_1 m_2} \quad , \quad g^2 = m_1 g_2^2 + m_2 g_1^2 \quad , \quad \vec{\Sigma} = g_1 \vec{\sigma}_1 + g_2 \vec{\sigma}_2 \ .$$

Note, however, that for $\underline{P} \neq 0$, there is a non-trivial term proportional to $\underline{P}$ and $\vec{r}/r^3$, unlike the Coulomb Case.

We shall now consider the simplest case of spin-orbit forces when we neglect the magnetic moment of particle 1 (i. e. $\mu_1 = 0$) and the Coulomb force. Then

$$H_{rel} = \frac{1}{2\mu} p^2 - \frac{e_1 e_2}{4m_1 m_2} g_2 \vec{p} \cdot \vec{\sigma}_2 \times \frac{\vec{r}}{r^3} + \frac{e_1^2 e_2^2}{16 m_1 m_2} \frac{g_2^2}{m_2} \frac{1}{r^4} + \frac{e_1 e_2}{r} \ . \quad (4)$$

The second term is of the form $\vec{S} \cdot \vec{L}/r^3$, $\vec{S} = \frac{1}{2}\vec{\sigma}_2$, $\vec{L} = \vec{r} \times \vec{p}$ hence the name "spin-orbit". The radial and angular parts of this Hamiltonian can be separated by introducing the total angular momentum J = L + S, and by expanding the wave function into spherical harmonics

$$\Psi(\vec{r}) = \sum_{J,L} \frac{1}{r} U_{JL} Y^m_{LSJ} \ .$$

We are interested in the radial equation which is then of the form

$$\left\{ - \frac{1}{2m} \frac{1}{r^2} \frac{\partial}{\partial r} \left(r^2 \frac{\partial}{\partial r} \right) + \frac{\ell(\ell + 1)}{2mr^2} + \frac{e_1 e_2}{r} \right.$$

$$\left. - \frac{e_1 \mu_2}{2m} \frac{[J(J + 1) - \ell(\ell + 1) - S(S + 1)]}{r^3} + \frac{e_1^2 \mu_2^2}{4m^2} \frac{1}{r^4} - E \right\} \frac{U_{J\ell}}{r} = 0,$$

or, finally, the Sturm-Liouville equation

$$\left\{ - \frac{d^2}{dr^2} + V(r, J, \ell, S) - k^2 \right\} U_{J\ell}(r) = 0 \quad (5)$$

with

$$k^2 = 2mE, \ V = \frac{\ell(\ell + 1)}{r^2} + \frac{e_1 e_2}{r} + \frac{v_3}{r^3} + \frac{v_4}{r^4} \ .$$

In this form the Hamiltonian is well-known in nuclear and atomic physics. However, when we come to the solutions of the equation (5), the universal procedure has been perturbation theory[1]. The Coulomb problem (or, in nuclear and particle physics, some spin-independent potential V(r)) is first exactly solved, and then the remaining terms are treated as perturbations to this solutions. But the term $1/r^4$ is too singular at the origin and cannot be treated as a perturbation, so this term is dropped altogether, the reason given is that its coefficient is too small. When this term is dropped, the first order perturbation integrals of the term $1/r^3$ barely converge, but nothing is said about higher order terms. On the other hand with $1/r^4$ - term dropped, the remaining Hamiltonian is, as is well known, not well-defined and not essentially self-adjoint. Fortunately, in atomic physics, the rigorous considerations will a posteriori, justify, as we shall see, the smallness of the spin effects relative to the Coulomb force. However, in nuclear and particle physics, the perturbative treatment outlined above is completely inadequate and more over does not reveal the remarkable new phenomena associated with the equation (5). All these difficulties can be avoided if one solves the problem nonperturbatively taking into account both the $1/r^3$ - and $1/r^4$ - terms from the beginning. Now since the $1/r^4$ - term is always repulsive and dominates as $r \to 0$, the wave function has a well-defined behavior at the origin. The Hamiltonian is in fact essentially self-adjoint. This is an instructive example, where a term, although its coefficient is very small, is nevertheless essential for the proper definition of the problem; both physically and mathematically, it describes a dominant effect at short distances and cannot be neglected. In addition to solving the difficulties mentioned above, the nonperturbative treatment will show a new phenomenon: the complete electromagnetic two-body problem (including magnetic effects) contains besides the well-known Coulomb bound states, new narrow high energy quasi-bound or resonance states whose wave functions are localized at much shorter distances (i. e. at magnetic radius α/m), then the Bohr radius $(1/\alpha m)$ of Coulomb orbits.

It is interesting to note that these magnetic resonance states also exist in classical and semiclassical theories of magnetic forces. For example, trapped charged particles in the dipole magnetic field of the earth (Störmer's problem)[2]. It is not a quantum effect, but an effect of attractive magnetic forces. The Bohr-Sommerfeld semi-classical quantization also leads to discrete set of magnetic pairing states.

3. EXACTLY SOLUBLE MODEL

In this section we show that complex eigenvalue problems (corresponding to resonances) can be formulated and solved in a similar fashion as the real eigenvalue problem (corresponding to bound states). This method provides than a precise time-independent description of resonances in quantum theory.

Consider the Sturm-Liouville equation

$$\left[\frac{d}{dr^2} + \lambda^2 - V(\lambda, r) \right] \phi(r) = 0, \qquad r \geq 0. \tag{6}$$

in the space of functions

$$L = \left\{ \phi(r) \mid \phi(0) = 0 \text{ and } \frac{\phi'(r)}{\phi(r)} \xrightarrow[r \to \infty]{} + i\lambda, \ \lambda \varepsilon \mathbb{C} \right\}. \tag{7}$$

This space includes the bound-states as a special case when λ is pure imaginary, $\lambda = i\kappa$. If we write the asympotic form of the solution as

$$\phi(r) \xrightarrow[r \to \infty]{} N \left[e^{-i\lambda r} - S(\lambda) e^{i\lambda r} \right], \tag{8}$$

where $S(\lambda)$ is the S-matrix, then it follows that a solution $\phi_n(r) \varepsilon L$, corresponding to $\lambda + \lambda_n$, is associated with the a pole of the S-matrix. It is well-known that the poles of the S-matrix in the positive imaginary axis in the λ-plane are bound-states, the poles on the negative imaginary axis are the so-called anti-bound states, and a pair of conjugate complex poles in the lower half plane correspond to a resonance.[3]

We now indicate the proof of the following result:[4]

<u>Theorem</u>: The complex eigenvalue problem (6) with

$$V = \frac{v_2}{r^2} + \frac{v_3}{r^3} + \frac{1}{r^4} \tag{9}$$

(the coefficient of r^{-4} can always be normalized to 1 without loss of generality) has a discrete set of (M + 1) in general complex, eigenvalues if $v_3 = -2(M + 1)$, M = 0, 1, 2, 3, ... These eigenvalues are given by the vanishing of the following tridiagonal determinant of order (M + 1):

$$\Delta = \begin{vmatrix} D & 1 & 0 & 0 & 0 & 0 & \cdot \\ -2i\lambda M & D-2M & 4 & 0 & 0 & 0 & \cdot \\ 0 & -2i\lambda(M-1) & D+2(1-2M) & 6 & 0 & 0 & \cdot \\ 0 & 0 & -2i\lambda(M-2) & D+3(2-2M) & \cdot & \cdot & \cdot \\ \cdot & \cdot & \cdot & \cdot & \cdot & \cdot & \cdot \\ \cdot & \cdot & \cdot & \cdot & \cdot & -2i\lambda & D+M(-M-1) \end{vmatrix} = 0$$

where

$$D = M^2 + M + 2i\lambda - v_2 \qquad . \tag{10}$$

The proof consists, as in the case of bound state problems, to separate the behavior of the solution at the singular points, r = 0 and ∞ , and make a polynomial expansion for the remainder.

Let

$$\phi(r) = F(r) \exp g(r) \ ,$$

where

$$g(r) = -\frac{1}{r} + i\lambda r + \nu \ln r \text{ with } 2\nu - 2 = v_3 \ , \qquad (11)$$

and

$$F(r) = \sum_n a_n r^n ;$$

we then obtain for the coefficients a_n the 3-term recursion relations

$$2(n+1)a_{n+1} + \left[D + n(n+2\nu-1) \right] a_n + \left[2i\lambda(n+\nu-1) \right] a_{n-1} = 0$$

where

$$D \neq \nu^2 - \nu + 2i\lambda - v_2 \ .$$

For $\nu = -M$, $M = 0, 1, 2, \ldots$ the power series for $F(r)$ terminates; i. e. we get the best possible asymptotic behavior. The quantization condition, $\Delta = 0$, eq. (10), is just this condition.

As an example, for $M = 1$, Δ in eq. (10) is a 2×2 determinant and we obtain a pair of eigenvalues

$$\lambda = \tfrac{1}{2} \left[-i(v_2 - 2) \pm \sqrt{2(v_2 - 2)} \right].$$

The coefficient v_2 is arbitrary in the whole problem. Thus for $v_2 > 0$ we have a pair of conjugate complex poles with negative imaginary part (i. e. a resonance); for $v_2 = 0$, a zero-eigenvalue solution, and for $v_2 < 0$, one bound state and one anti-bound state.

Discussion

The radial potential (5) has two widely separated minima, separated by a centrifugal barrier, one the Coulomb well, and the other the well due to $1/r^3$ - magnetic interaction. Thus perturbation theory around each well cannot lead to the other well. If the state is localized in the Coulomb minimum, then the magnetic effects are small perturbations. Conversely for the new magnetic resonances the Coulomb effects are small perturbations.

Once the exact solution in the magnetic potential well is found, perturbations around this solution can be performed, in order to take into account the effect of spin-spin interaction.

4. RELATIVISTIC MODELS

The same phenomenon can be studied in relativistic models: The Dirac equation with or without anomalous magnetic moment in the field of a charged dipole, and other relativistic equations including magnetic interaction.[5] In all cases similar radial potentials as eq. (5) result.

In particular, the limit of a Dirac particle with $e \to 0$, $m \to 0$, but $e/m = \kappa =$ const. (i. e. a 4-component neutrino with an anomalous magnetic moment) has an exact zero-energy bound-state solution under spin-orbit potential,[6] which can further become a resonances with the addition of spin-spin terms.

Numerical analysis has confirmed the existence of resonances in all these cases.

5. PHYSICAL APPLICATIONS

Do the magnetic resonances indeed occur in nature? Because this phenomenon has been overlooked, the formation of narrow high energy resonances between stable particles has been given a new name: strong interactions or nuclear forces. In fact, the potentials of the type of eq. (5), have all the properties of nuclear forces: attractive at short distances with a repulsive core at even shorter distances, spin dependence, correct size, energy and lifetime. It has been possible to construct all unstable lepton and hadron states (including nuclei) from the three absolutely stable particles: proton, electron and neutrino (and their antiparticles). This is a dynamical theory based on electromagnetism alone, hence an already unified theory, in which absolute values of the masses of leptons and hadrons can be calculated for the first time.

Pairing of electrons (Cooper pairs) and magnetic resonances also occur in solid state physics. Thus the further mathematical and physical study of this phenomenon is of great significance for physics.

REFERENCES

1. Cf. e. g. A. Messiah, Quantum Mechanics. Vol. 2. p. 541, (North Holland Publ. Co., 1962).
2. C. Størmer, The Polar Aurora, Clarendon Press, Oxford, 1955.
3. A. O. Barut, Resonance Scattering, in Lectures in Theoretical Physics, Vol. IV (Interscience, 1962), edited by W. E. Brittin.
4. A. O. Barut, M. Berrondo, C. Garcia-Calderon, J. Math. Phys. 21, 1851, (1980).
5. A. O. Barut and J. Kraus, J. Math. Phys. 17, 506 (1976) Phys. Rev. D16, 161 (1977); A. O. Barut and R. Rączka, Acta Phys. Polonica B10, 687 - 703 (1979).
6. A. O. Barut, J. Math. Phys. 21, 568 - 571 (1980).
7. A. O. Barut, Surveys in High Energy Physics, Vol. 1, (2), 113 - 140 (1980), and references therein.

DYONS SOLUTIONS TO YANG-MILLS EQUATIONS(*)

M. Schechter

Division of Natural Sciences and Mathematics 2495·
Amsterdam Avenue
New York, N.Y. 10033

R. Weder
Instituto de Investigación en Matemáticas Aplicadas y
en Sistemas. Universidad Nacional Autónoma de México
A.P. 20-726.
México 20, D.F.

Abstract

We prove the existence of dyon solutions to Yang-Mills equations, providing a rigorous model for the particles having electric and magnetic charge conjectured by Schwinger.

An abstract theory is developed to prove existence of stationary points for functionals that cannot be studied with the standard methods of the Calculus of Variations and is applied to the case of Dyons. It is also proved that the solutions are real analytic on R^+ and C^∞ at zero.

(*) Research partially supported by CONACYT under grant PCCBNAL 790025.

During the last years the Yang-Mills equations have received a great deal of attention. However it seems that there is no proof in the literature of the existence of dyon solutions.

In this talk we will consider a proof of the existence of dyon solutions satisfying the Julia-Zee [1] Ansatz for SU(2) gauge groups, and the generalization by Horváth and Palla [2] to SU(N) gauge groups.

The system of radial equations for Dyons ((6) - (8) and (20) - (22)) (22)) are highly singular, non linear, and defined in an unbounded set, and the Euler-Lagrange functional is not bounded below. To study this problem introducing Lagrange multipliers is difficult, because there is always a solution with the opposite sign of the multiplier.

We proved a new theorem [3] that allows to study constrained variational problems without introducing Lagrange multipliers. Our theorem is based in the concept of "natural constraint" introduced by Poincaré many years ago [4]. Since the system ((6) - (8) and (20) - (22)) are singular we also needed to introduce a new notion of derivative.

This kind of situation appears in many areas of physics and our theorem provides a strong tool to deal with them.

Dyons would be particles having both electric and magnetic charge. The existence of these particles has been conjectured by J. Schwinger [5]. He derived a conceivable dynamical interpretation of the subnuclear world on the basis of them. He obtained a picture in which hadronic matter was viewed as a magnetically neutral composite of Dyons, that provided a physical realization for the constituents used in empirical models of hadrons.

Our rigorous proof of the existence of Dyon solutions for SU(2) gauge group with one unit of magnetic charge and of solutions with one to N-1 units of magnetic charge in case of SU(N) gauge group puts Schwinger's conjecture on a firm theoretical basis.

We first introduce a new notion of derivative relative to a domain.

Definition 1

Let X and Y be Banach spaces and let G be a map from a set $D \subset X$, into Y. Let Q be a topological vector space, $Q \subset X$. We say that $G'(\mu) = A$ is the derivative of G relative to Q, at μ if:

(a) A is a linear operator from X to Y with $Q \subset D(A)$.

(b) For each $q \in C^1 [(-1, 1), Q]$ with $q(0) = 0$ and $q'(0) = q$ there exists a $t_o > 0$ such that $u + q(t) \in D$ for $-t_o \leqslant t \leqslant t_0$ and $\| G(u + q(t)) - G(u) - t A q \| = o(t)$, $t \to 0$.

A C^1 map will be, as usual, a map that has continous Frechét derivative.

The abstract theorem is as follows.

Theorem 2

Let N be a closed subspace of a Banach space X, and F(u) be a C^1 map from X into N. Denote $M = \{u \in X | F(u) = 0\}$. Let G(u) be a real functional defined on a set D contained in X. Denote $S = D \cap M$. Assume that $G'(u)$ exists relative to Q for all $u \in S$, $N \subset Q$, with the topology of X stronger than that of Q on N.

Suppose that there exists a $\mu_0 \in S$ such that $G(\mu_0) = \min_S G(\mu)$, $F(\mu_0)$ is bijective on N, $G'(\mu_0)\, g = 0$ for every $g \in N$, and finally for every $x \in Q, \mu_0 + x \in M$, implies $\mu_0 + x \in S$.

Then $\dot{G}(\mu_0)\, x = 0$, for every $x \in Q$.

Let us consider now the Yang-Mills-Higgs equations. The Lagragian is given by

$$L = -\frac{1}{4}(F_{u\nu}, F^{u\nu}) - \frac{1}{2}(D_u\phi, D^u\phi) - V(\phi) \tag{1}$$

the field strength, $F_{u\nu}$, is given by

$$F_{u\nu} = \partial_u A_\nu - \partial_\nu A_u + A_u \times A_\nu \tag{2}$$

where A_u belongs to the Lie algebra of the gauge group, that is assumed to be compact. **x** denotes Lie bracket and (,) denotes the natural inner product in the Lie algebra of the gauge group. The Higgs field, ϕ, belongs, by simplicity, to the adjoint representation. The potential $V(\phi)$ is a non negative C^1 function of $|\phi| = (\phi,\phi)^{1/2}$.

Julia and Zee [1] proposed the following Ansatz for a time independent solution on the case of SU(2) gauge group.

$$A_i^a = \varepsilon_{aij}\, \hat{x}_j \frac{1 - k(r)}{r} \tag{3}$$

$$A_o^a = \hat{x}_a J(r) \tag{4}$$

$$\phi^a = \hat{x}_a H(r) \tag{5}$$

where $1 \leqslant a, i \leqslant 3$, $\hat{x}_a = \frac{x_a}{|\bar{x}|}$.

The Yang-Mills-Higgs equations reduce to the following system on $(0,\infty)$

$$\ddot{k} = \frac{k}{r^2}(k^2 - 1) + (H^2 - J^2)\, k, \tag{6}$$

$$\ddot{J} + \frac{2}{r}\dot{J} = \frac{2J}{r^2} k^2 + V'(J) \tag{7}$$

$$\ddot{H} + \frac{2}{r}\dot{H} = \frac{2H}{r^2} k^2 \tag{8}$$

$\cdot$ denotes derivative with respect to r, and $V'(J) = \frac{d}{dJ} V(J)$.

The energy of the solution is given by

$$E = 4\pi\int_0^\infty dr\, [\dot{k}^2 + \frac{(k^2 - 1)^2}{2\, r^2} + \frac{1}{2} r^2 \dot{H}^2 + \frac{1}{2} r^2 \dot{J}^2 + k^2 (H^2 + J^2) + $$

$$+ r^2 V] \tag{9}$$

Equations (6) - (8) are the Euler-Lagrange equations from the following functional

$$G = \int_0^\infty dr [\dot{k}^2 + \frac{(k^2 - 1)^2}{2\, r^2} + \frac{1}{2} r^2 \dot{H}^2 - \frac{1}{2} r^2 \dot{J}^2 + $$

$$+ k^2 (H^2 - J^2) + r^2 V] \tag{10}$$

Notice that G is different from E.

From the finite energy requirement it follows the existence of the following limits

$$b = \lim_{r\to\infty} J(r) \tag{11}$$

$$c = \lim_{r\to\infty} H(r) \tag{12}$$

We introduce new variables

$$\beta(r) = J(r) - b \tag{13}$$

$$\gamma(r) = H(r) - c \tag{14}$$

and we require

$$\lim_{r\to\infty} \beta(r) = 0, \tag{15}$$

$$\lim_{r\to\infty} \gamma(r) = 0. \tag{16}$$

Equations (6) - (8) becomes

$$\ddot{k} = \frac{k}{r^2}(k^2 - 1) + (\beta + b)^2 - (\gamma + c)^2)k, \tag{17}$$

$$\ddot{\beta} + \frac{2}{r}\dot{\beta} = \frac{2(\beta + b)}{r^2} k^2 + V'(\beta + b), \tag{18}$$

$$\ddot{\gamma} + \frac{2}{r}\dot{\gamma} = \frac{2(\gamma + c)}{r^2} k^2. \tag{19}$$

The main theorem is

Theorem 3

Assume $V(\beta)$ is a C^1 function on R, $V(\beta) \geqslant 0$, and $V(b) = 0$ for some $b \neq 0$. Then for every $c \in R$, $|c| < |b|$, there are functions $k \in C^4(0,\infty)$, $\beta \in C^2(0,\infty)$ and $\gamma \in C^6(0,\infty)$ satisfying the system (6) - (8), and such that $\lim_{r\to\infty} \beta(r) = 0$, $\lim_{r\to\infty} \gamma(r) = 0$, $(1 + r)^{-\varepsilon}\beta(r)$, $(1 + r)^{-\varepsilon}\gamma(r) \in L^2$, for every $\varepsilon > 0$, and the energy (9) is finite. Moreover if $V \in C^m(R)$ $m \geqslant 1$: $\beta \in C^{m+1}(0,\infty)$, $k \in C^{m+3}(0,\infty)$ and $\gamma \in C^{m+5}(0,\infty)$, and if V is real analytic k, β, γ are real analytic on $(0,\infty)$.

We have also proven that the solutions are smooth at $r = 0$.

Corollary 4

If $V(\beta) \in C^\infty(R)$ the functions k, β, γ given by Theorem 3 are C^∞ at $r = 0$ provided that $V(\beta) = F(\beta)\beta$, where $|F(\beta)| \leqslant K|\beta|^\alpha$, for some α, $0 \leqslant \alpha < 4$, and some constant K.

We consider now the Horváth-Palla [2] Ansätze for SU(N) gauge groups. For each integer M, $1 \leqslant M \leqslant N - 1$, they considered a radial Ansatz consisting of functions $k_i(r)$, $1 \leqslant i \leqslant M$, $J_j(r)$, $H_j(r)$, $1 \leqslant j \leqslant N - 1$ satisfying the following equations*.

* In the notation of [2] $k_i = f_k$, where $k = i + 1$, $J_j = \frac{g_\ell}{r}$, and $H_j = \frac{h_\ell}{r}$, where $\ell = j + 1$, and $M = m - 1$.

$$\ddot{k}_i = \frac{1}{r^2} k_i \,[\, i(M + 1 - i)k_i^2 - \frac{(i + 1)(M - i)}{2} k_{i+1}^2 - \frac{(i - 1)(M - i + 2)}{2} k_{i-1}^2] - \frac{k_i}{r^2} - k_i \,[\,(H_i - \frac{i - 1}{i} H_{i-1})^2 - (J_i - \frac{i - 1}{i} J_{i-1})^2], \quad 1 \leqslant i \leqslant M \tag{20}$$

$$\ddot{J}_j + \frac{2}{r} \dot{J}_j = \frac{(j + 1)(M - j)}{r^2} k_{j+1}^2 (\frac{j}{j+1} J_j - J_{j+1}) - \frac{(j + 1)(M + 1 - j)}{r^2} \cdot k_j^2 (\frac{j-1}{j} J_{j-1} - J_j) + \frac{j + 1}{j} \frac{\partial V}{\partial J_j} \tag{21}$$

$$\ddot{H}_j + \frac{2}{r} H_j = \frac{(j + 1)(M - j)}{r^2} k_{j+1}^2 (\frac{j}{j+1} H_j - H_{j+1}) - \frac{(j + 1)(M + 1 - j)}{r^2} k_j^2 (\frac{j-1}{j} H_{j-1} - H_j) \tag{22}$$

in equations (21) and (22) $k_j \equiv 0$ for $j > M$.

Varying M from 1 to N - 1 we obtain Ansätze having 1 to N - 1 units of magnetic charge (see [2]).

The energy of a solution to (20) - (22) is given by (taking e = 1):

$$E = \pi \int_0^\infty dr [\sum_{i-1}^{M} (2i(M + 1 - i) \dot{k}_i^2 + \frac{1}{2} \frac{(i + 1)i}{r^2} \{(M + 1 - i) \cdot (k_i^2 - 1) + (M - i) (1 - k_{i+1}^2)\}^2) + \sum_{j=1}^{N-1} \frac{2j}{(j+1)} r^2 (\dot{J}_j^2 + \dot{H}_j^2) + \sum_{j=1}^{M} 2j(M + 1 - j)k_j^2 \cdot \{(J_j - \frac{j-1}{j} J_{j-1})^2 + (H_j - \frac{j-1}{j} H_{j-1})^2\} + 4 r^2 V(J_j)] \tag{23}$$

V is a nonnegative C^1 function of the J_j, $1 \leq j \leq N - 1$.

Equations (22) imply that for

$j \geq M + 1$

$$\ddot{H}_j + \frac{2}{r} \dot{H}_j = 0, \qquad j \geq M + 1 . \tag{24}$$

Then

$$(r^2 \dot{H}_j)^{\bullet} = 0, \; j \geq M + 1, \text{ then} \tag{25}$$

the finiteness of the energy implies

$$H_j \equiv 0, \quad j \geq 1 + M. \tag{26}$$

For (23) to be finite we must have

$$\| r\dot{J}_j \| < \infty, \qquad 1 \leq j \leq N - 1, \tag{27}$$

$$\| r\dot{H}_j \| < \infty, \qquad 1 \leq j \leq M. \tag{28}$$

This implies the existence of the following limits

$$b_j = \lim_{r\to\infty} J_j (r) \quad , \quad 1 \leq j \leq N - 1 \quad , \tag{29}$$

$$c_j = \lim_{r\to\infty} H_j (r) \quad , \quad 1 \leq j \leq M. \tag{30}$$

We introduce the new variables

$$\beta_j(r) = J_j(r) - b_j , \tag{31}$$

$$\gamma_j(r) = H_j(r) - c_j, \tag{32}$$

and we require $\lim_{r\to\infty} \beta_j(r) = \lim_{r\to\infty} \gamma_j(r) = 0$. In terms of the new variables the equations (20) - (22) becomes

$$\ddot{k}_i = \frac{1}{r^2} k_i \, [i(M + 1 - i) \, k_i^2 - \frac{(i + 1)(M - i)}{2} \, k_{i+1}^2 -$$

$$- \frac{(i - 1)(M - i + 2)}{2} \, k_{i-1}^2] - \frac{b_i}{r^2} + k_i \, [(\beta_i - \frac{i-1}{i} + \beta_{i-1} +$$

$$+ b_i - \frac{i-1}{i} b_{i-1})^2 - (\gamma_i - \frac{i-1}{i} \gamma_{i-1} + c_i - \frac{i-1}{i} c_{i-1})^2],$$

$$1 \leqslant i \leqslant M. \tag{33}$$

$$\ddot{\beta}_j + 2 \frac{\dot{\beta}_j}{r} = \frac{(j+1)(M-j)}{r^2} k_{j+1}^2 (\frac{j}{j+1} \beta_j - \beta_{j+1} + \frac{j}{j+1} b_j - b_{j+1}) -$$

$$\frac{(j+1)(M+1-j)}{r^2} k_j^2 (\frac{j-1}{j} \beta_{j-1} - \beta_j + \frac{j-1}{j} b_{j-1} - b_j) + \frac{j+1}{j} \cdot$$

$$\frac{\partial V}{\partial \beta_j}, \quad 1 \leqslant j \leqslant N-1, \tag{34}$$

$$\ddot{\gamma}_j + \frac{2}{r} \dot{\gamma}_j = \frac{(j+1)(M-j)}{r^2} k_{j+1}^2 (\frac{j}{j+1} \gamma_j - \gamma_{j+1} + \frac{j}{j+1} c_j - c_{j+1}) -$$

$$- \frac{(j+1)(M+1-j)}{r^2} k_j^2 (\frac{j-1}{j} \gamma_{j-1} - \gamma_j + \frac{j-1}{j} c_{j-1} - c_j), \tag{35}$$

$$1 \leqslant j \leqslant M,$$

where we have taken (26) into account and in (34), (35) $k_j \equiv 0$ for $j \geqslant M + 1$.

The energy is given by

$$E = \pi \int_0^\infty dr \, \{ \sum_{i=1}^{M} [2i(M+1-i) \dot{k}_i^2 + \frac{1}{2} \frac{(i+1) i}{r^2} [(M+1-i)$$

$$(k_i^2 - 1) + (M-i). (1 - k_{i+1}^2)]^2 + 2i(M+1-i) k_i^2 \cdot$$

$$[(b_i - \frac{i-1}{i} b_{i-1})^2 + (c_i - \frac{i-1}{i} c_{i-1})^2] + \sum_{j=1}^{N-1} \frac{2j}{j+1} r^2 \dot{\beta}_j^2 +$$

$$\sum_{i=1}^{M} 2i(M+1-i) k_i^2 [(\beta_i - \frac{i-1}{i} \beta_{i-1})^2 + (\gamma_i - \frac{i-1}{i} \gamma_{i-1})^2 +$$

$$+ 2(\beta_i - \frac{i-1}{i} \beta_{i-1})(b_i - \frac{i-1}{i} b_{i-1}) + 2(\gamma_i - \frac{i-1}{i} \gamma_{i-1}) \cdot$$

$$- (c_i - c_{i-1} \frac{i-1}{i}] + \sum_{i=1}^{M} \frac{2i}{i+1} r^2 \dot{\gamma}_i^2 + 4r^2 V(\beta_j + b_j)\}. \quad (36)$$

Equations (33) - (35) are the Euler-Lagrange equations associated with the following functional: denote by μ the vector valued function

$$\mu = \{k_1, \ldots, k_M, \beta_1, \ldots, \beta_{N-1}, \gamma_1, \ldots, \gamma_M\}. \quad (37)$$

Define:

$$G(\mu) = I(\mu) + J_{\overline{b}}(\overline{\beta}) - J_c(\overline{\gamma}) \quad (38)$$

where

$$I(\mu) = \sum_{i=1}^{M} \{2i(M + 1 - i) \|\dot{k}_i\|^2 + \frac{1}{2} i(i + 1) \|(M + 1 - i)$$

$$\frac{(k_i^2 - 1)}{r} + (M - i) \frac{(1 - k_{i+1}^2)}{r} \|^2 + 2i(M + 1 - i) (b_i - \frac{i-1}{i} \cdot$$

$$b_{i-1})^2 - (c_i - \frac{i-1}{i} c_{i-1})^2) \|k_i\|^2\} + 4\| r V^{1/2} (\beta_j + b_j) \|^2, \quad (39)$$

$$J_{\overline{b}}(\overline{\beta}) = \sum_{j=1}^{N-2} \frac{2j}{j+1} \|r\dot{\beta}_j\|^2 + \sum_{i=1}^{M} 2i(M + 1 - i) [\|k_i(\beta_i - \frac{i-1}{i} \cdot$$

$$\beta_{i-1})\|^2 + 2(b_i - \frac{i-1}{i} b_{i-1})(k_i, (\beta_i - \frac{i-1}{i} \beta_{i-1}) k_i)], \quad (40)$$

where we denote

$$\overline{\beta} = (\beta_1, \beta_2, \ldots, \beta_{N-1}), \text{ and} \quad (41)$$

$\bar{b} = (b_1, b_2, \ldots, b_{N-1})$. Finally (42)

$$J_c(\bar{\gamma}) = \sum_{i=1}^{M} \frac{2i}{i+1} \| r\, \dot{\gamma}_i \|^2 + \sum_{i=1}^{M} 2i(M+1-i)\Big[\| k_i(\gamma_i - \frac{i-1}{i} \gamma_{i-1}) \|^2 +$$

$$+ 2(c_i - \frac{i-1}{i} c_{i-1})\, (k_i, (\gamma_i - \frac{i-1}{i} \gamma_{i-1}) k_i)\Big], \tag{43}$$

where $(\cdot, \cdot)$ denotes the L^2 scalar product on R^+, $\| \ \|$ the L^2 norm on R^+, and we denote

$$\bar{c} = (c_1, c_2, \ldots c_M) \tag{44}$$

$$\bar{\gamma} = (\gamma_1, \gamma_2, \ldots \gamma_M) \tag{45}$$

For simplicity we assume that there are constants b and c such that $b = b_i - \frac{i-1}{i} b_{i-1}$, $1 \leq i \leq N-1$ (46)

$$c = c_i - \frac{i-1}{i} c_{i-1} \quad , 1 \leq i \leq M \tag{47}$$

Using Theorem 2 we prove

Theorem 5

Suppose $V(\beta_j)$ is a nonnegative C^1 function on R^{N-1}, and that $V(b_j) = 0$, for some b_j, $1 \leq j \leq N-1$, such that $b_j - \frac{j-1}{j} b_{j-1} = b$, for some $b \in R$. Then for every $\{c_i\}_{i=1}^{M}$, $c_i \in R$, $1 \leq M \leq N-1$, such that $c_i - \frac{i-1}{i} c_{i-1} = c$, for some $c \in R$, with $|c| < |b|$, there exist functions $k_i \in C^4(0,\infty)$, $\gamma_i \in C^6(0,\infty)$, $1 \leq i \leq M$, and $\beta_j \in C^2(0,\infty)$, $1 \leq j \leq N-1$, satisfying the system (33) - (35). The energy (36) is finite, $\lim_{r\to\infty} \beta_j(r) = 0$, $\lim_{r\to\infty} \gamma_i(r) = 0$, and $(1+r)^{-\varepsilon}\beta_j(r) \in L^2$, $(1+r)^{-\varepsilon}\gamma_i(r) \in L^2$, for every $\varepsilon > 0$. Moreover if $V \in C^m(R^{N-1})$: $k_i \in C^{m+3}(0,\infty)$, $\beta_j \in C^{m+1}(0,\infty)$ and $\gamma_i \in C^{m+5}(0,\infty)$. If $V \in C^\infty(R^{N-1})$, $k_i, \beta_j, \gamma_i \in C^\infty(0,\infty)$, and if V is real analytic k_i, β_j, γ_i are real analytic on $(0,\infty)$.

References

[1] Julia and A. Zee. Phys. Rev. D11 (1975) 2227.

[2] Z. Horváth and L. Palla. Nuclear Physics B116 (1976) 500.

[3] M. Schechter and R. Weder. A Theorem on the Existence of Dyon Solutions. Preprint.

[4] H. Poincaré. Trans. Amer. Math. Soc. 6, (1905) 237-274.

[5] J. Schwinger. Science 165, 3895 (1969) 757.

ASYMPTOTIC RESONANCE PROPERTIES OF THE FINITE-DIMENSIONAL FRIEDRICHS MODEL

Hellmut Baumgärtel
Zentralinstitut für Mathematik und Mechanik
der Akademie der Wissenschaften der DDR
DDR - 1080 Berlin

The Friedrichs model $H=H_0+\Gamma+\Gamma^*$ with an embedded eigenvalue λ_0 of H_0 of finite multiplicity is considered. Put $B(\lambda):=\Gamma^* E_0(d\lambda)\Gamma/d\lambda$, where $E_0(\cdot)$ denotes the spectral measure of H_0. Let $B(\lambda)>0$. Further put $\sigma(\lambda;B)=\|T(\lambda)\|^2_{2,\lambda}$ (σ scattering cross-section, $T(\lambda)$ scattering amplitude). Sufficient conditions on B are given for "asymptotic resonance", that is for $\sigma(\lambda;B)\to 0$ if $\lambda\neq\lambda_0$ and $\liminf \sigma(\lambda_0;B)>0$, if B tends to zero in some sense. If $B=B_\varepsilon$, the result can be interpreted in the sense of yielding the existence of a simple "resonance trajectory" in the $\{\lambda,\varepsilon\}$-plane tending to $\{\lambda_0,0\}$. This result is supplemented by an example which shows that already in very simple cases other "trajectories" must be used in order to obtain the resonance property at $\{\lambda_0,0\}$.

Let $\mathcal{H}$ be a separable Hilbert space, let H_0 be a selfadjoint operator acting on $\mathcal{H}$. The spectral measure of H_0 is denoted by $E_0(\cdot)$. The following properties are assumed:

(i) spec H_0=spec ${}^{ac}H_0 = [a,b]$, $-\infty < a < b < \infty$.

(ii) There is exactly one eigenvalue λ_0 of H_0 which is embedded into (a,b) and of finite multiplicity n. We put $E_0(\{\lambda_0\}) = P_0$, $\dim P_0 = n$.

(iii) $P_0^{\perp} = P^{ac}(H_0)$.

Let $\Gamma: P_0\mathcal{H} \mapsto P_0^{\perp}\mathcal{H}$ be a linear operator, which is non singular, that is dim ima Γ = n. The extension of Γ onto $\mathcal{H}$ given by $\Gamma \restriction P_0^{\perp}\mathcal{H} = 0$ is also denoted by Γ, $\Gamma P_0^{\perp} = 0$. Let

$$H = H_0 + \Gamma + \Gamma^* . \tag{1}$$

We may assume

$$P_0^{\perp}\mathcal{H} = \text{clo spa}\{E_0(\Delta)x,\ x \in \text{ima}\,\Gamma,\ \Delta \text{ Borel set}\},$$

because on $\mathcal{H} \ominus (P_0\mathcal{H} \oplus \text{clo spa}\{E_0(\Delta)x,\ x \in \text{ima}\,\Gamma,\ \Delta$ Borel set$\})$ we obtain $H = H_0$. Let

$$[k,k']_{\lambda} = (E_0(d\lambda)k,k')/d\lambda, \quad k,k' \in \text{ima}\,\Gamma . \tag{2}$$

and

$$[k]_{\lambda} = [k,k]_{\lambda}^{1/2} .$$

Then ima Γ mod ker $[\cdot]_{\lambda}$ may be equipped with the scalar product $[k,k']_{\lambda}$. The space obtained in this way is denoted by $\mathcal{K}_{\lambda}$. Hence $P_0^{\perp}\mathcal{H}$ may be identified with $L^2([a,b],d\lambda,\mathcal{K}_{\lambda})$. For simplicity we put

$$P_0^{\perp}\mathcal{H} = L^2([a,b],d\lambda,\mathcal{K}_{\lambda}). \tag{3}$$

Further let

$$B(\lambda) = \Gamma^* E_0(d\lambda)\Gamma/d\lambda , \quad B(\lambda) \geq 0 \tag{4}$$

The eigenspace $P_0\mathcal{H}$ may be equipped with the prescalar product $(B(\lambda)f,g)$; $f,g \in P_0\mathcal{H}$. Hence Γ is isometric as an operator from $P_0\mathcal{H}$ mod ker $(B(\lambda)\cdot,\cdot)$ onto $\mathcal{K}_\lambda$ because of

$$[\Gamma x,\Gamma y]_\lambda = \frac{(E_0(d\lambda)\Gamma x,\Gamma y)}{d\lambda} = \left(\frac{\Gamma^* E_0(d\lambda)\Gamma}{d\lambda} x,y\right);$$

$$x,y \in P_0\mathcal{H} .$$

The following property is assumed:

(iv) $\quad B(\lambda) > 0$ on $[a,b]$.

In this case we obtain $\mathcal{K}_\lambda = \text{ima}\,\Gamma$ as a set. Hence the multiplicity of the absolutely continuous spectrum of H_0 is constant, namely n.

By a well-known theorem the wave operators $W_\pm(H,H_0)$ for the pair $\{H_0,H\}$, which is called the finite-dimensional Friedrichs model, are existent and complete. Then $S = W_+^* W_-$ is unitary in $P_0^\perp \mathcal{H}$. We put $T = S-1$. Let $S(\lambda)$, $T(\lambda)$ be the scattering matrix and the scattering amplitude, respectively, which are $\mathcal{L}(\mathcal{K}_\lambda)$-valued functions, that is nxn-matrix-valued functions, defined a. e. on $[a,b]$. The scattering cross-section is given by

$$(5) \qquad \sigma(\lambda) = \|T(\lambda)\|_{2,\lambda}^2$$

where $\|\cdot\|_{2,\lambda}$ denotes the Hilbert-Schmidt norm on $\mathcal{K}_\lambda$. For convenience $\sigma(\lambda)$ is calculated:

We denote by $J(\lambda)$ the identification operator $J(\lambda)$: $\text{ima}\,\Gamma \mapsto \mathcal{K}_\lambda$. Then we obtain the following well-known formula for the scattering amplitude (see for instance J.S. Howland [1]):

$$(6) \quad T(\lambda) = J(\lambda)\Gamma P_0 R(\lambda+i0)P_0\Gamma^* J(\lambda)^* ,$$

that is $T(\lambda)$ is acting on $\mathcal{R}_\lambda$ as multiplication by the $n \times n$-matrix $\Gamma R(\lambda+i0)\Gamma^*$, where $R(z) = (z-H)^{-1}$. The operator $P_0R(z)P_0 \restriction P_0\mathcal{H}$ is called the partial resolvent and for this operator the well-known formula

$$(7) \qquad P_0R(z)P_0 \restriction P_0\mathcal{H} = H(z)^{-1}$$

is valid, where

$$(8) \qquad H(z) = (z-\lambda_0)P_0 - \int_a^b \frac{1}{z-\mu} B(\mu)d\mu \quad ,$$

which is called the Livšic-matrix (see J.S. Howland [1]). From (6)

$$(9) \qquad \| T(\lambda)\|^2_{2,\lambda} = \| B(\lambda)^{1/2} H(\lambda+i0)^{-1} B(\lambda)^{1/2} \|^2_2$$

is obtained, where $\| \cdot \|_2$ denotes the Hilbert-Schmidt norm on $P_0\mathcal{H}$ with regard to the original scalar product. Now we have, as is well-known,

$$H(\lambda+i0) = (\lambda-\lambda_0)1_{P_0\mathcal{H}} + i\pi B(\lambda) - \int_a^b \frac{1}{\lambda-\mu} B(\mu)d\mu \quad ,$$

hence

$$(10) \qquad B(\lambda)^{-1/2} H(\lambda+i0) B(\lambda)^{-1/2}$$
$$= (\lambda-\lambda_0)B(\lambda)^{-1} + i\pi + \int_a^b \frac{B(\lambda)^{-1/2}B(\mu)B(\lambda)^{-1/2}}{\mu-\lambda} d\mu .$$

In the following we use the notation $\sigma(\lambda) = \sigma(\lambda;B)$. Now the following questions arise. The general question asks for the structure of T and σ. A more special question asks: How does the eigenvalue λ_0 influence $T(\lambda)$ and $\sigma(\lambda)$, in particular, if the perturbation is small in some sense.

A partial answer to this question can be given within the framework of analytic continuation. Assume that $B(\lambda)$ is holomorphic in some region G, where $[a,b] \subset G$.

Then also $H(z)$ is holomorphic in G. Now we call a complex number ζ a *virtual pole*, if $\det H(\zeta) = 0$ or, equivalently, if ζ is a pole of $H(z)^{-1}$. Then one obtains: $T(\lambda)$ is analytic continuable into G. The virtual poles are exactly the poles of $T(\lambda)$ in G.

If for simplicity a parameter ε is introduced, $\Gamma_\varepsilon = \varepsilon\Gamma$, $B_\varepsilon = \varepsilon^2 B$, then according to Rouche's theorem the existence of virtual poles can be proved, if ε is small. In this case the resonance properties of $\sigma(\lambda;B_\varepsilon)$ can be expressed in terms of the virtual poles $\zeta(\varepsilon) = \xi(\varepsilon) + i\,\eta(\varepsilon)$, $\eta(\varepsilon) < 0$, for example we obtain

$$\liminf_{\varepsilon \to 0} \sigma(\xi(\varepsilon), B_\varepsilon) > 0.$$

Now we consider a more general case.

Let $H_\beta(a,b)$ be the Banach space of Hölder-continuous operator-functions φ on $P_0\mathcal{H}$ equipped with the norm

$$\|\varphi\|_{2,\beta} = \sup_{x\ (a,b)} \|\varphi(x)\|_2 + \sup_{\substack{x_1,x_2\ (a,b)\\ x_1 \neq x_2}} \frac{\|\varphi(x_1)-\varphi(x_2)\|_2}{|x_1-x_2|^\beta},$$

$$0 < \beta < 1,$$

($\varphi(x)$ is considered as a linear operator acting on $P_0\mathcal{H}$). As is well-known, $\int_a^b (\mu-\lambda)^{-1}\varphi(\mu)\,d\mu$, $a < \lambda < b$, is a bounded operator acting on $H_\beta(a,b)$. Now we can formulate the result. Recall that $\sigma(\lambda;0) \equiv 0$.

Theorem.

(i) *If* $B \in H_\beta(a,b)$ *and* $\|B\|_{2,\beta} \to 0$, *then* $\sigma(\lambda;B) \to 0$ *for all* $\lambda \neq \lambda_0$.

(ii) *If, additionally,*

(11) $$\sup_{\mu\in(a,b)} \| B(\mu) \| \cdot \| B(\lambda_o)^{-1} \| < K$$

(K *independent of* B) *is assumed, then*

$$\liminf_B \sigma(\lambda_o;B) > 0.$$

(iii) *If* (11) *is violated, then* $\sigma(\lambda_o;B) \to 0$ *is also possible.*

Proof.

(i) Let $e_1,\ldots,e_n$ be an orthonormal base in $P_o\mathcal{H}$. Then

$$\|B(\lambda)^{1/2}(\lambda-\lambda_o+ i\pi B(\lambda) + \int_a^b \frac{B(\mu)}{\mu-\lambda}\, d\mu)^{-1} B(\lambda)^{1/2}\|_2^2$$

$$= \sum_{i,j=1}^{n} |((\lambda-\lambda_o+ i\pi B(\lambda) + \int_a^b \frac{B(\lambda)}{\mu-\lambda}\, d\mu)^{-1} B(\lambda)^{1/2}e_i, B(\lambda)^{1/2}e_j)|^2$$

$$\leq \sum_{i,j=1}^{n} \|(\lambda-\lambda_o + i\pi B(\lambda) + \int_a^b \frac{B(\mu)}{\mu-\lambda}\, d\mu)^{-1}\|^2 \cdot (B(\lambda)e_i,e_i) \cdot (B(\lambda)e_j,e_j).$$

Now we have

(12) $$\| \int_a^b \frac{B(\mu)}{\mu-\cdot}\, d\mu\|_{2,\beta} \leq K \|B\|_{2,\beta} ,$$

hence

$$\| \int_a^b \frac{B(\mu)}{\mu-\lambda}\, d\mu\|_2 \leq K \|B\|_{2,\beta} \quad \text{and} \quad \|B(\lambda)\|_2 \leq \|B\|_{2,\beta}.$$

Let $\lambda \neq \lambda_o$. We put $\lambda - \lambda_o = \alpha \neq 0$. Now we obtain easily the estimation

$$\|(\alpha + i\pi B(\lambda) + \int_a^b \frac{B(\mu)}{\mu-\lambda}\,d\mu)^{-1}\|_2 \leq (|\alpha| - (\pi+K)\|B\|_{2,\beta})^{-1},$$

which is valid, if $\|B\|_{2,\beta}$ is sufficiently small. Noting $\|\cdot\| \leq \|\cdot\|_2$, (i) is proved.

(ii) Let $\lambda = \lambda_o$. Then

$$\sigma(\lambda_o;B) = \|(i\pi + B(\lambda_o)^{-1/2} \int_a^b \frac{B(\mu)}{\mu-\lambda_o}\,d\mu\, B(\lambda_o)^{-1/2})^{-1}\|_2^2.$$

Hence it is sufficient to show that

$$\|B(\lambda_o)^{-1/2} \int_a^b \frac{B(\mu)}{\mu-\lambda_o}\,d\mu\, B(\lambda_o)^{-1/2}\|_2$$

is bounded if $\|B\|_{2,\beta} \to 0$ and if (11) is valid. Because of (12) it is sufficient to show that $\|B(\lambda_o)^{-1/2}B(\mu) B(\lambda_o)^{-1/2}\|_2$ is bounded. Hence it is sufficient to show that $\|B(\lambda_o)^{-1/2}B(\mu)B(\lambda_o)^{-1/2}\|$ is bounded (the norms $\|\cdot\|$ and $\|\cdot\|_2$ are equivalent in our case). Now

$$\|B(\lambda_o)^{-1/2}B(\mu)B(\lambda_o)^{-1/2}\| \leq \|B(\mu)\| \cdot \|B_o^{-1}\|,$$

which is bounded because of (11). (ii) is proved.

(iii) Let $\lambda_o = 0$, $[a,b] = [-1,1]$, $B(\mu) = B_o + \mu B_o^{1/2}$. Let β_{max}, β_{min} be the maximal, minimal eigenvalue of B_o, respectively. Then

$$\sup\|B(\mu)\| = \beta_{max} + \beta_{max}^{1/2}, \quad \|B_o^{-1}\| = \beta_{min}^{-1},$$

hence

$$\sup\|B(\mu)\| \cdot \|B_o^{-1}\| = \beta_{min}^{-1}(\beta_{max} + \beta_{max}^{1/2}).$$

This may tend to infinity, if $B \to 0$, for example in the case $\beta_{min} = \beta_{max} = \gamma$, where $\beta_{min}^{-1}(\beta_{max} + \beta_{max}^{1/2}) = 1 + \gamma^{-1/2} \to \infty$. Further it is to calculate

$$B_0^{1/2}\left(\int_{-1}^{1}\left(\frac{B_0}{\mu} + B_0^{1/2}\right)d\mu\right)^{-1}B_0^{1/2} = \frac{1}{2}B_0^{1/2},$$

hence

$$\sigma(0;B) = \|(i\pi + 2B_0^{-1/2})^{-1}\|_2^2 \leq \left(\frac{1}{2}\|B_0^{1/2}\|_2 \cdot \right.$$

$$\left.(1 - \frac{\pi}{2}\|B_0^{1/2}\|_2)^{-1}\right)^2.$$

(iii) is proved. ∎

Special cases:

1. The case $n = 1$. Let e be a normed eigenvector corresponding to λ_0. Clearly, $\mathcal{K}_\lambda$ may be chosen as $\mathbb{C}$, $\Gamma e = \psi(\cdot)$, $\psi \in L^2([a,b], d\lambda, \mathbb{C})$; $B(\lambda) = |\psi(\lambda)|^2$. (iv) means $\psi(x) \neq 0$ on $[a,b]$. Then

$$\sigma(\lambda;B) = B(\lambda)^2\left(\pi^2 B(\lambda)^2 + \left(\lambda - \lambda_0 + \int_a^b \frac{B(\mu)}{\mu - \lambda}\,d\mu\right)^2\right)^{-1}.$$

If $\psi(x)$ is Hölder-continuous on $[a,b]$ and $\|B\|_{2,\beta} \to 0$, then the conclusion (i) of the theorem is valid. (11) means that $\sup_x |\psi(x)/\psi(\lambda_0)|^2$ must be bounded, independent of ψ. In this case also the conclusion of (ii) is valid.

2. A special case is that of a fixed Γ (respectively fixed B), where a small parameter ε is introduced: $\Gamma \mapsto \varepsilon\Gamma$, $B \to \varepsilon^2 B$. Then: if B is Hölder-continuous on $[a,b]$ and $\varepsilon \to 0$, then the conclusions of the theorem are valid, because (11) is satisfied in this case.

Now we discuss some remarks with respect to the interpretation of the result. The properties (i) and (ii) of the theorem may be considered as some kind of resonance behaviour of the scattering cross-section, due to the embedded unstable eigenvalue λ_0 of the unperturbed system, if the interaction is governed by a small perturbation. We introduce the interaction parameter ε as above, $\Gamma = \Gamma_\varepsilon$, $\varepsilon > 0$. Then $\sigma = \sigma(\lambda;\varepsilon)$. The spectral parameter λ and the interaction parameter ε span a plane. We consider trajectories $\{\lambda,\varepsilon\} = \{\lambda(\varepsilon), \varepsilon\}$ in this plane, which tend to $\{\lambda_0, 0\}$ for $\varepsilon \to +0$. The theorem says: Under certain assumptions the trajectories $\lambda(\varepsilon) \equiv \lambda$ have the property

$$\lim \sigma(\lambda(\varepsilon), \varepsilon) = 0 \quad , \quad \lambda \neq \lambda_0,$$

$$\liminf \sigma(\lambda(\varepsilon), \varepsilon) > 0 \quad , \quad \lambda = \lambda_0.$$

Therefore the following definition may be useful:

DEFINITION. *A trajectory* $\{\lambda(\varepsilon), \varepsilon\}$ *tending to* $\{\lambda_0, 0\}$ *is called a resonance trajectory, if*

$$\liminf_{\{\lambda(\varepsilon),\varepsilon\}} \sigma(\lambda(\varepsilon), \varepsilon) > 0$$

Then the theorem says: The n-dimensional Friedrichs model, under certain simple assumptions, has the resonance trajectory $\lambda(\varepsilon) \equiv \lambda_0$. There are simple examples, which show, that other trajectories than $\{\lambda_0, \varepsilon\}$ must be used in order to obtain a resonance trajectory for the point $\{\lambda_0, 0\}$. In other words, in such an example $\{\lambda_0, \varepsilon\}$ is not a resonance trajectory.

Example. Let $\mathcal{H} = L^2(0,\infty)$,

$$(H_\varepsilon f)(x) = -f''(x) + \varepsilon \chi_{[a,a+\ell]}(x) f(x),$$

where $\chi_{[a,b]}(x)$ denotes the characteristic function of $[a,b]$, with boundary condition $f(0) = 0$ (potential threshold). Let $\varepsilon - \lambda = \mu^2$, $\lambda = k^2$ and look at $\mu \to \infty$, λ fixed (that is $\varepsilon \to \infty$). We obtain in this case for $T_\varepsilon(\lambda) = T(k;\mu)$, as is well-known (for instance see S. Flügge [2], G.G. Emch and K.B. Sinha [3])

$$|T(k;\mu)|^2 = 4\,\frac{(\sin ka + k\mu^{-1}\cos ka \ \tanh \mu l)^2}{(\mu k^{-1}\sin ka \ \tanh \mu l + \cos ka)^2 + (\sin ka + k\mu^{-1}\cos ka \ \tanh \mu l)^2}$$

The square of the scattering amplitude takes its maximal value 4, if

$$\sin ka \cdot \tanh \mu l + k\mu^{-1} \cos ka = 0,$$

that is, if μ is large, one obtains because of $\tanh \mu l \approx 1$

$$\tan ka = -k\mu^{-1}, \tag{13}$$

which is an equation for resonance trajectories $\{k(\mu), \mu\}$ tending to $\{\lambda_0^{1/2}, \infty\}$ formally in the $\{k, \mu\}$-plane, where λ_0 denotes an eigenvalue of the unperturbed system. Now these eigenvalues are characterized by the equation

$$\sin ka = 0, \tag{14}$$

that is if $k(\mu) \equiv k$, where (14) is satisfied, then we obtain for the scattering amplitude

$$|T(k;\mu)|^2 = \frac{k^2(\tanh \mu l)^2}{\mu^2 + k^2(\tanh \mu l)^2} \to 0, \tag{15}$$

if $\mu \to \infty$ (or $\varepsilon \to \infty$), that is $k(\mu) \equiv k$ is not a resonance trajectory.

References

(1) J.S. Howland, The Livšic Matrix in Perturbation Theory, J. Math. Anal. Appl. 50(1975), 415-437.

(2) S. Flügge, Practical Quantum Mechanics, Springer Verlag, New York-Heidelberg-Berlin 1974.

(3) G.G. Emch and K.B. Sinha, Weak Quantization in a Non Perturbative Model. Preprint/FUSP/P-151, Universidade de São Paulo, Instituto de Fisica, July 1978.

SELF-CONSISTENT FIELD METHODS FOR METASTABLE ELECTRONIC STATES: A PROMISING EXTENSION OF THE COMPLEX COORDINATE TECHNIQUE

C. William McCurdy

Department of Chemistry
The Ohio State University
Columbus, OH 43210

ABSTRACT

The application of the complex coordinate technique to metastable electronic states of many-electron systems requires that ways be found to adapt the most successful methods of bound-state electronic structure calculations for this use. We describe the generalization of self-consistent field theory to the case of a complex coordinate Hamiltonian and present numerical results for the ^2P shape resonance of Be^-. The modifications necessary for molecular calculations are described for which the theory is more appropriately described as the complex basis function method rather than the complex coordinate method. Numerical results are presented for the $^2\Pi_g$ shape resonance of N_2^-. Finally, some speculations regarding other applications are offered.

I. INTRODUCTION

The audience to which this paper was originally delivered was a group representing a diverse range of backgrounds in mathematics and physics, so it seemed appropriate to begin with an informal outline of the problem at hand so as to at least attempt to make clear some of the motivation of this work. In view of the eclectic title of this volume it is certainly a good idea to begin again with an informal description of our starting point.

Over the past ten years a substantial body of literature has appeared dealing with either mathematical[1-5] or computational[6,9] aspects of the complex coordinate technique in quantum mechanics. For the mathematician, complex coordinate transformations provide powerful tools for studying the spectral properties of operators. For the physicist or chemist computational methods based on the complex coordinate idea can provide direct treatments of metastable states in scattering and can also be used in the calculation of photoionization and photodissociation cross sections[7]. This idea has been an attractive one for computational physicists, but it is only now, nearly ten years after the first numerical applications of the complex coordinate technique, that the practical difficulties have been surmounted to the extent that meaningful calculations on interesting systems — namely metastable molecular negative ions with more than three electrons — have become possible.

To understand why computational methods based on the complex coordinate technique have seemed so promising, it is only necessary to look briefly at the results of the early mathematical papers on the subject by Aguilar, Balslev and Combes[2]. The principal result of this work can be stated in a pedestrian manner as follows. Under the dilatation transformation, $\vec{r}_j \rightarrow \theta\vec{r}_j$, of the coordinates of the particles, the Hamiltonian, $H(\{\vec{r}_i\})$, of a system of particles interacting via Coulomb potentials becomes a non-Hermitian operator, H_θ, whose spectrum consists of a series of rays in the complex energy plane, which make an angle of $-2 \arg(\theta)$ with the real axis, together with a set of discrete eigenvalues. As shown in figure 1, the discrete eigenvalues are at the bound state energies of the untransformed Hamiltonian, and at complex values of the energy (sometimes called Siegert eigenvalues) which correspond to the metastable resonance states. This result becomes particularly interesting when we observe that the eigenfunctions corresponding to the complex resonance eigenvalues, as well as the bound state eigenvalues, are square integrable. Thus the dilatation transformation turns one aspect of the scattering problem, resonances, into a problem which has the essential features of a bound state problem.

The state of the computational art in bound state problems is far in advance of that in scattering calculations. There is now a vast array of practical and powerful variational methods and pertubation approaches for computing the energies and wavefunctions of the bound states of atoms and molecules[10]. So bringing the resonance problem within range of these computational tools constitutes an advance in itself. However, the application of bound state methods to the complex coordinate description of resonances has been hindered by computational and theoretical difficulties, particularly in the case of molecular resonances, which have only recently been surmounted to the extent that molecular calculations are now becoming feasible[8,9,11]. We will discuss some of those difficulties and how they are currently being overcome as the occasion arises in our description of self-consistent field calculations in the following sections.

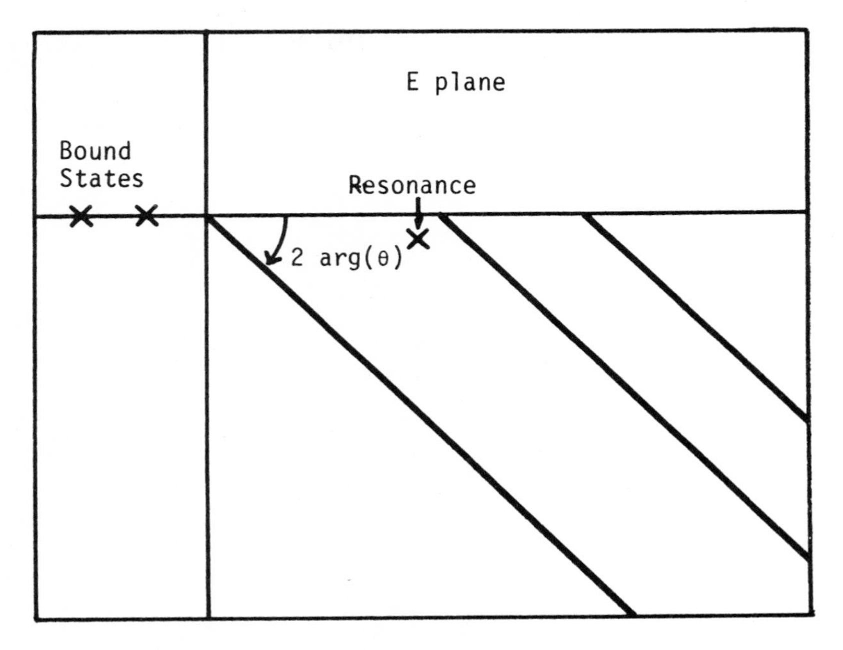

Figure 1. Spectrum of the scaled Hamiltonian H_θ.

In almost all of the previous calculations using the complex coordinate technique to describe resonances in many-electron systems the problem has been addressed by using configuration interaction (CI) methods as opposed to the other available methods of electronic structure calculations. In other words, workers in this area have usually persisted in simply diagonalizing the complex Hamiltonian using many-electron basis functions, thereby employing the most costly and cumbersome, albeit powerful, tool of bound state calculations. Particularly conspicuous by its absence from all applications of the complex coordinate technique to date is the use of Hartree-Fock theory or self-consistent field (SCF) theory in any form. Self-consistent field calculations have played a central role in the development of quantum chemistry and certainly seem to provide the best starting point for more accurate calculations. One expects that the basic mathematics of SCF theory (in some form) will be applicable to the square integrable eigenstates of the complex Hamiltonian. The question is whether or not the SCF approximation provides a reasonable physical description of electronic resonances.

The work presented here provides an initial answer to that question. We have found that simple SCF theory, i.e., open-shell restricted Hartree-Fock theory, is appropriate for shape resonances in atoms and molecules, but that this theory gives no information about the width of a Feshbach resonance. Multiconfiguration SCF theory, already known in several forms in electronic structure calculations, may however, provide a simple way to treat Feshbach resonances.

II. THE COMPLEX COORDINATE SCF EQUATIONS[12]

For the moment we will restrict ourselves to the case of an atomic resonance because the dilatation transformation itself must be modified to treat the molecular case.

The basic idea underlying our discussion of the SCF equations for resonances is that by beginning with the complex atomic Hamiltonian, $H(\{\theta\vec{r}_j\})$, and following (essentially) the same variational arguments used to derive the bound state SCF equations[13], we can derive complex SCF equations appropriate for resonance states. This procedure is successful because the resonance eigenfunctions of the complex Hamiltonian are square integrable. In that sense these eigenfunctions are sufficiently like bound state wavefunctions to allow treatment by the SCF methods currently employed in many bound state calculations. The extent to which the resonance eigenfunctions are unlike the bound state wavefunctions of an ordinary SCF calculation (they may oscillate at short distances) is manifested mostly in the numerical behavior of the complex SCF calculation and not in the formal mathematical statement of the problem.

We will specialize our discussion still further to the simple case of a shape resonance which corresponds to an electronic configuration with one electron outside a closed shell. The ^2P shape resonance in e-beryllium scattering is an example of such a case and the ^3P autoionizing state of helium (2s2p), for which we will also present numerical results demonstrating the failure of the method for Feshbach resonances, is simpler still. More general situations corresponding to open-shell target atoms and molecules lead to complications analogous to those in open-shell bound state calculations which need not concern us here.

The trial variational wavefunction[13] is a single Slater determinant of spin orbitals

$$\phi = \det \begin{vmatrix} \psi_1(r_1)\bar{\psi}_1(r_1)\psi_2(r_1)\bar{\psi}_2(r_1) & \cdots & \psi_k(r_1)\bar{\psi}_k(r_1)\psi_\mu(r_1) \\ \psi_1(r_1)\bar{\psi}_1(r_2)\psi_2(r_2)\bar{\psi}_2(r_2) & \cdots & \psi_k(r_2)\bar{\psi}_k(r_2)\psi_\mu(r_2) \\ \vdots & & \vdots \\ \psi_1(r_{2k+1})\bar{\psi}_1(r_{2k+1}) & \cdots & \psi_\mu(r_{2k+1}) \end{vmatrix} \quad (1)$$

Each spin orbital, ψ_i, is the product of a spatial orbital ϕ_i (depending on the coordinates of one electron) and a spin function, and $\bar{\psi}_i$ denotes the spin orbital with opposite spin. To derive the SCF equations for the orbitals, ϕ_i, we set to zero the first variations with respect to those orbitals of a particular functional. For the ordinary bound state problem that functional consists of the expectation value of the Hamiltonian with respect to Φ plus some Lagrange multiplier terms which serve to apply the constraint that the orbitals, ϕ_i, remain orthonormal. In our case we must generalize this expression slightly. Since the Hamiltonian, $H(\{\vec{r}_j\theta\})$, is not Hermitian, its eigenfunctions form a biorthogonal set, and the complex conjugate of the wavefunction which appears in the usual expectation value is not appropriate. Thus the functional we have chosen is

$$I = \int \Phi H(\{r_j \theta\}) \Phi d\tau_1 d\tau_2 \ldots d\tau_{2k+1}$$

$$- \sum_{i,j=1}^{k} \lambda_{ij} \int \phi_i(r)\phi_j(r) d^3r \qquad (2)$$

$$- 2 \sum_{i=1}^{k} \lambda_{i\mu} \int \phi_i(r)\phi_\mu(r) d^3r$$

$$- \varepsilon_\mu \int \phi_\mu(r)\phi_\mu(r) d^3r$$

where the matrix, λ_{ij}, of Lagrange multipliers is complex symmetric and the redundant terms in the first sum are for convenience only. Although this form of the functional is adequate for discussing the simple case at hand, it is probably not satisfactory for the general problem. As the scale factor, θ, approaches unity the SCF equations must approach the usual Hartree-Fock equations in which the orbital Hamiltonians are hermitian. The equations we derive using eq. (2) have this property only if the orbitals, ϕ_i, in the trial wavefunction are real when θ=1, because otherwise the coulomb and exchange operators as defined below are nonhermitian. We claim that this is no restriction at all in practice because it simply means that we are constructing the correct analytic continuation of the working equations of the ordinary real valued SCF calculations of quantum chemistry. Note that in defining the functional in equation (2), we have in mind a class of problems which, in the $\theta \rightarrow 1$ limit, can be formulated in terms of real orbitals. But, if for example, the orbitals of Eq. 1 were expressed as products of radial functions and spherical harmonics, the functional we would then use would have the complex conjugate of the angular variables of the left hand orbitals appearing in all the scalar products, but not the radial variables.

By setting the functional derivatives, $\delta I/\delta\phi_i$ and $\delta I/\delta\phi_\mu$, of equation (2) with respect to the open and closed shell orbitals to zero, we obtain the SCF equations[13]. If we define the Fock operators by

$$\hat{F}_o = h_\theta + \theta^{-1} \sum_{j=1}^{k} (2\hat{J}_j - \hat{K}_j) + \theta^{-1}(\hat{J}_\mu - \tfrac{1}{2}\hat{K}_\mu) \tag{3a}$$

$$\hat{F}_\mu = h_\theta + \theta^{-1} \sum_{j=1}^{k} (2\hat{J}_j - \hat{K}_j) \tag{3b}$$

where the one electron Hamiltonian, h_θ, is (atomic units)

$$h_\theta = \frac{\theta^{-2}}{2} \nabla^2 - \frac{Z\theta^{-1}}{r} \tag{4}$$

and the coulomb and exchange operators are

$$\hat{J}_j = \int \frac{\phi_j(\vec{r}_2)\phi_j(\vec{r}_2)}{r_{12}} d^3r_2 \tag{5a}$$

$$\hat{K}_j\phi_i = \phi_j(r_1)\int \frac{\phi_j(\vec{r}_2)\phi_i(\vec{r}_2)}{r_{12}} d^3r_2 \tag{5b}$$

the complex SCF equations can be written

$$2\hat{F}_o\phi_i = \sum_{j=1}^{k} \lambda_{ji}\phi_j + \lambda_{\mu i}\phi_\mu \tag{6a}$$

$$\hat{F}_\mu\phi_\mu = \sum_{j=1}^{k} \lambda_{\mu j}\phi_j + \varepsilon_\mu\phi_\mu \tag{6b}$$

$$\lambda_{ij} = \lambda_{ji} \tag{6c}$$

The usual SCF procedure is to find a set of equivalent matrix equations by multiplying equations (6a) and (6b) by the orbitals and integrating. Additional practical details about the matrix formulation can be found in references 13 and 14. The only difference between the usual SCF matrix equations and the ones we solve here is that the Fock operators in the complex SCF calculations are defined as in equations (3) and (4) with appropriate complex scaling. All the matrices appearing in these calculations are complex symmetric. We will discuss what is now known about the numerical behavior of the complex SCF equations in the next section. Finally, we note that the value of the complex resonance energy from this procedure is the complex SCF energy, E_{CSCF}

$$E_{CSCF} = \int \Phi H(\{\vec{r}_j, \theta\})\Phi d\tau_1 \ldots d\tau_{2k+1} \tag{7}$$

$$= \sum_{i=1}^{k} \int \phi_i (h_\theta + \hat{F}_o)\phi_i d^3r$$

$$+ \tfrac{1}{2} \int \phi_\mu (h_\theta + \hat{F}_\mu)\phi_\mu d^3r$$

This completes the formulation of the SCF problem for the simplest case of an atomic shape resonance: one electron outside of a closed shell. The same prescription is unfortunately not successful in the case of a Feshbach resonance because the complex SCF equations so derived have solutions which yield a real value for E_{CSCF} and consequently give no information about the width of the resonance. The reason for this failure is that the decay of a Feshbach resonance is a correlation effect which is not present in the SCF description of the resonance state.

Consider the simple case of the ^3P autoionizing state of helium with configuration 2s2p. Proceeding as before we can derive complex SCF equations for the 2s and 2p orbitals. The trial wavefunction can be written as a single Slater determinant, and after some algebra we find

$$\hat{F}_s \phi_{is} = \varepsilon_{is}\phi_{is} \tag{8a}$$

$$\hat{F}_p \phi_{ip} = \varepsilon_{ip}\phi_{ip} \tag{8b}$$

where the Fock Hamiltonians are given by

$$\hat{F}_s = h_\theta + \theta^{-1}(\hat{J}_{2p} - \hat{K}_{2p}) \tag{9a}$$

$$\hat{F}_p = h_\theta + \theta^{-1}(\hat{J}_{2s} - \hat{K}_{2s}) \tag{9b}$$

with h_θ and the coulomb and exchange operators defined as in equations (4) and (5). The important point to note is that the second root of equation (8a) (i=2) is used in the construction of F_p, equation (9b). Note that in equation (9a), $\hat{J}_{2p}$ and $\hat{K}_{2p}$ must be spherically averaged as in any restricted Hartree-Fock calculation on this state. This resonance decays to a continuum background which can be represented by the configuration 1skp where ϕ_{kp} is a continuum orbital of energy $k^2/2$. But this configuration is absent entirely from the space of variations being considered when we set to zero the first functional derivatives of the functional, I, defined in equation (2) for 2s2p single determinant trial function. Because the coupling to the 1skp continuum is absent from the SCF description of the 2s2p state. In fact equations (8) and (9) have a well defined solution corresponding to the 2s2p state even when θ is unity and they are not complex at all. The SCF energy computed for that state

$$E_{CSCF} = \tfrac{1}{2}(\varepsilon_{2s} + \varepsilon_{2p} + \int \phi_{2s} h_\theta \phi_{2s} d^3r + \int \phi_{2p} h_\theta \phi_{2p} d^3r) \qquad (10)$$

is therefore real and independent of θ. Numerical verification of this assertion is given in the next section.

Another way to see that the 1skp continuum is missing from the SCF description is to think of the equivalent configuration interaction (CI) calculation[15]. The open-shell version of Brillouin's theorem states[16], in this case, that matrix elements of the Hamiltonian between the 2s2p Slater determinant and any determinant corresponding to a single excitation to an orthogonal orbital from 2s2p all vanish. The SCF calculation is equivalent to a CI calculation in this space of single excitations from 2s2p. The configuration 1skp is a double excitation from 2s2p and does not appear in this space. The resonance therefore is prevented from decaying and the energy of the 2s2p state computed in this manner is real valued.

The possibility of employing multi-configuration SCF theory using the complex scaled Hamiltonian to describe Feshbach resonances immediately suggests itself. We will offer some speculations about such a theory in a later section.

III. ATOMIC EXAMPLES: THE Be^- SHAPE RESONANCE AND A SIMPLE FESHBACH RESONANCE

The ^2P shape resonance in e-beryllium scattering is an example of a resonance which the complex SCF approach should describe well. We have performed two sets of calculations on this system, the first of which employs a (14s/16p) basis of real valued Gaussians chosen as follows. The 10s Huzinana[17] basis for beryllium was augmented by four diffuse s functions with exponents given by $0.052/2^n$ with n+1,2 ..., and 16 p functions with exponents given by $15.46/(2.26)^n$ with n=0,1 .-. . The complex SCF equations (eqs (3) - (6)) were solved in matrix form and E_{CSCF} from this calculation with $\theta = \exp(i\alpha)$ is plotted in the main portion of figure 2 for a range of α values between .3 and .5 radians. We were unable to converge the SCF equations for values of α less than .3. The trajectory of E_{CSCF} as a function of α is somewhat surprising in view of what has been observed in complex coordinate CI calculations.

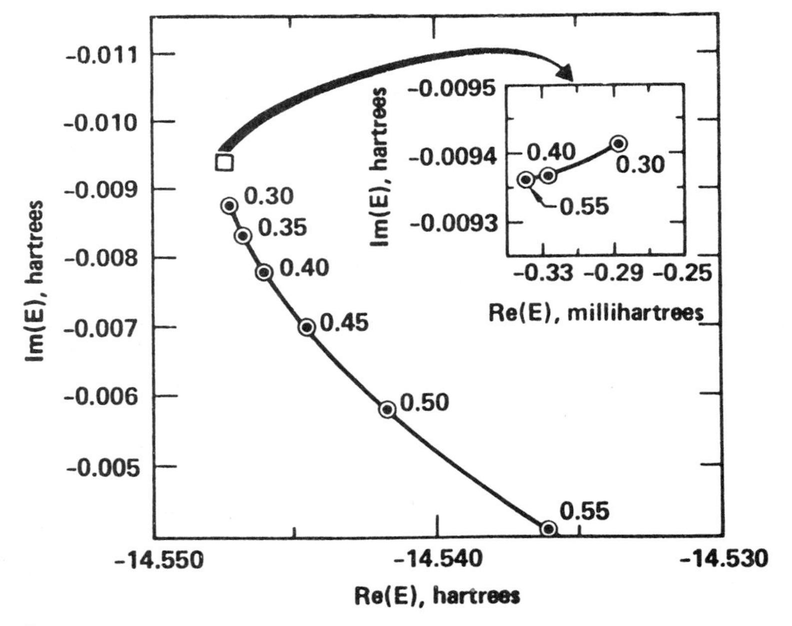

Figure 2. E_{CSCF} for Be as a function of α in the scaling parameter $\theta=\exp(i\alpha)$. The main curve is calculated with real basis functions, and the insert is the complex basis function result for the same range of α. The real part of the energy scale in the insert is relative to -14.547 Hartrees.

In a configuration interaction calculation using the complex Hamiltonian, $H(\{\theta\vec{r}_j\})$, there is a well known behavior to be expected of the complex resonance eigenvalue as a function of α. That behavior, which can be shown to be a consequence of the complex version of the virial theorem[18], is that for a given basis set the resonance eigenvalue, E_{res}, has a stationary point with respect to variation of θ so that, at some value of θ, $dE_{res}/d\theta$ vanishes. The stationary point does not necessarily occur with $|\theta| = 1$, but if a CI calculation is performed with $\theta=\exp(i\alpha)$ a sharp cusp in the complex value of E_{res} is often found as α is varied. The curve in figure 1 corresponding to the (14s/16p) real basis function calculation does not show such a cusp, nor is E_{CSCF} particularly stable as α is varied. This is a troublesome point because the stationary value of E_{res} is usually taken to be the best approximation to the resonance energy.

Thus the disturbing feature of the main curve in figure 2 is the fact that it shows little evidence of stationary behavior with respect to variation of α. This effect is due to a particular sort of inadequacy of the basis set and is an example of the numerical difficulties in the complex coordinate method to which we alluded in the introduction. The problem is not due to an inadequate description of the resonance part of the problem (specifically the 2p orbital in which the scattered electron is "trapped"), but instead it is due to a poor description of the orbitals which make up the Be target -- particularly the tightly bound 1s orbital. To see the origin of this problem consider a one particle system for which we are interested in finding the bound state eigenfunctions of the scaled Hamiltonian, $H(\theta\vec{r})$. If the bound eigenfunctions, $\phi_b(\vec{r})$, satisfy the unscaled equation

$$H(\vec{r})\phi_b(\vec{r}) = E_b\phi_b(\vec{r})$$

then the bound eigenfunctions of $H(\theta r)$ are found by a simple change of variable

$$H(\theta\vec{r})\phi_b(\theta\vec{r}) = E_b\phi_b(\theta\vec{r})$$

because, provided $\arg(\theta)$ is not too large (bound states covered by the cut), $\phi_b(\theta\vec{r})$ is square integrable. Now, the 1s orbital of beryllium in our problem behaves essentially as a one particle bound state. A simple approximation to the 1s orbital which has the right asymptotic form in the real coordinate case is a single exponential, $\exp(-\beta\theta\vec{r})$. For complex θ this function shows damped oscillatory behavior as a function of r, and the larger the value of β the more oscillatory this function becomes. It is exactly these oscillations which our basis of real valued Gaussian functions, $\exp(-\zeta_i\vec{r}^2)$, do not approximate well. This problem

Table I. Calculated Resonance Parameters for ^2P Be^-

	E_R(eV)	Γ(eV)
Rescigno, McCurdy, and Orel[19] (static-exchange)	0.76	1.11
Hunt and Moiseiwitsch[21] (static-exchange plus polarization)	0.60	0.22
Kurtz and Öhrn[22]		
Static-exchange	0.77	1.61
Static-exchange plus polarization	0.20	0.28
Donnelly and Simons[23] (Second order perturbation theory)	0.57	0.99
Present complex SCF	0.70	0.51

Finally, there are some comments to be made regarding practical aspects of these calculations. First of all, the complex SCF equations have continuum as well as resonance solutions and one may reasonably ask how convergence to the resonance solution, rather than to a poorly approximated continuum solution, can be assured. The solution to that problem lies in the fact that the resonance wavefunction is spatially very different from continuum wavefunctions. Thus among the eigenvectors of the orbital Hamiltonian, which have p symmetry, in our case, one may expect to find one which has the appearance of a bound wavefunction and which should be chosen as the occupied orbital. In figure 3 we plot the modulus of ϕ_{2p} for the resonant root and that of one of the continuum roots. The contrast is obvious with the 2p orbital density largely confined to small r values within the centrifugal barrier. In addition the orbital eigenvalue of the resonance orbital can be expected to lie closer to the real axis than those of nearby continuum orbitals. With only these simple notions of the qualitative behavior of resonance wavefunctions we were able to easily converge our calculations to the proper state in the cases we report here.

becomes worse with the more tightly bound inner orbitals of the target (characterized by larger values of β in the simple exponential approximation) which have little to do with the formation of resonances. So it would be worse still, for example, for the 1s orbital of calcium in the Ca^- p-wave resonance. The answer to this problem is to use complex Gaussian basis functions. If the set of our real valued Gaussian basis functions can adequately approximate the bound orbitals of the unscaled problem then complex Gaussian functions, $\exp(-\zeta_i\theta^2r^2)$, will be adequate for the bound orbitals in the complex scaled problem.

So we performed a second set of calculations on Be^- with only the s function exponents in the (14s/16p) basis scaled by θ^2. It should be noted that although the basis functions are now complex the entire calculation becomes real valued as θ approaches unity as we indicated should be the case in section II. The results of this calculation are plotted in the inset to figure 2. The stability of E_{CSCF} as a function of θ is remarkable in this calculation, but still no sign of clear cusp behavior was observed. Fortunately E_{CSCF} is so stable that it is unnecessary to find a stationary point of E_{CSCF} in order to find the resonance energy. Choosing a value of α near the center of the range plotted in figure 1 (α=0.4) we find E_{CSCF} = -14.54733 - .00937i (a.u.). The value for the width of the resonance is therefore 0.51 eV or about half the value from a static exchange calculation.[19] To get an estimate of the position of the resonance it seems most reasonable to subtract the Hartree-Fock value[20] of the ground state energy of Be from E_{CSCF} for Be^-, because one of the solutions of the complex SCF equations (a nonresonant, continuum solution) is the Hartree-Fock wavefunction for Be with the remaining electron in a continuum p orbital with (in a limiting sense) zero energy. The resonance position so obtained is 0.70 eV, only slightly lower than the static exchange value. In table I we compare these results with those of some other calculations. The principal source of disagreement among these results is the way in which the response of the target atom was treated in each of them. Unfortunately no experiments are currently available to resolve this problem. The situation is somewhat better in the case of N_2^- discussed in the following section, and it will be seen that the complex SCF results for simple shape resonances are indeed reliable. The case of ^2P Be^- seems to warrant more study.

Another important practical note concerns the use of complex basis functions. In an earlier publication[19] we made use of the identity, written here for the radial part of a one electron

$$\int \phi_i(r)H(r\theta)\phi_j(r)r^2dr = \theta^{-3}\int \phi_i(r\theta^{-1})H(r)\phi_j(r\theta^{-1})r^2dr \qquad (11)$$

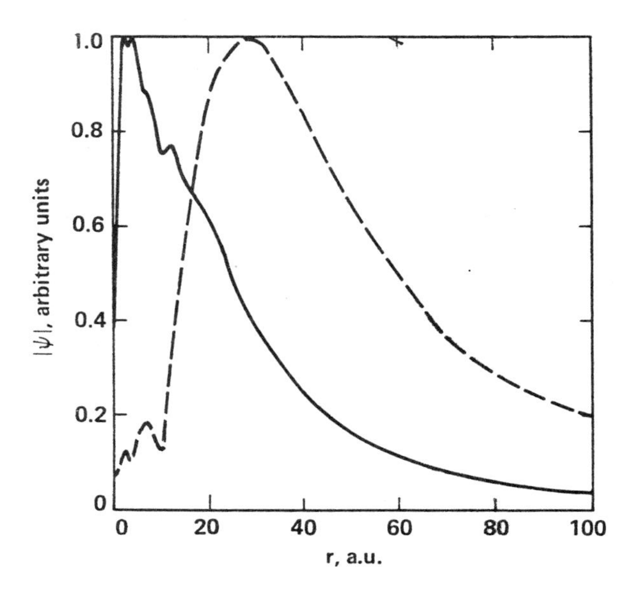

Figure 3. Modulus of the resonance ψ_{2p} orbital for Be^- (solid) curve) and the modulus of a virtual continuum p orbital (dashed curve) both at $\alpha=0.4$ from the complex basis function calculation.

problem, to perform a rotated coordinate calculation by diagonalizing the real Hamiltonian, $H(\vec{r})$, using complex basis functions, $\phi_i(\vec{r}\theta^{-1})$. It is in fact more convenient to use the same device in complex SCF calculations. Thus the second of our calculations on Be^- was actually performed with unscaled Fock operators and complex p basis functions of the form $\phi_i(\vec{r}\theta^{-1})$ and real s basis functions, instead of scaled Fock operators with real p and complex s basis functions as described above. These two forms of the calculation are completely equivalent, but the prescription of scaled Fock operators plus complex basis functions requires fewer complex two electron integrals and, more importantly, is a more convenient starting point for the extension to molecular problems. More will be said about the latter point in the next section.

The assertion that the complex SCF equations give no information about the width of a Feshbach resonance can be verified by performing a calculation on the 2s2p (^3P) autoionizing state of helium. We have performed this calculation using a (20s/18p) basis of uncontracted Gaussians consisting of the Huzinaga[17] 10s basis for helium augmented to 10 diffuse s functions with exponents given by $0.108/)2.25)^n$ with n=1,2 ..., and 18 p functions with exponents given by $30/(2.25)^n$ with n=0,1 ... . The complex Fock operators are given in equations (8) and (9). The value of E_{CSCF} from equation (10) for this state with the scale factor, θ, set to unity was found to be -0.753606 atomic units and for a substantially complex value of the scale factor, $\theta = \exp(i\ 20\ \pi/180)$, E_{CSCF} was $-0.753598 + 6.4 \times 10^{-6}\ i$ (a.u.). These two results would be identical in a complete basis and the extent to which they agree in the present calculation is evidence that the single determinant SCF description is inappropriate for Feshbach resonances. E_{CSCF} computed in this way does, however, give a fair approximation for the position of the resonance. E_{CSCF} minus the Hartree-Fock ground state energy for helium is 57.3 eV and E_{CSCF} minus the exact value of Pekeris[24] for the ground state energy is 58.4 eV. These values are to be compared with the value of 58.3 eV computed by more accurate methods[25].

IV. A MOLECULAR EXAMPLE: N_2^- [36]

In applying complex coordinate techniques to molecular problems we encounter a fundamental problem. Calculations on molecules are generally carried out in the Born-Oppenheimer approximation (fixed nuclei) and the problem of nuclear motion is solved subsequent to performing the electronic calculation for fixed nuclear positions. Molecular calculations would become impossibly complicated if we were to simultaneously treat nuclear and electronic motion in a single calculation. The problem we encounter as a result is that the elegant theorems of Aguilar, Balslev and Combes[2] are no longer applicable to molecules once we remove the nuclear motion. To see why this is the case, consider the Coulomb interaction potential, $V_\alpha(r)$, between an electron and one of the nuclei in a molecule

$$V_\alpha(\vec{r}) = \frac{1}{|\vec{R}_\alpha - \vec{r}|} \tag{12}$$

In this equation the nuclear coordinates, $\vec{R}_\alpha$ are fixed parameters. In a complex scaling calculation, since we are excluding the nuclear motion, we must now ask how this potential behaves if we scale the electronic coordinates, $\vec{r}$, while leaving those of the nuclei untouched.

$$
\begin{aligned}
V_\alpha(\theta r) &= \frac{1}{[(\vec{R}_\alpha - \theta\vec{r})^2]^{1/2}} \\
&= \frac{\theta^{-1}}{[(\theta^{-1}\vec{R}_\alpha - \vec{r})^2]^{1/2}}
\end{aligned} \tag{13}
$$

The square root branch point of the resulting complex nuclear attraction potential is the source of trouble in this attempt to analytically continue the Hamiltonian for the Born-Oppenheimer problem. Note that if we were including the complete dynamics of electronic and nuclear motion, we could scale $\vec{R}_\alpha$ by θ in equation (12) and then the resulting potential would be the real valued potential scaled by θ^{-1}. That scaled potential would be an analytic function of θ whereas the one in equation (13) is clearly not.

A solution to this dilemma was originally proposed by McCurdy and Rescigno,[11] and to date we have found no better approach. The procedure is most easily expressed as a generalization of the approach for atomic calculations described in the preceding section in which we use complex basis functions of the form $\phi_i(r\theta^{-1})$ together with a real valued Hamiltonian to effect the complex coordinate transformation. For molecular resonance calculations we use a real valued Hamiltonian again and Gaussian basis functions in which the exponents are scaled by θ^{-2}. These have the form

$$
\phi_{\ell mn}(\vec{r},\zeta\theta^{-2},A) = N(\zeta\theta^{-2})(x-A_x)^\ell(y-A_y)^m(z-A_z)^n\exp[-\zeta\theta^{-2}(\vec{r}-\vec{A})^2] \tag{14}
$$

where $\vec{A}$ denotes the coordinates of a nuclear center and $N(\zeta\theta^{-2})$ is the normalization constant. This prescription may be surprising at first glance, but it has been shown that it can be viewed[8,27] as a variant of a rigorous theory proposed by Simon[26] for the Born-Oppenheimer problem. Simon's theory is called exterior scaling, and as the name implies it calls for scaling the electronic coordinates only in the region outside a sphere which encloses the molecule. A particularly severe test of this numerical method has been given in an application to the calculation of the photoionization cross section of a molecule.[8] It was argued in that work that these complex basis functions can approximate the cusp behavior of the complex wavefunction at the nuclei and avoid the numerical problems which arise if that critical region of space is poorly described by the basis set. The numerical stability which the complex basis function approach provides is the principal reason for choosing it over alternative complex coordinate methods for the molecular problem.

To test the applicability of the complex basis function SCF theory on a molecular case, we chose the well studied case of the low energy $^2\Pi g$ resonance in electron $-N_2$ scattering. This is the first application of complex coordinate or complex basis function methods to a complicated many-electron molecular problem. The open-shall SCF equations for $N_2^-(^2\Pi g)$, which has the electronic configuration $(1\sigma_g^2 1\sigma_u^2 2\sigma_g^2 2\sigma_u^2 1\sigma_u^4 1\pi_g))\,^2\Pi_g$, were solved in a mixed basis set of real and complex cartesian Gaussian functions. The non-resonant core orbitals (all but $1\pi_g$) were expanded solely in terms of real functions. The physical reasons motivating such a choice are discussed at length in reference 19. We also carried out several numerical tests to check the adequacy of the core-orbital basis. These will be discussed later.

The $1\pi_g$ orbital was expanded in terms of both real and complex Gaussians. The latter were chosen following the prescription of McCurdy and Rescigno[11] by complex-scaling the orbital exponents of the Gaussian basis functions by a phase factor, $\theta^{-2}= e^{-i\phi}$. The nuclear centers associated with these functions are kept real as discussed above.

It is important to note that, in contrast to the atomic case, the one-and-two electron matrix elements involving complex functions cannot be obtained by any simple scaling of real matrix elements and must be recalculated for each new value of ϕ. These matrix elements are all analytic functions of the scale factor[8], however, and were obtained by appropriately modifying a standard molecular integral program.

The core orbitals were expanded in a nuclear-centered (9s 5p)/[5s 3p] contracted basis[28], augmented with two d-polarization functions (α = 1.0 and .4). Several choices for the π_g orbital space were tested. A preliminary set of calculations was done using a (4p) basis, augmented with additional diffuse d_π functions placed at the center-of-mass. All π_g functions were complex scaled. These calculations produced a complex total energy for $N_2^-(^2\Pi_g)$ which depended strongly on and varied monotonically with the rotation angle ϕ, the resonance width varying by roughly 40% over the range $15° < \phi < 25°$. A simple test was devised to demonstrate that this instability was due to the inadequacy of the π_g orbital basis and not the core orbital basis. The total SCF energy for $N_2^-(^2\Pi_g)$ can be written as $E = E_{core} + \varepsilon_{\pi_g}$ where E_{core} is the core-orbital contribution to the total energy and ε_{π_g} is the orbital energy of $1\pi_g$. A series of "static exchange" calculations was performed at different angles with the core orbitals frozen. It was found that changes in ε_{π_g} were precisely equal to the variations noted earlier in the SCF calculations.

The addition of several real p_π functions to the π_g orbital space greatly reduced the sensitivity of the energy to the rotation angle. The final π_g basis we used consisted of the (5p 2d)/[3p 2d] set of real functions, four nuclear-centered p_π functions with exponents .6, .26, .125, and .05 and six complex d_π functions placed at the center-of-mass with exponents ranging from 12 to .002 in a geometric series. The total energy as a function of rotation is shown in Figure 4 for an internuclear separation of 2.068 a.u., the equilibrium bond distance of N_2. The res-nance width is found to vary by roughly 6% within a range of angles between $28° < \phi < 38°$. An average of the data over this range gives a resonance width of .44 eV and an energy of 3.19 eV, when referenced to the SCF energy of ground state N_2 in the same basis. Table II compares these results to those of other recent theoretical calculations.

Table II. Comparison of calculated electronic resonance parameters for $N_2^-(^2\Pi_g)$. Energies, ε_r, and widths Γ are in electron volts and R_o = 2.068 a.u.

	$\varepsilon_r(R_o)$	$\Gamma(R_o)$
This work	3.19	.44
Krauss and Mies[29] (stabilization)	3.26	.8 ± .3
Schneider et. al.[30] (R-matrix)	2.15[a]	.34
Hazi et. al.[31]	3.23 (2.16[a])	.42
Levin and McKoy[32] (T-matrix)	2.19[a]	.37

[a] Resonance energy relative to the energy of a fictitious N_2 neutral core made up with N_2^- orbitals.

In all of the calculations in table II including those of Krauss and Mies[29], the N_2 target was allowed to respond to the presence of the scattered electron. Because the various calculations treated this polarization of the target somewhat differently, the agreement of the calculated widths is quite gratifying. There is some difficulty in reporting the real part of the resonance energy because it must be reported relative to that of the neutral N_2 molecule. The SCF method makes different overall errors in the energy N_2 and N_2^- because correlation effects (which are not treated in the SCF approximation) are quite different in molecules with different numbers of particles. This problem is an old one encountered in the calculation of electron affinities and ionization potentials. Although there are several possible dodges which involve the calculation of a reference energy of a fictitious N_2 neutral core made up of N_2^- orbitals[31], the only truly satisfactory answer is to perform more accurate calculations on both the neutral molecule and the anion. It is nevertheless true that the width of the resonance calculated by the complex SCF approach is the same as that used in calculations of vibrational excitation cross sections which are in agreement with experiment[31]. So our difficulties with the real part of the resonance energy are in some sense due to the use of an inconsistent reference energy for the neutral molecule.

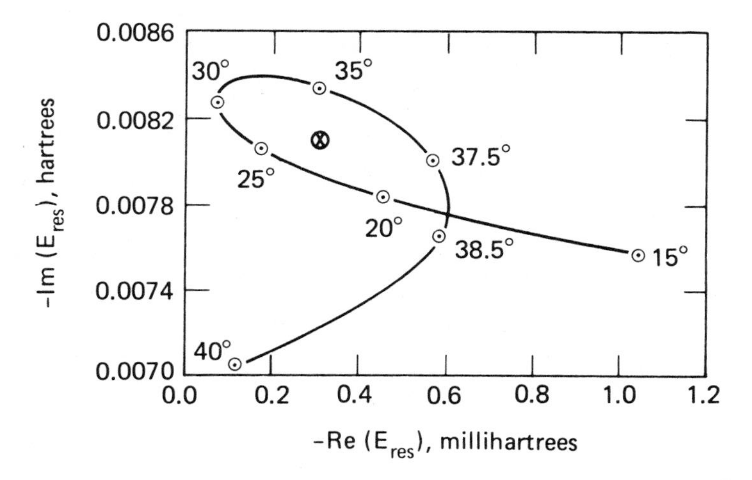

Figure 4. SCF resonance energy of $N_2^-(^2\Pi_g)$ as a function of rotation angle. The real part of the energy scale is given relative to -108.8494 hartrees. The internuclear separation is 2.068 a.u.

V. SPECULATIONS ON OTHER APPLICATIONS

In section II we discussed the reason why the complex coordinate version of restricted Hartree-Fock theory is unable to treat Feshbach resonances. Perhaps a convenient way to describe Feshbach resonances will be to use a complex coordinate version of multiconfiguration SCF theory[33] (MCSCF). For example, we would treat the 2s2p triplet autoionizing the state of helium by choosing a two configuration trial function of the form

$$\Phi = a|\psi_{2s}\psi_{2p}| + b|\psi_{1s}\psi_{kp}| \quad . \tag{15}$$

All four orbitals and the ratio of coefficients, a/b, would then be determined variationally. Although a number of practical problems could arise with convergence, one clearly expects that the basic mathematics of MCSCF theory (in some form) will be applicable to the states of the complex scaled Hamiltonian. The real question is whether or not there is enough physics in a simple wavefunction consisting of only a few configurations like that in equation (15) to accurately describe a Feshbach resonance in general.

The complex coordinate or complex basis function generalization of MCSCF theory, if it proves feasible, will be useful in calculations on shape resonances as well. Complex coordinate Hartree Fock calculations neglect effects of electron correlation which may strongly influence the widths and positions of even shape resonances in some cases. Examples of such systems can easily be found in electron scattering from target molecules in which electronic correlation is an important effect such as F_2 and HF. In fact, when one examines the multi-determinant trial functions for these cases it becomes apparent that the distinction between Feshbach and shape resonances is somewhat arbitrary for highly correlated systems. Strong configuration interaction cannot be clearly distinguished from virtual excitation of the target.

Another interesting possibility is the application of complex coordinate or complex basis function methods to calculations on virtual states in electron or position scattering from atoms or molecules. A virtual state[34] corresponds to a pole of the S-matrix on the negative imaginary axis of the momentum plane (k plane). Such states are known to occur in s-wave scattering from spherical potentials and cause a dramatic enhancement of the elastic scattering cross section near threshold. The interest in such states in electron scattering is largely due to the speculation that the enhancement of the vibrational excitation cross section near threshold in collisions of electrons with polar molecules (such as e^--HCℓ) is due to the presence of a virtual state[35].

It is easy to see that, in potential scattering, complex coordinates can be used to calculate the position of a virtual state. This fact is of some importance in itself because other methods applicable to the resonance problem which rely on localization (increased amplitude near the target) of the scattering wavefunction at the resonance energy cannot be applied to virtual states. There is in general no significant localization of the scattering wavefunction at low energies in the presence of a virtual state. To see that complex coordinates can be used to find virtual states consider the natural example of s-wave potential scattering for which the asymptotic form of the wavefunction is [34]

$$\psi(r) \sim \frac{i}{2} \left[f_o(k)e^{-ikr} - f_o(-k)e^{+ikr}\right] \tag{16}$$

where $f_o(k)$ is the s-wave Jost function. At the momentum corresponding to a virtual state, $k = -i\gamma$, the Jost function $f_o(-i\gamma)$ is zero and only the second term in equation (16) remains. If we scale r according to $e^{i\phi}r$ with ϕ real valued and greater than $\Pi/2$, the asymptotic form of the wavefunction at the virtual state momentum becomes

$$\psi_v(re^{i\phi}) \sim \frac{-i}{2} f_o(-i\gamma)e^{-i\gamma re^{i\phi}} \xrightarrow[r\to\infty]{} 0$$

So the complex coordinate transformation can be used to render the virtual state wavefunction square integrable, and it can then be approximated by basis set methods. All this is not surprising of course because all we have done is to rotate the continuum branch cut by more than 180° in the energy plane and therefore exposed the virtual state pole.

The problem with this idea is that if we apply it to a many-electron system we find that the inner orbitals of the target cease to be square integrable as the rotation angle exceeds 180°. There are several possible ways to evade this problem one of which is to find an effective one-body potential (an "optical potential") which adequately describes the scattering of electrons by the target. The complex coordinate transformation could then be applied to the effective single particle problem. The whole question of virtual states in electron and position scattering is an interesting one which hopefully will receive more attention in the future.

In this paper we have described one particular advance in the effort to find practical and accurate methods to exploit complex coordinate ideas in resonance calculations. It will certainly be necessary to explore other approaches as well. Reducing a scattering calculation to a "bound state" problem is only useful if we can solve the resulting complex bound state problem. Even ordinary bound states of many-electron molecules are difficult to describe accurately so it is to be expected that we will be a long time in refining complex coordinate calculations to the point that they finally become routine.

ACKNOWLEDGMENT

The author's research on this topic and his attendance at the Boulder conference were supported by National Science Foundation grant No. CHE-7907787.

REFERENCES

1. J. Nuttall and H.L. Cohen, Phys. Rev. 188, 1542 (1969).

2. J. Aguilar and J. Combes, Commun. Math. Phys. 22, 269 (1971); E. Balslev and J. Combes, Commun. Math. Phys. 22, 280 (1971).

3. B. Simon, Commun. Math. Phys 27, 1 (1972); B. Simon, Ann. Math. 97, 247 (1973).

4. J. Avron and I. Herbst, Commun. Math. Phys. 52, 239 (1977); I. Herbst and B. Simon, Phys. Rev. Lett. 41, 67 (1978).

5. For a brief overview of the mathematical aspects of complex coordinates see: B. Simon, Int. J. Quantum Chem. 14, 529 (1978). We give only a few examples from the mathematical literature. Further references can be found in Int. J. Quantum Chem 14 (4), (1978) which is an issue devoted entirely to complex scaling.

6. To give only a few examples of computational applications of complex coordinate ideas: G. Doolen, J. Nuttall and R. Stagat, Phys. Rev. A10, 1612 (1974); G. D. Doolen, J. Phys. B 8, 525 (1975); R. A. Bain, J. N. Bardsley, B. R. Junker and C. V. Sukumar, J. Phys. B 7, 2189 (1974); T. Rescigno and V. McKoy, Phys. Rev. A12, 552 (1975); A. P. Hickman, A. D. Isaacson, and W. H. Miller, Chem. Phys. Lett. 37, 63 (1976). See also reference 5.

7. T. Rescigno and V. McKoy, Phys. Rev. A12, 552 (1975); C. V. Sukamar and K. C. Kulander, J. Phys. B 11, 4155 (1978).

8. C. W. McCurdy, Phys. Rev. A 21, 464 (1980); C. W. McCurdy and T. N. Rescigno Phys. Rev. A 21, 1499 (1980).

9. For a brief review see: C. W. McCurdy in Electron-Molecule and Photon-Molecule Collisions, edited by T. Rescigno, V. McKoy, and B. Schneider (Plenum, New York 1979).

10. A collection of reviews appears in: Methods of Electronic Structure Theory, Modern Theoretical Chemistry, Vol. 3, edited by H. F. Schaefer (Plenum, New York 1976).

11. C. W. McCurdy and T. N. Rescigno, Phys. Rev. Lett. 41, 1364 (1978).

12. Most of the theory described in this section has appeared in: C. W. McCurdy, T. N. Rescigno, E. R. Davidson, and J. G. Lauderdale, J. Chem. Phys. 73, 3268 (1980).

13. SCF methods in bound state problems are reviewed by: R. McWeeny and B. T. Suteliff, Methods of Molecular Quantum Mechanics, Academic Press, New York (1979) and in reference 10.

14. E. R. Davidson and L. Z. Stenkamp, Int. J. Quant. Chem. 10 S, 21 (1976).

15. Equivalent CI for SCF calculations on excited states is discussed in reference 6 and by H. Hsu, E. R. Davidson and R. M. Pitzer, J. Chem. Phys. 65, 609 (1976).

16. R. Lefebvre in Modern Quantum Chemistry, part I. ed. O. Sinanoglu (Academic Press, New York, 1965) p. 125.

17. S. Huzinaga, J. Chem. Phys. 42, 1293 (1965).

18. See reference 5 and: P. Froelich and E. Brandas, Phys. Rev. A16, 2207 (1977); R. Yaris and P. Winkler, J. Phys. B 11, 1475 (1978).

19. T. N. Rescigno, C. W. McCurdy and A. E. Orel, Phys. Rev. A 17, 1931 (1978).

20. The value of the Hartree-Fock ground state energy of Be used here is that obtained using the Slater basis given by E. Clementi, Tables of Atomic Functions (IBM Corp., Armonk, N.Y., 1965).

21. J. Hunt and B. L. Moiseiwitsch, J. Phys. B 3, 892 (1970).

22. H. A. Kurtz and Y. Öhrn, Phys. Rev. A 19, 43 (1979).

23. R. A. Donelly and J. Simons, J. Chem. Phys. 73, 2858 (1980).

24. C. L. Pekeris, Phys. Rev. 115, 1216 (1959).

25. A. U. Hazi, in Electron-Molecule and Photon-Molecule Collisions, Edited by T. N. Rescigno, V. McKoy and B. Schneider (Plenum, New York, 1979) p. 281.

26. B. Simon, Phys. Lett. 71A, 211 (1979).

27. B. Simon and J. Morgan (preprint).

28. T. H. Dunning, J. Chem. Phys. 53, 2023 (1970).

29. M. Krauss and F. H. Mies, Phys. Rev. A 1, 1592 (1970).

30. B. Schneider, M. LeDourneuf and Vo Ky Lan, Phys. Rev. Letts. 43, 1926 (1979).

31. A. U. Hazi, T. N. Rescigno, and M. Kurilla, Phys. Rev. A (submitted).

32. D. A. Levin and V. McKoy, Phys. Rev. A (submitted).

33. For a review see: A. C. Wahl and G. Das, in Methods of Electronic Structure Theory, Modern Theoretical Chemistry, Vol. 3, edited by H. F. Schaefer, (Plenum, New York 1976) p.51.

34. J. R. Taylor, Scattering Theory (Wiley, New York 1972) p.246.

35. R. K. Nesbet, J. Phys. B 10, L739 (1977); L. Dube' and A. Herzenberg, Phys. Rev. Lett. 38, 820 (1977).

36. The calculation presented in this section is described in: T. N. Rescigno, A. E. Orel, and C. W. McCurdy, J. Chem. Phys. (to appear in Dec. 1980).

TIMESTEP CONTROL FOR THE NUMERICAL SOLUTIONS OF INITIAL-BOUNDARY-VALUE PROBLEMS

H. Tadjeran, K. Gustafson, and J. Gary

Department of Mathematics and Computer Science

University of Colorado, Boulder, Colorado 80309

Abstract. The process of dynamic time-step selection for fixed spatial resolution based on the idea of balancing different local truncation errors will be presented. Some theoretical justification and numerical results will also be given. The application of iterative improvement to ADI methods will also be discussed.

INTRODUCTION AND GENERAL REMARKS.

The question of optimalizing the time step for the numerical solution of initial-boundary-value problems is not an easy one. Even for linear problems the sensitivity of existing schemes to the type of boundary and initial conditions and their interplay can be surprising. One need only take a textbook example (for example, start with Ames [1, Problem 2-12]) and experiment with different mesh ratios, even on stable schemes, to find this out.

For partial differential equations in higher dimensions with spatial mesh already of inherently large size, taking a time-step relatively too small will greatly increase the amount and cost of computing without significantly improving the accuracy of the numerical solution. On the other hand, a relatively large time step may adversely affect the accuracy. As domain and dimension size increase, these considerations become of increasing importance, even for schemes such as ADI where efficiencies of calculation have already been introduced.

For nonlinear problems of partial differential equations little is known about time step control. Even for the erratic trajectories being numerically found in other talks at this conference attempting to distinguish stochastic from quasi-periodic regions in intramolecular dynamics one is placing great faith on the time step control features in the numerical schemes and package being employed. Should one attempt true three dimensional trajectory tracing rather than two dimensional sectioning these considerations would become increasingly important.

In a number of problems of mathematical physics, S. Ulam (see, e.g , [12, 13]) became concerned with the possibility that on the microscopic level mathematical structures other than the usual Euclidean geometry might better describe the real space time dynamics taking place. Even within the usual framework one can see problems of time step control becoming possibly of paramount importance when working with a dynamical system in which there is great variance in the rate of transition in the flow of phase space from one region to another. And there are always questions of whether there are fundamental critical time increment limits in approximating quantum behavior by numerical studies based on a semiclassical approach.

TIME STEP CONTROL

To set the optimal stepsize to meet a given error tolerance in the numerical solution of ordinary differential equations (ODE's), one usually obtains two approximate solutions. The difference between these approximate solutions (or the appropriate norm of that difference) estimates the error in the less accurate solution. This estimate is then used to set the optimal stepsize, using some properties of the numerical solution, such as the order of convergence. See Shampine and Gordon [7], Shampine, et al. [8], Gear [5] . Many production codes implement this idea.

One may combine the above idea with the method of lines (see Gary [3]) in order to solve initial-boundary-value problems (IBVP) . In this procedure, the IBVP is transformed into a system of ODE's by discretizing the problem in the spatial variable(s) only. One can then use the ODE methods (or solvers) to obtain the solution of this system and therefore that of the IBVP . The error produced by the ODE solver is viewed as an error in time only, and "interplay" between the errors due to spatial and temporal discretization is now allowed. (In other words, the error due to temporal discretization is controlled without concern for the error present due to spatial discretization.) See Warming and Beam [11] .

Swartz and Wendroff [9] , using an idea of Douglas [2], pose the problem of optimal time step as follows:

Given a finite difference method in one space dimension with time step Δt and spatial mesh h , assume that the global error ε has the form:

$$\varepsilon \approx C_1(\Delta t)^q + C_2(h)^q \qquad (p,q > 0)$$

Then assume that the work required to obtain a solution for $0 \leq t \leq T$ is:

$$W = \frac{K}{\Delta t \cdot h}$$

Now choose Δt and h such that a fixed error ε is achieved with a minimum W. The constant K will depend on the scheme and the problem involved.

Gary [3] generalizes the above treatment to the case where

$$\varepsilon \approx h^q e(x,t,\lambda) \quad , \quad \lambda = \frac{\Delta t}{h^{q/p}}$$

and

$$W = \frac{K}{\Delta t \cdot h^d} \quad ,$$

d the dimension of the space of integration. The problem is then reduced to the choice of the optimal value of λ (which yields the minimum value of the work function). In [3] two approximate solutions using two different mesh spacings (with the same value of λ) estimate the error function. This is then repeated for different values of λ, and the optimal value of λ is estimated. Thus, once the spatial mesh is specified, the optimal (constant) time step can be determined.

These results can also be used to compare the relative effectiveness of different finite difference schemes.

However, theoretically the determination of the work function requires the knowledge of the exact solution. Thus although the above procedure yields a theoretical analysis of the time step selection, in many problems it cannot be employed as a practical tool.

A PRACTIAL ALGORITHM FOR OPTIMAL TIME STEP.

We now investigate a practical process of dynamically selecting the "optimal" time step for the numerical solution of the initial-boundary-value problems. By "optimal" time step we mean the following.

Given the spatial resolution(s) Δx on a uniform grid, the global (or total) error in integration with a time step of Δt will be partly due to spatial discretization and partly due to temporal discretization. At each step of the integration, we would like to choose Δt so that the contribution to the global error by the spatial and temporal discretizations are balanced.

Thus we define the optimal time step to be the time step for which the temporal truncation error is balanced by the spatial truncation error. For a general finite difference scheme let ϵ_t denote the temporal truncation error, ϵ_s the spatial truncation error, and ϵ_{st} the total truncation error (in the next section we will consider a particular scheme). We may then describe the general algorithm as follows.

Let the spatial mesh be fixed and assume that we can estimate the spatial truncation error ϵ_s . The latter may be obtained for example by known methods such as those using convergence rates or by varying Δt near the initial values until there is no change in ϵ_s . Suppose we expect a quadratic ratio between Δt and Δx in a parabolic problem, we then set up an algorithm which may be essentially described as:

$$\epsilon_t = (\text{const})((\Delta t_{est.})^2 = (\text{const})((\Delta t)_{opt.})^2 = \epsilon_s \quad .$$

Requiring the constant to be invariant then produces $(\Delta t)_{opt.} = (\epsilon_s/\epsilon_t)^{\frac{1}{2}}(\Delta t)_{est}$. The quantity $(\Delta t)_{est.}$ may be found in terms of the particular scheme employed and the approximate solutions to it (see the next section).

ITERATIVE IMPROVEMENT FOR TIME STEP CONTROL

In this section we will show the application of iterative improvement to the dynamic determination of an optimal time step in a marching problem. To illustrate the method we will briefly examine a simple example of linear parabolic equations. For details of theoretical considerations and some numerical results refer to Tadjeran [10] .

Consider

$$u_t = A(x,t)u_x + B(x,t)u_x + C(x,t)u + f(x,t) \quad ,$$

$$0 < x < 1 \quad , \quad 1 < t \leq T \quad ,$$

with appropriate initial and boundary conditions (say, Dirichlet boundary cnnditions). We assume that the problem is properly-posed.

Suppose now that one uses the implicit (or backward) Euler scheme to numerically solve this problem.

Let $h = \frac{1}{I}$, $x_i = ih$, $0 \leq i \leq I$ and $\Delta t = \frac{T}{N}$, $t_n = nh$, $0 \leq n \leq N$.

We use lower-case u_i^n to denote $u(x_i, t_n)$, and use upper-case U_i^n to denote the (approximate) numerical solution. Then the scheme becomes:

$$\frac{U_i^{n+1} - U_i^n}{\Delta t} = A_i^{n+1} \frac{U_{i+1}^{n+1} - 2U_i^{n+1} + U_{i-1}^{n+1}}{h^2} + B_i^{n+1} \frac{U_{i+1}^{n+1} - U_{i-1}^{n+1}}{2h} + C_i^{n+1} U_i^{n+1} + f_i^{n+1}$$

and we set the boundary and the initial conditions accordingly.

This scheme is stable, first order accurate in time and second order accurate in space.

The error in discretization is partly due to discretization in time and partly due to discretization in space. We attempt to balance those errors by selecting a suitable time step.

We refer to $\tau t_i^{n+1} = \dfrac{u_i^{n+1} - u_i^n}{\Delta t} - \left(\dfrac{\partial u}{\partial t}\right)_i^{n+1}$ as the temporal truncation error, and to

$$\tau x_i^{n+1} = A_i^{n+1} \frac{u_{i+1}^{n+1} - 2u_i^{n+1} + u_{i-1}^{n+1}}{h^2} + B_i^{n+1} \frac{u_{i+1}^{n+1} - u_{i-1}^{n+1}}{2h} - \left(A\frac{\partial^2 u}{\partial x^2} + B\frac{\partial u}{\partial x}\right)_i^{n+1}$$

as the spatial truncation error.

The idea of error estimation in our algorithm is an extension of the Error Estimation and Iterative Improvement given by Lindberg [6] .

To estimate the spatial truncation error, let $\Delta t = k \cdot \Delta x^2$ (or some such relation). Let U_i^n be the solution obtained from the implicit Euler scheme. Then, under appropriate conditions (see Tadjeran [10]):

$$A_i^n \cdot \frac{U_{i+1}^n - 2U_{i+1}^n U_{i-1}^n}{} + B_i \cdot \frac{U_{i+1}^n - U_{i-1}^n}{2h} - A_i^n \cdot \frac{-U_{i+2}^n + 16U_{i+1}^n - 30U_i^n + 16U_{i-1}^n - U_{i-2}^n}{12h^2} - B_i^n \cdot \frac{-U_{i+2}^n - 8U_{i+1}^n - 8U_{i-1}^n + U_{i-2}^n}{12h}$$

approximates τx_i^n, up to terms of $O(h^4)$, for $2 \le i \le n - 2$. For $i = 1$ or $n - 1$, the five-point approximations for u_{xx} and u_x should be replaced with corresponding one-sided

differences in the above calculations. Moreover
$\frac{U_i^n - U_i^{n-1}}{\Delta t} - \frac{U_i^{n+1} - U_i^{n-1}}{2\Delta t}$ is an $O(h^2)$ correct estimate of the temporal truncation error τt_i^n, for $1 \leq n \leq N - 1$. For $n = 1$ or $N - 1$, the formula $\frac{U_i^{n+1} - U_i^{n-1}}{2\Delta t}$ which is a second order approximation to $(\frac{\partial u}{\partial t})_i^n$ should be replaced by the corresponding second order accurate one-sided differences.

We may then use the numerical properties of the approximate solution to set the optimal time step, (see e.g., Shampine and Gordon [7]). The above algorithm has been satisfactorily applied to some linear test problems of the form $u_t = Au_{xx} + Bu_x + Cu + f(x,t)$ with Dirichlet boundary conditions. The theory is valid for non-linear problems and we are planning to apply it to those problems.

Final Note: We should remark that iterative improvement theory of Lindberg [6] applies to the ADI method for the two dimensional heat equation to produce a fourth order solution. This will appear in a separate paper, or see Tadjeran [10] Chapters II and III

References

1. Ames, W.F., Numerical Methods for Partial Differential Equations, 2nd Edition, Academic Press, New York, 1977.
2. Douglas, Jr., J., "A survey of numerical methods for parabolic differential equations," in Advances in Computers, Vol. 2, edited by F. Alt, New York, Academic Press, 1961, pp. 1-54.
3. Gary, J.M., "On the optimal time step and computational efficiency of difference schemes for PDE," J. of Computational Physics, Vol. 16 (1974), pp. 298-303.
4. __________, "The method of lines applied to a simple hyperbolic equation," J. of Computational Physics, Vol. 22 (1976), pp.. 131-149.
5. Gear, C.W., Numerical Initial Value Problems in Ordinary Differential Equations, Englewood Cliffs, N.J., Prentice Hall, 1971.
6. Lindberg, B., "Error estimation and iterative improvement for the numerical solution of operator equations," UIUCDSD-R-76-820, Dept. of Computer Science, University of Illinois at Urbana-Champaign (July 1976).
7. Shampine, L.F., and M.K. Gordon, Computer Solution of Ordinary Differential Equations, San Francisco, W.H. Freeman and Company, 1975.

8. Shampine, L.F., H.A. Watts, and S.M. Davenport, "Solving non-stiff ordinary differential equations-- the state of the art," SIAM Review, Vol. 18 (1976), pp. 376-410.

9. Swartz, B., and B. Wendroff, "The relative efficiency of finite difference and finite element methods, I: Hyperbolic problems and splines," SIAM J. of Numerical Analysis, Vol. 11 (1974), pp. 979-993.

10. Tadjeran, Hamid, Ph.D. Dissertation, University of Colorado at Boulder (1980).

11. Warming, R.F., and R.M. Bean, "Factored, A-stable, linear multistep methods -- an alternative to the method of lines for multidimensions," Conference Working Paper, 1979 SIGNUM Meeting on Numerical ODEs, Champaign Illinois, (April, 1979).

12. Ulam, S., A Collection of Mathematical Problems, Wiley-Interscience, New York, 1960.

13. ________, How to formulate mathematically problems of rate of evolution, Mathematical Challenges to the Neo-Darwinian Interpretation of Evolution, Symposium Held at the Wistar Institute of Anatomy and Biology, 1966, Ed.: P. Moorhead and M. Kaplan, Wistar Institute Symposium Monograph no. 5, Philadelphia, 1967, 21-23.

QUANTUM MECHANICAL ANGULAR DISTRIBUTIONS AND GROUP REPRESENTATIONS ON BANACH SPACES

Michael P. Strand[1] and R. Stephen Berry[2]

[1]Joint Institute for Laboratory Astrophysics
University of Colorado and National Bureau of
Standards, Boulder, Colorado 80309

[2]Department of Chemistry, The University of Chicago
Chicago, Illinois 60637

ABSTRACT

In large classes of scattering experiments in atomic, molecular, and nuclear physics the angular distribution of a product of the reaction can be described by simple trigonometric formulae. For example, the angular distribution of a particle ejected from an unpolarized target due to the absorption of a photon via a dipole process takes the phenomenological form $A + B \cos^2 \theta$. The microscopic parameters describing the detailed structure of the target determine the values of the macroscopic parameters A and B but do not change the phenomenological form. We show that the phenomenological form of an angular distribution may be simply derived from physically induced representations of the rotation group on the space of states and on the space of observables which are represented, respectively, by the trace class operators T(H) and the bounded operators L(H) on a Hilbert space H. Specifically we show that with appropriate topologies the representations on T(H) and L(H) are completely reducible and that Schur's lemma applies. These results lead to the appropriate phenomenological form for the angular distribution and also provide straightforward procedures for evaluating the macroscopic parameters in terms of microscopic parameters and geometrical factors.

I. INTRODUCTION

The application of group representation theory to quantum mechanical systems which possess some sort of symmetry has a long history. Its utility for simplifying both concepts and calculations in quantum theory may be seen by examining any of the vast number of books on the subject by mathematicians, chemists and physicists.

Traditionally group theory has entered into quantum mechanics through the use of unitary group representations on the Hilbert space of pure quantum mechanical states. By applying results from the well-developed mathematical theory of unitary group representations many strong results of physical importance may be derived. These include conservation laws for the values of observables associated with the symmetries of the system, selection rules for the possible changes in the values of these observables when two or more systems interact with each other, and limits to the number of independent parameters needed to describe the dynamics of such interacting systems. All of these results are expressed in terms of transition amplitudes. Utilization of group theory provides significant economies in the description and calculation of such processes.

In this paper we consider quantum mechanical applications of group representations on Banach and more general topological vector spaces. The utility of considering these more general representations arises because the physical systems involved in most experiments are not in pure quantum mechanical states; rather they are in mixed states which are most conveniently represented by density operators. Of course the theory of unitary group representations may still be applied to such systems by representing the mixed state as a weighted average of pure states and applying group theory to the pure states involved. In this procedure group theory is applied at the wave function (i.e., pure state) level and makes direct statements about transition amplitudes only while the restrictions on the physically observed transition probabilities that are imposed by symmetry are frequently obtained only after detailed calculations are performed. In contrast to this we have recently shown[1-3] that group representation theory may be applied at the mixed state (i.e., density operator) level directly. Thus applied, group theory makes statements directly about transition probabilities and is consequently more closely related to quantum theory at the observational level. The price paid for this advantage is that the representation spaces involved are no longer Hilbert spaces so that the representations are no longer unitary. Consequently many of the familiar results from unitary representation theory which lead to useful physical conclusions need not be true in this more general representation theory. Since the theory of group representations on Banach spaces (and more general topological vector spaces) is not well developed these results must be individually examined and verified or rejected. We have shown that for compact Lie groups many of the familiar results from unitary representation theory do in fact generalize to the physically induced (non-unitary) representations at the density operator level.

In Sec. II we review some aspects of the theory of unitary representations of compact Lie groups and show how they lead immediately to conservation laws, selection rules, and the Wigner-Eckart theorem when the representation space is the Hilbert space of pure states. This is done so that when nonunitary representations are considered we will have a model for both the mathematics and the physical consequences of representation theory immediately available. Section III contains a brief description of the angular distribution experiments which prompted our investigation of group representation theory at the mixed state level. In Sec. IV the Liouville representation of mixed state quantum theory is reviewed. Since the fundamental vector spaces in this representation -- the State Space and the Observable Space -- are not Hilbert spaces we are led to the consideration of group representations on topological vector spaces which have at most a Banach space structure. It is shown that when State Space and Observable Space are given physically appropriate topologies the physically induced representations of compact Lie groups are sufficiently well behaved that conservation laws, selection rules, and the Wigner-Eckart theorem may be derived. These three results are stronger than the corresponding Hilbert space results since they are expressed directly in terms of transition probabilities. The application of these results to the study of angular distributions leads to considerable simplification in the theoretical analysis of such experiments.

No proofs are given here as they have been presented elsewhere. Here a review is given of those results which have thus far been obtained. While this work was prompted by the need to efficiently calculate angular distributions we believe that these methods will prove useful for treating many other experiments where correlations in mixed state systems are involved.

II. UNITARY REPRESENTATIONS ON THE HILBERT SPACE OF PURE STATES

In this section we consider some of the implications of symmetry when a system is in a pure state. The Hilbert space formulation of von Neumann is used, together with the notation of Dirac. All of the results quoted in this section are well known; they are presented here only to serve as a model for the consequences of symmetry. In the next section when mixed states are considered this model will be followed for both mathematical and physical results.

Using the standard Hilbert space formulation of quantum mechanics we represent a pure state of the physical system by a vector $|\psi\rangle$ in a separable Hilbert space H. The vector $|\psi\rangle \in H$, when regarded (through the natural isomorphism of a Hilbert space and its dual) as a linear functional on H is denoted by $\langle\psi|$. The Hermitian product of $|\phi\rangle$ and $|\psi\rangle$ is denoted by $\langle\phi|\psi\rangle$ which satisfies $\langle a\phi|b\psi\rangle = a^*b\langle\phi|\psi\rangle$ for complex a and b. Dynamics of an isolated system is represented by a continuous one parameter group of unitary operators U(t) on H; if $|\psi\rangle$ is the vector representing the system at time t = 0 then

$$|\psi(t)\rangle = U(t)|\psi\rangle \tag{1}$$

is the vector representing the system at time t. Finally, an observable is represented by a self-adjoint operator O on H. If the system is in the state $|\psi\rangle$ the expectation value for the measurement of O is given by

$$\langle O\rangle_\psi = \langle\psi|O|\psi\rangle \quad . \tag{2}$$

A particularly important application of this rule is when the observable O corresponds to asking the question: Is the system in state $|\phi\rangle$? Then O is the projection operator $O = |\phi\rangle\langle\phi|$ and the probability that the system, initially in state $|\psi\rangle$ is found at time t to be in state $|\phi\rangle$, is

$$\langle\psi(t)|\phi\rangle\langle\phi|\psi(t)\rangle = \langle\psi|U^+(t)|\phi\rangle\langle\phi|U(t)|\psi\rangle \quad . \tag{3}$$

Equation (3) gives the expression for the transition probability from state $|\psi\rangle$ to state $|\phi\rangle$, while the factor $\langle\phi|U(t)|\psi\rangle$ is termed the transition amplitude.

A symmetry operation on a physical system may be defined as an invertible operation which commutes with time development. It is known[4] that every symmetry operation may be represented by a unitary operator U on H. (We neglect here the possibility of time reversal.) If we have a group G of symmetry operations it follows from the definition of a symmetry operation that U(t) commutes with the action of the group G for all times t, i.e., that

$$U(t)U(g) = U(g)U(t) \quad \text{for all} \quad g\varepsilon G \quad \text{and all} \quad t\varepsilon R \quad . \tag{4}$$

From the preceding it is seen that symmetry groups lead naturally to the study of unitary group representations within the conventional formulation of pure state quantum theory. A few classical results from the theory of unitary group representations are now listed. These results have been chosen because of their physical implications and in order that they may be contrasted with results for representations on topological vector spaces which are not Hilbert spaces. At this time we restrict attention to representations of compact topological groups. All representations U will be assumed to be continuous unitary representations of a compact topological group G on a Hilbert space H. It is known that[5]:

U1 If H_1 is a closed subspace of H invariant under the representation U then there exists a complementing closed invariant subspace H_{1c} such that H is the direct sum $H = H_1 \oplus H_{1c}$ of H_1 and its complement, H_{1c}. H_{1c} may be chosen to be the subspace of all vectors orthogonal to H_1.

U2 (Peter-Weyl) - a) If U is irreducible then H is finite dimensional. b) The right regular representation of G is completely reducible into a direct sum of irreducible subrepresentations. c) Every irreducible unitary representation of G occurs in the right regular representation with a multiplicity equal to its dimension.

U3 The unitary representation U is completely reducible into a direct sum of irreducible representations of G. That is H is a direct sum

$$H = \sum_{a\hat{g}} \oplus H_{a\hat{g}} \quad , \tag{5}$$

where $\hat{g} \varepsilon \hat{G}$ (where the dual object $\hat{G}$ is the collection of irreducible unitary representations of G) labels which irreducible representation $H_{a\hat{g}}$ transforms as, and a is an index that distinguishes the various subspaces which transform according to the same irreducible representation $\hat{g}$. Equation (5) means that there exists a set of self-adjoint projection operators $P_{a\hat{g}}$ on H with range $H_{a\hat{g}}$ such that

$$\sum_{a\hat{g}} P_{a\hat{g}} = I \tag{6}$$

with the sum converging to the identity operator I in the strong operator topology on L(H). Furthermore each $P_{a\hat{g}}$ commutes with the action of the group.

U4 (Schur's lemma) Let U_1 be an irreducible representation of G on H_1 and U_2 be an irreducible representation of G on H_2. Let $R(U_1,U_2)$ be the vector space of intertwining operators from H_1 to H_2, i.e., the set of all operators S mapping H_1 into H_2 which commute with the actions of the group:

$$SU_1(g) = U_2(g)S \quad \text{for all} \quad g\varepsilon G \quad . \tag{7}$$

Then if the representations U_1 and U_2 are inequivalent $R(U_1,U_2)$ consists of the zero operator while if U_1 and U_2 are equivalent $R(U_1,U_2)$ is isomorphic to the vector space of the complex numbers. In the latter case there exists a unitary intertwining operator U_{21} from H_1 to H_2 such that each S in $R(U_1,U_2)$ has the form

$$S = sU_{21} \tag{8}$$

where s is a complex number and $\|S\| = |s|$.

The application of these four results to quantum theory is straightforward. If a physical system admits a compact symmetry group G then by U3 the Hilbert space H of pure states is a direct sum

$$H = \sum_{a\hat{g}} \oplus H_{a\hat{g}} \tag{9}$$

of irreducible subspaces. For example, if G = SO(3) then $\hat{G}$ is the set of nonnegative integers, $\hat{g}$ is conventionally denoted as the angular momentum quantum number ℓ, and each $H_{a\ell}$ is $2\ell+1$ dimensional. A basis in $H_{a\ell}$ may be chosen which transforms under rotations as the spherical harmonics $Y_{\ell m}(\theta,\phi)$ for integer m satisfying $-\ell \leqslant m \leqslant \ell$.

Conservation laws, selection rules, and the Wigner-Eckart theorem now follow from the observation that the time development operator U(t) is an intertwining operator for the action of the symmetry group. [See Eq. (4).] Consequently the projections $P_{a'\hat{g}'}\, U(t)\, P_{a\hat{g}}$ of U(t) are intertwining operators from the irreducible representation spaces $H_{a\hat{g}}$ to the irreducible representation spaces $H_{a'\hat{g}'}$. Assume then that at time t = 0 the system is described by a vector $|\psi_{a\hat{g}}\rangle \in H_{a\hat{g}}$. Then by Schur's lemma the transition amplitude

$$\langle\phi_{a'\hat{g}'}|U(t)|\psi_{a\hat{g}}\rangle = \langle\phi_{a'\hat{g}'}|P_{a'\hat{g}'}U(t)P_{a\hat{g}}|\psi_{a\hat{g}}\rangle \tag{10}$$

for being found in state $|\phi_{a'\hat{g}'}\rangle \in H_{a'\hat{g}'}$ at time t is zero unless $\hat{g} = \hat{g}'$. Thus the group theoretically derived quantum number $\hat{g}$ does not change as the system evolves. In the case where G is the rotation group SO(3) Eq. (10) states that angular momentum is conserved.

Assume now that two subsystems interact in such a way that their dynamics is invariant under G. If initially subsystem one is in state $|\psi_{a_1\hat{g}_1}\rangle \in H_{a_1\hat{g}_1}$ while subsystem two is in state $|\psi_{a_2\hat{g}_2}\rangle \in H_{a_2\hat{g}_2}$ then the initial state of the composite system is represented by the vector $|\psi_{a_1\hat{g}_1} \otimes \psi_{a_2\hat{g}_2}\rangle = |\psi_{a_1\hat{g}_1}\rangle \otimes |\psi_{a_2\hat{g}_2}\rangle$ in the tensor product space $H_{a_1\hat{g}_1a_2\hat{g}_2} = H_{a_1\hat{g}_1} \otimes H_{a_2\hat{g}_2}$ and the probability amplitude for being found in state $|\phi_{a\hat{g}}\rangle$ at time t is

$$\langle\phi_{a\hat{g}}|U(t)|\psi_{a_1\hat{g}_1}\otimes\psi_{a_2\hat{g}_2}\rangle =$$

$$\sum_{a_{12}\hat{g}_{12}} \langle\phi_{a\hat{g}}|P_{a\hat{g}}U(t)P_{a_{12}\hat{g}_{12}}|\psi_{a_1\hat{g}_1}\otimes\psi_{a_2\hat{g}_2}\rangle \tag{11}$$

where the $P_{a_{12}\hat{g}_{12}}$ are the projection operators associated with the direct sum decomposition

$$H_{a_1\hat{g}_1 a_2\hat{g}_2} = \sum_{a_{12}\hat{g}_{12}} \oplus H_{(a_1\hat{g}_1 a_2\hat{g}_2)a_{12}\hat{g}_{12}} \tag{12}$$

of the representation space $H_{a_1\hat{g}_1 a_2\hat{g}_2}$ into its irreducible subspaces. Now $P_{a\hat{g}}U(t)P_{a_{12}\hat{g}_{12}}$ restricted to $H_{(a_1\hat{g}_1 a_2\hat{g}_2)a_{12}\hat{g}_{12}}$ is an intertwining operator from the irreducible representation space $H_{(a_1\hat{g}_1 a_2\hat{g}_2)a_{12}\hat{g}_{12}}$ to the irreducible representation space $H_{a\hat{g}}$ so Schur's lemma implies that it is zero if $\hat{g}_{12} \neq \hat{g}$. Consequently, the transition amplitude (11) must be zero if $\hat{g}$ is not found in the direct sum decomposition of the tensor product representation $\hat{g}_1 \otimes \hat{g}_2$. This provides a selection rule for which transitions are possible when two systems interact. For the rotation group this implies that

$$\langle\phi_{a\ell}|U(t)|\psi_{a_1\ell_1}\otimes\psi_{a_2\ell_2}\rangle = 0 \quad \text{unless} \quad |\ell_1-\ell_2| \leqslant \ell \leqslant \ell_1+\ell_2 \quad . \tag{13}$$

This of course has important implications in, for example, spectroscopic experiments. If an atom or molecule initially has angular momentum ℓ_1, then when it absorbs a photon of angular momentum $\ell_2 = 1$ (a "dipole photon") its angular momentum will change by at most one (i.e., the final angular momentum ℓ satisfies $|\ell_1-1| \leqslant \ell \leqslant \ell_1+1$). These group theoretically derived selection rules may be generalized to the interaction of more than two subsystems.

The above derived selection rule states that many transition amplitudes must be zero but says nothing about the numerical values of those transition amplitudes which are not required to vanish because of symmetry. However Schur's lemma also leads to statements about these numerical values. Thus if $\hat{g} = \hat{g}_{12}$ the intertwining operator $P_{a\hat{g}}U(t)P_{a_{12}\hat{g}_{12}}$ has the form

$$P_{a\hat{g}}U(t)P_{a_{12}\hat{g}_{12}} = \langle a\hat{g}\|U(t)\|(a_1\hat{g}_1 a_2\hat{g}_2)a_{12}\hat{g}\rangle U_{a(a_1\hat{g}_1 a_2\hat{g}_2)a_{12}\hat{g}} \tag{14}$$

where $U_{a(a_1\hat{g}_1 a_2\hat{g}_2)a_{12}\hat{g}}$ is an intertwining operator from $H_{(a_1\hat{g}_1 a_2\hat{g}_2)a_{12}\hat{g}}$ to $H_{a\hat{g}}$ and $\langle a\hat{g}\|U(t)\|(a_1\hat{g}_1 a_2\hat{g}_2)a_{12}\hat{g}\rangle$ is a complex function of t. Thus (11) implies that

$$\begin{aligned}&\langle\phi_{a\hat{g}}|U(t)|\psi_{a_1\hat{g}_1}\otimes\psi_{a_2\hat{g}_2}\rangle \\ &= \sum_{a_{12}} \langle a\hat{g}\|U(t)\|(a_1\hat{g}_1 a_2\hat{g}_2)a_{12}\hat{g}\rangle\langle\phi_{a\hat{g}}|U_{a(a_1\hat{g}_1 a_2\hat{g}_2)a_{12}\hat{g}}|\psi_{a_1\hat{g}_1}\otimes\psi_{a_2\hat{g}_2}\rangle \; .\end{aligned} \tag{15}$$

Equation (15) has important consequences. First of all it provides a division of the dynamics into factors which are universal for all systems that have the same symmetry group, and other factors which differentiate between systems with the same symmetry group. Specifically each factor $\langle\phi_{a\hat{g}}|U_{a(a_1\hat{g}_1a_2\hat{g}_2)a_{12}\hat{g}}|\psi_{a_1\hat{g}_1}\otimes\psi_{a_2\hat{g}_2}\rangle$ is a "geometrical" factor determined by the properties of the irreducible representations $\hat{g}_1$, $\hat{g}_2$ and $\hat{g}$. It gives the "projection" of $P_{a_{12}\hat{g}}|\psi_{a_1\hat{g}_1}\otimes\psi_{a_2\hat{g}_1}\rangle$ onto $|\phi_{a\hat{g}}\rangle$ and is independent of the details of the dynamics. The details of the dynamics are given by the reduced transition amplitudes $\langle a\hat{g}\|U(t)\|(a_1\hat{g}_1a_2\hat{g}_2)a_{12}\hat{g}\rangle$ which depend on the microscopic parameters and on the vector space labels a, a_1, $\hat{g}_1$, a_2, $\hat{g}_2$, a_{12} and $\hat{g}$ but are independent of the particular vectors $|\psi_{a_1\hat{g}_1}\rangle$, $|\psi_{a_2\hat{g}_2}\rangle$ and $|\phi_{a\hat{g}}\rangle$ under consideration. This leads to the important conclusion that the evolution of the system from $H_{a_1\hat{g}_1}\otimes H_{a_2\hat{g}_2}$ into $H_{a\hat{g}}$ is determined by the N functions

$$\langle a\hat{g}\|U(t)\|(a_1\hat{g}_1a_2\hat{g}_2)a_{12}\hat{g}\rangle \quad ,$$

where N is the multiplicity of $\hat{g}$ in $\hat{g}_1\otimes\hat{g}_2$. Equation (15) is a coordinate free statement of the Wigner-Eckart theorem generalized to unitary representations of arbitrary compact topological groups.

The mathematical basis of these physical results -- conservation laws, selection rules, and the Wigner-Eckart theorem -- is clearly provided by Schur's lemma together with the complete reducibility of an arbitrary unitary representation of a compact topological group into finite dimensional irreducible representations.

III. THE NEED FOR A NEW APPROACH

In the preceding section we have reviewed some of the consequences of group theory in the standard Hilbert space formulation of quantum mechanics. As powerful as these results are there are many experimental situations in which symmetry imposes restrictions and simplifications which are not directly obtainable through these Hilbert space group theoretic methods.

The standard Hilbert space formulation of quantum theory has the disadvantage that a vector in H can only represent a pure state of the quantum mechanical system whereas most physical systems are in mixed (or impure) states. (Recall that a pure state is characterized by the existence of an experiment that gives a result predictable with certainty when performed on a system in that state and in that state only, while other states are mixed states.[6] Polarized light is in a pure state, while unpolarized or partially polarized light is in a mixed state.) While a mixed state may not be represented by a single vector $|\psi\rangle \in H$ it may be represented through a weighted average (a mixture) of different vectors $|\psi_i\rangle$ (pure states).

A more convenient representation of mixed states is provided by a density operator formulation.[6] An arbitrary pure or mixed state may be represented by a density operator ρ on Hilbert space H which satisfies

$$\rho^{+} = \rho \qquad \text{i.e., } \rho \text{ is Hermitian} \tag{16}$$

$$\langle\psi|\rho|\psi\rangle \geq 0 \qquad \text{i.e., } \rho \text{ is non-negative} \tag{17}$$

$$\mathrm{tr}\rho = 1 \qquad \text{i.e., } \rho \text{ has unit trace} \quad . \tag{18}$$

When the system is in the state ρ the expectation value of an observable represented by the self-adjoint operator 0 is given by

$$\langle 0\rangle_{\rho} = \mathrm{tr}(0\rho) \quad . \tag{19}$$

The density operators form a convex set (i.e., if ρ_1 and ρ_2 are density operators and if $0 \leq \lambda \leq 1$ then $\rho = \lambda\rho_1 + (1-\lambda)\rho_2$ is a density operator). An extreme point in this convex set represents a pure state, and every pure state is represented by a Hermitian projection operator $\rho_{\psi} = P_{\psi} = |\psi\rangle\langle\psi|$.

The motivation of the following application of group theory at the density operator level is provided by the results of certain classes of experiments performed on mixed states. Although there are many different types of experiments in atomic, molecular, and nuclear physics which exhibit the desired behavior, for concreteness we will consider only one such type of experiment -- a photoionization experiment.

Consider the photoionization of an atom or molecule through the absorption of a single linearly polarized "dipole" photon. Then, provided the atomic or molecular target is initially unpolarized (i.e., has no preferred orientation in space) the angular distribution of the ejected electron is always found to have the form

$$A + B\cos^2\theta \tag{20}$$

where θ is the angle between the polarization vector of the photon and the propagation direction of the electron (see Fig. 1). This phenomenological form for the angular distribution is universal in that it is independent of the detailed nature of the target. For different targets and different photon energies the value of the coefficients A and B are different, but the form remains the same.

Again, consider the ionization of an unpolarized atomic or molecular target through the absorption of two linearly polarized "dipole" photons. In this case the angular distribution will have the phenomenological form

$$\sum_{\substack{mn \\ 0\leq m+n\leq 4}} C_{mn}(\eta)\sin^{m}\theta\cos^{n}\theta \tag{21}$$

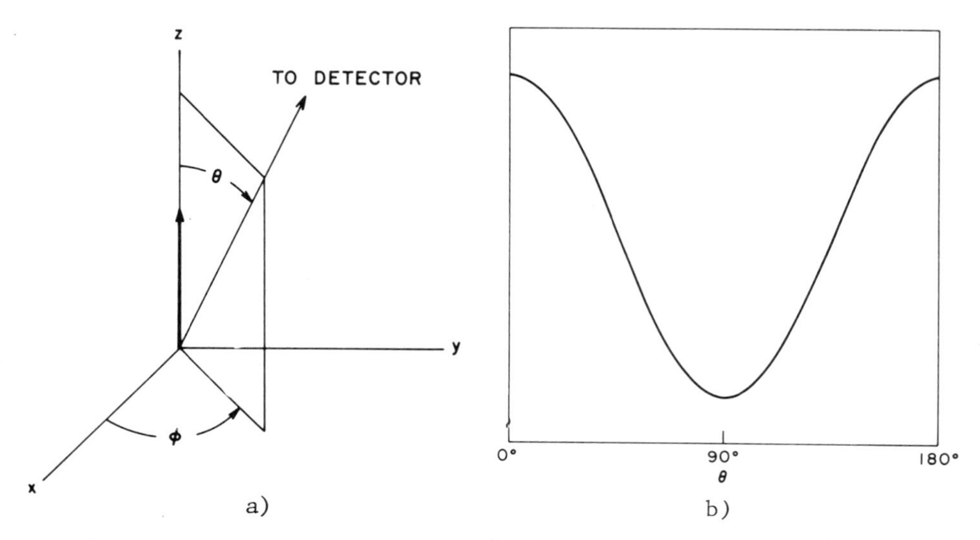

Fig. 1. One photon ionization. a) The position of the electron detector is specified by the spherical angles (θ,ϕ) where the z axis coincides with the polarization vector of the photon. b) The angular distribution of the electron may always be described by the form $A + B\cos^2\theta$ provided the target is initially unpolarized and the ionization proceeds through the absorption at a single linearly polarized dipole photon.

(see Fig. 2). Again this phenomenological form is independent of the detailed nature of the target but is universal for all such processes. Again the microscopic details determine the numerical values of the parameters $C_{mn}(\eta)$ but do not change the generic form of the angular distribution.

The feature common to all such phenomenological equations is that they place restrictions on the possible complexity of the angular distribution. They are quite analogous to the spectroscopic selection rules which restrict, for example, the possible changes in the value of the angular momentum [see, e.g., (13)] but cannot be derived from them. They also differ from the Hilbert space selection rules in that they refer to observed transition probabilities rather than to transition amplitudes. The analogy, however, suggests that they may have a group theoretic origin. Since the phenomenological equations refer explicitly to mixed states (an unpolarized state being in general a mixed state) this suggests that group theory should be applied at the density operator level directly. In the next section we show that when this is done one obtains conservation laws, selection rules, and a Wigner-Eckart theorem expressed in terms of transition probabilities. The phenomenological form for angular distributions follows from the selection rules.

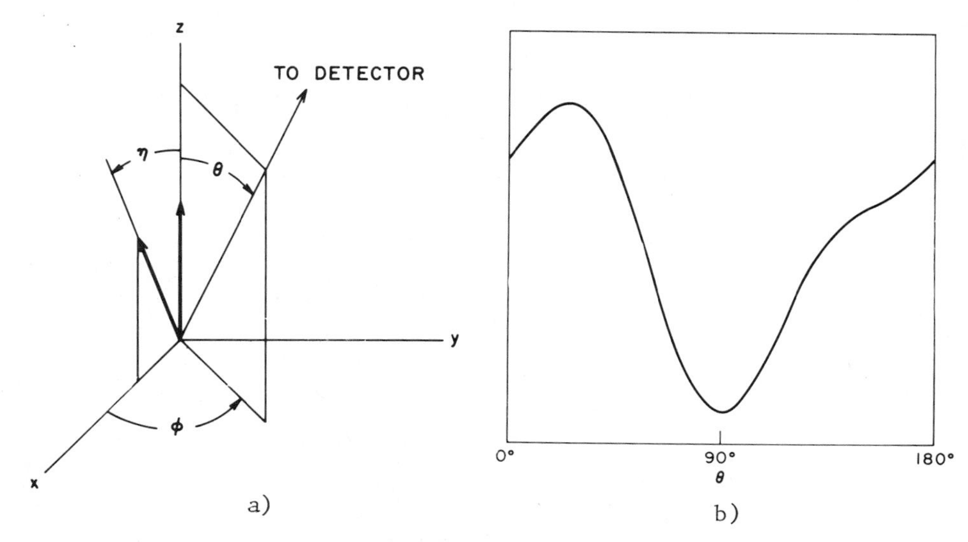

Fig. 2. Two photon ionization. a) The position of the electron detector is specified by the spherical angles (θ,ϕ) where the polarization vector of the ionizing photon coincides with the z-axis, the polarization vector of the exciting photon lies in the x-z plane, and η is the angle between the two polarization vectors. b) The angular distribution in the x-z plane (i.e., $\phi = 0$) may always be described by Eq. (21) provided the target is initially unpolarized and the ionization proceeds through the absorption of two linearly polarized dipole photons.

IV. GROUP THEORY ON STATE SPACE AND OBSERVABLE SPACE

To apply group representation theory on mixed state systems we employ the Liouville representation of quantum theory. Here the states are represented by vectors in the complex vector space generated by the set of density operators. This is the space T(H) of trace class operators on H. T(H) is a Banach space with the trace class norm

$$\|A\|_T = \mathrm{tr}[(A^+A)^{1/2}] \tag{22}$$

and is called the State Space of the system. If U: G×H → H is a continuous unitary representation of G on H then the adjoint and product representations

$$U_a: G\times T(H) \to T(H) \qquad U_a(g)\rho = U(g)\rho U^+(g) \tag{23}$$

and

$$U_p: G\times G\times T(H) \to T(H) \qquad U_p(g,h)\rho = U(g)\rho U^+(h) \tag{24}$$

are continuous isometric representations of G and G×G, respectively, on State Space.[1]

Similarly the bounded self-adjoint operators generate the complex vector space L(H) of bounded operators on H. L(H) is a Banach space with the norm

$$\|O\|_L = \sup_{\substack{\psi\varepsilon H \\ \psi\neq 0}} \|O\psi\|/\|\psi\| \tag{25}$$

and is called the Observable Space of the system. If $U:G\times H \to H$ is a continuous unitary representation of G on H then the adjoint and product representations

$$\tilde{U}_a : G\times L(H) \to L(H) \qquad \tilde{U}_a(g)O = U(g)OU^+(g) \tag{26}$$

and

$$\tilde{U}_p : G\times G\times L(H) \to L(H) \qquad \tilde{U}_p(g,h)O = U(g)OU^+(h) \tag{27}$$

are isometric representations of G and G×G, respectively, on Observable Space. In general the representations $\tilde{U}_a$ and $\tilde{U}_p$ are not norm continuous.[1]

The norms $\| \ \|_T$ and $\| \ \|_L$ on State Space and Observable Space, respectively, have been provided by mathematics, but they also have direct physical relevance. Their physical interpretation follows from the fact that Observable Space is the dual space of State Space with the duality being provided by the expectation value (19). In order to clearly exhibit this duality it is convenient to employ a notation that is analogous to Dirac's notation for Hilbert space quantum theory. Thus if each trace class operator ρ is written as a "ket" $|\rho\rangle\rangle$ and each bounded operator O is written as a "bra" $\langle\langle O|$ then the "bracket"

$$\langle\langle O|\rho\rangle\rangle = tr(O^+\rho) \tag{28}$$

defines a continuous linear functional on T(H) for each $O\varepsilon L(H)$ and each continuous linear functional on T(H) arises in this manner from some $O\varepsilon L(H)$. Clearly if O is self-adjoint (i.e., if O represents an observable) and ρ is a density operator the bracket $\langle\langle O|\rho\rangle\rangle$ coincides with the expectation value (19):

$$\langle O\rangle_\rho = \langle\langle O|\rho\rangle\rangle \quad . \tag{29}$$

Now since L(H) is the dual space of T(H) we know that the norm $\| \ \|_L$ is alternatively given by

$$\|O\|_L = \sup_{\substack{\rho\varepsilon T(H) \\ \|\rho\|=1}} |\langle\langle O|\rho\rangle\rangle| \quad . \tag{30}$$

Consequently if O represents an observable $\|O\|_L$ is the maximum expectation value $\langle O\rangle_\rho$ of O as ρ is varied over all states of the system:

$$\|O\|_L = \sup_{\rho\varepsilon DO} |\langle\langle O|\rho\rangle\rangle| \quad . \tag{31}$$

This implies that two observables O_1 and O_2 are "close" in the norm topology if their expectation values are "close" for all states of the system:

$$\|O_1-O_2\|_L = \sup_{\rho\varepsilon DO} |\langle\langle O_1|\rho\rangle\rangle - \langle\langle O_2|\rho\rangle\rangle| \quad . \tag{32}$$

A similar interpretation is obtained for the norm topology on State Space: two states ρ_1 and ρ_2 are "close" in the norm topology if their expectation values are "close" for all "normalized" observables on the system

$$\|\rho_1-\rho_2\|_T = \sup_{\substack{O\varepsilon L(H)\\ O^+=O\neq 0}} |\langle\langle O|\rho_1\rangle\rangle - \langle\langle O|\rho_1\rangle\rangle|/\|O\| \quad . \tag{33}$$

Dynamics is given by the one parameter group of isometric operators

$$U_a(t):T(H) \rightarrow T(H) \qquad U_a(t)\rho = U(t)\rho U^+(t) \tag{34}$$

where U(t) is the one parameter unitary group (1) governing pure state dynamics. Thus if at time t = 0 the system is in state ρ then at time t the expectation value of observable O is

$$\langle\langle O|\rho(t)\rangle\rangle = \langle\langle O|U(t)\rho U^+(t)\rangle\rangle = \langle\langle O|U_a(t)|\rho\rangle\rangle \quad . \tag{35}$$

Assume now that the compact topological group G is a symmetry group for the system. Then for any time t the time development operator $U_a(t)$ commutes with the action of the group, i.e.,

$$U_a(t)U_a(g) = U_a(g)U_a(t) \quad \text{for all} \quad t\varepsilon R \quad \text{and} \quad g\varepsilon G \quad . \tag{36}$$

Thus we have a situation for mixed state quantum theory which is quite similar to the situation for pure state quantum theory. If State Space and Observable Space have direct sum decompositions

$$T(H) = \sum_{a\hat{g}} \oplus T_{a\hat{g}} \tag{37}$$

and

$$L(H) = \sum_{a\hat{g}} \oplus L_{a\hat{g}} \tag{38}$$

into finite dimensional irreducible subspaces under the adjoint representations U_a and $\tilde{U}_a$, respectively, of G then Schur's lemma applies and the procedure used for proving conservation laws, selection rules, and the Wigner-Eckart theorem in the pure state case may be followed with very little change to provide similar conclusions in the mixed state case. The soundness of this procedure rests on the validity of (37) and (38) which we now investigate.

Since the representations U_a and $\tilde{U}_a$ are defined on Banach spaces it is natural to ask if the general results U1-U4 for unitary representations on Hilbert spaces generalize in some way to non-unitary representations.[1,3]

TVS1 If V_1 is a closed subspace of topological vector space V which is invariant under the continuous representation $U:G\times V \rightarrow V$ of the compact topological group G, does there exist a complementing closed invariant subspace V_{1c} of V such that V is the direct sum $V = V_1 \oplus V_{1c}$ of V_1 and its complement V_{1c}? Not always. Let $V = L(H)$ and $U(g) = \tilde{U}_a(g)$ the adjoint representation (26) associated with a unitary representation $U:G\times H \rightarrow H$ of G. A group G and representation $U:G\times H \rightarrow H$ may be found such that $\tilde{U}_a(g)$ is continuous on L(H) (e.g., U could be the trivial representation $U(g) = I$, the identity operator). Then the subspace $V_1 = \mathrm{Com}(H)$ of compact operators on H is a closed subspace of L(H) which is invariant under $\tilde{U}_a$. But Com(H) has no complementing subspace whatsoever, so it certainly has no complementing subspace invariant under $\tilde{U}_a$.

TVS2 a) Is every continuous irreducible representation $U:G\times V\rightarrow V$ of compact topological group G on the topological vector space V finite dimensional? This is known to be true if V has a nontrivial dual, i.e., if there exists at least one nonzero continuous linear functional on V. We can conclude that since Observable Space L(H) is the dual of State Space T(H), any irreducible subspace of L(H) or T(H) must be finite dimensional. b) and c) are results specifically about the (unitary) right regular representation and thus need not be generalized. However, because each irreducible subrepresentation of G occurring in L(H) and T(H) is finite dimensional we know that it also appears in the right regular representation of G. Thus no "new" irreducible subrepresentation will appear in L(H) and T(H) which are not already known from the theory of unitary group representations.

TVS3 Is every continuous representation $U:G\times V \rightarrow V$ of the compact topological group G on topological vector space V a direct sum of irreducible representations of G? No. Examples will be given after the consideration of TVS4.

TVS4 Does Schur's lemma generalize? Yes, at least for the irreducible subrepresentations of L(H) and T(H) which are known to be finite dimensional.

The theory of Fourier Series provides examples of continuous group representations which are not completely reducible. Let G be the circle group, i.e., the multiplicative group of complex numbers of modulus 1. G acts on the Banach spaces $L^1(G)$ and $L^\infty(G)$ through the action $[U(\theta)f](\phi) = f(\theta+\phi)$. In both $L^1(G)$ and $L^\infty(G)$ the one dimensional subspaces generated by the functions $e_n = e^{in\theta}$ are irreducible. It is known, however, that[7]

a) $U:G\times L^1(G)\to L^1(G)$ is not completely reducible, i.e., if $f\in L^1(G)$ has Fourier coefficients $\hat{f}(n) = 1/2\pi \int_0^{2\pi} f(\theta)\, e^{-in\theta}d\theta$, $\sum_n \hat{f}(n)e_n$ need not converge to f in L^1 norm.

b) $U:G\times L^\infty(G)\to L^\infty(G)$ is not completely reducible, i.e., if $g\in L^\infty(G)$ has Fourier coefficients $\hat{g}(n) = 1/2\pi \int_0^{2\pi} g(\theta)\, e^{-in\theta}d\theta$, $\sum_n \hat{g}(n)e_n$ need not converge to g in L^∞ norm.

For our purposes these examples are quite interesting. Since $L^\infty(G)$ is the dual space of $L^1(G)$, the spaces $L^1(G)$ and $L^\infty(G)$ are analogous to state space T(H) and observable space L(H), respectively, with integration playing the role of the trace operation. With this interpretation a) suggests that an arbitrary "state" $f\in L^1(G)$ cannot be reconstructed by summing its irreducible components while b) makes a similar statement about an "observable" $g\in L^\infty(G)$. From a physical point of view one might argue that it does not matter if a "state" f or an "observable" g can be reconstructed as long as the "expectation value" [or dual action $g(f) = 1/2\pi \int_0^{2\pi} g^*(\theta)f(\theta)d\theta$] may be reconstructed from the irreducible components of f and g. But it is known[7] that if $g\in L^\infty(G)$ and $f\in L^1(G)$ then $\sum_n g^*(n)f(n)$ need not converge to g(f). These examples indicate that one does not have to search for pathological cases to find representations on topological vector spaces which are not completely reducible.

Since there are no general theorems which assert that group representations on Banach spaces (or more general topological vector spaces) are completely reducible, we consider only the physically induced representations (23) and (26) on State Space and Observable Space, respectively.

We have shown elsewhere[1,3] that for these representations the direct sum decompositions (33) and (38) of State Space and Observable Space do, in fact, exist. These direct sum decompositions are constructed with the aid of the corresponding direct sum decompositions (5) of the underlying Hilbert space H. Thus we define operators

$$\Pi_{a'\hat{g}'a''\hat{g}''}:T(H) \to T_{a'\hat{g}'a''\hat{g}''} \quad ; \quad \Pi_{a'\hat{g}'a''\hat{g}''}|\rho\rangle\rangle = |P_{a'\hat{g}'}\rho P_{a''\hat{g}''}\rangle\rangle \tag{39}$$

where the $P_{a'\hat{g}'}$ are the projection operators (6) associated with the direct sum decomposition (5) of H. It may be shown that each $\Pi_{a'\hat{g}'a''\hat{g}''}$ is a projection operator which is continuous when T(H) is given the norm topology (33). Furthermore,

$$\sum_{a'\hat{g}'a''\hat{g}''} \Pi_{a'\hat{g}'a''\hat{g}''} = I \tag{40}$$

with the sum converging to the identity operator on T(H) in the strong operator topology [again with T(H) having the norm topology]. This means that for any $|\rho\rangle\rangle \varepsilon T(H)$ the series $\Sigma\ |\rho_{a'\hat{g}'a''\hat{g}''}\rangle\rangle$ (where $|\rho_{a'\hat{g}'a''\hat{g}''}\rangle\rangle = \Pi_{a'\hat{g}'a''\hat{g}''}|\rho\rangle\rangle$) converges to $|\rho\rangle\rangle$ in the $\|\ \|_{T(H)}$ norm topology. Since each $T_{a'\hat{g}'a''\hat{g}''}$ is irreducible under U_p (24) we have a direct sum decomposition

$$T(H) = \sum_{a'\hat{g}'a''\hat{g}''} \oplus\ T_{a'\hat{g}'a''\hat{g}''} \tag{41}$$

of State Space into finite dimensional irreducible representations of the product representation. What is needed in (37) is a direct sum decomposition of state space into irreducible representations of the adjoint representation (23). But the adjoint representation is the product representation restricted to the diagonal $U_a(g) = U_p(g,g)$ so each $T_{a'\hat{g}'a''\hat{g}''}$ is invariant, if not irreducible, under the adjoint representation. Since each $T_{a'\hat{g}'a''\hat{g}''}$ is finite dimensional it may be reduced into a direct sum of representations irreducible under the adjoint representation using familiar finite dimensional techniques, producing the desired direct sum decomposition (37) of state space. One may attempt to obtain the direct sum decomposition (38) of Observable Space through a similar procedure by defining the projection operators

$$\tilde{\Pi}_{a'\hat{g}'a''\hat{g}''}:L(H) \to L_{a'\hat{g}'a''\hat{g}''} \quad ; \quad \langle\langle \tilde{\Pi}_{a'\hat{g}'a''\hat{g}''}O| = \langle\langle P_{a'\hat{g}'}OP_{a'\hat{g}''}| \tag{42}$$

on L(H). This does not succeed, however, since each $\langle\langle P_{a'\hat{g}'}OP_{a''\hat{g}''}|$ is of finite rank and the norm closure of the set of finite rank operators is the set of compact operators. Consequently $\langle\langle \tilde{\Pi}_{a'\hat{g}'a''\hat{g}''}O|$ does not converge to $\langle\langle O|$ for a noncompact operator $O\varepsilon L(H)$. However, this procedure does succeed if the topology on L(H) is weakened to the weak-* topology. That is the series $\langle\langle \tilde{\Pi}_{a'\hat{g}'a''\hat{g}''}O|$ converges to $\langle\langle O|$ in the weak-* topology on L(H). While the physical meaning of the norm topology on L(H) has already been established [see (32)], we see that it does not lead to complete reducibility. For mathematical reasons we are led to the weak-* topology on L(H). We must ask if this new topology is physically reasonable. The weak-* topology is the weakest topology on L(H) such that for each fixed $\rho\varepsilon T(H)$ the expectation value mapping

$$E_\rho:L(H) \to C \quad ; \quad E_\rho(\langle\langle O|) = \langle\langle O|\rho\rangle\rangle \tag{43}$$

is continuous. Clearly this topology, being expressed in terms of measured quantities (the expectation values), is physically reasonable.

We thus have the direct sum decompositions (37) and (38) of State Space and Observable Space into finite dimensional irreducible subspaces. Since the irreducible representations are finite dimensional, Schur's lemma applies and conservation laws, selection rules, and the Wigner-Eckart theorem may be proven by simple transcriptions of the Hilbert Space (pure state) proofs given in Sec. II. The interpretations of these results, however, are different. We will illustrate these results by specializing to the rotation group and using them to derive the phenomenological equations (20) and (21) describing angular distributions.

When G = SO(3) the direct sum decompositions (37) and (38) become

$$T(H) = \sum_{a\ell} \oplus\, T_{a\ell} \tag{44}$$

and

$$L(H) = \sum_{a\ell} \oplus\, L_{a\ell} \tag{45}$$

where ℓ ranges over the non-negative integers. Equation (44) means that an arbitrary state $|\rho\rangle\rangle$ has an expansion

$$|\rho\rangle\rangle = \sum_{a\ell} |\rho_{a\ell}\rangle\rangle \tag{46}$$

which converges in the $\|\ \|_{T(H)}$ norm topology while (45) means that an arbitrary observable $\langle\langle O|$ has an expansion

$$\langle\langle O| = \sum_{a\ell} \langle\langle O_{a\ell}| \tag{47}$$

which converges in the weak-* topology. The State Space quantum number ℓ in (44) and (46) is derived from the action of the adjoint representation of the rotation group on State Space and characterizes the transformational properties of the ensemble under rotations; the component $|\rho_{a\ell}\rangle\rangle$ is invariant under rotations for $\ell = 0$, transforms as a dipole for $\ell = 1$, transforms as a quadrupole for $\ell = 2$ and, in general, transforms as a 2^{ℓ}-pole for arbitrary ℓ. The observable space quantum number ℓ has a "dual" interpretation. Since $L_{a'\ell'}$ is "orthogonal" to $T_{a\ell}$ if ℓ' does not equal ℓ in the sense that

$$\langle\langle O_{a'\ell'}|\rho_{a\ell}\rangle\rangle = 0 \quad \text{if} \quad \ell' \neq \ell \quad , \tag{48}$$

we see that the observable $\langle\langle O_{a\ell}|$ measures the unpolarized component of $|\rho\rangle\rangle$ if $\ell = 0$, the dipole moment of $|\rho\rangle\rangle$ for $\ell = 1$, and the 2^{ℓ}-pole moment for arbitrary ℓ. Because these quantum numbers ℓ refer to ensemble properties they are called the polarization quantum numbers of the system.

Conservation laws, selection rules, and the Wigner-Eckart theorem are now expressed in terms of these polarization quantum numbers. Thus for an isolated system we have

$$\langle\langle O_{a'\ell'}|U_a(t)|\rho_{a\ell}\rangle\rangle = 0 \quad \text{if} \quad \ell' \neq \ell \tag{49}$$

which expresses the conservation for polarization. The interaction of two systems leads to the selection rule on polarization quantum numbers

$$\langle\langle O_{a\ell}|U_a(t)|\rho_{a_1\ell_1} \otimes \rho_{a_2\ell_2}\rangle\rangle = 0 \quad \text{unless} \quad |\ell_1-\ell_2| \leqslant \ell \leqslant \ell_1+\ell_2 \ . \tag{50}$$

If the selection rule is satisfied then the Liouville representation Wigner-Eckart theorem states

$$\langle\langle O_{a\ell}|U_a(t)|\rho_{a_1\ell_1} \otimes \rho_{a_2\ell_2}\rangle\rangle$$
$$= \langle\langle a\ell||U_a(t)||(a_1\ell_1a_2\ell_2)\ell\rangle\rangle\langle\langle O_{a\ell}|U_{a(a_1\ell_1a_2\ell_2)\ell}|\rho_{a_1\ell_1} \otimes \rho_{a_2\ell_2}\rangle\rangle \tag{51}$$

where

$$\langle\langle O_{a\ell}|U_{a(a_1\ell_2a_2\ell_1)\ell}|\rho_{a_1\ell_1} \otimes \rho_{a_2\ell_2}\rangle\rangle$$

is a geometrical factor determined by the properties of the irreducible representations, and the details of the dynamics are given by the reduced transition probabilities $\langle\langle a\ell||U_a(t)||(a_1\ell_1a_2\ell_2)\ell\rangle\rangle$.

The phenomenological equations describing angular distributions may be easily derived from the selection rules for the polarization quantum numbers in the following way. It can be shown quite generally that an observable $\langle\langle O(\theta,\phi)|$ which responds to (i.e., counts) electrons traveling into a solid angle about the polar angle (θ,ϕ) (see Fig. 1) has a direct sum expansion of the form

$$\langle\langle O(\theta,\phi)| = \sum_{a\ell m} Y_{\ell m}(\theta,\phi)O_{a\ell}\langle\langle a\ell m| \tag{52}$$

Now consider an experiment in which a target interacts with a photon. Let L_1 be the set of polarization quantum numbers occurring in the direct sum decomposition

$$|{}^1\rho\rangle\rangle = \sum_{a_1\ell_1} |{}^1\rho_{a_1\ell_1}\rangle\rangle \tag{53}$$

of the density operator of the target (i.e., $\ell_1 \varepsilon L_1$ if $|{}^1\rho_{a_1\ell_1}\rangle\rangle$ $\neq 0$ for at least one a_1) and let L_2 be the set of polarization quantum numbers occurring in the direct sum decomposition

$$|^2\rho\rangle\rangle = \sum |^2\rho_{a_2\ell_2}\rangle\rangle \tag{54}$$

of the density operator of the photon. The angular distribution of the photoelectron is

$$\langle\langle O(\theta,\phi)|U_a(t)|^1\rho \otimes {}^2\rho\rangle\rangle$$

$$= \sum_{\substack{maa_1a_2 \\ \ell_1\varepsilon L_1 \\ \ell_2\varepsilon L_2 \\ \ell\varepsilon L_1\times L_2}} Y_{\ell m}(\theta,\phi)O_{a\ell}\langle\langle a\ell||U_a(t)||(a_1\ell_1a_2\ell_2)\ell\rangle\rangle \times \langle\langle a\ell m|U_{a(a_1\ell_1a_2\ell_2)\ell}|^1\rho_{a_1\ell_1} \otimes {}^2\rho_{a_2\ell_2}\rangle\rangle \tag{55}$$

from which it is clear that the angular distribution can contain spherical harmonics of order ℓ only if ℓ satisfies the triangle rule $|\ell_1-\ell_2| \leq \ell \leq \ell_1+\ell_2$ for some $\ell_1\varepsilon L_1$ and $\ell_2\varepsilon L_2$. For an isotropic target $L_1 = \{0\}$ and for a linearly polarized "dipole" photon $L_2 = \{0,2\}$. Thus the angular distribution will only contain spherical harmonics of order 0 and 2. For the particular geometry shown in Fig. 1 it is easy to see that the angular distribution reduces to the form given in Eq. (20). Similar arguments show that for a two-photon ionization of an unpolarized target the angular distribution will contain spherical harmonics of order 4 or less from which the phenomenological form (21) follows.

We have sketched how the use of group theory at the ensemble level leads to simple derivations of the phenomenological selection rules describing angular distributions. No discussion has been given as to how to compute the reduced transition probabilities in terms of microscopic parameters. We simply comment here that many identities satisfied by the reduced transition amplitudes in the pure state, Hilbert space theory generalize to the Liouville representation. These identities provide straightforward procedures for expressing the reduced transition amplitudes, and thus the angular distributions, in terms of microscopic parameters. This provides an important calculational tool which is discussed in detail elsewhere.[2,3]

V. CONCLUDING REMARKS

Density operator methods have long been used to calculate angular distributions. Group theoretic arguments have until this time been confined to the wave function level. In this way certain of the consequences of symmetry are arrived at only through the use of detailed analysis, and not as a result of symmetry principles. We have shown that, despite certain mathematical difficulties which have been surmounted, group theoretic methods may be applied at the ensemble level. As one application of this method we have shown that easy derivations, based only on symmetry arguments, of the phenomenological equations describing angular distributions are obtained. Moreover, the theory leads to straightforward procedures for evaluating the macroscopic parameters occurring in the angular distributions in terms of microscopic parameters and geometrical factors.

Beyond its applications to angular distributions this work may be regarded as a contribution to a growing body of research devoted to treating mixed state quantum theory in a mathematically rigorous fashion.[8] The necessity of paying attention to questions of rigor is clearly illustrated by the example of group representations on Banach spaces which are not completely reducible; an intuition based upon Hilbert space experience may lead one far astray.

This work was supported in part by National Science Foundation grants CHE76-17821, CHE80-11442 and PHY79-04928.

REFERENCES

1. M. P. Strand and R. S. Berry (submitted to J. Math. Phys.).
2. M. P. Strand and R. S. Berry (to be submitted).
3. M. P. Strand, Doctoral Thesis, University of Chicago (1979).
4. E. P. Wigner, Group Theory (Academic Press, New York, 1959).
5. G. W. Mackey, Unitary Group Representations in Physics, Probability and Number Theory (Benjamin/Cummings, Reading, Mass., 1978).
6. U. Fano, Rev. Mod. Phys. 29, 74 (1957).
7. R. E. Edwards, Fourier Series, Vol. 1, 2nd Ed. (Holt, Reinhardt, and Winston, New York, 1979).
8. See, e.g., E. B. Davies, Quantum Theory of Open Systems (Academic Press, New York, 1976).

TENSOR PRODUCT DECOMPOSITION OF HOLOMORPHICALLY INDUCED REPRESENTATIONS AND CLEBSCH-GORDAN COEFFICIENTS

Tuong Ton-That and William H. Klink

The University of Iowa
Iowa City, Iowa 52242

ABSTRACT

A formalism of decomposing tensor product of irreducible representations of compact semisimple Lie groups is presented, using holomorphic induction techniques. The case of SU(n) is examined in detail. Irreducible representation spaces are realized as polynomial functions over $GL(n,\mathbb{C})$ group variables and it is shown how to generate invariant spaces labelled by double cosets, with one double coset subspace isomorphic to the original tensor product space. Global and noninductive procedure for constructing orthogonal polynomial bases is presented and is compared with the Gelfand-Žetlin procedure. Clebsch-Gordan (Wigner) coefficients are computed and are used to provide a resolution of the multiplicity problem occurring in the tensor product decompositions of U(n) groups.

INTRODUCTION

Concrete decompositions of tensor product of irreducible representations of groups are of great interest in mathematical physics. In this article we will give a general and unified method of decomposition of all irreducible representations of groups which can be realized by holomorphic induction, these include the class of all finite dimensional representations of the classical groups. To achieve our goal we first realize such G-modules, where G are complex reductive groups, as vector spaces of all polynomial functions over $\mathbb{C}^{n\times n}$ which transform covariantly with respect to holomorphic characters of Borel subgroups B of G. By restriction to the compact real forms G_0 of G (the "Unitarian trick") and by equipping these vector spaces with the "differentiation" inner product we obtain irreducible unitary representations of the compact groups G_0 on Hilbert spaces. Tensor products of such representations are then considered. Using double cosets which arise naturally in this holomorphic induction procedure we define double coset spaces and double coset maps, these in turn will give concrete decompositions of tensor products. A formalism of the resolution of the multiplicity problem is also given. Finally, we show how to construct orthogonal bases for these G_0-modules, in fact, an orthogonal polynomial basis for $U(n)$-modules is exhibited. Using these bases we indicate how to compute matrix coefficients (D-functions) and Clebsch-Gordan coefficients of these representations. Many concrete realizations of irreducible finite dimensional representations of classical groups have appeared in the literature, the two closest to ours are found in the Boson calculus (see, for example, [1]) and in Ref. [2]. However, in the Boson calculus these spaces are too "big" (i.e., no covariant condition is imposed, thus in general one has to extract irreducible subspaces out of these spaces by a complicated procedure), and in Ref. [2] these spaces consist of polynomial over coset spaces G/B, thus the convenient "differentiation" inner product is lost. Finally, this "differentiation" inner product is used implicitly in the Boson calculus; however, it should be mentioned that it arises naturally in our context from the work on G-invariant polynomials and differential operators of Harish-Chandra [3], B. Kostant [4], and S. Helgason [5]. (See also [6].)

1. TENSOR PRODUCTS OF HOLOMORPHICALLY INDUCED REPRESENTATIONS

Let G be a complex reductive Lie group with a compact real form G_0. By the Iwasawa decomposition $G = BG_0$, where B is a Borel subgroup, that is, a maximal connected solvable subgroup of G. Let $\pi : B \longrightarrow \mathbb{C}^*$ be a holomorphic homomorphism (or character) of B onto $\mathbb{C}^*$. Set

$$\mathrm{Hol}(G,\pi) = \{f : G \longrightarrow \mathbb{C} \mid f(bg) = \pi(b)f(g), \forall (b,g) \in B \times G,\ f \text{ polynomial}\}$$

and let $R_\pi : G \longrightarrow \mathrm{Aut}(\mathrm{Hol}(G,\pi))$ be the representation of G which is obtained by right translation on $\mathrm{Hol}(G,\pi)$. In this context we have the following

THEOREM (Borel-Weil). (1) <u>The space</u> $\mathrm{Hol}(G,\pi)$ <u>is nonzero iff</u> π <u>is dominant</u>.

(2) <u>If</u> π <u>is dominant then</u> R_π <u>is irreducible and the restriction</u> $R_\pi|G_0 = R_\pi^0$ <u>is irreducible with highest weight</u> π (cf. [7]).

If G is a subgroup of the general linear group, e.g., $G = GL(n,\mathbb{C})$, $SO(n,\mathbb{C})$, or $Sp(2n,\mathbb{C})$, etc., then without loss of generality $Hol(G,\pi)$ can be realized as a space of polynomial functions on $\mathbb{C}^{n\times n}$. This space can be in turn equipped with the following "differentiation" inner product

$$(p,q) = p(D)\overline{q(\bar{x})}\Big|_{x=0}, \quad x \in \mathbb{C}^{n\times n}$$

where $p(D)$ is obtained from the polynomial $p(x)$ by replacing the entries x_{ij} by the partial derivatives $\frac{\partial}{\partial x_{ij}}$. Then it can be shown that

$$(R_\pi^0(u)p, R_\pi^0(u)q) = (p,q), \ \forall\, u \in G_0,$$

thus R_π^0 is irreducible and unitary (cf. [7]).

Our main problem can then be formulated as follows:

<u>Set</u> $V^{(\pi')\otimes(\pi'')} = \left\{ \begin{matrix} F: G\times G \to \mathbb{C} \mid F \text{ holomorphic, } F(b'g', b''g'') = \\ \pi'(b')\pi''(b'')F(g',g''),\ \forall\, b',b'' \in B,\ \forall\, g',g'' \in G \end{matrix} \right\}$

<u>and if</u> ρ <u>denotes the representation of</u> G <u>in</u> $V^{(\pi')\otimes(\pi'')}$ <u>obtained by right translation, then give an explicit decomposition of</u> $V^{(\pi')\otimes(\pi'')}$.

Assuming the knowledge of the spectral decomposition of $V^{(\pi')\otimes(\pi'')}$, i.e., the Clebsch-Gordan series for $(\pi')\otimes(\pi'')$, which is given by many well-known formulae (cf. [8]), and for most concrete cases is not too difficult to compute, our main task consists then of resolving the two main difficulties:

(a) Give concrete bases for the spaces $\mathcal{H}(G,\pi)$ for which the decomposition of $V^{(\pi')}\otimes V^{(\pi'')}$ can be easily handled.

(b) Give a canonical resolution of the multiplicity problem occurring in this spectral decomposition.

We shall give a sketch of how this can be achieved.

For the remaining part of this exposition we will restrict ourselves to the cases where G is a classical group. For $G = GL(n,\mathbb{C})$, B can be taken as the subgroup consisting of all lower triangular matrices. A holomorphic character of B is given by $\pi^{(m)}(b) = b_{11}^{m_1} b_{22}^{m_2} \cdots b_{nn}^{m_n}$ where (m) belongs to the integer lattice $\mathbb{Z}^n$; $\pi^{(m)}$ is dominant iff $m_1 \geq m_2 \geq \cdots \geq m_n$. If we realize

$Sp(2n,\mathbb{C})$ and $SO(2n,\mathbb{C})$ (a slight modification is needed for the case $SO(2n+1,\mathbb{C})$) as follows:

$$Sp(2n,\mathbb{C}) = \{h \in GL(2n,\mathbb{C}) \mid h\sigma_- h^t = \sigma_-\}$$

$$SO(2n,\mathbb{C}) = \{h \in GL(2n,\mathbb{C}) \mid h\sigma_+ h^t = \sigma_+,\ \det h = 1\}$$

where

$$\sigma_\pm = \begin{pmatrix} 0 & \pm I_n \\ I_n & 0 \end{pmatrix}, \text{ and } h^t \text{ denotes the transpose of } h.$$

Then Borel subgroups of $Sp(2n,\mathbb{C})$ and $SO(2n,\mathbb{C})$ can be taken as intersection of these groups with appropriate conjugates of the lower triangular subgroup of $GL(2n,\mathbb{C})$, and holomorphic characters of the Borel subgroups thus obtained can be also appropriately defined. For $G = GL(n,\mathbb{C})$ set

$$V^{(m)} = \{f:\mathbb{C}^{n\times n} \to \mathbb{C} \mid f \text{ polynomial},\ f(bx) = \pi^{(m)}(b)f(x), \forall (b,x) \in B \times \mathbb{C}^{n\times n}\},$$

set $R_{\pi^{(m)}} = R_{(m)}$, and $V^{(\pi^{(m')})\otimes(\pi^{(m'')})} = V^{(m')\otimes(m'')}$. For this case we also have that $R^0_{(m)} \mid SU(n)$ is irreducible. A double coset map can be used to obtain irreducible polynomial representations of $Sp(2n,\mathbb{C})$ and $SO(2n,\mathbb{C})$. A general element of $Sp(2n,\mathbb{C})$ and $SO(2n,\mathbb{C})$ can be respectively written as

$$s_1\hat{g}s_2 \in Sp(2n,\mathbb{C}) \quad \text{and} \quad a_1\hat{g}a_2 \in SO(2n,\mathbb{C})$$

where

$$\hat{g} = \begin{pmatrix} g & 0 \\ 0 & (g^{-1})^t \end{pmatrix},\ g \in GL(n,\mathbb{C}),\ s_1 = \begin{pmatrix} I_n & 0 \\ s & I_n \end{pmatrix},$$

$$s_2 = \begin{pmatrix} I_n & s' \\ 0 & I_n \end{pmatrix},\ a_1 = \begin{pmatrix} I_n & 0 \\ a & I_n \end{pmatrix}, \text{ and } a_2 = \begin{pmatrix} I_n & a' \\ 0 & I_n \end{pmatrix}.$$

Here s,s' are symmetric $n \times n$ matrices, while a,a' are antisymmetric, and I_n is the identity matrix of order n.

Given this parametrization of $Sp(2n,\mathbb{C})$ and $SO(2n,\mathbb{C})$, we define maps from $V^{(m)}$ to representation spaces of $Sp(2n,\mathbb{C})$ and $SO(2n,\mathbb{C})$ via the double cosets representatives of $B\backslash G/Sp(2n,\mathbb{C})$ and $B\backslash G/SO(2n,\mathbb{C})$ with $G = GL(2n,\mathbb{C})$, and B the Borel subgroup of G (cf. [9]). Thus, irreducible representations of $Sp(2n,\mathbb{C})$ are concretely realized on polynomial spaces as with $GL(n,\mathbb{C})$;

moreover, the restrictions to the corresponding compact real forms become irreducible and unitary with respect to the "differentiation" inner product defined above.

For the sake of concision we will consider the case $G = GL(n,\mathbb{C})$ only, but it is clear that the procedure described below applies to the cases of the symplectic and orthogonal groups as well. Let T be a maximal torus of G_0 ($U(n)$ in this case), let $N(T)$ be the normalizer of T. Then the Weyl group $N(T)/T \equiv W(G)$ can be identified with the symmetric group S_n. We have

$$G = \cup_{p \in W(G)} BpB \quad \text{(Bruhat decomposition [7])}$$

and there exists a unique element $p_0 \in W(G)$, $p_0 = \begin{pmatrix} 0 & & 1 \\ & \cdot^{\cdot^{\cdot}} & \\ 1 & & 0 \end{pmatrix}$ such that Bp_0B is dense in G (C. C. Moore, [7]). Define

$p_i \cdot (m) = (m_{p_i^{-1}(1)}, \dots, m_{p_i^{-1}(n)})$ and $B_i = \{b \in B \mid p_i b p_i^{-1} \in B\}$,

$W_i = \{w \in B \mid p_i w p_i^{-1} \in Z\}$ where Z is the subgroup of all upper triangular unipotent matrices of G. Set

$$H^{(m')+p_i\cdot(m'')} = \left\{ \begin{array}{l} f : G \to \mathbb{C} \mid f \text{ holomorphic}, f(b_i g) = \\ \pi^{(m')+p_i\cdot(m'')}(b_i) f(g), \forall (b_i, g) \in B_i \times G \end{array} \right\}.$$

Then the $H^{(m')+p_i\cdot(m'')}$ are "double coset spaces" and we define $\Phi_i : V^{(m')\otimes(m'')} \longrightarrow H^{(m')+p_i\cdot(m'')}$ by $\Phi_i F(g) = F(g, p_i g)$.

THEOREM (D. King). *If* $(m') + p_i \cdot (m'')$ *is dominant then* $V^{(m')+p_i\cdot(m'')} \subset H^{(m')+p_i\cdot(m'')}$ *occurs in* $V^{(m')\otimes(m'')}$ *with multiplicity one* (cf. [10]).

(This theorem is a particular case of a conjecture made by B. Kostant and Ranga Rao (cf. [10]).)

The pair $(\Phi_0, H^{(m')+p_0\cdot(m'')})$ will play an important role in the tensor product decomposition problem. Indeed, we have the following

THEOREM I. *The mapping* Φ_0 *defined by* $(\Phi_0 F)(g) = F(g, p_0 g)$, *for all* $g \in G$ *and* $F \in V^{(m')\otimes(m'')}$ *is an isomorphism* (*of* G-*modules*) *of* $V^{(m')\otimes(m'')}$ *into the "dense double coset" space*

$H^{(m')+p_0\cdot(m'')}$ (cf. [9]).

Since by Moore's theorem almost all $(g',g'') \in G \times G$ can be uniquely written as $g' = b'g$ and $g'' = b''p_0g$ with $b',b'' \in B$ and $g \in G$ we can define a restricted inverse map

$$\Psi_0 : \Phi_0(V^{(m')\otimes(m'')}) \longrightarrow V^{(m')\otimes(m'')} \quad \text{by}$$

$$(\Psi_0 f)(g',g'') = \pi^{(m')}(b')\pi^{(m'')}(b'')f(g) \quad ([9]).$$

To decompose the tensor product $V^{(m')\otimes(m'')}$ it is necessary to find maps that send elements of an irreducible space $V^{(m)}$ (where (m) includes those representations that occur in the Clebsch-Gordan series of $(m')\otimes(m'')$) into $V^{(m')\otimes(m'')}$. When $(m) = (m') + p_i\cdot(m'')$ is dominant, hence multiplicity free, as shown above, this is achieved by the following

THEOREM II. _If_ $(m) = (m') + p_i\cdot(m'')$ _is dominant then the irreducible representation space_ $V^{(m)}$ _is injected isomorphically into the tensor product_ $V^{(m')\otimes(m'')}$ _via the mapping_

$$\Psi_i h^{(m)}_{[k]}(g',g'') = \sum_{[k'],[k'']} C^{(m)(m')(m'')}_{[k][k'][k'']} h^{(m')}_{[k']}(g') h^{(m'')}_{[k'']}(g'')$$

where the constants C^{--}_{--} _are Clebsch-Gordan coefficients computed with respect to the nonnormalized bases_ $\{h^{(m)}_{[k]}\}$, $\{h^{(m')}_{[k']}\}$, _and_ $\{h^{(m'')}_{[k'']}\}$ _of_ $V^{(m)}$, $V^{(m')}$, _and_ $V^{(m'')}$, _respectively. The map_ Ψ_i _extends by linearity to all elements of_ $V^{(m)}$ _and satisfies_ $\Phi_i\Psi_i = \mathrm{Id}_{V^{(m)}}$, _where_ $\mathrm{Id}_{V^{(m)}}$ _is the identity map on_ $V^{(m)}$ (cf. [9]).

In Section 2 we will show how to construct the bases $\{h^{(m)}_{[k]}\}$ of $V^{(m)}$ and to compute these Clebsch-Gordan coefficients. For other irreducible representations which are not of the form $(m) = (m') + p_i\cdot(m'')$, especially the ones that occur with multiplicity > 1, our strategy is to map them into $H^{(m')+p_0\cdot(m'')}$ and then send them to $V^{(m')\otimes(m'')}$ via the mapping Ψ_0. This procedure gives also a resolution of the multiplicity problem. Indeed, set $(M) = (M_1,\ldots,M_n) = (m') + p_0\cdot(m'')$ and $(M_i) = (M_i,0,\ldots,0)$, $1 \le i \le n$

then we have the following

THEOREM III. *The space* $H^{(m')+p_0\cdot(m'')}$ *is isomorphic to the tensor product* $V^{(M_1)}\otimes\cdots\otimes V^{(M_n)}$ (cf. [11]).

Now it is well-known that the tensor product of any irreducible representation with an irreducible representation of signature $(p,0,\ldots,0)$ is multiplicity free. Therefore, if the tensor product $V^{(M_1)}\otimes\cdots\otimes V^{(M_n)}$ is computed in a stepwise fashion, $(M_1)\otimes(M_2)$, $[(M_1)\otimes(M_2)]\otimes(M_3)$, etc., at each stage the reduction will be multiplicity free, but more importantly, the intermediate irreducible representation labels may be used as labels that distinguish equivalent representations of $V^{(m)}$. In connection with this theorem it is of interest to formulate the following

CONJECTURE (Fröbenius reciprocity theorem for $H^{(M)}$ and $V^{(m)}$). *The number of times that* $V^{(m)}$ *is contained in* $H^{(M)}$ *is equal to the number of times the restriction* $V^{(m)}|_D$ *contains the one-dimensional character* $\pi^{(m)}$ *of the diagonal subgroup* D *of* G.

2. CONSTRUCTION OF BASES FOR $V^{(m)}$ AND COMPUTATION OF CLEBSCH-GORDAN COEFFICIENTS

In this section we shall exhibit a method of construction of orthogonal polynomial bases for the representation spaces $V^{(m)}$ of $G = GL(n,\mathbb{C})$. Using these bases we shall show how to compute matrix coefficients and Clebsch-Gordan coefficients of representations of G. The strategy we follow is to construct all irreducible representations of G from the fundamental ones. Thus, in this section we first turn our attention to an analysis of the fundamental representations of G. The fundamental representations of G are those representations of G with highest weight indexed by $(\underbrace{1,\ldots 1}_{s},\underbrace{0,\ldots,0}_{n-s})$, $0\le s\le n$. For y in $\mathbb{C}^{n\times n}$ let $\Delta^{1\ldots s}_{k_1\ldots k_s}(y)$ denote the minor of y formed from the rows $1,\ldots,s$, and the columns $k_1,\ldots,k_s$ where $(k_1,\ldots,k_s)$ is an s-shuffle, i.e., $1\le k_1<\cdots<k_s\le n$. If $\{[k]\}$ denotes a set of Gelfand-Žetlin labels (cf. [12]) needed to distinguish different orthogonal vectors for a fixed representation with signature (m), then for a fundamental representation the $k_1,\ldots,k_s$ are related to the pattern $[k]$ in the following way.

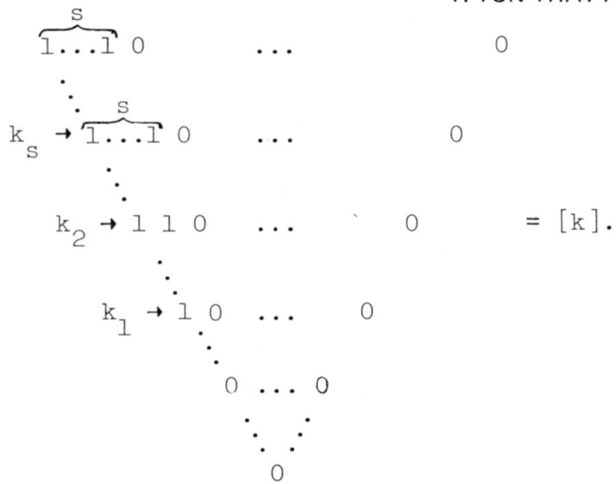

Let $h^{(m)}_{[k]}(y) = \Delta^{1\ldots s}_{k_1\ldots k_s}(y)$ then when $[k]$ varies over all possible tableaus written above, the $h^{(m)}_{[k]}$'s form an orthogonal basis for $V^{(m)}((m) = (\underbrace{1,\ldots,1}_{s},\underbrace{0,\ldots,0}_{n-s}))$ with respect to the "differentiation" inner product $(\cdot,\cdot)$. The matrix coefficients of the fundamental representations are obtained by right translating the basis elements:

$$R_{(m)}(g_0)h^{(m)}_{[k]}(g) = \Delta^{1\ldots s}_{k_1\ldots k_s}(gg_0)$$
$$= \sum_{j_1<\cdots<j_s} \Delta^{1\ldots s}_{j_1\ldots j_s}(g)\Delta^{j_1\ldots j_s}_{k_1\ldots k_s}(g_0).$$

Thus the matrix coefficients are given by

$$D^{(m)}_{[j][k]}(g_0) = \Delta^{j_1\ldots j_s}_{k_1\ldots k_s}(g_0), \quad (m) \text{ a fundamental representation.}$$

To obtain a basis for a general irreducible representation of G with signature $(m) = (m_1,\ldots,m_n)$, $m_1 \geq \cdots \geq m_n \geq 0$, we make use of the fact that any dominant weight (m) can be viewed as the highest weight in an r fold tensor product of fundamental representations. Indeed, set $m_{n+1} = 0$, write

$$(m_1,\ldots,m_n) = \sum_{s=1}^{n} (m_s - m_{s+1})(\underbrace{1,\ldots,1}_{s},\underbrace{0,\ldots,0}_{n-s})$$

and consider the tensor product

$$\underbrace{V^{(1,0,..,0)} \otimes \cdots \otimes V^{(1,0,..,0)}}_{m_1-m_2} \otimes \underbrace{V^{(1,1,0,..,0)} \otimes \cdots \otimes V^{(1,1,0,..,0)}}_{m_2-m_3} \otimes$$

$$\cdots \otimes \underbrace{V^{(1,1,..,1)} \otimes \cdots \otimes V^{(1,1,..,1)}}_{m_n}.$$

If $r = m_1-m_n$ and F is an element of this r-fold tensor product, then under the identity double coset Φ_e defined by

$$(\Phi_e F)(g) = F(g,\ldots,g)$$

F is mapped into $V^{(m)}$. So the identity double coset map generates an irreducible representation, the highest weight representation obtained by adding all the fundamental representations together. Now for a general representation with signature (m) a Gelfand-Žetlin label $[k]$ is given by the following tableau

$$[k] = \begin{bmatrix} m_{1n} & \cdots & m_{nn} \\ & m_{1n-1} \cdots m_{n-1,n-1} & \\ & \ddots \quad \cdot^{\cdot^{\cdot}} & \\ & m_{11} & \end{bmatrix}$$

where $m_{in} = m_i$, $1 \le i \le n$, and each integer m_{ij} in row j is subject to the constraint $m_{i,j+1} \ge m_{ij} \ge m_{i+1,j+1}$. Let $\{h^{(m)}_{[k]}\}$ denote the orthogonal polynomial basis of $V^{(m)}$ that we want to construct, with $[k]$ ranges over all tableaus defined above, then Ref. [13] shows that

$$h^{(m)}_{[k]}(g) = \sum_{[k^{(1)}]+\cdots+[k^{(r)}]=[k]} \Phi_e(h^{(m^{(1)})}_{[k^{(1)}]} \otimes \cdots \otimes h^{(m^{(r)})}_{[k^{(r)}]})(g)$$

$$= \sum_{[k^{(1)}]+\cdots+[k^{(r)}]=[k]} h^{(m^{(1)})}_{[k^{(1)}]}(g)\cdots h^{(m^{(r)})}_{[k^{(r)}]}(g).$$

Note that if we want our basis elements to have the same norms as those given abstractly by Gelfand we must multiply each $h^{(m)}_{[k]}$ by the factor

$$\left[\prod_{j=2}^{n}\left(\prod_{i=1}^{j-1}\binom{m_{ij}-m_{i+1,j}}{m_{i,j-1}-m_{i+1,j}}\right)\right]^{-1}$$

where $m_{i,j}$ are entries of the tableau $[k]$ and $\binom{p}{q}$ denotes the binomial coefficient $\frac{p!}{p!(p-q)!}$. Before computing matrix coefficients and Clebsch-Gordan coefficients of $U(n)$, it is useful to review the relationship between orthonormal bases, Wigner coefficients, and matrix coefficients, using the more familiar ket notation of Dirac. Let $|(m)[k]\rangle$ denote an orthonormal basis, with $[k]$ a Gelfand labeling as before. Then the highest weight in the tensor product $(m^{(1)})\otimes\cdots\otimes(m^{(r)})$ is given by $(m) = (m^{(1)}+\cdots+(m^{(r)})$ with

$$|(m)[k]\rangle = \sum_{[k^{(1)}]\cdots[k^{(r)}]}\langle(m^{(1)})[k^{(1)}]\cdots(m^{(r)})[k^{(r)}]|(m)[k]\rangle \times |(m^{(1)})[k^{(1)}]\rangle\cdots|(m^{(r)})[k^{(r)}]\rangle$$

where $\langle(m^{(1)})[k^{(1)}]\cdots(m^{(r)})[k^{(r)}]|(m)[k]\rangle$ is a highest weight Wigner coefficient. A distinction is being made between Wigner and Clebsch-Gordan coefficients in that Wigner coefficients are defined as reducing an ortho<u>normal</u> basis in an r-fold tensor product space to an ortho<u>normal</u> direct sum basis, whereas Clebsch-Gordan coefficients merely reduce an r-fold orthogonal basis to an orthogonal basis. The reason for this distinction is that the various Φ maps do not in general preserve norms. Thus, it is convenient not to fix the norms of various bases until the very end of a calculation. Now we have

$$R_{(m)}(g)|(m)[k]\rangle = \sum_{[k^{(1)}]\cdots[k^{(r)}]}\langle(m^{(1)})[k^{(1)}]\cdots(m^{(r)})[k^{(r)}]|(m)[k]\rangle \times R_{(m)}(g)(|(m^{(1)}[k^{(1)}]\rangle\cdots|(m^{(r)})[k^{(r)}]\rangle),$$

$$\sum_{[k']} D^{(m)}_{[k'][k]}(g)|(m)[k']\rangle = \sum_{\substack{[k^{(1)}]..[k^{(r)}]\\ [k'^{(1)}]..[k'^{(r)}]}}\langle(m^{(1)}[k^{(1)}]\cdots(m^{(r)})[k^{(r)}]|(m)[k]\rangle$$

$$\times D^{(m^{(1)})}_{[k'^{(1)}][k^{(1)}]}(g)\cdots D^{(m^{(r)})}_{[k'^{(r)}][k^{(r)}]}(g)|(m^{1})[k'^{(1)}]\rangle\cdots|(m^{(r)})[k'^{(r)}]\rangle.$$

It follows that

$$D^{(m)}_{[k'][k]}(g) = \sum_{\substack{[k^{(1)}]..[k^{(r)}] \\ [k'^{(1)}]..[k'^{(r)}]}} \langle (m)[k'] | (m^{(1)})[k'^{(1)}] \cdots (m^{(r)})[k'^{(r)}] \rangle$$

$$\times D^{(m^{(1)})}_{[k'^{(1)}][k^{(1)}]}(g) \cdots$$

$$\cdots D^{(m^{(r)})}_{[k'^{(r)}][k^{(r)}]}(g) \langle (m^{(1)})[k^{(1)}] \cdots (m^{(r)})[k^{(r)}] | (m)[k] \rangle .$$

Thus, to compute the matrix coefficients $D^{(m)}_{[k'][k]}(g)$ it is necessary to know the matrix coefficients of the fundamental representations $D^{(m^{(i)})}_{[k'^{(i)}][k^{(i)}]}(g)$ and the highest weight Wigner coefficients realting $(m) = \sum_{i=1}^{r} (m^{(i)})$ to the fundamental representations. But the matrix coefficients of the fundamental representations were computed in the beginning of this section. It remains to compute the highest weight Wigner coefficients. Below we will actually give a method of computing Clebsch-Gordan coefficients of all multiplicity free representations occurring in the tensor products of r arbitrary irreducible representations. This in particular enables us to calculate the coefficients $C^{(m)(m')(m'')}_{[k][k'][k'']}$ in Theorem II and the highest weight Wigner coefficients after proper normalization. Define the maps $\Lambda^{(m)}_{\eta}$ as

$$(\Lambda^{(m)}_{\eta} h^{(m)}_{[k]})(g_1,\ldots,g_r)$$

$$= \sum_{[k^{(1)}]\ldots[k^{(r)}]} C\eta^{(m)(m^{(1)})\ldots(m^{(r)})}_{[k][k^{(1)}]\ldots[k^{(r)}]} h^{(m^{(1)})}_{[k^{(1)}]} \otimes \cdots \otimes h^{(m^{(r)})}_{[k^{(r)}]}$$

carrying basis elements from $V^{(m)}$ to $V^{(m^{(1)})} \otimes \cdots \otimes V^{(m^{(r)})}$, where η is a degeneracy parameter. Notice that when $(m) = (m') + p_i \cdot (m'')$ occurs with multiplicity one in the tensor product of two representations (m') and (m''), $\Lambda^{(m)}$ is just a map Ψ_i. It follows that

$$h^{(m^{(1)})}_{[k^{(1)}]} \otimes \cdots \otimes h^{(m^{(r)})}_{[k^{(r)}]} = \sum_{(m)\eta[k]} K\eta^{(m)(m^{(1)})\ldots(m^{(r)})}_{[k][k^{(1)}]\ldots[k^{(r)}]} h^{(m)}_{[k]}$$

where K^{---}_{---} are the coefficients inverse to the Clebsch-Gordan coefficients C^{---}_{---}.

If now the map Φ_i (generalized to r-fold tendor product) is applied to both sides of the equation above it follows that

$$(\Phi_i h^{(m^{(1)})}_{[k^{(1)}]} \otimes \cdots \otimes h^{(m^{(r)})}_{[k^{(r)}]})(g) = \sum_{[k]} K^{(m)(m^{(1)})\ldots(m^{(r)})}_{[k][k^{(1)}]\ldots[k^{(r)}]} h^{(m)}_{[k]}$$

Using the orthogonality of the $h^{(m)}_{[k]}$ gives finally

$$K^{(m)(m^{(1)})\ldots(m^{(r)})}_{[k][k^{(1)}]\ldots[k^{(r)}]} = \frac{1}{\|h^{(m)}_{[k]}\|^2} (h^{(m)}_{[k]}, \Phi_e(h^{(m^{(1)})}_{[k^{(1)}]} \otimes \cdots \otimes h^{(m^{(r)})}_{[k^{(r)}]})).$$

It is clear that when we have a two-fold tensor product or an r-fold tensor product of fundamental representations, these coefficients K^{---}_{---} (and thus the Clebsch-Gordan coefficients C^{---}_{---}) can be easily computed since the norms $\|h^{(m)}_{[k]}\|$ are well-known and the "differentiation" inner product $(\cdot,\cdot)$ of the quantities involved are simple.

REFERENCES

1. L. C. Biedenharn, A. Giovannini, and J. D. Louck, J. Math. Phys. t.8:691 (1967), and references cited therein.
2. D. P. Želobenko, Compact Lie groups and their representations, "Nauka", Moscow, 1970; English transl., Transl. Math. Monographs, vol. 40, A.M.S., Providence, R. I., 1973, and references cited therein.
3. Harish-Chandra, Differential operators on a semisimple Lie algebra, Amer. J. Math. 79:87-120 (1957).
4. B. Kostant, Lie group representations on polynomial rings, Amer. J. Math. 85:327-404 (1963).
5. S. Helgason, Invariants and fundamental functions, Acta Math. 109:241-258 (1963).
6. Tuong Ton-That, Lie group representations and harmonic polynomials of a matrix variable, Trans. Amer. Math. Soc. 216:1-46 (1976).
7. G. Warner, Harmonic Analysis on Semisimple Lie Groups, Vol. I, Springer-Verlag, Berlin (1972), and references cited therein.
8. J. Humphreys, Introduction to Lie Algebras and Representation Theory, 2nd ed., Springer-Verlag, Berlin (1972), and references cited therein.

9. W. H. Klink and T. Ton-That, Holomorphic induction and tensor product decomposition of irreducible representations of compact groups I. SU(n) groups, Ann. Inst. Henri Poincaré 31: 77-97 (1979).
10. D. King, The geometric structure of the tensor product of irreducible representations of a complex semisimple Lie algebra (preprint).
11. W. H. Klink and T. Ton-That, On the resolution of the multiplicity problem for U(n) (submitted for publication).
12. I. M. Gelfand and M. I. Graev, Finite dimensional irreducible representations of the unitary group and the full linear groups and related special functions, Izv. Akad. Nauk. SSSR Ser. Mat. t.29:1329-1356 (1965); English transl., Amer. Math. Soc. Transl., t.64:116-146 (1967).
13. W. H. Klink and T. Ton-That, Construction explicite non itérative des bases de GL(n,ℂ)-modules, C.R. Acad. Sci. Paris, Ser. B 289:115-118 (1979).

ORTHOGONAL LAURENT POLYNOMIALS AND GAUSSIAN QUADRATURE

William B. Jones and W.J. Thron

University of Colorado

Boulder, Colorado 80309

The purpose of this paper is to introduce a class of orthogonal functions similar in many respects to the classical orthogonal polynomials [6]. These functions are linear combinations of integral (positive, negative and zero) powers of a single complex variable z. Hence they are called Laurent polynomials. Historically the general theory of orthogonal polynomials had its origins in the investigations of certain types of continued fractions (the Stieltjes fractions and real J-fractions). In like manner, the orthogonal Laurent polynomials originated in the study of continued fractions known as positive T-fractions. The latter have been investigated recently [3] in connection with the following <u>strong Stieltjes moment problem</u>: For a given double sequence of real numbers $\{c_k\}_{k=-\infty}^{\infty}$, does there exist a real-valued bounded, non-decreasing function $\psi(t)$ with infinitely many points of increase on $[0,\infty)$ such that

$$(1) \qquad c_k = \int_0^\infty (-t)^k d\psi(t), \quad \text{for } k = 0,\pm1,\pm2,\ldots \ ?$$

The paper [3] contains (among other things) necessary and sufficient conditions for the existence of a solution to the strong Stieltjes moment problem. That paper is the starting point of the work reported here. Theorems 1,2 and 3 provide the basic existence theory for orthogonal Laurent polynomials. They also describe the uniqueness of these functions and connections with

Research supported in part by the United States National Science Foundation under Grant No. MCS 78-02152.

three-term recurrence relations and with positive T-fractions. Theorems 4,5 and 6 deal with Gaussian quadrature formulas and their convergence. For brevity these theorems are stated with only some ideas about proofs. Complete proofs will be given in a subsequent paper. All of the results stated here have analogues in the theory of classical orthogonal polynomials. Among these results the analogue of Theorem 2 is attributed to Favard [1], that of Theorem 4 (in a special case) to Gauss [2], that of Theorem 5 to Stieltjes [5] and that of Theorem 6 to Markoff [4]

A function $R(z)$ of the form

$$R(z) = \sum_{j=k}^{m} r_j z^j \quad , \quad r_j \in \mathbb{R} \quad , \quad -\infty < k \leq m < +\infty \quad ,$$

is called a (k,m) Laurent polynomial in the complex variable z. A sequence of Laurent polynomials $\{R_n(z)\}_0^\infty$ of the form

$$R_{2m}(z) = \sum_{j=-m}^{m} r_{2m,j} z^j \quad , \quad R_{2m-1}(z) = \sum_{j=-m}^{m-1} r_{2m-1,j} z^j$$

is called regular if, for all m,

$$r_{2m,m} \neq 0 \quad \text{and} \quad r_{2m-1,-m} \neq 0 \quad ;$$

it is called completely regular if, in addition,

$$r_{2m,-m} \neq 0 \quad \text{and} \quad r_{2m-1,m-1} \neq 0 \quad .$$

Let $\mathcal{R}$ denote the linear vector space of all Laurent polynomials over the field of real numbers $\mathbb{R}$ with respect to scalar multiplication and addition of elements in $\mathcal{R}$. Let $\mathcal{R}_{2m}$ and $\mathcal{R}_{2m-1}$ denote the subspaces of Laurent polynomials of type $(-m,m)$ and $(-m,m-1)$, respectively. Then clearly $\mathcal{R}_n \subset \mathcal{R}_{n+1} \subset \mathcal{R}$ for all $n \geq 0$.

For $0 \leq a < b \leq +\infty$, we denote by $\Phi(a,b)$ the set of all real-valued, bounded, non-decreasing functions $\psi(t)$ with infinitely many points of increase on $a \leq t \leq b$ such that all of the moments

$$(2) \qquad c_k = \int_a^b (-t)^k d\psi(t) \quad , \quad k = 0,\pm 1,\pm 2,\cdots$$

exist. If $b = +\infty$, then $\psi(t)$ is defined only on $a \leq t < +\infty$. For each $\psi(t) \in \Phi(a,b)$, an inner product $(\cdot,\cdot)$ on $\mathcal{R}$ is defined by

(3) $$(R,S) = \int_a^b R(t)S(t)d\psi(t) \quad , \quad R,S \in \mathcal{R} \quad .$$

The integrals in (1) , (2) , (3) and hereafter are of Riemann-Stieltjes type. A regular sequence of Laurent polynomials $\{R_n(z)\}_0^\infty$ is called a _sequence of orthogonal Laurent polynomials with respect to_ $\psi(t) \in \Phi(a,b)$ if

(4) $$(R_k,R_n) = \begin{cases} 0 \quad , & \text{if } 0 \le k < n < +\infty \\ \|R_n\|^2 \neq 0 \quad , & \text{if } 0 \le k = n < +\infty \end{cases} \quad .$$

Theorem 1. _Let_ $\psi(t) \in \Phi(a,b)$, $0 \le a < b \le +\infty$. _Then_:

(A) _There exists a sequence_ $\{Q_n(z)\}_0^\infty$ _of orthogonal Laurent polynomials with respect to_ $\psi(t)$ _of the form_

$$Q_{2m}(z) = \sum_{j=-m}^{m} q_{2m,j}z^j \quad , \quad q_{2m,-m} = 1 \quad , \quad q_{2m,m} > 0 \quad , \quad m = 0,1,2,\ldots \quad ,$$

$$Q_{2m-1}(z) = \sum_{j=-m}^{m-1} q_{2m-1,j}z^j \quad , \quad q_{2m-1,-m} = 1 \quad , \quad q_{2m-1,m-1} < 0 \quad , \quad m = 1,2,3,\ldots$$

(B) _Every sequence of orthogonal Laurent polynomials with respect to_ $\psi(t)$ _is completely regular_.

(C) (_Uniqueness_) _If_ $\{\hat{Q}_n(z)\}_0^\infty$ _is any sequence of orthogonal Laurent polynomials with respect to_ $\psi(t)$ _such that_

$$\hat{Q}_{2m}(z) = \sum_{j=-m}^{m} \hat{q}_{2m,j}z^j \quad , \quad \hat{q}_{2m,-m} = 1 \quad , \quad m = 0,1,2,\ldots \quad ,$$

$$\hat{Q}_{2m-1}(z) = \sum_{j=-m}^{m-1} \hat{q}_{2m-1,j}z^j \quad , \quad \hat{q}_{2m-1,-m} = 1 \quad , \quad m = 1,2,3,\ldots \quad ,$$

then

$$\hat{Q}_n(z) \equiv Q_n(z) \quad , \quad n = 0,1,2,\ldots \quad .$$

(D) _The_ $Q_n(z)$ _satisfy a system of three-term recurrence relations of the form_

(5a) $$Q_{2m}(z) = (1 - G_{2m}z)Q_{2m-1}(z) - F_{2m}Q_{2m-2}(z) \quad , \quad m = 1,2,3,\ldots \quad ,$$

(5b) $Q_{2m+1}(z) = (\frac{1}{z} - G_{2m+1})Q_{2m}(z) - F_{2m+1}Q_{2m-1}(z)$, $m = 1,2,3,\ldots$,

where

(5c) $G_{2m} = -\frac{q_{2m,m}}{q_{2m-1,m-1}} > 0$, $F_{2m} = -\frac{q_{2m-2,m-1}}{q_{2m-1,m-1}} \frac{\|Q_{2m-1}\|^2}{\|Q_{2m-2}\|^2} > 0$,

(5d) $G_{2m+1} = -\frac{q_{2m+1,m}}{q_{2m,m}} > 0$, $F_{2m+1} = -\frac{q_{2m+1,m}}{q_{2m,m}} \frac{\|Q_{2m}\|^2}{\|Q_{2m-1}\|^2} > 0$.

(E) Let $V_n(z)$ denote the nth denominator of the continued fraction

(6) $$\frac{F_1}{\frac{1}{z}+G_1} + \frac{F_2}{1+G_2z} + \frac{F_3}{\frac{1}{z}+G_3} + \frac{F_4}{1+G_4z} + \cdots$$

(called a positive T - fraction). Then

(7) $Q_{2m}(z) = (-1)^m V_{2m}(-z)$, $Q_{2m-1}(z) = (-1)^m V_{2m-1}(-z)$,

$m = 1,2,3,\ldots$.

(F) For each $n \geq 1$, the Laurent polynomial $Q_n(z)$ has exactly n zeros $t_j^{(n)}$, $j = 1,2,\ldots,n$. They are distinct positive real numbers satisfying

$$0 \leq a < t_1^{(n)} < t_2^{(n)} < \cdots < t_n^{(n)} < b$$

Ideas on proof of Theorem 1: An application of the Gram-Schmidt orthogonalization process to the sequence $1, 1/z, z, 1/z^2, z^2, \ldots$ would produce a sequence of orthogonal Laurent polynomials. However, it was not obvious (to the authors) that the resulting sequence would be completely regular. Our proof of (A) and (E) came indirectly from a long proof of sufficient conditions for the uniqueness of the solution to the strong Stieltjes moment problem [3, Theorem 6.2] . A more direct proof will be given in a subsequent paper. (C) follows from orthogonality by a standard argument and (B) is an immediate consequence of (A) and (C) (D) can be deduced from (7) and the recurrence relations for the denominators $V_n(z)$ of (6) . Finally, (F) can be proved by induction, using (D) .

The next two theorems are converses of Theorem 1 (A,D) and Theorem 1 (E), respectively. These theorems are direct consequences of results proved in [3, Theorems 3.1, and 6.1]

Theorem 2. *Let $\{Q_n(z)\}_0^\infty$ be a sequence of Laurent polynomials satisfying a system of three-term recurrence relations of the form* (5), *with* $Q_0(z) \equiv 1$. *Then there exists* $\Psi(t) \in \Phi(a,b)$ *with* $0 \le a < b \le +\infty$, *such that* $\{Q_n(z)\}_0^\infty$ *is the sequence of orthogonal Laurent polynomials with respect to* $\Psi(t)$, *normalized as in Theorem* 1 (A).

Theorem 3. *Let* $V_n(z)$ *denote the* nth *denominator of a positive* T-*fraction* (6) *and let* $\{Q_n(z)\}_0^\infty$ *be defined by* (7) *with* $Q_0(z) \equiv 1$. *Then there exists a* $\Psi(t) \in \Phi(a,b)$ *with* $0 \le a < b \le +\infty$, *such that* $\{Q_n(z)\}_0^\infty$ *is the sequence of orthogonal Laurent polynomials normalized as in Theorem* 1 (A)

Using orthogonality of the $Q_n(z)$ and Theorem 1 (F), one can derive the Gaussian quadrature formulas given by the following:

Theorem 4. *Let* $\Psi(t) \in \Phi(a,b)$ *with* $0 \le a < b \le +\infty$. *Let* n *be a positive integer and let* $Q_n(z)$ *denote the* nth *orthogonal Laurent polynomial with respect to* $\Psi(t)$ *normalized as in Theorem* 1 (A). *Let* $t_j^{(n)}$, $j = 1,2,\ldots,n$, *denote the zeros of* $Q_n(z)$. *Then for every* $F(z) \in \mathcal{R}_{2n-1}$,

$$\int_a^b F(t)d\Psi(t) = \sum_{j=1}^{n} w_j^{(n)} F(t_j^{(n)}) , \tag{9a}$$

where

$$w_j^{(n)} = \frac{1}{Q_n'(t_j^{(n)})}\int_a^b \frac{Q_n(t)}{t - t_j^{(n)}} d\Psi(t) , \quad j = 1,2,\ldots,n . \tag{9b}$$

The $w_j^{(n)}$ in (9b) are the *Christoffel numbers*. We will summarize some further properties of them. Let $U_n(z)$ denote the nth numerator of the positive T-fraction (6) associated with the $Q_n(z)$ as in Theorem 1 (E). Let $\{P_n(z)\}_0^\infty$ be defined by $P_0(z) \equiv 0$ and

$$(10)\quad P_{2m}(z) = (-1)^m U_{2m}(-z) \quad , \quad P_{2m-1}(z) = (-1)^m U_{2m-1}(-z) \quad , \quad m = 1,2,3,\ldots \quad .$$

By using the three-term recurrence relations (5) satisfied also by the $P_n(z)$, one can show that

$$(11)\qquad P_n(z) = z\int_a^b \frac{Q_n(z) - Q_n(t)}{z - t}\, d\Psi(t) \quad , \quad n = 0,1,2,\ldots \quad ,$$

and hence, for $j = 1,2,\ldots,n$,

$$(12)\quad P_n(t_j^{(n)}) = t_j^{(n)}\int_a^b \frac{Q_j(t)}{t - t_j^{(n)}}\, d\Psi(t) \quad \text{and} \quad w_j^{(n)} = \frac{P_n(t_j^{(n)})}{t_j^{(n)} Q_n'(t_j^{(n)})} \quad .$$

Moreover, it can be shown that, for $n = 1,2,3,\ldots$,

$$(13)\quad \frac{P_n(z)}{Q_n(z)} = \sum_{j=1}^{n} \frac{z w_j^{(n)}}{z - t_j^{(n)}} \quad , \quad w_j^{(n)} > 0 \quad \text{and} \quad \sum_{j=1}^{n} w_j^{(n)} = \frac{F_1}{G_1} \quad , \quad j = 1,2,\ldots,n \quad .$$

Using these results together with [6, Theorem 15.2.1] , we obtain

<u>Theorem</u> 5. <u>Let</u> $\Psi(t) \in \Phi(a,b)$ <u>with</u> $0 \le a < b < +\infty$. <u>Let</u> $t_j^{(n)}$ <u>and</u> $w_j^{(n)}$ <u>be defined as in Theorem</u> 4 . <u>Then</u>

$$(14)\qquad \int_a^b F(t)\,d\Psi(t) = \lim_{n\to\infty} \sum_{j=1}^{n} w_j^{(n)} F(t_j^{(n)})$$

<u>holds for every function</u> $F(t)$ <u>continuous on</u> $[a,b]$.

The authors conjecture that Theorem 5 can be extended to hold for all functions $F(t)$ for which the integral on the left side of (14) exists. Our final result is obtained as an application of Theorem 5 and (13) by taking $F(t) = 1/(z - t)$.

<u>Theorem</u> 6. <u>Let</u> $\Psi(t) \in \Phi(a,b)$ <u>with</u> $0 \le a < b < +\infty$. <u>Let</u> $U_n(z)/V_n(z)$ <u>denote the</u> nth <u>approximant of the positive</u> T-<u>fraction</u> (6) <u>associated with</u> $\Psi(t)$ <u>as in Theorem</u> 1 (E) . <u>Then, for every complex</u> z <u>not in the interval</u> $[-b,-a]$,

$$(15)\qquad \int_a^b \frac{z\,d\Psi(t)}{z + t} = \lim_{n\to\infty} \frac{U_n(z)}{V_n(z)} \quad .$$

The Markoff theorem stated above asserts that the function on the left side of (15) is represented by the convergent positive T-fraction in Theorem 1 (E) .

We conclude by noting that a great deal of further work must be done in order to investigate fully the ideas set forth in this paper. There are a number of theoretical questions that should be dealt with such as closure of systems of orthogonal Laurent polynomials and the conjecture mentioned in connection with Theorem 5 . There is also a need to study interesting special cases and to explore possible applications. Some special cases of orthogonal Laurent polynomials with respect to $\Psi(t) \in \Phi(a,b)$, where $d\Psi(t) = w(t)dt$, that appear to be of interest include the following:

$w(t)$	(a,b)
$e^{-t-\frac{1}{t}}$	$(0,\infty)$
$e^{-t^2-\frac{1}{t^2}}$	$(0,\infty)$
$e^{-\frac{1}{t}}$	$(0,1)$
e^{-t}	$(1,\infty)$
1	$(1,2)$

References

1. Favard, J., Sur les polynomes de Tchebicheff, C.R. Acad. Sci. Paris 200 (1935), 2052-2053.
2. Gauss, C.F., Methodus nova integralium valores per approximationem inveniendi (1814); Werke, vol. 3, Göttingen (1876), 165-196.
3. Jones, William B., W.J. Thron and H. Waadeland, A strong Stieltjes moment problem, Trans. Amer. Math. Soc., to appear.
4. Markoff, A., Deux démonstrations de la convergence de certaines fractions continues, Acta Math. 19 (1895).
5. Stieltjes, T.J., Quelques recherches sur la théorie des quadratures dites mécaniques, Ann. Sci. Éc. Norm. Paries (3), 1 (1884), 409-426.
6. Szegö, G., Orthogonal Polynomials, Amer. Math. Soc. Coll. Publ. Vol. XXIII, New York, revised edition (1959).

MEASUREMENT THEORY WITH INSTRUMENTS TREATED PARTIALLY QUANTUM MECHANICALLY

Franklin E. Schroeck, Jr.

Florida Atlantic University
Boca Raton, Florida 33432
University of Denver
Denver, Colorado 80208

ABSTRACT

Starting from a general Weyl system, projections onto minimum uncertainty states are derived. The corresponding conditionings (instruments) are defined and interpreted physically, giving a generalization of some previous results for Weyl systems by other authors. Sequential and simultaneous instruments are discussed. The observables corresponding to the instruments are discussed and a natural sense in which action-at-a-distance is an inherent consequence of the uncertainty principle is explored.

INTRODUCTION

Our objective is to treat instruments and the measurement process in an intrinsically quantum mechanical way. To this end, we take the point of view that instruments are composed of particles and that a particle is an element of an irreducible representation of the Weyl algebra. Using the formalisms and languages of the Davies and Lewis[1] measurement scheme and fuzzy phase space, we are able to treat instruments for the description of which it is unnecessary to consider coherent superpositions of states. Since the technical details and computations of these results have appeared[2] or will appear elsewhere[3], we shall concentrate at present on the general structure and philosophy of this approach.

ELEMENTARY WEYL PARTICLES

A Weyl system on a Hilbert Space, $\mathfrak{H}$, is a set of unitary operators $W(x,y)$ on $\mathfrak{H}$, x, y real, satisfying

$$W(x,y)\, W(a,b) = \exp\{-i\hbar(ay-xb)/2\}\, W(x+a,y+b) \ . \tag{1}$$

One can show $W(x,y) = \exp\{ixP+iyQ\}$ where $[P,Q] = -i\hbar\, \mathbb{1}$; so that P and Q may be interpreted as the momentum and position operators. As we are interested in functions of these Weyl operators, we consider

$$T_g \equiv \int_{\mathbb{R}^2} g(x,y)\, W(x,y)\, dxdy \tag{2}$$

where g is a complex-valued function. The integral exists in the strong sense for measurable g , and uniformly for g absolutely integrable in which case T_g is a bounded operator. Positivity and self-adjointness of T_g may also be discussed[3] in terms of properties of g . Our primary interest here is to take

$$g(x,y;a,b,c) = \frac{\hbar}{2\pi} \exp\{-\frac{\hbar c}{4}(x^2+y^2/c^2) - ibx - iay/c\} \tag{3}$$

in which case T_g becomes a projection operator satisfying

$$[P-icQ+i(a+ib)\mathbb{1}]\, T_g = 0 \ . \tag{4}$$

Suppose now that ρ is any positive trace class operator (a state) and we form from it the new state given by

$$\rho' = [\mathrm{Tr}T_g\rho T_g]^{-1}\, T_g\, \rho\, T_g \ . \tag{5}$$

From (3) we then obtain $a/c = \bar{Q}$, $b = \bar{P}$, $2\, c\, \mathrm{Var}\, Q = \hbar$, where $\bar{Q}$ and $\bar{P}$ are the expected values of Q and P respectively in the state ρ' and Var Q is the expected value of $(Q-\bar{Q})^2$ in state ρ' . One further obtains $(\mathrm{Var}\, P)(\mathrm{Var}\, Q) = (\hbar/2)^2$ so that ρ' describes a state of minimum uncertainty. Henceforth we shall treat c , i.e., Var Q and Var P , as a constant intrinsic to the system of particles under consideration. Then we denote $T_g = T_{\bar{Q},\bar{P}}$. It is useful to obtain by direct computation the following relations among the $T_{\bar{Q},\bar{P}}$ and $W(a,b)$ operators:

$$\begin{aligned} T_{\bar{Q},\bar{P}}\, T_{\bar{Q}',\bar{P}'} = {} & \exp\{-i(\bar{Q}'\bar{P}-\bar{Q}\,\bar{P}')/\hbar\} \\ & \exp\{-(c/4\hbar)[(\bar{Q}-\bar{Q}')^2 + (\bar{P}-\bar{P}')^2/c^2]\} \\ & W(\hbar^{-1}(\bar{Q}'-\bar{Q})\ ,\ \hbar^{-1}(\bar{P}-\bar{P}'))\, T_{\bar{Q}',\bar{P}'}\ ; \end{aligned} \tag{6}$$

$$T_{\bar{Q},\bar{P}}\ W(a,b)\ T_{\bar{Q},\bar{P}} = \exp\{i(a\bar{P}+b\bar{Q}) - \hbar c 4^{-1}(a^2+b^2/c^2)\}\ T_{\bar{Q},\bar{P}}\ ; \quad (7)$$

$$T_{\bar{Q},\bar{P}}\ T_{\bar{Q}',\bar{P}'}\ T_{\bar{Q},\bar{P}} = \exp\{-(c/2\hbar)[(\bar{Q}-\bar{Q}')^2 + (\bar{P}-\bar{P}')^2/c^2]\}\ T_{\bar{Q},\bar{P}}\ . \quad (8)$$

We emphasize here that from (6) we observe that the $T_{\bar{Q},\bar{P}}$ do not form a pairwise orthogonal family of projections.

We follow the procedure of J. von Neumann[4] to obtain a decomposition of our Hilbert space into irreducible representations of the Weyl system. This will enable us to give a physical interpretation to the $T_{\bar{Q},\bar{P}}$: we first observe that from (1) and (7), if f and g are in $T_{\bar{Q},\bar{P}}\mathcal{H}$, then

$$(W(a,b)f\ ,\ W(a,b)g) = K\ (f,g)$$

where K is a scalar function of (a,b) . From this we show that for $\{e_\alpha\}$ an orthonormal basis for $T_{\bar{Q},\bar{P}}$, and defining $\mathcal{H}_\alpha$ = closed linear span of $\{W(a,b)e_\alpha \mid a,b \in \mathbb{R}\}$, then $\mathcal{H} = \bigoplus_\alpha \mathcal{H}_\alpha$. By its very definition $\mathcal{H}_\alpha$ hosts an irreducible representation of the Weyl algebra. If we next follow the philosophy from particle physics, we may associate each vector from $\mathcal{H}_\alpha$ with an elementary (Weyl) particle. In particular, $e_\alpha = T_{\bar{Q},\bar{P}}\, e_\alpha$ represents an elementary particle with fuzzy coordinates $(\bar{Q},\bar{P})$ in fuzzy phase space. Furthermore, from (6) we have

$$T_{\bar{Q}',\bar{P}'}\ e_\alpha = K'\ W(\hbar^{-1}(\bar{Q}-\bar{Q}'),\ \hbar^{-1}(P'-P))\ e_\alpha \quad (9)$$

where K' is a scalar. Hence $W(\hbar^{-1}(\bar{Q}-\bar{Q}'),\ \hbar^{-1}(\bar{P}'-\bar{P}))\ e_\alpha$ represents, in $\mathcal{H}_\alpha$, a particle at fuzzy point $(\bar{Q}',\ \bar{P}')$. In the general representation space $T_{\bar{Q},\bar{P}}$ would project out the part of a state belonging to a collection of elementary particles at point $(\bar{Q},\bar{P})$ and any $W(a,b)$ would translate the collection to another fuzzy point.

Conditionings of States: Instruments, Expectations and Measurements

Let μ be a measure on phase space, B a bounded operator in $\mathcal{H}$ (perhaps a density operator) and define

$$\mathcal{E}(\mu,B) = \int T_{\bar{Q},\bar{P}}\ B\ T_{\bar{Q},\bar{P}}\ d\mu(\bar{Q},\bar{P}) \quad . \quad (10)$$

In the measurement scheme of Davies and Lewis[1] this defines both an "instrument" and an "expectation". In the terminology of Ali and Emch[5] it determines a "measurement". What is new here is that we may give an interpretation to (10) directly in terms of elementary particles. First of all, if ρ is a density operator, then $Tr(T_{\bar{Q},\bar{P}}\ \rho\ T_{\bar{Q},\bar{P}})$ represents the probability that ρ is a (collection of) particle(s) at fuzzy point $(\bar{Q},\bar{P})$. It follows that $Tr(\mathcal{E}(\mu,\rho))$ is the probability that ρ is a distribution of a particle (particles) with classical distribution μ.

In spite of the similarity of form of (10) with the von Neumann collapse scheme[6], the fact that $\{T_{\bar{Q},\bar{P}}\}$ is a non-orthogonal family of projectors leads to important differences, one of which is that if an experiment is repeated on a system "immediately after" the first experiment one does not recover the results of the first experiment. In particular, if μ is concentrated at more than a single point in fuzzy phase space, then

$$\mathcal{E}(\mu,\ \mathcal{E}(\mu,\rho)) \neq K\ \mathcal{E}(\mu,\rho)\ ,\ K \text{ a scalar .} \tag{11}$$

Equality in (11) is achieved if μ is concentrated at a point.

As an aside, we note that the results of Ali and Emch[5] and Davies and Lewis[1] in their treatment of the position of the photon may be recovered from (10) from the choices

$$d\mu(\bar{Q},\bar{P}) = d\bar{Q}\ dT(\bar{P}) \quad \text{or} \quad d\bar{P}\ dT(\bar{Q}) \tag{12}$$

for T some measure on one variable.

One may, using (6), (7), (8), compute the variances of P and Q in the state $\mathcal{E}(\mu,\rho)$ with the result[3] that if μ is positive and supported at more than a single point then the product of the variances of P and Q is strictly greater than $(\hbar/2)^2$. Hence the only states with minimum uncertainty that we may prepare with these conditionings are those associated with any single $T_{\bar{Q},\bar{P}}$; i.e., only an elementary Weyl particle at some given fuzzy point has minimum uncertainty.

The Observables Corresponding to the Instruments: A Joint Spectral Family for P and Q.

Following Davies and Lewis[1] we define the observable $A(\mu)$ corresponding to the instrument $\mathcal{E}(\mu,\cdot)$ by

$$T_r(\mathcal{E}(\mu,\rho)\mathbb{1}) \equiv T_r(\rho\ \ A(\mu))\ . \tag{13}$$

From this it follows that

$$A(\mu) = \int d\mu(\bar{Q},\bar{P})\ T_{\bar{Q},\bar{P}} \tag{14}$$

which has the interpretation of a distribution of particle(s) with classical distribution μ when μ is a probability distribution.

For special choices of μ one obtains the following:

$$\int \bar{P}\, T_{\bar{Q},\bar{P}}\, d\bar{Q}d\bar{P} = 2\pi\hbar P \,, \tag{15}$$

$$\int \bar{Q}\, T_{\bar{Q},\bar{P}}\, d\bar{Q}d\bar{P} = 2\pi\hbar Q \,, \tag{16}$$

$$\int T_{\bar{Q},\bar{P}}\, d\bar{Q}d\bar{P} = 2\pi\hbar\, \mathbb{1} \,. \tag{17}$$

More generally,

$$\int \bar{P}^n\, d\bar{P}d\bar{Q}\, T_{\bar{Q},\bar{P}} = n^{th} \text{ order polynomial in } P \,, \tag{18}$$

and similarly for Q. In fact "all" functions of P and "all" functions of Q as well as "all" functions of P and Q may be obtained from the $A(\mu)$ by a suitable choice of μ. We therefore lack only the pairwise orthogonality property for the $T_{\bar{Q},\bar{P}}$ to think of $\{A(\mu)\}$ as the "joint spectral family for P,Q"; since P and Q do not commute, some property of an ordinary joint spectral family had to be violated. As we have none-the-less obtained a formulation from which we may efficiently compute quantities of interest, we shall label the set $\{A(\mu)\}$ as "the joint spectral family of P and Q".

If we wish to know the probability of observing a distribution of particles with distribution σ on a state ρ conditioned by an instrument composed of particles in a detector distributed according to μ, we compute

$$\begin{aligned} &\mathrm{Tr}(\mathcal{E}(\mu,\rho)\, A(\sigma)) \\ &= \int d\mu(\bar{Q},\bar{P})\, d\sigma(\bar{Q}',\bar{P}')\, T_r(\rho\, T_{\bar{Q},\bar{P}}) \\ &\quad \exp\{-c\, 2^{-1}\hbar^{-1}[(\bar{Q}-\bar{Q}')^2 + (\bar{P}-\bar{P}')^2 c^{-2}]\} \end{aligned} \tag{19}$$

which does not vanish even if μ and σ have disjoint supports. The result does, however, tend to zero exponentially in the square of the distance between their supports. We may view (19) as a precise description of action-at-a-distance which devolves from the inherent fuzziness of the constituent particles of the apparatus and observed subjects.

Compositions of Instruments and Observables

We have already seen that the repetition of an experiment does not yield the same result as the original experiment, so that the von Neumann scheme of measurement fails. We may instead try the generalized spectral measure methods of Jauch and Piron[7] in which the composition of two elements A, B of a generalized spectral family are composed by the rule

$$A \cap B \equiv s\text{-}\lim_{n\to\infty} (AB)^n .$$

For the case A = B = projection we would have $A \cap A = A$ which would agree with (11) if μ were concentrated at a single point. For a less trivial example we set

$$d\mu(\bar{Q},\bar{P}) = \exp\{-2\hbar\beta c^{-1}(\bar{Q}^2 + \bar{P}^2 c^{-2}\} \, d\bar{Q}d\bar{P} \tag{20}$$

and compute

$$\lim_{n\to\infty} \frac{Tr([\mathfrak{E}(\mu,\cdot)]^n (\rho) \; W(a,b))}{Tr([\mathfrak{E}(\mu,\cdot)]^n (\rho))} = \exp\{-\hbar c 4^{-1}(1+2c(\beta + b_\infty)^{-1})(a^2 + b^2c^{-2})\} \tag{21}$$

where $[\mathfrak{E}(\mu,\cdot)]^n (\rho)$ is the n-fold application of $\mathfrak{E}(\mu,\cdot)$ on ρ , and b_∞ is a finite constant.[3] Since (21) is not of the form $T_r(\mathfrak{E}(\sigma,\rho) \, W(a,b))$ for any measure σ (the right hand side of (21) being independent of the initial state ρ) we may not use the Jauch and Piron scheme for composition of instruments, since in general we do not thereby recover an instrument. As an aside, (21) describes a non-Fock representation of the canonical commuta-ion relations.

It is useful to consider a particle traveling in the x direction and two detectors D_1, D_2 consisting of particles distributed according to μ_1 and μ_2 respectively with D_1 "over" D_2 in the y direction. The instrument corresponding to detecting the particle in either detector is $\mathfrak{E}(\mu_1 + \mu_2, \cdot)$(suitably normalized), and the corresponding observable is $A(\mu_1 + \mu_2)$. We thus define the operation $\vee$ by

$$\mathfrak{E}(\mu,\cdot) \vee \mathfrak{E}(\nu,\cdot) = \mathfrak{E}(\mu + \nu), \tag{22}$$

$$A(\mu) \vee A(\nu) = A(\mu + \nu) . \tag{23}$$

We may also ask for the instrument corresponding to registering the particle in both detectors. Because of the Gaussian spread of all the particles involved, (i) we should get a non-zero result even though μ_1 and μ_2 have disjoint supports, although (ii) the result should vanish as the separation between the detectors diverges. (iii) The result should also be insensitive to the interchange of μ_1 and μ_2 . Thus we search for an operation o from pairs of measures to measures such that i, ii, iii above hold and define

$$\mathcal{E}(\mu,\cdot) \wedge \mathcal{E}(\nu,\cdot) = \mathcal{E}(\mu \circ \nu) , \tag{24}$$

$$A(\mu) \wedge A(\nu) = A(\mu \circ \nu) . \tag{25}$$

The usual candidates for $\mu_1 \circ \mu_2$ given by the product of their densities, maximum of their densities, minimum of their densities, or convolution of their densities (assuming some kind of absolute continuity) all fail to satisfy all of i, ii, and iii . We present, however,

$$[\mathcal{E}(\mu_1,\cdot) \wedge \mathcal{E}(\mu_2,\cdot)] (\rho)$$

$$= \int d\mu_1(\bar{Q},\bar{P})\, d\mu_2(\bar{Q}',\bar{P}')\, E^2(\bar{Q}-\bar{Q}', \bar{P}-\bar{P}') \; T_{\frac{\bar{Q}+\bar{Q}'}{2}, \frac{\bar{P}+\bar{P}'}{2}}\, \rho\, T_{\frac{\bar{Q}+\bar{Q}'}{2}, \frac{\bar{P}+\bar{P}'}{2}} \tag{26}$$

where

$$E(x, y) = \exp\{-c/2\hbar(x^2 + y^2/c^2)\} \tag{27}$$

as well as an expression similar to (26) with E^2 replaced by E . These two expression for $\wedge$ both satisfy i, ii, iii, and are nice in the sense that they are the only two candidates which arise naturally[3] from a commutatuve operation that sends pairs of self adjoint T_f operators to a self-adjoint operator and also preserves positivity for products of $A(\mu_i)$, μ_i positive measures.

The Classical Limit

If in (1) we were to let $\hbar \to 0$, we would recover a commuting set of operators. We would expect, then, to recover classical measurement theory if we take the limit $\hbar \to 0$ in the various formulae presented so far. This indeed happens, as may easily be checked; the von Neumann collapse scheme is recovered in the classical limit, etc.

Dynamics and Imprimitivity

Since essentially all functions of P and Q may be written in the form $A(\mu)$ for some measure μ , then given some dynamics

α_t or any automorphism "α_t" of the bounded operators on $\mathcal{H}$, we have

$$\alpha_t[A(\mu)] = A(\mu_t) \tag{28}$$

for some new measure μ_t. As an example, if α_t is unitarily implemented with generator $P^2(2m)^{-1}$ then we obtain the supposed "free" dynamics. One obtains

$$\alpha_t^{\text{free}} (W(x, y)) = W(x + tym^{-1}, y) . \tag{29}$$

In fact this is not the appropriate free dynamics for a stable elementary particle, since it leads to "wave-packet spreading". The free dynamics would be that for which

$$\alpha_t[T_{\bar{Q},\bar{P}}] = T_{\bar{Q}+t\bar{P}m^{-1}, \bar{P}} . \tag{30}$$

The explicit computation of the form for the dynamics with the usual free Hamiltonian perturbed by an arbitrary number of harmonic oscillator potentials has also been carried out.[3]

In a previous paper[2] we have shown that the $\{A(\mu)\}$ under the action of the $W(a,b)$ form a system of imprimitivity. In particular

$$W(a,b)\ T_{\bar{Q},\bar{P}}\ W(a,b)^* = T_{\bar{Q}-\hbar a,\ \bar{P}+\hbar b} ; \tag{31}$$

so

$$W(a,b)\ A(\mu)\ W(a,b)^* = A(\mu_{a,b}) , \tag{32}$$

where

$$d\mu_{a,b}(\bar{Q},\bar{P}) = d\mu(\bar{Q}+\hbar a,\ \bar{P}-\hbar b) . \tag{33}$$

Implications for Axioms of Physics

By considering the variance of the observable $A(\mu)$ in state $\mathcal{E}(\sigma,\rho)$ it is seen that if $\mu = \sigma =$ point measure, then $A(\mu)$ is dispersion free on $\mathcal{E}(\sigma,\rho)$. If σ is supported at more than a single point then the variance is non-zero. It seems[3] that if σ is a point mass and μ is supported at more than one point then the dispersion is again non-zero. Thus the only observables with physically preparable dispersion free states are the $T_{\bar{Q},\bar{P}}$ But we cannot generate all of the $A(\mu)$ by taking limits of finite linear combinations of the $T_{\bar{Q},\bar{P}}$. Thus we have a contradiction with a basic axiom of the C* approach to physics[8] which requires the existence of a (large) number of dispersion free states for each observable. Similarly in the quantum logic framework one hypothesizes that to every physical proposition (observable) there is a state for which the probability of obtaining a "yes" is one. Such a state would therefore be dispersion-free for the observable, in contradiction with our results.

Using equations 22, 23, 24, 25 we may attempt to generate the quantum logic, L . However, o is not associative and has no identity so that $\wedge$ is non-associative and non-idempotent, and L has no top element. Without a top element, no complement can be defined on L either. It would seem that quantum logic has little to do with physical situations.

We recall that action-at-a-distance is an inherent consequence of the uncertainty principle. We have a dynamics which is causal in the sense of having explicit formulae, but because conditionings and observables may only be described in fuzzy phase space, we have a non-deterministic theory.

REFERENCES

1. Davis, E.B. & Lewis, J.T., "An Operational Approach to Quantum Probability, Comm. Math. Phys., 17, (1970), pp. 239-260.
2. Schroeck, F.E. Jr., "Measures with Minimum Uncertainty", in Mathematical Foundations of Quantum Theory, A.R. Marlow, ed., Academic Press, New York, (1978), pp. 299-327.
3. Schroeck, F.E. Jr., "A Quantum Mechanical Treatment of Measurement with a Physical Interpretation", preprint, (1979).
4. von Neumann, J., "Die eindentigheit der Schrödingerschen Operatoren, Math. Ann, 104, (1931), pp. 570-578.
5. Ali, S.T. & Emch, G.G., "Fuzzy Observables in Quantum Mechanics", J. Math. Phys, 15, (1974), pp. 176-182.
6. von Neumann, J., Mathematical Foundations of Quantum Mechanics, Princeton University Press, Princeton, N.J., (1955), Chapters V, VI.
7. Jauch, J.M. & Piron, C., "Generalized Localizability", Helv. Phys. Acta, 40, (1967), pp. 559-570.
8. Emch, G.G., Algebraic Methods in Statistical Mechanics and Quantum Field Theory, Wiley-Interscience, New York, (1972), n.b. pg. 40, xiom 4.

TRANSITION MAPS AND LOCALITY

Richard Mercer

Department of Mathematics
Wright State University
Dayton, Ohio 45435

1. INTRODUCTION

This is a summary of some of the results obtained in [1]. We will deal with quantum measurement theory and the notion of a quantum stochastic process, and the close relationship between the two. The unifying theme will be the concept of a local transition map (to be defined below). We will present the claim that local transition maps are generalized quantum measurements, and state results justifying this claim. Local transition maps in a different context will serve as propagation maps for quantum stochastic processes, leading to the result that quantum stochastic processes (as defined here) necessarily contain quantum measurements. For details and further references on this material, see [1].

We use the following notation and conventions:

$\mathcal{H}$ will always denote a Hilbert space; all Hilbert spaces are complex and separable with inner products conjugate linear in the first entry. $\mathcal{B}(\mathcal{H})$ denotes the von Neumann algebra of bounded linear operators on $\mathcal{H}$. $\Sigma(\mathcal{H})$ denotes the convex set of normal states on $\mathcal{B}(\mathcal{H})$, as represented by the positive trace-class operators ρ on $\mathcal{H}$ with $tr(\rho) = 1$. 1 will represent the identity of any operator algebra. The expression "$\mathcal{A}$ is a von Neumann algebra in $\mathcal{B}(\mathcal{H})$" will include the assumption that $\mathcal{A}$ contains the identity of $\mathcal{B}(\mathcal{H})$.

We assume the reader is familiar with the theory of bounded operators on a Hilbert space and the fundamental concepts of quantum mechanics.

2. QUANTUM STOCHASTIC PROCESSES

We present here a definition for quantum stochastic process which is at once simple, closely analogous to the classical concept, and quite general. It is also easy to add "gadgets" to this definition for further generalization or describing additional structure.

Definition: Let H be a separable complex Hilbert space, H_t a copy of H for each $t \in [0, \infty)$, and F the family of finite subsets of $[0, \infty)$. Let $H_F = \bigotimes_{t \in F} H_t$ for each $F \in F$. A quantum stochastic process (QSP) is a family $\{\rho_F\}_{F \in F}$, where ρ_F is a normal state in $B(H_F)$, which satisfies the following consistency conditions:

For each $F \in F$ there is a commutative von Neumann algebra C_F in $B(H_F)$ such that whenever $G \in F$ with $F \cap G = \emptyset$,

$$\text{(QCC)} \qquad \rho_{F \cup G}(x_F \otimes 1_G) = \rho_F(x_F) \qquad \text{for all } x_F \in C_F,$$

where 1_G is the identity in $B(H_G)$. **

The interpretation of ρ_F for $F = \{t_1, \ldots, t_n\}$ is that it is the joint state of the system under consideration when measurements are made at times $t_1, \ldots, t_n$. (For convenience it is often useful to assume that a measurement is not made at the final time t_n in F.) We assume that $C_F = C_{t_1} \otimes \cdots \otimes C_{t_n}$ where C_{t_i} is a commutative von Neumann algebra in $B(H_{t_i})$, and that the observables measured at time t_i together with the identity generate C_{t_i}.

If $x_i \in C_{t_i}$ for $i = 1, \ldots, n$, the interpretation of $\rho_F(\bigotimes_{i=1}^{n} x_i)$ is clear: it is the joint expectation of the observables x_i at the respective times t_i. If x_i is not in C_{t_i} for some i, the interpretation of $\rho_F(\bigotimes_{i=1}^{n} x_i)$ is not as clear, but it is not to be dismissed out of hand as meaningless; to do so would be to say that in principle all we can know about a system are the results of measurements. Such an assumption should not be built into the mathematical framework.

The conditions QCC ("quantum consistency conditions") are a generalization of the Kolmogorov consistency conditions for a classical stochastic process [2, p. 43]. The latter may be written as

(KCC) $\mu_{F\cup G}(f_F \otimes 1_G) = \mu_F(f_F)$ for all $f \in C_F$

with the following explanation:

if X is the measurable space serving as a phase space for our classical stochastic process and X_t is a copy of X for each t, define $X_F = \underset{t\in F}{\times} X_t$ for $F \in \mathcal{F}$. A classical stochastic process is then given as a collection $\{\mu_F\}_{F\in\mathcal{F}}$ where μ_F is a measure on X_F. Let $C_F = L^\infty(X_F)$. If 1_G is the function on X_G which is identically 1 (and is the identity of C_G), then (KCC) are the Kolmogorov consistency conditions.

We can then see that (QCC) are a direct generalization of (KCC) under the usual rules of correspondence between classical and quantum probability. An alternative generalization would be to require (QCC) to hold for all $x_F \in \mathcal{B}(\mathcal{H}_F)$. However, this would not be appropriate in a quantum situation as it is equivalent to the assumption that the measurements occurring at times in F do not disturb the state of the system. To see this, we restrict ourselves to the simple case where $F = \{t_1\}$ and $G = \{t_2\}$ with $t_1 < t_2$. Then we have

(QCC-2) $\rho_{1,2}(x_1 \otimes 1) = \rho_1(x_1)$ for all $x_1 \in C_1$.

This condition may be interpreted as follows:

Let ρ_1 be the state of the system at time t_1. Suppose the system is measured at time t_1 and then evolves further to time t_2, giving a joint state $\rho_{1,2}$. If we then take the partial trace over H_2 we obtain a state ρ_1', which we may interpret to be the state immediately following the measurement defined by $\rho_1'(x_1) = \rho_{1,2}(x_1 \otimes 1)$. To claim that (QCC-2) holds for all $x_1 \in B(H_1)$ would then be to claim that $\rho_1' = \rho_1$, i.e., the measurement has not disturbed the state of the system.

This is an unacceptable requirement in a quantum system. What (QCC-2) in fact claims is that the expectations of observables which are functions of those being measured are unaffected by the measurement, that is $\rho_1'(x_1) = \rho_1(x_1)$ for all $x_1 \in C_1$.

3. TRANSITION MAPS

If A, B are von Neumann algebras we may define a transition map from A to B to be a completely positive normal linear map $\Phi: A \to B$ with $\Phi(1) = 1$. Here "normal" means "continuous when A and B have their respective σ-weak topologies." In this article, we will only be concerned with the case $A = B(H_2)$ and $B = B(H_1)$ for Hilbert spaces H_1, H_2.

If $\rho \in \Sigma(H_1)$ and $\Phi : B(H_2) \to B(H_1)$ is a transition map, then $\rho \circ \Phi \in \Sigma(H_2)$.

Transition maps are useful to describe the following situations:

(A) Measurement: if $\rho \in \Sigma(H)$ is the state of the system prior to a measurement and $\rho' \in \Sigma(H)$ the state after measurement, the map $\rho \to \rho'$ may be given by a transition map $\Phi: B(H) \to B(H)$, with $\rho' = \rho \circ \Phi$. (In this paper we assume measurements to be instantaneous.)

(B) Evolution: if $\rho_0 \in \Sigma(H)$ is the state of the system at time 0, and $\rho_t \in \Sigma(H)$ the state at time t, the map $\rho_0 \to \rho_t$ may be given by a transition map $\Phi_t: B(H) \to B(H)$, with $\rho_t = \rho_0 \circ \Phi_t$.

(C) Propagation of QSP's: Suppose we have a quantum system on Hilbert space H with the following property:

> Let $F, G \in F$ with $F \subset G$ and such that $s \in F$, $T \in G \setminus F$ imply $s < t$. Then ρ_F uniquely determines ρ_G in the sense that all QSP's occurring in this system with F-joint state ρ_F have the same G-joint state ρ_G.

The map $\rho_F \to \rho_G$ may be given by a transition map $\Phi_{G,F}: \mathcal{B}(H_G) \to \mathcal{B}(H_F)$, with $\rho_G = \rho_F \circ \Phi_{G,F}$. Such maps $\Phi_{G,F}$ will be called propagation maps.

It is not hard to construct examples of systems where propagation maps do not exist, but for many situations, it is not unreasonable to assume that they do, at least for some pairs F, G. This would amount to assuming that one had enough information about the system and the influence of the outside universe to predict ρ_G, given ρ_F.

There is a nice result on the representation of transition maps. If $\Phi: \mathcal{B}(H_2) \to \mathcal{B}(H_1)$ is a transition map, then $\Phi x = \sum_i A_i^* x A_i$ for all $x \in \mathcal{B}(H_2)$, for some sequence $\{A_i\}$ in $\mathcal{B}(H_1, H_2)$. (We will abbreviate this by $\Phi = \sum_i A_i^* \cdot A_i$.) Since $\Phi(1) = 1$ we have $\sum_i A_i^* A_i = 1$.

4. LOCALITY

Let A_0 be a von Neumann algebra in $\mathcal{B}(H)$. Then a transition map $\Phi: \mathcal{B}(H) \to \mathcal{B}(H)$ is <u>local with respect to</u> A_0 if for each projection $p \in A_0$, $\Phi(p) \leq p$. More generally, if $\Phi: \mathcal{B}(H(\otimes H_1)) \to \mathcal{B}(H(\otimes H_2))$, where each tensor product is optional, then Φ is local with respect to A_0 if for each projection $p \in A_0$, $\Phi(\tilde{p}) \leq \tilde{p}$, where each $\tilde{p}$ is p or $p \otimes 1$ as appropriate. (For example, if $\Phi: \mathcal{B}(H \otimes H_1) \to \mathcal{B}(H)$ the requirement is $\Phi(p \otimes 1) \leq p$.)

Let $A \in \mathcal{B}(H)$. Then A is said to be local with respect to A_0 if for all projections $p \in A_0$, $pAp = Ap$ (or equivalently, $A \in A_0'$.) A bounded linear operator $A: H(\otimes H_1) \to H(\otimes H_2)$ is local with respect to A_o if for all projections $p \in A_o$, $\tilde{p}A\tilde{p} = A\tilde{p}$, where again each tensor product is optional and $\tilde{p} = p$ or $p \otimes 1$ as appropriate.

Proposition: Let $\Phi: \mathcal{B}(H(\otimes H_1)) \to \mathcal{B}(H(\otimes H_2))$ be a transition map given by $\Phi = \sum_i A_i^* \cdot A_i$ where $A_i: H(\otimes H_2) \to H(\otimes H_1)$. Then Φ is local with respect to a von Neumann algebra A_o in $\mathcal{B}(H)$ if and only if each A_i and A_i^* are local with respect to A_o. **

Proposition: If Φ is a transition map on $\mathcal{B}(H)$ which is local with respect to A_o, then $\Phi(axb) = a\Phi(x)b$ for $x \in \mathcal{B}(H)$, $a, b \in A_o$. **

Thus local transition maps may be considered as generalizations of conditional expectations [3, p. 101].

Proposition: Let $\Phi: \mathcal{B}(H_1 \otimes H_2) \to \mathcal{B}(H_1)$ be a transition map, and for each $\rho_1 \in \sum(H_1)$ define $\rho_{1,2} \in \sum(H_1 \otimes H_2)$ by $\rho_{1,2} = \rho_1 \circ \Phi$. Then (QCC-2) is satisfied for each $\rho_1 \in \sum(H_1)$ if and only if Φ is local with respect to C_1. **

5. MEASUREMENTS

We define a measurement on H to be a transition map $\Phi: \mathcal{B}(H) \to \mathcal{B}(H)$ together with a commutative von Neumann algebra C_o in $\mathcal{B}(H)$, such that Φ is local with respect to C_o.

We distinguish three levels of measurement. A transition map Φ is

a (weak) measurement if Φ is local with respect to C_o
a strong measurement if Φ is local with respect to C_o'
a complete measurement if $\Phi: \mathcal{B}(H) \to C_o$.

Weak or strong measurements which are not complete may be referred to as incomplete measurements. As an example, let $\{P_k\}$ be projections in C_0 with $\sum_k P_k = 1$. Then $\Phi = \sum_k P_k \cdot P_k$ is a strong measurement, as each $P_k \in (C_0')' = C_0$. If each P_k is one-dimensional, Φ is a complete measurement. It is known that complete measurements of this type increase entropy of a state [4], and they obviously decrease off-diagonal matrix elements of a state ρ if the matrix elements ρ_{ij} are defined by $\rho_{ij} = P_i \rho P_j$.

<u>Proposition</u>: [5, p. 275] Let C_0 be a commutative von Neumann algebra in $\mathcal{B}(H)$. There is a σ-finite standard measure space (Γ, μ) and a measurable field of Hilbert spaces $\{H(t): t\in\Gamma\}$ such that $H = \int_\Gamma^\oplus H(t)d\mu(t)$, the direct integral of the $H(t)$, and C_0 is the algebra of multiplication operators on H. **

(Recall that $A\in\mathcal{B}(H)$ is a multiplication operator if there exists $a(t) \in L^2(\Gamma, \mu)$ such that for $f = \int_\Gamma^\oplus f(t)d\mu(t)\in H$, $Af = \int_\Gamma^\oplus a(t)f(t)d\mu(t)$.)

Let C_0 be as above and let $H = \int_\Gamma^\oplus H(t)d\mu(t)$ be the corresponding direct integral decomposition of H. We say that an operator $x\in\mathcal{B}(H)$ has a kernel with respect to C_0 if there is a measurable field of operators [5, p. 272] $\{x(s,t); s, t\in\Gamma\}$ with $x(s,t): H(t) \to H(s)$ such that for $f, g \in H$

$$\langle f,xg\rangle = \int_{\Gamma\times\Gamma} \overline{f(s)}x(s,t)g(t)d\mu(s)d\mu(t).$$

<u>Proposition</u>: Let $x\in\mathcal{B}(H)$ be a Hilbert-Schmidt operator. Then x has a kernel with respect to any commutative von Neumann algebra C_0 in $\mathcal{B}(H)$. **

In particular, any state $\rho\in\sum(H)$ has a kernel with respect to any C_0. Assuming C_0 is given, we denote this kernel by $\rho(s,t)$.

If C_0 is a maximal commutative algebra in $B(H)$, then each $H(t)$ is one-dimensional for each $t \in \Gamma$ and $H = L^2(\Gamma, \mu)$. If ρ is a state on H, its kernel $p(s,t)$ is a complex-valued function on $\Gamma \times \Gamma$.

PROPOSITION: Suppose that C_0 is maximal commutative in $B(H)$. Let $\Phi: B(H) \to B(H)$ be a transition map which is local with respect to C_0. Let $\rho \in \Sigma(H)$, and let $\rho' \in \Sigma(H)$ be given by $\rho' = \rho \circ \Phi$. If $\rho(s,t)$ and $\rho'(s,t)$ are the respective kernels of ρ and ρ', then

$$|\rho'(s,t)| \leq |\rho(s,t)| \quad \text{a.e.} \quad (\mu \times \mu). \quad **$$

This proposition shows that when C_0 is maximal measurements "decrease matrix entries". The next proposition deals with the increase of entropy by measurements. The most widely accepted definition of the entropy of a quantum state is $S(\rho) = -\mathrm{tr}(\rho \ell n \rho)$. The results given here are valid as long as the entropy functional is of the form $S(\rho) = \mathrm{tr}(f(\rho))$ for some concave function f on [0,1]. (Taking $f(x) = -x \ell n x$ gives the above definition.)

PROPOSITION: In the situation of the above proposition, $S(\rho') \geq S(\rho)$. **

In order to get the same results when C_0 is not maximal commutative, we must assume Φ to be a strong measurement.

PROPOSITION: Let $\Phi: B(H) \to B(H)$ be a transition map which is local with respect to C_0'. Let $\rho \in \Sigma(H)$, and let $\rho' \in \Sigma(H)$ be given by $\rho' = \rho \circ \Phi$. Then

$$||\rho'(s,t)|| \leq ||\rho(s,t)|| \quad \text{a.e.} \quad (\mu \times \mu)$$

and $S(\rho') \geq S(\rho)$. **

One can obtain results similar to these in the general case of a weak measurement. However, they require a considerable amount of preliminary work and some new definitions; hence they will not be given here. But one can get a feeling for what happens in the general case by examining the case of uniform multiplicity. This occurs when each $H(t)$ has the same dimension, say n. In this case $H = \int_{\Gamma}^{\oplus} H(t)d\mu(t)$ is isomorphic to $H_0 \otimes K$, where $H = L^2(\Gamma,\mu)$ and K is a fixed Hilbert space of dimension n [6, p. 153].

We may define the <u>partial trace</u> over K of a state $\nu \in \Sigma(H)$ to be $tr_K \nu \in \Sigma(H_0)$, where for $x_0 \in B(H_0)$, $tr_K \nu(x_0) = \nu(x_0 \otimes 1)$.

PROPOSITION: Let Φ be a transition map on $B(H_0 \otimes K)$ which is local with respect to C_0, the algebra of multiplication operators on $H_0 = L^2(\Gamma,\mu)$. Let $\nu \in \Sigma(H_0 \otimes K)$ be given by $\nu = \rho \otimes \omega$ for $\rho \in \Sigma(H_0)$, $\omega \in \Sigma(K)$. Let $\nu' = \nu \circ \Phi$ and $\rho' = tr_K \nu$. Then

$$|\rho'(s,t)| \leq |\rho(s,t)| \qquad \text{a.e.} \quad (\mu \times \mu)$$

and $S(\rho') \geq S(\rho)$. **

6. PROPAGATION

Let $\rho_1 \in \Sigma(H_1)$ be the state of our system at time t_1. We assume that the propagation of ρ_1 to $\rho_{1,2} \in \Sigma(H_1 \otimes H_2)$ (the joint state for times t_1 and t_2) is given by a transition map $\Phi: B(H_1 \otimes H_2) \to B(H_1)$, i.e. $\rho_{1,2} = \rho_1 \circ \Phi$.

PROPOSITION: Let C_1 be a commutative von Neumann algebra in $B(H_1)$, and let $\Phi: B(H_1 \otimes H_2) \to B(H_1)$ be a transition map which is local with respect to C_1. If $\Phi_1: B(H_1) \to B(H_1)$ is defined by

$\Phi_1(x_1) = \Phi(x_1 \otimes 1)$ for $x_1 \in \mathcal{B}(H_1)$, then Φ_1 is also a transition map local with respect to C_1. Conversely, if Φ_1 is local with respect to C_1, so is Φ. Furthermore, if we define for each $\rho_1 \in \Sigma(H_1)$ a state $\rho_1' \in \Sigma(H_1)$ by $\rho_1' = tr_{H_2}\rho_{1,2} = tr_{H_2}(\rho_1 \circ \Phi)$, then $\rho_1' = \rho_1 \circ \Phi_1$. **

This proposition may be interpreted as follows: the propagation from ρ_1 to $\rho_{1,2}$ contains embedded in it a quantum measurement $\rho_1 \to \rho_1'$ occurring at time t_1. More generally, if $\{\rho_F\}_{F \in F}$ is a QSP and $F, G \in F$ with $F \subset G$ and $s \in F$, $t \in G \backslash F$ implying $s < t$, then the propagation $\rho_F \to \rho_G$ contains quantum measurements at the times in F. (The latter result is obtained from the former by replacing H_1 by H_F and H_2 by $H_{G \backslash F}$.) This result follows from just two assumptions: our definition of a QSP and the assumption that the propagation is given by a transition map.

Before we can be satisfied with this result, however, we must realize that the identity transition map on $\mathcal{B}(H)$ is local with respect to any algebra C_0 and hence always qualifies as a measurement. The following proposition rules out this possibility in virtually all cases and therefore guarantees that the quantum measurements occurring in a QSP actually change the state of the system.

The identity transition map on $\mathcal{B}(H_1)$ is given by $\Phi_1 = 1 \cdot 1$. This map is local with respect to $\mathcal{B}(H_1)$, and by the above proposition so is any $\Phi: \mathcal{B}(H_1 \otimes H_2) \to \mathcal{B}(H_1)$ from which Φ_1 is derived via $\Phi_1(x_1) = \Phi(x_1 \otimes 1)$.

PROPOSITION: Let $\Phi: \mathcal{B}(H_1 \otimes H_2) \to \mathcal{B}(H_1)$ be a transition map local with respect to $\mathcal{B}(H_1)$. Then for each $\rho_1 \in \Sigma(H_1)$, $\rho_{1,2} = \rho_1 \circ \Phi = \rho_1 \otimes \rho_2$, where $\rho_2 \in \Sigma(H_2)$ does not depend on ρ_1. **

From this we can conclude that if the identity measurement $\Phi_1 = 1 \cdot 1$ occurs at time t_1, then the state of the system and the results of measurements at future times are independent of the initial state. In most quantum systems this simply cannot be the case, and so the identity measurement is ruled out. (On the other hand, if such a situation does exist, it makes no difference whether or not the initial state is disturbed by the measurement!) Incidentally, this result also excludes the possibility of requiring (QCC) to hold for all $x_F \in \mathcal{B}(H_F)$ as then the propagation map $\Phi_{G,F}$ taking ρ_F to ρ_G would have to be local with respect to $\mathcal{B}(H_F)$.

7. DISCUSSION

The mathematical theory of QSP's first got off the ground with the works of E.B. Davies, collected in [7] with further references. Davies restricted his attention to Markov processes, however, and during the 1970's several different definitions for a QSP were proposed, with that of Lindblad [8] being in my opinion closest to the mark. I did not find any of them satisfactory in all respects and so have constructed my own, presented here.

The field of quantum stochastic processes is still in its infancy, with much work to be done both on theory and examples. I believe the ideas presented here can serve as a framework for a general theory of QSP's, although it's not yet clear precisely what role is to be played by transition maps.

Looking to the future, some of the extensions of this theory I hope to develop are:

(i) a "transition-map valued measure" approach, in the spirit of Davies' work;
(ii) an extension to von Neumann algebras other than $\mathcal{B}(H)$;
(iii) simultaneous measurement of non-commuting observables;
(iv) "generalized quantum processes" in the spirit of Gelfand-Shilov, volume 4, where measurements take place over an interval of time.

REFERENCES

[1] Mercer, R.: Ph.D. Thesis, University of Washington (1980).

[2] Gikhman, I.I., Skorohod, A.V.: The Theory of Stochastic Processes I (1970), Springer, Berlin.

[3] Sakai, S.: C* and W* Algebras (1971) Springer, Berlin.

[4] Lindblad, G.: Comm. Math. Phys., 33, 305-322 (1973). "Entropy, Information, and Quantum Measurements".

[5] Takesaki, M.: Theory of Operator Algebras I (1979), Springer New York.

[6] Dixmier, J.: Les Algebras D' Operateurs Dans L'Espace Hilbertien (1969) Gauthier-Villars, Paris.

[7] Davies, E.B.: Quantum Theory of Open Systems (1976), Academic Press, London.

[8] Lindblad, G.: Comm. Math. Phys., 65, 281-294 (1979): "NonMarkovian Quantum Stochastic Processes and their Entropy".

THE FEYNMAN-KAC FORMULA FOR BOSON WIENER PROCESSES

R.L. Hudson

Mathematics Department

University of Nottingham
Nottingham, NG7 2RD
England

P.D.F. Ion

Angewandte Mathematik

Universität Heidelberg
D-6900 Heidelberg
West Germany

K.R. Parthasarathy

Indian Statistical Institute

7, S J S Sansanwal Marg
New Delhi - 110029
India

ABSTRACT

A dilation of a self-adjoint contraction semigroup is constructed which consists of a unitary evolution in which the evolution operators for disjoint time intervals act in mutually orthogonal supplementary Hilbert spaces. On second quantizing, one obtains noncommutative Feynman-Kac formulae for perturbed semigroups which are related to the Boson noncommutative Wiener process in the same way as the original Feynman-Kac formula is related to the classical Wiener process.

In the case when the original semigroup is $t \to e^{-t}$, the Feynman-Kac formula can be expressed heuristicly as

$$\exp\{-(s-t)(a^{\dagger}a+V(q))\}$$
$$= \mathbb{E}\left[\exp\{-\int_t^s V(e^{-(s-r)}q \otimes I + \sqrt{2}\, I \otimes \int_r^s e^{-(r'-r)}\, dQ(r'))\, dr \prod_t^s e^{i\sqrt{2}(pdQ-qdP)}\right]$$

where $a^{\dagger} = 2^{-\frac{1}{2}}(p+iq)$ and a are creation and annihilation operators,

$V = V(q)$ is a perturbing potential, $\mathbb{E}$ denotes expectation over the Boson Wiener process (P,Q) and the continuous product is a "stochastic product integral".

1. INTRODUCTION

A simple Feynman-Kac formula is

$$e^{-(s-t)(\frac{1}{2}p^2+V(q))} = \mathbb{E}\left[\exp\{-\int_t^s V(q+X(s)-X(r))\,dr\}e^{i(X(s)-X(t))p}\right] \quad (1)$$

Here the semigroup $e^{-(s-t)(\frac{1}{2}p^2+V(q))}$ generated by perturbing with the potential $V(q)$ the free Hamiltonian $\frac{1}{2}p^2$ is expressed as the expectation over the auxilliary Wiener process X of the product of two terms. The second term

$$U_{st} = e^{i(X(s)-X(t))p} \quad (2)$$

may be regarded as a random unitary evolution, insofar as the operators (2) are random unitary operators satisfying the identity

$$U_{rs}U_{st} = U_{rt}$$

which inherit strong continuity in s and t from the sample-path continuity property of the Wiener process. The first term in the expectation in (1)

$$M_{st} = \exp\{-\int_t^s V(q+X(s)-X(r))\,dr\}$$

is a cocycle for the evolution (U_{st}), insofar as it satisfies the identity

$$M_{rs}U_{rs}M_{st}U_{rs}^{-1} = M_{rt}, \quad (3)$$

as is easily verified using the commutation relation

$$e^{iap}F(q)e^{-iap} = F(q+a).$$

In the free case $V = 0$, the right hand side of (1) becomes

$$A_{st} = \mathbb{E}[e^{i(X(s)-X(t))p}] = \mathbb{E}[U_{st}]$$

That these operators form an evolution,

$$A_{rs}A_{st} = A_{rt},$$

may be thought of as a consequence of the independent increments property of the Wiener process; that this evolution is a semigroup, that is $A_{st} = A_{s-t}$ depends only on the difference $s-t$, is a consequence of the stationarity of the process. That the semigroup is self-adjoint may be regarded as a consequence of the reflection property of the Wiener process; specifically that if $X_1(r) = X(r) - X(t)$ is a Wiener process on $[t,s]$ (with $X_1(t) = 0$), then $\tilde{X}_1$, defined by

$$\tilde{X}_1(r) = X_1(-r) - X_1(s)$$

is a Wiener process on $[-s,-t]$. That the right hand side of (1) defines a self-adjoint semigroup in the general case is a consequence of the fact that M inherits these invariance properties as well as satisfying the cocycle identity (3).

In this paper we shall describe a method of constructing unitary evolutions having analogous independence and stationarity properties and for which corresponding perturbation cocycles can be constructed, by means of second quantisation. Instead of being random unitary operators on a classical probability space, the constituent operators have an independence property which must be described in the language of noncommutative probability theory on von Neumann algebras; in the simplest case they are related to the (noncommutative) canonical (Boson) Wiener process [1,2] in a way analogous to the dependence (2) of U_{st} on the classical Wiener process. The resulting Feynman-Kac formula are thus in this sense noncommutative. In the special case referred to the resulting formula is essentially the "oscillator process" Feynman-Kac formula of [3] (see also [4,5], as was made clear to us by a comment of L van Hemmen, and it is anticipated that this way of deriving this formula will indicate some new generalisations, in particular involving non-Fock Boson Wiener processes and also Fermion second quantisation.

Our basic construction is of a certain dilation, consisting of a unitary evolution, of a given self-adjoint contraction semigroup acting in a Hilbert space h_0; which we call the time-orthogonal unitary dilation. The essential feature of this dilation is that for each time interval $]t,s]$ the corresponding unitary operator acts non-trivially only in a Hilbert space $h_0 \oplus h_{]t,s]}$ where, for disjoint intervals $]t,s]$, the corresponding spaces $h_{]t,s]}$ are orthogonal. Upon second quantization, this orthogonality property transforms into the independence property required. The "free" or unperturbed semigroup obtained is the second quantization of the originally given contraction semigroup.

2. TIME ORTHOGONAL UNITARY DILATIONS

Let $(A_t : t \geq 0)$ be a strongly continuous self-adjoint contraction semigroup acting in a Hilbert space h_0 and let H be its infinitesimal generator, so that H is a positive self-adjoint operator and $A_t = e^{-tH}$. Let h be the Hilbert space $L^2(\mathbb{R}, h_0)$ of vector-valued functions on $\mathbb{R}$ and let $h_1 = h_0 \oplus h$. Let J be the embedding $\zeta \mapsto (\zeta, 0)$ from h_0 into h. For $t \in \mathbb{R}$ denote by S_t the shift operator

$$S_t f(s) = f(s-t)$$

in h, and by S the reflection at the origin

$$Sf(s) = f(-s);$$

notice that (S_t) and S generate a unitary representation of the one dimensional Euclidean group, consisting of all distance-preserving transformations of $\mathbb{R}$. Finally, let h_A denote the subspace of h comprising functions which vanish outside the measurable subset A of $\mathbb{R}$.

Our objective is to construct a family of unitary operators $(U_{st} : s \geq t)$ in h_1 with the following properties

(i) $U_{rs} U_{st} = U_{rt}$.

(ii) U_{st} is strongly continuous in s and t.

(iii) In the direct sum decomposition $h_1 = (h_0 \oplus h_{]t,s]}) \oplus h_{]t,s]}^{\perp}$, where $h_{]t,s]}^{\perp}$ is the orthogonal complement of $h_{]t,s]}$ in h, U_{st} assumes the form

$$U_{st} = V_{st} \oplus I \tag{4}$$

for some operator V_{st} in $h_0 \oplus h_{]t,s]}$.

(iv) $(I \oplus S_r) U_{st} (I \oplus S_r)^{-1} = U_{r+s\ r+t}$

(v) $(I \oplus S) U_{st}^{*} (I \oplus S)^{-1} = U_{-t-s}$ (6)

(vi) $A_{s-t} = J^{*} U_{st} J.$ (7)

Such a family will be called a <u>time-orthogonal unitary dilation</u> of (A_t).

We may assume without loss of generality that (A_t) is uniformly continuous, equivalently that H is bounded. Indeed, if this is not the case we can decompose h_0 has an infinite direct sum,

$$h_0 = \oplus_n h_0^{(n)}$$

of components each of which reduces H and on each of which H is bounded, by using the spectral decomposition of H. If $\{U^{(n)}_{st}\}$ is a time-orthogonal unitary dilation for the restriction of (A_t) to $h_0^{(n)}$ and if V is the isomorphism from

$$h_1 = h_0 \oplus L^2(\mathbb{R}, h_0) = \left(\bigoplus_n h_0^{(n)}\right) \oplus \left(\bigoplus_n L^2(\mathbb{R}, h_0^{(n)})\right)$$

to $\bigoplus_n (h_0^{(n)} \oplus L^2(\mathbb{R}, h_0^{(n)}))$ which appropriately permutes components, then it is easily seen that

$$U_{st} = V^{-1}\left(\bigoplus_n U^{(n)}_{st}\right)V$$

defines a time orthogonal unitary dilation for (A_t).

Assuming then that H is bounded, we construct the operators U_{st} in $h_1 = h_0 \oplus h$ as matrices

$$U_{st} = \begin{pmatrix} A_{st} & B_{st} \\ C_{st} & I + D_{st} \end{pmatrix}.$$

where $A_{st} = A_{s-t}$ and the actions of $B_{st} : h \to h_0$, $C_{st} : h_0 \to h$ and $D_{st}\ h \to h$ are respectively

$$B_{st}f = -(2H)^{\frac{1}{2}} \int \chi_{]t,s]}(x) A_{sx} f(x)\, dx \tag{8}$$

$$(C_{st}\zeta)(x) = \chi_{]t,s]}(x) A_{xt} (2H)^{\frac{1}{2}} \zeta \tag{9}$$

$$(D_{st}f)(x) = -2 \int \chi_{]t,s]}(y) \chi_{]y,s]}(x) H A_{xy} f(y)\, dy \tag{10}$$

To show that U_{st} is unitary one must check that

$$A^*_{st}A_{st} + C^*_{st}C_{st} = I$$

$$A^*_{st}B_{st} + C^*_{st} + C^*_{st}D_{st} = 0$$

$$B^*_{st}B_{st} + D^*_{st} + D_{st} + D^*_{st}D_{st} = 0. \tag{11}$$

together with corresponding component equations for $UU^* = I$. We give the proof for (11). The actions of B^*_{st} and D^*_{st} are

$$(B^*\zeta)(x) = -\chi_{]t,s]}(x)A_{sx}(2H)^{\frac{1}{2}}\zeta.$$

$$D^*_{st}f(x) = -2\int \chi_{]t,s]}(x)\chi_{]x,s]}(y)HA_{yx}f(y)\,dy.$$

Thus $D^*_{st}D_{st}$ is the integral operator in $L^2(\mathbb{R}, h_0)$,

$$D^*_{st}D_{st}f(x) = \int K(x,y)f(y)\,dy,$$

whose kernel is

$$\begin{aligned} K(x,y) &= 4\int \chi_{]t,s]}(x)\chi_{]x,s]}(z)HA_{zx}\chi_{]t,s]}(y)\chi_{]y,s]}(z)HA_{zy}\,dz \\ &= 4\chi_{]t,s]}(x)\chi_{]t,s]}(y)He^{(x+y)H}\int_{\max\{x,y\}}^{s} He^{-2zH}\,dz \\ &= -2\chi_{]t,s]}(x)\chi_{]t,s]}(y)e^{-(s-x)H}He^{-(s-y)H} \\ &+ 2\chi_{]t,s]}(x)\chi_{]t,s]}(y)He^{-|x-y|H} \end{aligned} \tag{12}$$

In (12), the first term is the kernel of $-B^*_{st}B_{st}$, while the restrictions of the second term to the complementary regions $\{(x,y) : x \geq y\}$ and $\{(x,y) : x \leq y\}$ give the negatives of the kernels of D^*_{st} and D_{st} respectively, proving (11).

To show that (i) holds one verifies the component identities

$$\begin{aligned} A_{rs}A_{st} + B_{rs}C_{st} &= A_{rt} \\ A_{rs}B_{st} + B_{rs} + B_{rs}D_{st} &= B_{rt} \\ C_{rs}A_{st} + C_{st} + D_{rs}C_{st} &= C_{rt} \\ C_{rs}B_{st} + D_{rs} + D_{st} + D_{rs}D_{st} &= D_{rt}; \end{aligned} \tag{13}$$

for instance to prove (13) note first that $D_{rs}D_{st} = 0$ and secondly that the kernels of the integral operators $C_{rs}B_{st}$, D_{rs}, D_{st} are respectively

$$-2\chi_{]s,r]}(x)\chi_{]s,t]}(y)HA_{xy},$$

$$-2\chi_{]s,r]}(y)\chi_{]y,r]}(x)HA_{xy},$$

$$-2\chi_{]t,s]}(y)\chi_{]y,r]}(x)HA_{xy},$$

the sum of which is the kernel $-2\chi_{]t,r]}(y)\chi_{]y,s]}(x)HA_{xy}$ of D_{rt},

as may be seen by decomposing the triangular region on which $\chi_{]t,r]}(y)\chi_{]y,r]}(x)$ is non zero into the union of a rectangle and two further triangular regions.

In view of i) and the unitarity of (U_{st}), to verify ii) it suffices to establish weak continuity; this in turn is equivalent to weak continuity of the matrix components of U_{st} and is evident by inspection. iii) is equivalent to the statement that

$$U_{st} = \begin{pmatrix} I & 0 \\ 0 & E \end{pmatrix} U_{st} \begin{pmatrix} I & 0 \\ 0 & E \end{pmatrix} + \begin{pmatrix} 0 & 0 \\ 0 & I-E \end{pmatrix},$$

where E is the project or onto $h_{]t,s]}$ in h, and this in turn is equivalent to the easily checked component indentities

$$B_{st}E = B_{st}, \quad E\,C_{st} = C_{st}, \quad ED_{st} = D_{st}E = U_{st}.$$

To prove iv) and v) we again write the identities in component form; for instance (6) is equivalent to

$$\begin{aligned} A^*_{st} &= A_{-t-s}, \\ SB^*_{st} &= C_{-t-s}, \\ C^*_{st}S^{-1} &= B_{-t-s}, \\ SD^*_{st}S^{-1} &= D_{-t-s} \end{aligned} \tag{14}$$

and, for instance, (14) is verified by the computation

$$\begin{aligned} SD^*_{st}S^{-1}f(x) &= D^*_{st}S^{-1}f(-x) \\ &= -2\int \chi_{]t,s]}(-x)\chi_{]-x,s]}(y)He^{-(y+x)H}f(-y)\,dy \\ &= -2\int \chi_{]t,s]}(-x)\chi_{]-x,s]}(-y)He^{(y-x)H}f(y)\,dy \\ &= -2\int \chi_{]-s,-t]}(x)\chi_{]-s,x]}(y)He^{-(x-y)H}f(y)\,dy \\ &= -2\int \chi_{]-s,-t]}(y)\chi_{]y,-t]}(x)He^{-(x-y)H}f(y)\,dy \\ &= D_{-t-s}f(x). \end{aligned}$$

vi) holds evidently.

3. PERTURBATIONS OF SECOND QUANTIZED SEMIGROUPS

Let h be a Hilbert space. We denote by $\Gamma(h)$ the (symmetric) Fock space over h,

$$\Gamma(h) = \bigoplus_{n=0}^{\infty} \left(\bigotimes_{j=1}^{n} h \right)_s ,$$

and by $e(f)$ the exponential vector [6],

$$e(f) = (1, f, (2!)^{-\frac{1}{2}} f \otimes f, (3!)^{-\frac{1}{2}} f \otimes f \otimes f, \dots)$$

corresponding to $f \in h$. As is well known [6], the exponential vectors form a total, linearly independent set in $\Gamma(h)$. If $h = h_1 \oplus h_2$ is a direct sum then there is a Hilbert space isomorphism from $\Gamma(h)$ onto $\Gamma(h_1) \otimes \Gamma(h_2)$, under which each exponential vector $e(f_1, f_2)$ is mapped to $e(f_1) \otimes e(f_2)$, which we use to identify the two Hilbert spaces.

We denote by Ω the Fock vacuum vector, $\Omega = e(0)$, and by $W = (W(f) : f \in h)$ the Fock representation of the canonical commutation relations over h; the operators $W(f)$ act on exponential vectors as

$$W(f)e(g) = \exp\{-\tfrac{1}{4}\|f\|^2 + 2^{-\frac{1}{2}} i \langle g, f\rangle\} e(g + 2^{-\frac{1}{2}} if)$$

and satisfy

$$\langle W(f)\Omega, \Omega\rangle = \exp\{-\tfrac{1}{4}\|f\|^2\}.$$

If C is a linear contraction from h to a Hilbert space h', the second quantization $\Gamma(C)$ of C is a contraction from $\Gamma(h)$ to $\Gamma(h')$ whose action on exponential vectors is

$$\Gamma(C)e(f) = e(Cf).$$

Second quantization has the properties

$$\Gamma(C_1 C_2) = \Gamma(C_1)\Gamma(C_2)$$

$$\Gamma(C^*) = \Gamma(C)^*$$

$$\Gamma(I) = I;$$

moreover the map $C \mapsto \Gamma(C)$ is continuous for the strong operator topologies in $B(h, h')$ and $B(\Gamma(h), \Gamma(h'))$. If $C = C_1 \oplus C_2$ is a direct sum then we have the identification

$$\Gamma(C_1 \oplus C_2) = \Gamma(C_1) \otimes \Gamma(C_2) \tag{15}$$

Now let $(A_t : t \geq 0)$ be a strongly continuous self-adjoint contraction semigroup in a Hilbert space h_0 and write

$$A_{s-t} = J^* U_{st} J \tag{16}$$

where the time-orthogonal unitary dilation (U_{st}) acts in the Hilbert space $h_0 \oplus h$, $h = L^2(\mathbb{R}, h_0)$. Second quantizing (16), we obtain

$$\Gamma(A_{st}) = \Gamma(J)^* \Gamma(U_{st}) \Gamma(J) \tag{17}$$

The left hand side of (17) is a strongly continuous self-adjoint contraction semigroup, the <u>second quantized semigroup</u> of (A_t). To interpret the right hand side, we first note that $\Gamma(J)$ is the map $\psi \to \psi \otimes \Omega$ from $\Gamma(h_0)$ into $\Gamma(h_0 \oplus h) = \Gamma(h_0) \otimes \Gamma(h)$ and that consequently, for every operator $A \in B(\Gamma(h_0) \otimes \Gamma(h))$, the operator in $B(\Gamma(h_0))$

$$\mathbb{E}[A] = \Gamma(J)^* A \Gamma(J)$$

is the "vacuum conditional expectation of A, given $B(\Gamma(h_0))$", that is, the unique operator in $B(\Gamma(h_0))$ such that, for all $\psi, \Phi \in \Gamma(h_0)$

$$\langle \mathbb{E}[A]\psi, \Phi \rangle = \langle A\psi \otimes \Omega, \Phi \otimes \Omega \rangle$$

Secondly, observe that second quantization of (4), using (15), gives

$$\Gamma(U_{st}) \in B(\Gamma(h_0)) \otimes N_{]t,s]}$$

where $N_{]t,s]}$ is the von Neumann algebra in $\Gamma(h)$ of operators of the form $A \otimes I$ in the decomposition $\Gamma(h) = \Gamma(h_{]t,s]}) \otimes \Gamma(h_{]t,s]}^{\perp})$. For disjoint intervals $]t_j, s_j]$ the corresponding von Neumann algebras $N_j = N_{]t_j,s_j]}$ are <u>independent</u> [2] in the vacuum state Ω, meaning that given $A_j \in N_j$, $j = 1,2,\ldots$, the different A_j commute with each other and

$$\mathbb{E}[\prod_j A_j] = \prod_j \mathbb{E}[A_j]$$

where $\mathbb{E}[A] = \langle A\Omega, \Omega \rangle$. This is the "independent increments" property, responsible for the fact that the right hand side of (17) is an evolution. In this connection we note the following fact, of which the proof may be found in [7] : if $N_1, N_2, \ldots$ are von Neumann algebras which are independent in the state Ω, if K is a Hilbert space and if $A_1, A_2, \ldots$ are elements of $B(K_0) \otimes N_1$, $B(K_0) \otimes N_2, \ldots$ respectively, then

$$\mathbb{E}[A_1 A_2 \ldots] = \mathbb{E}[A_1]\mathbb{E}[A_2] \ldots \quad .$$

Finally, observe that second quantization of (5) and (6) together with the relations

$$(I \oplus S_t)J = (I \oplus S)J = J$$

yields properties analogous to the stationarity and reflection invariance of the Wiener process, ensuring the semigroup and self-adjointness properties of the evolution (17).

Now let K_0 be a conjugation in h_0, that is a conjugate-linear isometric map from h_0 to itself whose square is the identity, such that

$$K_0 A_t = A_t K_0 \qquad (t \geq 0).$$

Then if K is the conjugation in $h = L^2(\mathbb{R}, h_0)$ given by $(Kf)(x) = K_0(f(x))$ and $K_1 = K_0 \oplus K$, it is easy to check that

$$K_1 U_{ts} = U_{ts} K_1, \qquad K_1 J = JK_0.$$

For j = 1,2 let R_j be the real-linear subspace of h_j comprising vectors invariant under K_j. Then the von Neumann algebras N_0, N_1 generated by W(f), $f \in R_0$, R_1 respectively are commutative. Also

$$\Gamma(U_{st})W(\zeta) \otimes I\Gamma(U_{st})^{-1} = \Gamma(U_{st})W(J\zeta)\Gamma(U_{st})^{-1} = W(U_{st}J\zeta)$$

and, for $\zeta \in R_0$,

$$K_1 U_{st} J\zeta = U_{st} K_1 J\zeta = U_{st} JK_0 \zeta = U_{st} J\zeta,$$

showing that conjugation by $\Gamma(U_{st})$ maps $N_0 \otimes I$ into N_1. It follows that if V is a self-adjoint element of N_0, the family of operators $\Gamma(U_{st})V \otimes I\Gamma(U_{st})^{-1}$, $s \geq t$, is mutually commutative. Since $\Gamma(U_{st})$ inherit the strong continuity of U_{st} the strong integral $\int_t^s \Gamma(U_{sr})V \otimes I\Gamma(U_{sr})^{-1}\, dr$ is a well defined bounded self-adjoint operator. We set

$$M_{st} = \exp\left\{- \int_t^s \Gamma(U_{sr})V \otimes I\Gamma(U_{sr})^{-1}\, dr\right\}. \tag{18}$$

It is straightforward to verify that M_{st} is a cocyle for the evolution $\Gamma(U_{st})$, that is that

$$M_{rs}\Gamma(U_{rs})M_{st}\Gamma(U_{rs})^{-1} = M_{rt}.$$

Hence

$$(s,t) \to M_{st}\Gamma(U_{st})$$

is an evolution; moreover $M_{st}\Gamma(U_{st}) \in N_{]t,s]}$. Hence we may form the "reduced" evolution

$$Z_{st} = \Gamma(J)^{*}M_{st}\Gamma(U_{st})\Gamma(J)$$

in h_0. Using the second quantizations of (5) and (6) it is easily verified that M satisfies the identities

$$I \otimes \Gamma(S_r)M_{st}I \otimes \Gamma(S_r)^{-1} = M_{r+s\ r+t}$$

$$M^{*}_{st}U_{st}(I \otimes \Gamma(S)^{-1}) = U_{st}(I \otimes \Gamma(S)^{-1})M_{-t-s}$$

which, together with the invariance of Ω under $\Gamma(S_t)$, $\Gamma(S)$ imply that the reduced evolution is a semigroup, $Z_{st} = Z_{s-t}$, and that this semigroup is self-adjoint. Strong continuity is a consequence of continuity of second quantization. Differentiation at the origin shows that the infinitesimal generator H of (Z_t) is given by

$$H = H_0 + V$$

where H_0 is the infinitesimal generator of $(\Gamma(A_t))$.

4. AN EXAMPLE

The simplest illustration of our theory [7] is obtained by taking $h_0 = \mathbb{C}$, $A_t = e^{-t}$. Then $(\Gamma(A_t))$ is the semigroup in $\Gamma(\mathbb{C}) = \ell^2$ whose infinitesimal generator is the number operator

$$a^{\dagger}a(z_0, z_1, \ldots) = (0, z_1, 2z_2, \ldots)$$

or equivalently the renormalised harmonic oscillator Hamiltonian $\frac{1}{2}(p^2+q^2-1)$. The operator $\Gamma(U_{st})$ may be expressed formally as a "stochastic product integral",

$$\Gamma(U_{st}) = \prod_{t}^{s} e^{i\sqrt{2}(p \otimes dQ - q \otimes dP)} \tag{19}$$

where (P,Q) is the Fock space canonical (Boson) Wiener process of [1,2]. Thus (17) becomes

$$e^{-(s-t)(\frac{1}{2}(p^2+q^2-1))} = \mathbb{E}\left[\prod_{t}^{s} e^{i\sqrt{2}(p \otimes dQ - q \otimes dP)}\right] \tag{20}$$

which may be compared with the classical free case

$$e^{-(s-t)\frac{1}{2}p^2} = \mathbb{E}[e^{i(X(s)-X(t))p}] = \mathbb{E}\left[\prod_t^s e^{ipdX}\right];$$

because of the non-commutativity of the infinitesimal rotations $e^{i\sqrt{2}(p\otimes dQ - q\otimes dP)}$ the product integral (20) cannot be integrated directly.

Taking K_0 to be the usual conjugation in $\mathbb{C}$ (replacing it by $e^{i\alpha}K_0e^{-i\alpha}$ would give formulae for perturbations of form $V(\cos\alpha q + \sin\alpha p)$), the algebra N_0 is that generated by the position operator q and the perturbation V is a (bounded, measurable) function $V = V(q)$. We write the cocycle (18) in the form

$$M_{st} = \exp\{-\int_t^s V(q_r^s)\,dr\}$$

where q_t^s is defined by

$$\begin{aligned} e^{ixq_t^s} &= \Gamma(U_{ts})e^{ixq}\otimes I\,\Gamma(U_{ts})^{-1} \\ &= \Gamma(U_{ts})W(x,0)\Gamma(U_{ts})^{-1} \\ &= W(U_{ts}(x,0)) \\ &= W(e^{-(s-t)}x, f) \end{aligned}$$

using the definition of U_{ts}, where

$$f(y) = \chi_{]t,s]}(y)e^{-(y-t)}\sqrt{2}\,x.$$

Hence, using the stochastic integral notation for field operators of [1] we can write

$$q_t^s = e^{-(s-t)}q\otimes I + \sqrt{2}\,I\otimes\int_t^s e^{-(r-t)}\,dQ(r) \tag{21}$$

The final formula can then be expressed as

$$\exp\{-(s-t)(a^\dagger a + V(q))\}$$
$$= \mathbb{E}\left[\exp\{-\int_t^s V(e^{-(s-r)}q\otimes I + \sqrt{2}I\otimes\int_r^s e^{-(r'-r)}\,dQ(r'))dr\}\prod_t^s e^{i\sqrt{2}(p\otimes dQ - q\otimes dP)}\right] \tag{22}$$

The family of operators (21), together with its momentum counterpart

$$p_t^s = e^{-(s-t)} p \otimes I + \sqrt{2} I \otimes \int_t^s e^{-(r-t)} dP(r).$$

may be regarded as constituting a non-commutative analog of the Orstein-Uhlenbeck velocity process [8].

The elements (p_t^s, q_t^s) of the canonical Ornstein-Uhlenbeck velocity processes are canonical pairs, satisfying the fixed commutation relation $pq - qp = -i$. This process may thus be regarded as a (Gaussian) quantum stochastic process in the sense of [9], unlike the canonical Wiener process whose elements $(P(t), Q(t))$ satisfy the "time-dependent" commutation relation

$$P(t)Q(t) - Q(t)P(t) = -it.$$

The classical Ornstein Uhlenbeck velocity process has been used by previous authors [3,4,5] to construct formulae of Feynman-Kac type in which the "free" Hamiltonian is that of the harmonic oscillator Hamiltonian, and the "oscillator process" formula of [5], page 52, can be derived (in the case of bounded V) from (22). We are grateful to L van Hemmen for bringing this connection to our attention.

5. NON-FOCK SECOND QUANTIZATION

Let σ be a real number ≥ 1. Given a Hilbert space h we construct a Hilbert space $\Gamma_\sigma(h)$ and a representation of the canonical commutation relations $(W_\sigma(f) : f \in h)$ in $\Gamma_\sigma(h)$ as follows; we set

$$\Gamma_\sigma(h) = \Gamma(h) \otimes \Gamma(\bar{h})$$

where $\Gamma(h)$ is Fock space over h and $\Gamma(\bar{h})$ is Fock space over the dual Hilbert space of h, which we identify with $\overline{\Gamma(h)}$, and define

$$W_\sigma(f) = W(\cosh \alpha f) \otimes W(\sinh \alpha \bar{f})$$

where α is a real number such that $\cosh 2\alpha = \sigma^2$ and, for $f \in h$, $\bar{f}$ is the corresponding element $g \mapsto \langle g, f \rangle$ of $\bar{h}$. We set $\Omega_\sigma = \Omega \otimes \bar{\Omega}$ (where Ω is the Fock vacuum); then

$$\langle W_\sigma(f)\Omega_\sigma, \Omega_\sigma \rangle = \exp\{-\tfrac{1}{4}\sigma^2 \|f\|^2\}.$$

If $h = h_1 \oplus h_2$ there is a natural identification, derived from the corresponding identification for the Fock case, between $\Gamma_\sigma(h)$ and $\Gamma_\sigma(h_1) \otimes \Gamma_\sigma(h_2)$. If C is a contraction from h to a second Hilbert space h' then $\Gamma_\sigma(C) = \Gamma(C) \otimes \Gamma(\bar{C})$ (where $\bar{C}\bar{f} = \overline{Cf}$) defines a contraction from $\Gamma_\sigma(h)$ to $\Gamma_\sigma(h')$ and Γ_σ has *-morphism and continuity properties similar to those of Γ; also

$$\Gamma_\sigma(C_1 \oplus C_2) = \Gamma_\sigma(C_1) \otimes \Gamma_\sigma(C_2).$$

If U is a unitary operator in h the operator $\Gamma_\sigma(U)$ does not in general belong to the von Neumann algebra generated by the representation $(W_\sigma(f))$. However, using the well known theorem of Shale [10] on implementability of linear canonical transformations in the Fock representation, it can be shown [11] that a necessary and sufficient condition that $\Gamma_\sigma(U)$ factorise into a product $\Lambda(U)\,\Lambda'(U)$ of unitary elements of this algebra and its commutant (and thus that conjugation by $\Gamma_\sigma(U)$ be an inner automorphism of the algebra) is that U differ from the identity by a Hilbert Schmidt operators, moreover the map $U \to \{c\,\Lambda(U) : |c| = 1\}$ is a continuous unitary ray representation of the group G of such unitary operators, topologized by the metric $\rho(U_1, U_2)$ = Hilbert Schmidt norm of $U_1 - U_2$.

Analysis of the structure of (U_{st}) [11] shows that a necessary and sufficient condition that the elements of the time-orthogonal unitary dilation (U_{st}) of a strongly continuous one-parameter semigroup (A_t) belong to the group G is that the infinitesimal operator of A_t be of trace class, and that if this is the case, U_{st} depends continuously on s and t is the sense of the topology of G. It can then be shown [11] that there is a unique choice of representatives Λ_{st} of the unitary rays $\{c\,\Lambda(U_{st}) = |c| = 1\}$ having the properties

i) $\quad \Lambda_{st} \in B(\Gamma_\sigma(h_0)) \otimes N_{]t,s]}$

where now $N_{]t,s]}$ is the von Neumann algebra generated by the $W_\sigma(f)$ with f supported by $]t,s]$.

ii) $\quad \Lambda_{rs}\,\Lambda_{st} = \Lambda_{rt}$

iii) $\quad \Gamma_\sigma(I \oplus S_r)\,\Lambda_{st}\,\Gamma_\sigma(I \oplus S_r)^{-1} = \Lambda_{r+s,\,r+t}$

iv) $\quad \Gamma_\sigma(I \oplus S)^*\Lambda_{st}\,\Gamma_\sigma(I \oplus S)^{-1} = \Lambda_{-t-s}.$

These relations imply that the formula

$$C_{s-t} = \Gamma_\sigma(J)^*\,\Lambda_{st}\,\Gamma_\sigma(J)$$

defines a strongly continuous self-adjoint contraction semigroup. Cocycles of the type (18) may be constructed exactly as in the Fock case. In the case $h_0 = \mathbb{C}$, $A_t = e^{-t}$, Λ_{st} is given formally by the stochastic product integral (19) in which the canonical Wiener process is now of variance σ^2 [1]. The semigroup C_t is given by

$$C^t = e^{-t\left(\frac{\sigma^2}{2}(p^2+q^2-1) + \frac{1}{2}(\sigma^2-1)\right)}.$$

References

[1] A.M. Cockcroft and R.L. Hudson, Quantum mechanical Wiener processes, J. Multivariate Anal. 7, 107-24 (1977).
[2] R.L. Hudson, The strong Markov property for canonical Wiener processes, J. Functional Anal. 34, 266-81 (1979).
[3] B. Simon, "Functional Integration and Quantum Physics," Academic Press, New York (1979).
[4] J. Glimm, Boson fields with nonlinear selfinteraction in two dimensions, Commun.math.Phys. 8, 12-25 (1968).
[5] E. Nelson, A quartic interaction in two dimensions, in "Mathematical theory of elementary particles," R. Goodman and I.E. Segal, eds., M.I.T. Press, Cambridge, Mass. (1966).
[6] A Guichardet, "Symmetric Hilbert spaces and related topics," Lecture notes in Mathematics 261, Springer, Berlin (1972).
[7] R.L. Hudson and P.D.F. Ion, The Feynman-Kac formula for a canonical quantum mechanical Wiener process, to appear in Proceedings of the Colloquium "Random fields : rigorous results in statistical mechanics and quantum field theory," Esztergom, 1979.
[8] L. Breiman, "Probability," Addison-Wesley, New York (1968).
[9] L. Accardi, A. Frigerio and J.T. Lewis, Quantum stochastic processes, preprint.
[10] D. Shale, Linear symmetries of free Boson fields, Trans. Amer.Math.Soc. 103, 149-167 (1962).
[11] R.L. Hudson, P.D.F. Ion and K.R. Parthasarathy, Time-orthogonal unitary dilations and non commutative Feynman-Kac formulae, preprint.

The third author wishes to thank the British Council for offering him financial support under the Commonwealth Visiting Professors' Programme during the period January 15 to March 15, 1979.

ON NONUNITARY EQUIVALENCE BETWEEN UNITARY GROUP OF DYNAMICS AND CONTRACTION SEMIGROUPS OF MARKOV PROCESSES

B. Misra

Chimie-Physique II, C.P. 231
Université Libre de Bruxelles
1050 Bruxelles, Belgium

ABSTRACT

We discuss the problem of deriving an exact Markovian Master equation from dynamics without resorting to approximation schemes such as the weak coupling limit, Boltzmann-Grad limit etc. Mathematically, it is the problem of the existence of suitable positivity preserving operator Λ such that the unitary group U_t induced from dynamics satisfies the intertwining relation:

$$\Lambda U_t = W_t^* \Lambda$$

with the contraction semigroup W_t^* of a strongly irreversible stochastic Markov process. Two cases are of special interest: (i) Λ is a projection operator, (ii) Λ has densely defined inverse Λ^{-1}. Our recent work, which we summarize here, shows that the class of (classical) dynamical systems for which a suitable projection operator satisfying the above intertwining relatin exists is identical with the class of K-flows or K-systems. As a corollary of our consideration it follows that the function $\int \hat{\rho}_t \ln \hat{\rho}_t d\mu$ with $\hat{\rho}_t$ denoting the coarse-grained distribution with respect to a K-partition obtained from $\rho_t = U_t\rho$ is a Boltzmann type H-function for K-flows. This is not in contradiction with the time reversal (velocity inversion) symmetry of dynamical evolution as it is the operation of coarse-graining with respect to K partition that breaks the symmetry.

INTRODUCTION

The study of the possible connections that may exist between deterministic dynamics and probabilistic processes is of obvious importance for the foundation of nonequilibrium statistical mechanics. As it is well known, stochastic Markov processes provide suitable models to represent irreversible evolution admitting a Liapounov functional or H-function. The important question, thus, is how is the passage from deterministic dynamics to probabilistic Markov processes to be achieved.

Closely related to this is the much discussed question of the relation between the chaotic nature of dynamical motion and the instability of (phase space) trajectories. Our work, to be described below, shows that in the presence of suitable instability of motion, described by the condition of K-flow, the dynamical evolution is indeed similar to the stochastic evolution of an irreversible Markov process.

Let us recall that the procedure for obtaining a Markovian Master Equation from dynamics starts with an initial "contraction of description" or coarse-graining brought about by a projection operator P. The operation of "coarse-graining" alone however generally leads to the so-called generalized Master equation which is non-Markovian in character [1]. To obtain a Markovian evolution equation one needs to consider special asymptotic limits (e.g. the weak coupling limit etc.). Thus, even when this program succeeds, the resulting Master equation is only an approximation. To lay a more satisfactory foundation of nonequilibrium statistical mechanics it seems desirable to investigate the possibility of establishing exact Markovian Master Equations whose validity does not depend on special approximation schemes. This paper summarizes our recent work in this direction [2-6]. We shall discuss the problem for classical dynamical systems. Our main result is that the class of dynamical system for which an exact Markovian master equation follows from a suitable operation of "coarse-graining" alone is not empty but is precisely the class of so-called K-flow. The condition of K-flow is thus seen to play the same role in the foundation of nonequilibrium statistical mecahnics as that of ergodicity in the foundation of equilibrium statistical mechanics.

Let us, however, mention that just as the method of replacing time average by ensemble average can be justified for a special class of functions on phase space even if the system as a whole is not ergodic, one may be able to derive exact Markovian Master Equation for special subclass of initial distribution even if the system is not a K-flow. But it is only for K-flows that one can derive an exact master equation through a projection operator for all initial distribution function in $L^2_\mu \cap L^1_\mu$.

MATHEMATICAL FORMULATION OF THE PROBLEM

Consider an abstract dynamical system $[\Gamma, B, \mu, T_t]$. Here Γ denotes the phase space of the system equipped with a σ-algebra B of measurable subsets, T_t a group of measurable transformations mapping Γ onto itself and preserving the measure μ. For example, Γ could be the energy surface of a classical dynamical system, T_t the group of dynamical evolution and μ the invariant measure whose existence is assured by Liouville's theorem. For convenience we shall assume the measure to be normalized: $\mu(\Gamma) = 1$. As is well known, the evolution $\rho \to \rho_t$ of density function under the given deterministic dynamics is described by the unitary group U_t induced by T_t:

$$\rho_t(\omega) = (U_t \rho)(\omega) = \rho(T_{-t}\omega) \quad .$$

The generator L of the unitary group U_t is called Liouvillian operator of the system: $U_t = e^{-iLt}$. It is given by

$$L\rho = i[H,\rho]_{P.B.}$$

if the evolution is generated by the Hamiltonian function H. Here $[\ ]_{P.B.}$ denotes the usual Poisson bracket.

Every measure preserving deterministic evolution T_t thus defines a unitary group. Conversely (under the mild assumption that (Γ, B, μ) is a standard measure space), every unitary group which preserves positivity (i.e. maps nonnegative functions to nonnegative functions) and leaves the constant functions on Γ unchanged is induced by a group T_t of measure preserving transformations on Γ (cf. [7]).

On the other hand, stochastic Markov processes on the state space Γ, preserving μ, are associated with contraction semigroups [8]. In fact, let $P(t,\omega,\Delta)$ denote the probability of transition from the point $\omega \in \Gamma$ to the region Δ in time t. Then the operators W_t defined by:

$$(W_t f)(\omega) = \int_\Gamma f(\omega')P(t,\omega,d\omega')$$

$$f \in L^2_\mu$$

form a contraction semigroup for $t \geqslant 0$.

Moreover, W_t has the following properties:

i) W_t preserves positivity (i.e. $f \geqslant 0$ implies $W_t f \geqslant 0$ for $t \geqslant 0$);

ii) $W_t .1. = 1$.

The evolution of the distribution functions $\hat{\rho}$ under the Markov process is described now by the adjoint semigroup W_t^* which also preserves positivity since W_t does: $\hat{\rho} \to \hat{\rho}_t = W_t^*\hat{\rho}$. Since the measure μ is an invariant measure for the process (or equivalently the micro canonical distribution function 1 is the equilibrium state of the process) we also have

iii) $W_t^* .1 = 1.$

Every Markov process on Γ with stationary measure μ is thus associated with a contraction semigroup satisfying the conditions (i) through (iii). Conversely every contraction semigroup W_t on L^2_μ satisfying the above conditions comes from a stochastic Markov process, the transition probabilities $P(t,\omega,\Delta)$ being given by

$$P(t,\omega,\Delta\} = (W_t \phi_\Delta)(\omega) \quad .$$

Here ϕ_Δ denotes the characteristic (or indicator) function of the set Δ.

In the following we are interested in a special class of Markov processes whose semigroups W_t satisfy (in addition to the condition i-iii) the condition:

iv) $\| W_t\hat{\rho} - 1 \|$ decreases strictly monotonically to 0 as $t \to +\infty$; for all states $\hat{\rho}$ (i.e. for all nonnegative distribution functions $\hat{\rho}$ with ($\int \hat{\rho} d\mu = 1$). This condition expresses the requirement that any initial distribution $\hat{\rho}$ tends strictly monotonically in time to the equilibrium distribution 1. For such processes the functional

$$\|\hat{\rho}_t\|^2 = \int \hat{\rho}_t^2 \, d\mu \quad ; \quad \hat{\rho}_t = W_t^*\hat{\rho}$$

and indeed any other convex functional including the usual expression for negative entropy

$$\int \hat{\rho}_t \, \ell n \, \hat{\rho}_t d\mu$$

is an H-function. Such Markov processes thus provide the best possible model of irreversible evolution obeying the law of monotonic increase of entropy. Semigroups satisfying the conditions (i) through (iv) will be called strongly irreversible Markov semigroups.

The problem before us is to determine the class of dynamical systems for which one can construct a bounded operator Λ having the following properties:

i) Λ preserves positivity;

ii) $\int_\Gamma \Lambda\rho d\mu = \int_\Gamma \rho d\mu$;

iii) $\Lambda 1 = 1$;

iv) The dynamical group $U_t = e^{-i\varepsilon L}$ satisfies the intertwining relation: $\Lambda U_t = W_t^* \Lambda$ (for $t \geq 0$) with a strongly irreversible Markov semigroup W_t.

We shall consider two cases:

(1) Λ has a densely defined inverse Λ^{-1}. In this case Λ may be interpreted as defining a "change of representation" of dynamics $\rho_t \to \Lambda\rho_t = \hat{\rho}_t$. Condition (iv) then means that the evolution $\hat{\rho}_o \to \hat{\rho}_t$ of transformed states obeys Master equation of the Markov process W_t. For dynamical systems admitting an operator Λ satisfying conditions (i-iv) and (1) the dynamical group U_t is similar to a strongly irreversible Markov semigroup: $\Lambda U_t \Lambda^{-1} = W_t^*$ for $t \geq 0$.

Note that the demanded invertibility of Λ assures that the passage $\rho \to \Lambda\rho$ involves no "coarse-graining" or "loss of information." Dynamical systems admitting such a Λ may, hence, be said to be <u>intrinsically random</u>.

The other case we consider is: Λ is a projection operator P. Such projection operators (i.e. projections P satisfying conditions i-iii, with Λ replaced by P) correspond to operations of "coarse-graining." The existence of such a projection P satisfying the intertwining relation (iv) thus imply an <u>exact</u> Markovian Master equation for the system which does not depend on special asymptotic approximations but result solely from the coarse-graining operation.

As described below, it is a remarkable fact that there is a rather general class of dynamical systems for which an exact master equation holds in this sense.

DYNAMICAL SYSTEMS ADMITTING EXACT MARKOVIAN MASTER EQUATIONS

A K-flow is, by definition [9], an abstract dynamical system (Γ, B, μ, T_t) for which there exists a distinguished (measurable) partition ξ_o of the phase space into disjoint cells having the following properties:

i) $\xi_t \leq \xi_s$ if $t \leq s$.

Here $\xi_t = T_t \xi_o$, the partition into which the original partition ξ_o is transformed in time under the dynamical evolution. The notation $\xi_t \leq \xi_s$ signifies that the partition ξ_s is "finer" than ξ_t (i.e. every cell of ξ_s is entirely contained in one of the cells of ξ_t).

ii) The (least fine) partition $V\xi_t$ which is finer than each ξ_t, $-\infty < t < +\infty$ is the partition of the phase space into distinct phase points.

iii) The (finest) partition $\cap\xi_t$ which is less fine than every ξ_t is a trivial partition consisting of a cell of measure 1.

A partition ξ_o with the above stated properties is called a K-partition. Many systems of physical interest have been recently found to be K-flows: For instance, the motion of hard spheres within a finite box [10], the geodesic flow on space of constaint negative curvature [11], the Lorentz gas and infinite gas and hard rod systems etc. Generally, a K-partition consists of an uncountable number of cells, each of null μ-measure. The notion of "coarse-graining" with respect to a K-partition cannot, therefore, be defined directly as the operation of taking averages over the cells of the partition. The appropriate extension of the usual concept of coarse-graining is provided by the projection operator P_o of L^2_μ onto the subspace $L^2(A(\xi_o),\mu)$. $A(\xi_o)$ denotes the σ-subalgebra of B consisting of only those measurable subsets of Γ that are unions of cells in ξ_o and $L^2(A(\xi_o),\mu)$ is the subspace of all $f \in L^2_\mu$ that are measurable with respect to $A(\xi_o)$.

In fact, it is clear that for any nonnegative density function $\rho \in L^2_\mu$ the function $P_o\rho$ has the following characteristic properties of coarse-grained distribution with respect to the partition ξ_o

i) $P_o\rho = \hat{\rho} \geqslant 0$;

ii) $\int \hat{\rho} d\mu = \int \rho d\mu$;

iii) $\hat{\rho}$ being measurable with respect to $A(\xi_o)$, can only assume constant values on individual cells of ξ_o

iv) $\int_\Delta \hat{\rho} d\mu = \int_\Delta \rho d\mu$ for any measurable Δ that is a union of cells in ξ_o.

The (self adjoint) projection operator P_o of "coarse-graining" with respect to a partition ξ_o, obviously preserves positivity and maps the constant function (microcanonical ensemble) onto itself. It is interesting that the converse is also true: For standard measure spaces (Γ,B,μ) with $\mu(\Gamma) = 1$ every self adjoint projection operation P_o of L^2_μ that preserves positivity and maps the constant function onto itself is the projection operator onto $L^2(A(\xi_o),\mu)$ for some measurable partition ξ_o of Γ. The operations of coarse-graining may thus be identified with (self-adjoint) projection operators P satisfying:

i) $P\rho \geqslant 0$ if $\rho \geqslant 0$ and;

ii) $P1 = 1$.

The following theorem tells that the condition of K-flow is both necessary and sufficient for the existence of an operation of "coarse-graining" that converts the dynamical evolution into that of a strongly irreversible stochastic Markov process.

THEOREM 1.

Suppose a dynamical system with induced unitary group U_t of dynamical evolution admits a positivity preserving projection P mapping the function 1 to itself such that

i) $P_o U_t \rho = P_o U_t \rho'$ for all t
$\Rightarrow \rho = \rho'$;

ii) $P_o U_t = W_t^* P_o$ for $t \geqslant 0$, where W_t is a strongly irreversible Markov semigroup (see §2 for definition). Then the dynamical system is necessarily a K-flow. Conversely, every K-flow admits an operation of "coarse-graining" P_o (namely, the "coarse-graining" operation with respect to a K-partition) which satisfies the conditions (i) and (ii) above.

Remark: condition (i) means that the "coarse-graining" under consideration is sufficiently fine so that a knowledge of the "coarse-grained" distribution $P_o U_t \rho$ during the entire history, both past and future, of the systems evolution is equivalent to a knowledge of the original distribution.

We shall not stop here for a proof of the theorem which may be found elsewhere. But let us mention an important corollary of this result.

COROLLARY 2

For K-flows the "coarse-grained" (negative) entropy functional

$$\int \hat{\rho}_t \ln \hat{\rho}_t d\mu \quad , \quad \hat{\rho}_t = P_o U_t \rho$$

with respect to a K-partition is a monotonically decreasing function (H-function) of t which attains the "fine grained" value $\int \rho \ln \rho d\mu$ at $t \to -\infty$ and the equilibrium value (0, due to our normalization) at $t \to +\infty$.

This result, which has been recently proved by a different method [12], follows immediately from theorem 1 because $\hat{\rho}_t = W_t^* \rho_o$ for $t \geq 0$ with W_t^* a strongly irreversible Markov process. It is well known that for such processes the functional (2), and indeed any convex functional of $\hat{\rho}_t$, is an H-function.

Finally, let us mention that for K-flow one can not only construct a coarse-graining operation that leads from the deterministic and irreversible evolution U_t to that of a strongly irreversible stochastic process but one can construct also an invertible and positivity preserving transformation Λ such that $\Lambda U_t \Lambda^{-1}$ is a strongly irreversible Markov semigroup for $t \geq 0$. In other terms, the unitary group U_t induced from every K-flow is nonunitarily equivalent through a positivity preserving similarity Λ, to the contraction semigroup of a stochastic Markov process. A detailed discussion of this construction is described in our previous publications [2-6]. Let us only mention that the operator Λ establishing nonunitary equivalence between U_t and a stochastic Markov process W_t may be constructed as suitable function of operator time T

$$\Lambda = h(T) + P_o$$

$$T = \int_{-\infty}^{+\infty} \lambda dF_\lambda \quad , \quad F_\lambda = U_\lambda P_o U_\lambda^* - P_{-\infty} \quad .$$

Here P_o denotes the projection of coarse-graining with respect to the K-partition and $P_{-\infty}$, the projection to the equilibrium ensemble.

REFERENCES

[1] I. Prigogine, Nonequilibrium Statistical Mechanics (Wiley, New York, 1962).
[2] I. Prigogine, C. George, F. Henin and L. Rosenfeld, Chemica Scripta 4, 5 (1973).
[3] B. Misra, Proc. Natl. Acad. Sc. U.S.A. 75, 1627 (1978).
[4] B. Misra, I. Prigogine and M. Courbage, Physica 98A, 1 (1979).
[5] S. Goldstein, B. Misra and M. Courbage, to appear in J. Stat. Phys.
[6] B. Misra and I. Prigogine, to appear.
[7] K. Goodrich, K. Gustafson and B. Misra, Physica 102A, 379 (1980). See also the article in this proceedings.
[8] E. G. Dynkin, Markov Process (Springer, New York, 1965).
[9] V. I. Arnold and A. Avez, Ergodic Problems in Classical Mechanics (Benjamin, New York, 1968).
[10] Y. Sinai, Sov. Math. Dolk. 4, 1818 (1963).
[11] D. Anosov, Proc. Steklov Inst. n°90 (1967).
[12] S. Goldstein and O. Penrose, Univ. of Rutgers, preprint.

INDEX